COMPARATIVE PHYSIOLOGY:
LIFE IN WATER
AND ON LAND

COMPARATIVE PHYSIOLOGY: LIFE IN WATER AND ON LAND

Edited by

Pierre Dejours
  Centre National de la Recherche
  Scientifique, Strasbourg, France

Liana Bolis
  University of Milan, Italy

C. Richard Taylor
  Harvard University, Bedford,
  MA, USA

Ewald R. Weibel
  University of Bern, Switzerland

FIDIA
RESEARCH
SERIES

Volume 9

LIVIANA PRESS
Padova

SPRINGER VERLAG
Berlin - Heidelberg
New York - Tokyo

FIDIA RESEARCH SERIES

An open-end series of publications on international biomedical research, with special emphasis on the neurosciences, published by LIVIANA Press, Padova, Italy, in cooperation with FIDIA Research Labs, Abano Terme, Italy.

The series will be devoted to advances in basic and clinical research in the neurosciences and other fields.

The aim of the series is the rapid and worldwide dissemination of up-to-date, interdisciplinary data as presented at selected international scientific meetings and study groups.

Each volume is published under the editorial responsibility of scientists chosen by organizing committees of the meetings on the basis of their active involvement in the research of the field concerned.

DISTRIBUTION

Sole distribution rights outside Italy granted to Springer - Verlag.

All orders for the Fidia Research Series should be sent to the following addresses

Italy:
LIVIANA EDITRICE S.p.A. - Via Luigi Dottesio, 1, 35138 Padova, Italia

North America:
SPRINGER - VERLAG New York, Inc. - 175 Fifth Avenue, New York, N.Y. 10010, USA

Japan:
SPRINGER - VERLAG - 37-3 Hongo 3-chome, Bunkyo-Ku, Tokyo 113, Japan

Rest of the World:
SPRINGER - VERLAG Berlin - Heidelberger Platz 3, 1000 Berlin 33, FRG

Printed in Italy

ISBN   88-7675-428-8:   Liviana Editrice
ISBN   3-540-96515-7:   Springer - Verlag Berlin Heidelberg New York Tokyo
ISBN   0-387-96515-7:   Springer - Verlag New York Heidelberg Berlin Tokyo

LIVIANA Editrice S.p.A. - via Luigi Dottesio 1, 35138, Padova, Italy.

CONTENTS

PART VIII - WATER, IONIC EXCHANGES, OSMOREGULATION MECHANISMS

PREFACE

This volume examines the extent to which the design and function of terrestrial and aquatic animals are determined by the physiochemical properties of the media in which they live. The topic is addressed from the viewpoint of scientists representing a variety of disciplines and approaches. Anatomists, biochemists, biophysicists, physiologists and zoologists each contribute their perspectives. The general topics examined include: respiration; acid base balance; osmoregulation; water and ionic exchanges; nutrient acquisition and absorption; nitrogen and sulfur metabolism; locomotion; sensory information and behavior; energy metabolism; temperature and evolution. Four or five papers deal with each of these general topics.

The papers contained in this volume were presented at the Eighth International Conference on Comparative Physiology held in Crans-sur-Sierre, Switzerland, in June 1986. The conference was sponsored by the Interunion Commission on Comparative Physiology representing the International Unions of Biological Sciences, Physiological Sciences, and Pure and Applied Biophysics. It brought together outstanding scientists from various fields to discuss the design and function of terrestrial and aquatic animals.

Pierre Dejours
Liana Bolis
C. Richard Taylor
Ewald R. Weibel

INTRODUCTION

WATER AND AIR
PHYSICAL CHARACTERISTICS
AND THEIR PHYSIOLOGICAL CONSEQUENCES

Pierre Dejours

Laboratoire d'Etude des Régulations Physiologiques
associé à l'Université Louis Pasteur, C.N.R.S.
23 rue Becquerel, 67087 Strasbourg, France

The physical properties of water and air are strikingly different and, consequently, biochemical, anatomical and physiological features of aquatic and terrestrial animals differ
markedly. Whether the aquatic animals are invertebrates or
vertebrates, the problems raised for those which invade the
land are common to all. Many biochemical and physiological
traits are convergent, are related to the environmental
characteristics and suggest an ecophysiological transphyletic
division of the animal kingdom.

In Ancient Greece, Empedocles contended that water and air together with earth and
fire were the four elements which, associated in various proportions by the action
of love or strife, accounted for the matters and forms of the real world as man
saw it. This cosmology has inspired philosophers, artists and scientists. Water
and air have long been recognized as basic entities.

The physical properties of water and air are markedly different. Table 1 gives
various properties of water and air with their physiological consequences. The
headings are organized to follow approximately the sequence of chapters in this
book, but this sequence has no particular logical justification. Actual numerical
values are not given because they vary, especially in water, and there are many
types of water, with differing temperature, salinity, pressure and so on. Thus the
ratio air/water are only indicative, although they are exact for certain given
physical conditions. Actual values may be found in Dejours, 1981, p. 35; Little,
1983, p. 3; Schmidt-Nielsen, 1983, p. 27.

As said above, the properties of the two milieus are very different except for
one. The CO_2 capacitance, at 18°C, is identical in distilled water and air.
However, if the water contains some buffer, its CO_2 capacitance is lower than the
value in air, as is illustrated in the chapter by Rahn and Dejours of this volume.
This is one example among many that the numerical value of any property is correct

Comparative Physiology: Life in Water and on Land. P. Dejours, L. Bolis, C.R. Taylor, E.R. Weibel
(eds.) Fidia Research Series, IX-Liviana Press, Padova © 1987

Table 1

Main differences between water and air and their physiological consequences

	WATER	AIR	AIR/WATER	CONSEQUENCES
O_2 and CO_2 diffusivities	+	++++	$\simeq$ 8000	O_2 & CO_2 tensions
O_2 capacitance	+	++	$\simeq$ 30	
CO_2 "	++	++	$\simeq$ 1	acid-base balance
NH_3 "	++++	++	$\simeq$ 1/700	N end products
Viscosity	++	+	$\simeq$ 1/60	work of breathing
				circulation
Density	+++	+	$\simeq$ 1/800	skeleton
				locomotion
Kinematic viscosity	+	++	$\simeq$ 13	buoyancy, gravity
Water availability	+++	very variable		water turnover
				osmoregulation
Ionic environment	very variable			ionoregulation
Sound velocity	++	+	$\simeq$ 1/4	
				audition
" absorption	++	+	$\simeq$ 1/4	
Light refractive index	1.33	1	$\simeq$ 0.75	
				vision
" absorption	++	+	$\simeq$ 1/12	
Dielectric constant	+++	+	$\simeq$ 80	electroreception
Solubility of molecules	variable			
				distance
Volatility " "		+		
				chemoreception
Diffusivity " "	+	++++		
Heat capacity	++++	+	$\simeq$ 1/3500	
				heat dissipation
" conductivity	++	+	$\simeq$ 1/24	
				body temperature
" of evaporation	$\simeq$ 2450 kJ·L^{-1}			

only for a given set of conditions.

While one can consider two main groups of *waters*: the oceanic waters and the fresh
waters, one should remember that (1) briny waters are saltier than open sea water;
(2) there is a continuum of brackish waters, often estuarine, between open sea
water and fresh water; (3) every fresh water has its own properties. The composition of each water must be known in order to make certain predictions, in
particular as to respiratory and ionic exchanges.

By contrast, *air*, except for a few instances, has a constant fractional composition; it is sufficient to know the ambient temperature, the barometric pressure
and the humidity to define the exact composition of the milieu and the changes it
can undergo in the process of respiration.

Soil, if it is very dry, may be considered as a collection of interconnecting gas
pockets. if it is completely wet, it should be treated as some kind of water. But
if it is partially wet, it may be viewed as a collection of water-saturated gas
pockets *and* water pockets. This special milieu has been termed *porosphere*
(Vannier, 1983). The animal life of soils has been less studied than that of the
aquatic and land mediums.

The broad division of the animal kingdom into aquatic and terrestrial animals is
of course very old. It was first a distinction made by anatomists and zoologists
who in the past inferred the nature of the functions from the knowledge of the
structures. Later, with the development of physiology and biochemistry, more
contrasts between aquatic and terrestrial life became evident.

Schmidt-Nielsen, in "Animal Physiology: Adaptation and Environment" (1975, 1979,
1983) emphasized many contrasts between terrestrial and aquatic life. Dejours
(1979) concluded that it may be more important for the physiologist to consider
animal features as functions of the two main milieus they live in, namely water
and air. Little (1983) developed a systematic approach to this problem in a
little publicized book, "The Colonisation of Land. Origins and Adaptations of
Terrestrial Animals" (1983), which is today the most complete attempt to contrast
terrestrial and aquatic life in the animal kingdom.

With respect to the contrast of terrestrial and aquatic animals, certain physiological functions have been studied prior to others. There is some tendency,
perhaps justified, to think of one or another function as the main one; this,
however, suggests that the rest may be subordinate. The fact is that all functions
are interrelated.

Nonetheless, the history of the contrasted features of terrestrial and aquatic animals should be undertaken for each function. *Locomotion* is well known because, at first glance, the differences between both groups of animals is striking. As for the end products of *nitrogen metabolism*, it was pointed out in the late twenties that most wholly aquatic animals are ammoniotelic whereas most wholly terrestrial animals are uricotelic or ureotelic. For aquatic animals, the exceptions to the ammoniotelic rule are readily explained (see Schmidt-Nielsen, Animal Physiology). But the crocodilians excrete ammonium salts and uric acid; their important ammoniotelism remains to be explained. Regarding *respiration*, the opposition between air breathers and water breathers has been schematically outlined by Rahn (1966) and the consequence for acid-base balance of body fluids immediately envisaged (Rahn, 1967).

Many functions are completely different in terrestrial animals on the one hand, and aquatic animals on the other. Table 2 is a schematic résumé of the main differences in completely aquatic and completely terrestrial animals. The fact that these differences exist in many groups (Table 3) which are totally unconnected by evolution indicates that some *convergences of functions* are related to the characteristics of the milieus.

However, many books of comparative physiology do not attempt such a distinction. One reason may be that the contrasts may not be general. Whereas in some terrestrial animals, *e.g.* insects, all functions, respiration, excretion, water economy, etc., are different from those of aquatic animals, *e.g.* cephalopods, some animals are *intermediate*. (1) Some are amphibious, because they spend a part of their lifetime in water and the rest in air, as intertidal animals or some *amphibians*. (2) Some annelids and arthropods are intermediate because they live in a more or less wet soil. (3) Other animals as turtles and cetaceans are descendants of terrestrial animals, and they keep some traits of their terrestrial origins. For example, the whales and porpoises are air-breathers and have the respiratory and excretory characteristics of their ancestors, but because their gross body shape is fishlike, the cetaceans were classified until the 18th century among the Pisces. Animals like the aquatic insects, reptiles and mammals are certainly aquatic by their abode, but physiologically they are intermediate; certain functions - locomotion, circulation, sensory information - are of an aquatic nature; some others - as respiration, nitrogen metabolism - are terrestrial.

But the existence of intermediate animals is not in itself an insuperable difficulty. All animal classifications meet similar problems of borderline, intermediate animals. We easily contrast:

Table 2

Some characteristics of typical aquatic and terrestrial animals

	A Q U A T I C	T E R R E S T R I A L
Respiration	skin and/or gills	tracheae or lungs
P_{CO_2}	small P_{CO_2} between body fluids and ambient water	large P_{CO_2} between body fluids and ambient air
$[HCO_3^-]$	low	high
N end products	mainly ammonia	mainly urea and/or uric acid and other purine derivatives
Water turnover	high or very high	low or very low
Locomotion	swimming	running, flying
Temperature	poikilothermy (most of them)	homeothermy (some of them)

- Invertebrates and vertebrates, but we have the non-vertebrate chordates.
- Poikilotherms and homeotherms, but some poikilotherms have features of homeotherms (*e.g.* tuna, social insects) and in some higher vertebrates homeothermy may
be imperfect or intermittent.
- Osmoconformers and osmoregulators, but many animals do not enter into either
category and may be considered as imperfect osmoregulators.

On the contrary, the existence of intermediate animals may be instructive because
their study may shed light upon the evolutionary process by which some aquatic
animals could invade land, and particularly how some terrestrial animals could
become homeothermic.

The amphibians, particularly the anurans, are most interesting because in this
group we find all the intermediaries between the completely aquatic life of the
tadpoles to the completely aerial life of the treefrogs which have excellent
water-conserving mechanisms and are uricotelic, like the reptiles. Also, among
adult amphibians, one finds some who are ammoniotelic in water, and ureotelic when
they are out of water, *e.g. Xenopus*, and these animals may oscillate between
ammoniotelism and ureotelism as a function of the water availability. It is rather

Table 3

Main animal groups with aquatic and terrestiral forms

Phylum	Aquatic	Terrestrial
Flatworms	+	few
Annelids	+	+
Molluscs	+	+ (only snails)
Arthropods	+	+ { Arachnids Myriapods Insects
Vertebrates	+	+

common to say that the amphibians offer a recapitulation of the *evolution* from water to land colonization.

In the programme of this symposium, the various functions which have been mentioned above are considered separately as if they were independent of each other. This may be necessary, but is obviously unsatisfactory. Comparative anatomists know very well the correlation between the various parts of the body. Certainly, some physiologists do, but maybe not as systematically. In most studies presented here, terrestrial life is compared to aquatic life. This seems logical if one accepts that terrestrial animals are the heirs of aquatic animals. But to understand the problems raised by the recolonization of waters the terms of comparison are the terrestrial animals.

Very little has been done to consider *simultaneously* all the anatomical and physiological problems raised by the transition from aquatic to terrestrial life, an evolutionary question. One may suggest the following scenario. The *major* problem for an aquatic animal leaving water and entering air is not a problem of dessication, since there are many microenvironments which are completely water-saturated. It is not a problem of respiration because air is rich in oxygen and accepts carbon dioxide as easily as waters. The major problem is the catabolism of the nitrogenous compounds. Most aquatic animals are ammoniotelic. They may be since water is a sink for ammonia due to its extreme solubility, whereas the capacitance of air for ammonia is 1/700th that of water. That is to say that it is impossible for ammoniotelic animals to remain so if they leave water since ammonia is highly toxic. There are two exceptions: some arthropods and the crocodilians.

Certainly *some* ammonia can go into air through the tegument or via ventilation, or into urine and feces, but the majority of the organic nitrogen catabolism yields uric acid which needs little water since it precipitates (reptiles, birds), or some purine derivatives (insects), or urea which is relatively soluble in water (mammals).

The tracheal system allows insects to use the oxygen of the atmosphere without the need of a blood carrier. The clearance of CO_2 does not raise particular problems since its diffusive coefficient is only a little lower than that of oxygen.

Air being very rich in oxygen, the animals can breathe much less as long as they possess organs for air breathing, and can solve the problem of hypercapnia and potential acidosis. Modified gills, special mucosal diverticula, modified gas bladders, and *stricto sensu* lungs provide the surface through which oxygen enters the blood or the hemolymph, and since air is rich in oxygen, the animals can breathe much less, as they actually do. But any decrease of ventilation automatically increases P_{CO_2} over the respiratory surface and in the body fluids. This hypercapnia leads to acidosis, unless there is a proportional increase of $[HCO_3^-]$. A common feature of all air breathing animals is hyperconcentration of HCO_3^-, as compared to the aquatic animal's $[HCO_3^-]$ (see Rahn and Dejours, this volume). But an increase of $[HCO_3^-]$ immediately raises the problem of ionoregulation. Necessarily the sum of anions is equal to the sum of cations. HCO_3^- is a weak anion. Its concentration may be 3-4 meq$\cdot$L^{-1} in aquatic animals and HCO_3^- fills most of the strong ion difference (Stewart, 1978). $[HCO_3^-]$ may reach 50 meq$\cdot$L^{-1} in reptiles, that is the strong ion difference may be increased more than ten times. Of the two logical possible solutions to keep the ionic balance, namely an increase of the cations Na^+, K^+, Mg^{2+}, Ca^{2+}, or a decrease of a strong anion, namely Cl^-, it is the second solution which is mainly observed. In this way, there are small changes in osmotic pressure and in the individual concentrations of cations, whose interstitial activities are very important for the cellular metabolism.

Obviously what is said above cannot be considered independently of the water economy. A frog may live in a damp atmosphere and use its lungs, become hypercapnic, and increase its bicarbonate to keep a normal pH. Actually it may enter a relatively dry atmosphere for a short time, but to avoid dessication it will have to return from time to time to some water pool. However, there exist some tree-frogs who are able to live in a very dry atmosphere; their skin is no more permeable to water than the reptilian skin. Furthermore they produce uric acid which precipitates in the cloaca. As adults they may be considered as perfect terrestrial animals, although they start their life as aquatic tadpoles (see Shoemaker, in this volume).

But it is among reptiles, birds and mammals, that are found the most terrestrial vertebrates. The end product of the nitrogen catabolism is mainly uric acid in reptiles and birds, mainly urea in mammals. In reptiles and birds, the kidney does not form hypertonic urine but water is absorbed in the lower intestine. In mammals the kidney forms hypertonic urine, the concentration of which may exceed 9 osmol·L^{-1} in one species of Australian desert hopping mouse.

Insects, reptiles, birds and mammals may not lose much water by evaporation, thus minimizing the cooling effect of vaporization. Then they are able to have a body temperature higher than ambient temperature. Some insects and reptiles show some kind of thermal regulation by basking in the sun, or by increased energy metaboliqm (see Schmidt-Nielsen, 1983, pp. 296-301). But complete homeothermy is observed only in birds and mammals and results from behavorial tricks, from variations in energy metabolism and from the control of the heat excess by varying the water evaporation (panting, gular flutter, sweating). Some birds and mammals are permanent homeotherms. Some are heterotherms: the bats are circadian heterotherms; the hibernators are circannual heterotherms. Some accept a variable body temperature and, as the camels, make good use of thermal inertia.

If one adds to the problems of N catabolism, respiration, ionoregulation, water economy, and eventual homeothermy, the problems of communication, nutrition, reproduction (in terrestrial animals the fecundation is internal), locomotion, skeleton development and circulation (because of the strong effect of gravity in terrestrial animals), it is obvious that the transition from aquatic life to terrestrial life is very complex; all functions are different.

The topic of this conference thus requires the contribution of scientists of different fields: zoology, morphology, ecology, ecophysiology, comparative physiology, biochemistry, biophysics. The presence of botanists would have been justified for their appreciation of the possible contrasts in design and function of aquatic and terrestrial plants and, because plants are the primary foodstuff of the animals, one must consider the coevolution of the two kingdoms.

REFERENCES

Dejours P (1979) La vie dans l'eau et dans l'air (Life in water and in air). Pour la Science No. 20: 87-95.

Dejours P (1981) Principles of comparative respiratory physiology, second edition, Elsevier, Amsterdam, 264 p. First edition (1975), 253 p.

Little C (1983) The colonisation of land. Origins and adaptations of terrestrial animals, Cambridge University Press, Cambridge; 290 p.

Rahn H (1966) Aquatic gas exchange. Theory. Respir Physiol 1: 1-12.

Rahn H (1967) Gas transport from the external environment to the cell. In: de Reuck AVS, Porter R (eds): Development of the lung. Ciba Foundation Symposium, J & A Churchill, London; pp. 3-23.

Schmidt-Nielsen K (1983) Animal physiology. Adaptation and environment, Third edition, Cambridge University Press, Cambridge, 619 p. First edition (1975) 699 p., Second edition (1979), 560 p.

Stewart PA (1978) Independent and dependent variables of acid-base control. Respir Physiol 33: 9-26.

Vannier G (1983) The importance of ecophysiology for both biotic and abiotic studies of the soil. In Lebrun P et al. (eds): New trends in soil biology. Dieu-Brichart, Ottignies-Louvain-la-Neuve; pp. 289-314.

Part I

RESPIRATION AND ACID-BASE BALANCE

RESPIRATORY EXCHANGES IN AIR BREATHERS VERSUS WATER BREATHERS: DESIGN CONTRASTS

José Eduardo P.W. Bicudo and Ewald R. Weibel

Department of Anatomy
University of Berne
Berne, Switzerland

The design of gas exchangers for water and air breathing is determined by (a) anatomic and functional features of circulation, and (b) physical properties of the respiratory medium, specifically the relative solubility and diffusivity of oxygen in water and air. Gills are evaginated and have parallel lamellae allowing complete convective ventilation. Lungs are invaginated to prevent water loss and counteract mechanical forces. Microstructure of the gas exchangers depends on capillary blood pressure. Secondary adaptation to air breathing in lungfishes follows the same design principles, but with other means.

INTRODUCTION: THE EVOLUTION OF GAS EXCHANGERS

To supply oxygen to the metabolically active cells in the depth of the body and to remove carbon dioxide efficiently is one of the fundamental problems of survival of all animals too large to exchange gases directly with the environment by simple diffusion. The problems are different in water and in air for two main reasons: (1) the air is rich in O_2, whereas at the same partial pressure the O_2 content of water is 30 times lower because of poor solubility; (2) O_2 diffuses about 300 000 times faster through air than through water. Additional problems occur in air breathing, however, because of the mechanical forces that act at air-fluid interfaces, i.e. at the interface between the body or its gas exchanger and air. This required the development of different designs of respiratory organs for air or water breathing.

Primary adaptation to air breathing in invertebrates

It appears that terrestrial arthropods evolved primarily in air and have taken advantage of the high solubility and diffusivity of O_2 in air in building an internal system of air conduits that penetrate into the tissues. In insects and arachnids the air breathing organ consists of a system of noncollapsible chitinous hollow tubes, called tracheae, opening onto the surface through special apertures, the spiracles, which are more or less open according to the acitivity of the animals and to the ambient conditions. The tracheae originate as ectodermal invaginations and the system may be ramified, branching into the tissues and even penetrating into the muscle cells where the terminal tracheoles come to lie in the immediate vicinity of mitochondria (Fig. 1). The tracheal system may be simple in small, relatively inactive insects and arachnids, but in others it may be quite complex, with many spiracles, atria, and even air-sacs; longitudinal and transversal connections may occur between tracheae.

Renewal of gas along the tracheae is often adequately insured by diffusion through the air, although some convection could be produced by various body movements. The chitin spirals form a skeleton that counteracts surface forces and keep the tracheal system open.

Comparative Physiology: Life in Water and on Land. P. Dejours, L. Bolis, C.R. Taylor, E.R. Weibel (eds.) Fidia Research Series, IX-Liviana Press, Padova © 1987

The tracheal system of aquatic insects is also air-filled and they take O_2 either from an air-bubble that they maintain around their abdomen and periodically replenish at the water surface, or from water through a gas exchanger membrane that separates the tracheal system from the water. It is likely that the respiratory system of insects evolved in air as a primary adaptation to terrestrial life, as attested by their probable predecessors, the onychophorans, which are terrestrial invertebrates and possess a tracheal system similar to that of insects. Aquatic insects have therefore adapted to their wet environment secondarily.

The development of special air breathing organs in some arachnids, namely tracheal or book lungs, depended on the development of an efficient circulation of hemolymph between the lung and the tissues, and of an O_2 carrying pigment, mostly hemocyanin, which allowed to increase O_2 content of hemolymph. Similar simple lungs develop in terrestrial molluscs, in the pulmonate snails.

Primary adaptation to water breathing in invertebrates

Aquatic molluscs, like the octopus and squid, have well developed gills or ctenidia specially designed for gas exchange, while in bivalves, for instance, the gills play a dual role, i.e. respiration and ciliary feeding. Cephalopods, in general, are quite active animals requiring a more effective gas exchanger, which is further accomplished by a capillary circulation in the gills and the presence of a respiratory pigment in the blood, usually hemocyanin. Aquatic gastropods, although less active molluscs than cephalopods, also possess gills and circulating hemocyanin.

Aquatic crustaceans present gills which are usually ventilated by the paddle-like movements of special appendages. In the larger forms they are well vascularized with circulating respiratory pigments (hemocyanin) in the blood, although they do not develop capillaries such as are found in the cephalopods.

Primary adaptation to water breathing in vertebrates

Vertebrates have evolved primarily in water and thus they needed to adapt to water breathing. A set of branchial vessels first evolved at the interface between the foregut and skin - already developed in Amphioxus. The slits that form between the branchial arches are perfused by water which is pumped outwards by the pharynx and into a gill chamber which opens externally. The branchial arches, which are provided with a skeleton, increase their gas exchanging surface by forming long gill filaments which extend outwards into the gill chamber; on these filaments an actual gas exchanger is mounted in form of 30 - 40 densely spaced leaf-like secondary lamellae (Fig. 2). Water flows between these lamellae in the outward direction. Each filament has an afferent and efferent vascular supply derived from branchial arteries in the gill arches. In the capillaries of the secondary lamellae blood flows in the opposite direction to the external respiratory water current. This countercurrent design greatly increases the efficiency of gills as gas exchangers so that the low O_2 content of water can be better exploited.

Secondary adaptation to air breathing in vertebrates: Phylogenetic aspects

Evaginated gills are not suitable for breathing in air, particularly since surface tension would make it very difficult to keep them separated; they would tend to stick together so that the surface available for gas exchange is very much reduced. Eventhough, a few fish like Synbranchus and also some land crabs are able to practice air breathing with their gills, already engineered to such remarkable efficiency. In most of the cases, however, air breathing requires the gas exchangers to be housed in invaginated pouches with the capillaries contained in thin walls that can be kept extended by tension. Such internal gas exchangers have not evolved from the skin, as was the case with gills, but

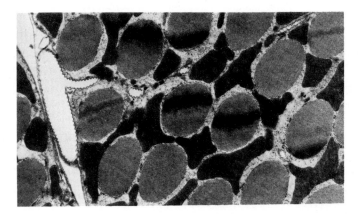

Fig. 1. Transverse section of flight muscle of bumblebee demonstrating the close relation between tracheoles and mitochondria; X 6400 (courtesy of J. Gabriel, Cambridge Univ. and H. Hoppeler, Univ. of Berne).

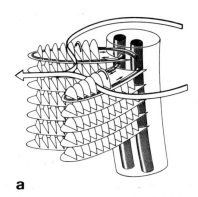

Fig.2. Gill arch general arrangement (a) with filaments in paired rows and thin lamellae mounted on the filaments. (b) SEM of a gill filament with regularly spaced lamellae mounted on both sides (X 130).

rather from the gut where the integument is thinner a priori, being designed to allow the uptake of nutrients, and thus offers a distinct advantage for gas exchange. In air breathing the greatest danger is dessication from the large surface and this can be minimized by controlling the humidity in the invaginated lung, so that the barrier separating air and blood can be made very thin. Most fishes that depend on air breathing, will also develop lungs from outpouchings of the dorsal lateral wall of the foregut, in addition to the gills they need for water breathing, as will be described below.

The lungs which are the evolutionary antecedants of the gas exchanger of terrestrial vertebrates are formed from ventral invaginations of the foregut into the chest cavity. The principal design feature of these lungs is that their gas exchange surface is made large by forming multiple walls that subdivide the air space and contain dense capillary networks (Fig. 3). The large numbers of air chambers resulting from this partitioning all communicate with the outside air through a system of usually branched tubes. These air chambers or alveoli are ventilated either with the use of a pressure pump, as in amphibians, or a suction pump, as in most reptiles, birds, and mammals. In either case the air in the lung is renewed periodically by cyclic inspiration and expiration movements. The lung serves as well as gas exchanger and as ventilator bellow.

While the mammalian lungs derived probably from the homogeneous lungs of therapsid reptiles, the avian lung differentiated from reptilian lungs with heterogeneous lung partitioning (Duncker, 1978). In the bird, ventilation is achieved by means of air sacs, which originate ontogenetically out of bronchi, whereas the volume-constant lung develops a set of parabronchi surrounded by an air capillary network intimately associated with blood capillaries. The long and only poorly interconnected blood capillaries arise from arterioles at the outside of the parabronchi, the functional unit of the lung, and are collected mainly by venules at the luminal side. This special arrangement of the capillaries and that of the long, unidirectional parabronchi are responsible for the high gas exchange capacity of the avian lung, and can be best described as a cross-current type of system.

Ontogenetic aspects - Amphibians

The transition of amphibians from aquatic eggs via tedpoles to air breathing adults is accompanied by changes in the respiratory processes. The amphibian egg does not exhibit structural specializations for gas exchange and therefore obtains ample gas exchange through the general surface. The tadpole shows, as a general pattern, first external gills, then internal gills and, later in the metamorphosis to adult frog, also lungs.

DESIGN PRINCIPLES FOR GAS EXCHANGERS

In principle, the main function of a gas exchanger, whether the organism is a water breather or an air breather, is to supply continuously enough O_2 to cover the demands, and also to eliminate CO_2, the gaseous end product of combustion, from the interior compartments of the body. The efficiency of the gas exchanger is importantly determined by certain structural properties. Thus gas exchange between water or air and blood is improved the greater the surface area of the gas exchanger, and the smaller the tissue barrier thickness. Also, the volume of capillary blood exposed to either water or air influences gas exchange.

The fact that gas exchanger surface area must be large enough to provide adequate gas exchange is well expressed in fish. Highly active fish have correspondingly the largest relative gill areas. The fast swimming mackerel's gill surface area, for example, expressed per unit body weight, is some 50 times than that of the sluggish, bottom-living goosefish (Gray, 1954).
The obvious environmental difference presented by aquatic and terrestrial

habitat is associated with a generally higher metabolic rate in animals that live on land, and one sees a progressive increase in the surface area of the lungs as one goes from amphibians and reptiles to mammals and birds (Czopek, 1965; Perry, 1983).

Cutaneous gas exchange, typically associated with branchial or pulmonary gas exchange, also contributes significantly to tissue respiration, mostly CO_2 elimination, in various fishes, nearly all amphibians, many reptiles and certain mammals (Feder and Burggren, 1985).

DESIGN CONTRASTS

Why are gills evaginated?

Because the O_2 content of water is approximately 30 times lower than that of air and also because of the poor diffusivity of O_2 in water, organisms breathing in water have to renew the medium on the outside of the gas exchanger at a greater rate than the blood that perfuses it. Ventilation is tidal in lungs, but usually continuous and unidirectional in gills. The basic design principle of gills is therefore to provide a set of slits for laminar water flow, the branchial clefts, separated by perfused branchial arches. This is the situation in Amphioxus. As the demand for more respiration increases, the ventilation slits are made more complex by adding secondary (filaments) and tertiary structures (lamellae) to the primary branchial arches. This is achieved by building these elements of the gas exchanger about a hierarchically structured vascular system (Fig. 4a). Thus the gill structures form as evaginations of branchial arches into the gill chamber. This construction principle serves water ventilation by continuous through-flow particularly well; no water pockets are formed through which O_2 would have to diffuse, and also, the gills have virtually no dead space.

It is important to note that, in the fish, the heart has a single chamber which pumps only deoxygenated blood. The gill vasculature is inserted between ventral and dorsal aorta; it thus is in series with the tissue vessels which it precedes, and is therefore exposed to high blood pressures. This will have important effects on the construction of the secondary lamellae which must resist this high pressure and still must serve gas exchange efficiently (see below).
Gill structures are, therefore, evaginated to allow smooth uninhibited ventilation by water which needs to flow through at high rate and meet a low resistance, whereas blood perfusion at high pressure is easy. This design eliminates pockets of "residual water" whose PO_2 would fall rapidly because of slow diffusion of O_2 in water.

Why are lungs invaginated?

In air breathers the rate of air flow (ventilation) over the respiratory surfaces practically matches the rate of blood flow, mainly because the O_2 content of air is high (similar to that of blood in mammals) and also because of the high rate of diffusion of O_2 in it, which at the same partial pressure, is some 300 000 times as rapid as in water. Accordingly, ventilation of the respiratory surface must not be achieved exclusively by convection, as the PO_2 in lung air can be easily equilibrated by diffusion of O_2 in the air phase, even into pockets of residual air. On the other hand, one of the major dangers in air breathers is water loss. To prevent undue evaporation from the respiratory surfaces, these should not be directly exposed to the outside air. The gas exchange surfaces are therefore located in specialized respiratory cavities, the lungs, which are connected to outside air through restricted pathways, a single trachea from which the intrapulmonary airways arise by branching.
The alveolar lungs of air breathing vertebrates are designed from the airways

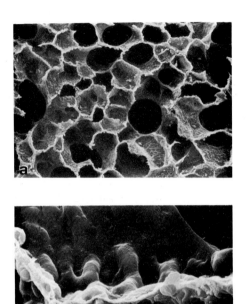

Fig. 3. (a) Alveolar lung gas exchange surface attached to airway ducts (X 50). (b) Walls between alveoli containing a dense network of blood capillaries, separated from air by thin tissue barrier (X 900).

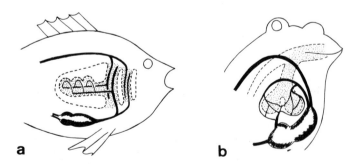

Fig. 4. (a) Gills are evaginations of branchial arches into gill chamber, where the elements of the gas exchanger are built about a hierarchically structured vascular system. (b) Lungs are invaginated structures derived from the gut anlage.

which derive directly from the gut anlage (Fig. 4b). Thus, the pattern of lung development is determined by the branching pattern of the airway tubes, but this depends on a close interaction with the mesenchyme. The pulmonary arteries grow into the lung anlage from the sixth branchial artery and form a capillary network around the proliferating bronchial tubes. The pulmonary blood circuit of terrestrial vertebrates is characterized as being a low pressure system, permitting, consequently, the construction of a very thin tissue barrier between the circulating blood and the outside air. The thin barrier of alveolar lungs is suspended on a fibrous skeleton in form of a continuous fiber network which is kept under tension, a feature which allows the surface forces to be better counteracted (Fig. 5b).

In alveolar lungs all air chambers are apparently ventilated as a pool, whereas in the lungs of birds the parabronchi are ventilated by a continuous stream of air from large abdominal air sacs that are periodically filled with fresh air; the outflowing air is collected into a separate set of anterior air sacs.

Lungs are therefore invaginated to allow mechanical stability in spite of adverse surface forces, and to prevent excessive water loss. The ensuing problems of ventilation are unimportant because of rapid diffusive O_2 equilibration in air.

Construction of gill lamella versus alveolar wall

The major differences between the gill lamella and the alveolar walls are, on the one hand, the high pressure of blood flow through the gills and the low pressure of pulmonary blood flow, and, on the other hand, the circumfusion of gill lamellae by rapidly flowing water and the exposure of alveolar walls to an air-water interface with ensuing surface forces. The design of the gas exchangers will have to account for these differences.

The gill lamella must resist the shear forces of water flow. They are accordingly lined by a relatively thick two-layered epithelium provided with a thick basement membrane reinforced by a felt of collagen fibrils (Fig. 6a). In order to resist the high vascular pressure the "capillaries" of the lamella are spanned and lined by pillar cells; these are transformed fibroblasts and are associated with collagen columns which brace the two strong basement membranes. In addition, pillar cells contain cytoplasmic bundles of actin filaments (Bettex-Galland and Hughes, 1973) and thus have contractile properties similar to smooth muscle cells. This construction of the gill lamella results in a relatively thick tissue barrier, but this may not be a major disadvantage because the counter current flow of blood and water allows for efficient gas exchange. The high blood pressure prevailing in the gill vasculature is also responsible for maintaining the gills turgid, thus allowing them to stand erect and to resist the shear forces of water.

In contrast, alveolar walls are very delicate structures; the low blood pressure allows the gas exchange barrier to be made very thin. The structural backbone of the alveolar lung is a continuous system of fibers that are anchored at the hilum and are kept under tension by the negative intrapleural pressure that tugs on the visceral pleura (Weibel, 1984). The presence of the surfactant lining of alveoli reduces the surface forces and can vary its surface tension. As a result of this feature, alveoli (small bubbles 1/4 mm in diameter) do not behave like soap bubbles; the alveolar complex is astonishingly stable and the interplay between these two features allows the gas exchange surface to remain adequately expanded.

Fiber tension and surface forces also affect the configuration of capillaries in the alveolar septa. The capillary network is expanded in the septal plane due to the extension of the septal fibers with which it is interlaced (Fig. 5b). The low blood pressure in the pulmonary circuit is possible because of low surface

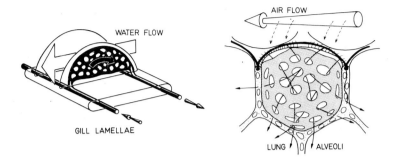

Fig. 5. Comparison between the designs of (a) gill lamella and (b) lung
alveolus. Gill lamellae are rigid structures supported by pillar cells and
perfused by blood at high pressure, whereas lung alveoli are delicate structures
supported by an interlacing fiber network kept under tension.

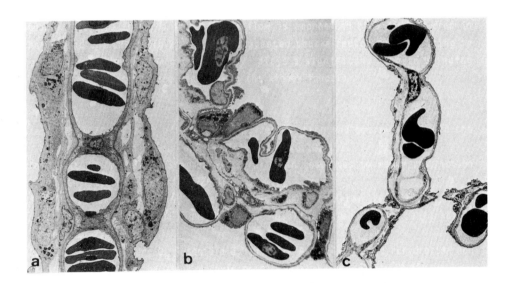

Fig. 6. (a) Gill lamella microstructure showing a thick two-layered epithelium
and a strong basement membrane reinforced by collagen fibrils. The capillaries
are lined by pillar cells (X 1300). (b) Frog alveolar septum with double
capillary network (X 750). (c) Dog alveolar septum with a single capillary
network. The tissue barrier is made very thin because the epithelial and
endothelial basement membranes are fused over half of the barrier surface (X
2000).

forces prevailing on the alveolar wall. The design of the alveolar septum is the result of balancing the various micromechanical forces that act on the thin air-blood tissue barrier: tissue tension, blood pressure, and surface tension (Weibel, 1984).

Keeping the air-blood barrier dry and thin is a further mechanical problem, important to insure adequate conditions for efficient gas exchange. It is solved by giving the two cellular linings of epithelium and endothelium different permeability properties, and by providing the septal interstitial space with features that allow rapid drainage of lymph fluid. Another important feature is that the epithelial and endothelial basement membranes are fused over half the barrier surface, thus eliminating an interstitial space in these regions and minimizing the thickness of the tissue barrier.
It is interesting to notice that there is a systematic decrease in the harmonic mean thickness of the tissue barrier as well as in capillary loading (Vc/Sa) from amphibians to mammals and birds (Meban, 1980; Perry, 1983). One of the reasons for the differences in capillary loading is because the alveolar surface is only found on one side of the capillaries in amphibians as opposed to what is found in adult mammals, for example (Fig. 6c). This together with the increasing lung surface areas mentioned before has led to progressively increased lung diffusing capacities (D_LO_2) as one goes from amphibians to mammals and birds (Perry, 1983).

In the mammalian lung, phylogeny is in a way recreated during development. At birth the lung is still in its saccular stage, similar to the design found in amphibia, but it is adequately designed to take up gas exchange. The saccules can be expanded with air upon the first breath if the secretory cells have produced enough surfactant. The barrier is thin and O_2 can diffuse into the blood, but the septa still have two capillary networks as the ones observed in the lower terrestrial vertebrates. In the weeks that follow, the saccules transform into alveolar ducts by forming a very large number of alveoli, small pouches in the wall of the saccule. This causes the gas exchange surface to increase drastically; now the septa between alveoli have only a single capillary network which is in contact with alveolar air on both sides (Burri, 1985).

SECONDARY AIR BREATHING IN FISHES

Shortage of O_2 and the high cost of extracting it from water compelled some fish to select means of absorbing O_2 directly from the atmosphere. Seasonal droughts further aggravated the conditions for respiratory exchange in some water breathers. Environmental conditions, thus, called for direct absorption of atmospheric O_2 in some fish, and fossil records and extant fish suggest that diverse solutions for obtaining O_2 directly from the atmosphere did occur and that air breathing in fish evolved repeatedly in many different ways. A few selected examples will suffice to demonstrate the different designs of the air breathing organs in fishes.

The air breathing organ of Heteropneustes fossilis has been derived by modification of the gills (last branchial arch) (Hughes and Munshi, 1973). The suprabranchial chambers extend backwards into the trunk region in the form of a tube deeply embedded into the trunk myotomes. Air enters first the suprabranchial chamber through the inhalant aperture guarded by the second and third gill fans and then into the posterior tubular part. The respiratory surface is derived from transformed secondary lamellae which lie flat on the surface of the respiratory sac, having two distinct regions: the non-respiratory basal and the respiratory distal parts (Fig. 7a). The lamellae retain their relative position in relation to the afferent and efferent primary filamentar arteries. Pillar cells with columns containing collagen fibrils, as found in the gills of fish, are also found in the lamella-like structures of the respiratory islets of the air sacs (Fig. 7b), in both the respiratory and non-respiratory

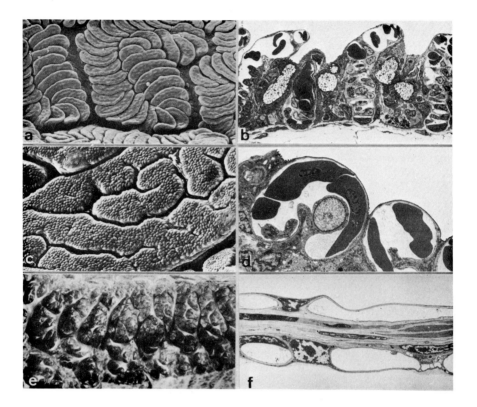

Fig. 7. Air breathing organs in fishes. (a) Respiratory surface (X 180) and (b) lamella-like structures (X 750) of the respiratory islets of the air sacs in _Heteropneustes fossilis_. (c) Respiratory islets (X 100) and (d) vascular papillae (X 2000) in _Monopterus cuchia_. (e) Inside view of _Lepidosiren paradoxa_ lung with septa and trabeculae and (f) lung septum with double capillary network (X 680).

regions, which suggests that they are structures homologous to gill secondary lamellae (Munshi et al., 1986a).

The mud eel Monopterus cuchia is an obligatory air breather, highly terrestrial and can survive indefinitely on land if it is kept moist. Air breathing is mostly accomplished with a pair of pharyngeal inflatable sacs, which receive their blood supply from the heart via afferent branchial vessels having lost nearly all functional gill filaments. The air sacs are postero-dorsal extensions of the pharynx; the air sacs are lined by a membrane having vascular respiratory islets (Fig. 7c) and non-vascular areas (Munshi et al., 1986b). Also the pharynx, the hypopharynx and the pharyngeal region of the branchial arches participate in air breathing, and are lined by the same type of respiratory islets. The entire respiratory membrane is covered by a protective mucous layer which is anchored by microbridges on the outer epithelial surface. The respiratory islets consist of vascular papillae, which are specialized capillaries which take the form of spirals (Fig. 7d). This provides an efficient mechanism to bring every individual red blood cell into intimate contact with the respiratory surface for gas exchange. It also provides enough resistance to increase their residence time and ensure proper oxygenation (Munshi et al., 1986b).

In the South-American lungfish, Lepidosiren paradoxa, the air breathing organ consists of two long dorsal tubes, originating from the foregut. The main airway which opens into the posterior region of the pharynx, continues throughout the whole length of the lung. The internal lung surface is much folded, especially in the dorsal regions (Fig. 7e). Folding takes the form of septa and trabeculae (Hughes and Weibel, 1976). In some ways the lungs of Lepidosiren resemble the lungs of amphibians (Goniakowska-Witalinska, 1980). The surface of the lung is provided with a lining film which contains phospholipids, similar to surfactant in mammalian and amphibians lungs, except for the absence of the tubular myelin figures characteristic of true lung surfactant (Hughes and Weibel, 1978). Also the capillaries do not continue at the edges of some of the septa, and consequently non-respiratory surfaces are present. The septa are relatively thick, containing much connective tissue and smooth muscle, and there are capillaries on both sides, as in the African lungfish (Maina and Maloiy, 1985) and amphibian lung septa (Fig. 7f).

BREATHING AIR VERSUS WATER: DESIGN CONTRASTS

The gas exchangers that have evolved for air and water breathing, respectively, have adapted their design to the boundary conditions for gas exchange imposed by the respiratory medium. Gills of invertebrates and fishes show basically the same design pattern: the vasculature is contained in parallel lamellae which are circumfused with rapidly flowing water. In fishes, the construction of a counter-current ventilation-perfusion system greatly improves efficiency and counteracts some of the problems, for example the need for a thick barrier because of the high perfusion pressure.

In air breathing organs the major problems to overcome are the danger of water loss and the mechanical stresses generated at the air-water surfaces. Invaginating the gas exchangers contributes to the solution of both problems, and is no disadvantage for ventilation. The mechanical forces are overcome in insects by providing the tracheoles with a rigid chitin spiral. In lungs these forces are compensated by mounting the gas exchange elements (alveolar septa) on a continuous fiber system kept under tension, and by providing the surface with a surfactant lining. It is interesting that in the secondary development of lung-like organs in air breathing fishes design features very similar to those encountered in amphibian and mammalian lungs evolve.

AKNOWLEDGEMENTS: The technical assistance of K. Babl, F. Doffey and W. Herrmann are gratefully appreciated. This work was supported by Swiss National Foundation grant 3.036.84, and by Sao Paulo State Research Foundation (FAPESP), Brazil.

REFERENCES

Bettex-Galland M & Hughes GM (1973) Contractile filamentous material in pillar cells of fish gills. J Cell Sci 13: 359-366.

Burri, PH (1985) Development and growth of the human lung. In: Fishman, AP, Fisher, AB (eds): Handbook of Physiology, vol 4. American Physiological Society, Washington; pp. 1-46.

Czopek, J (1965) Quantitative studies on the morphology of respiratory surfaces in amphibians. Acta Anat 62: 296-323.

Duncker, H-R (1978) General morphological principles of amniotic lungs. In: Piiper, J (ed): Respiratory function in birds, adult and embryonic. Springer-Verlag, Berlin, Heidelberg; pp. 2-15.

Feder, ME & Burggren, WW (1985) Cutaneous gas exchange in vertebrates: design, patterns, control and implications. Biol Rev 60: 1-45.

Goniakowska-Witalinska, L (1980) Ultrastructural and morphometric changes in the lung of Triturus crisatus carnifex Laur. during ontogeny. J Anat 130: 571-583.

Gray, IE (1954) Comparative study of the gill area of marine fishes. Biol Bull 107: 219-225.

Hughes, GM & Munshi, JSD (1973) Nature of the air-breathing organs of the Indian fishes Channa, Amphipnous, Clarias and Saccobranchus as shown by electron microscopy. J Zool (Lond) 170: 245-270.

Hughes, GM & Weibel, ER (1976) Morphometry of fish lungs. In: Hughes, GM (ed): Respiration of amphibious vertebrates. Academic Press, London, New York; pp. 213-232.

Hughes, GM & Weibel, ER (1978) Visualization of layers lining the lung of the South American (Lepidosiren paradoxa) and a comparison with the frog and rat. Tissue & Cell 10: 343-353.

Maina, JN & Maloiy, GMO (1985) The morphometry of the lung of the African lungfish (Protopterus aethiopicus): its structural-functional correlations. Proc R Soc Lond B 224: 399-420.

Meban, JN (1980) Thickness of the air-blood barriers in vertebrate lungs. J Anat 131: 299-307.

Munshi, JSD, Weibel, ER, Gehr, P, Hughes, GM (1986a) Structure of the respiratory air sac of Heteropneustes fossilis (Bloch), Proc Ind Nat Sci Acad, in press.

Munshi, JSD, Hughes, GM, Gehr, P, Weibel, ER (1986b) Structure and morphometry of the air breathing organs of a swamp mud eel, Monopterus cuchia (Ham.), J Zool (Lond), in press.

Perry, SF (1983) Reptilian lungs: functional anatomy and evolution. Adv Anat Embryol Cell Biol 79: 1-81.

Tenney, SM & Tenney, JB (1970) Quantitative morphology of cold-blooded lungs: amphibia and reptilia. Respir Physiol 9: 197-215.

Weibel, ER (1984) The Pathway for Oxygen. Harvard University Press, Cambridge.

RESPIRATORY TRANSITION AND ACID-BASE BALANCE FROM WATER TO AIR

Hermann Rahn and Pierre Dejours[1]

Department of Physiology, State University of New York at Buffalo, Buffalo, NY, U.S.A., and [1] Laboratoire d'Etude des Régulations Physiologiques, C.N.R.S., 67087 Strasbourg, France

Because O_2 in water is about 30 times less soluble than in air, water breathing animals must ventilate considerably more than air breathers, and as a consequence their CO_2 tensions are maintained below 3-4 Torr. With the initiation of air breathing, ventilation could be reduced, giving rise to an increase in CO_2 tensions and blood bicarbonate level. Air breathing fish and amphibians today serve as examples of how the transition from water to land was accomplished. The unimodal gas exchange system of water breathing gills was exchanged for a bimodal gas exchange system where the primitive air organ or lung provided about 65% of the oxygen needs, while the gills and/or skin served for the elimination of about 75% of the CO_2. Not until the advent of reptiles do we again find a unimodal gas exchange system, the modern lung, which took care of all the O_2 and CO_2 exchange so that the naked skin of the transition animals could be replaced by scales to resist dehydration. In contrast to water breathers, the modern lung-air breather lives in a compensated state of respiratory acidosis with a reduced ventilation, high CO_2 tension and bicarbonate level. However, throughout the long transition and whatever the body temperature, a normal acid-base balance was maintained, not by regulating a constant blood pH, but by regulating a constant relative alkalinity.

INTRODUCTION

Transition from water to land eventually led to replacement of water breathing gills by air breathing lungs. This change, however, was not accomplished easily, judging from the large number of air breathing fish and amphibians which were caught at the air-water interface, where they can still be found today. These extant forms allow one to interpret the various structural and functional changes

Comparative Physiology: Life in Water and on Land. P. Dejours, L. Bolis, C.R. Taylor, E.R. Weibel (eds.) Fidia Research Series, IX-Liviana Press, Padova © 1987

that must have occurred which permitted other more successful ancestral animals to make the complete transition from gills to modern lungs of reptiles, birds, and mammals.

This transition was constrained not only by morphological changes imposed by the physico-chemical differences of the respiratory medium, but also by internal constraints imposed by laws of acid-base balance. It is of interest to note that it was not the adaptation to the greater availability of O_2 which presented a problem, but rather the adaptation to a mandatory increase in CO_2 reserves and tension. As shown in Figure 1, the low solubility of O_2 in water required a very large gill ventilation and as a consequence an acid-base equilibrium balanced around a very low P_{CO_2} tension. With access to air the ventilation could be reduced (and still maintain an adequate blood O_2 tension) but of necessity required a large increase in blood-bicarbonate to maintain the same relative alkalinity (or OH^-/H^+ ratio) at a higher CO_2 tension. In other words, compared with their water breathing ancestors, air breathing vertebrates live in a chronic state of compensated respiratory acidosis.

The gross changes in this transition are illustrated in Figure 2 where most or all the exchange of O_2 and CO_2 was handled by the typical gills of fish (shown on the left-hand side). The transition, however, required the development of a bimodal gas exchange system because the primitive air organ or primitive lungs were unable to ventilate sufficiently to take care of the metabolic CO_2 excretion. These structures were able to take in a "gulp" of air and slowly extract the O_2 to very low levels, in other words, a prolonged breath-hold. The CO_2, on the other hand, was excreted continuously through the gills and/or the skin. This is illustrated in the center of Figure 2 and describes the events in air breathing fish and amphibians. While the skin was able to get rid of CO_2 but presented a large barrier to the transport of O_2, its respiratory exchange ratio was very large. On the other hand, the primitive lung took care of most of the O_2 demand, but because of the low ventilation could excrete only a small amount of CO_2. Its respiratory exchange ratio was, therefore, very low. Average values of these partitions and the exchange ratio in air breathing fish and terrestrial amphibians are shown in Table 1. In general we see that about 75 to 80% of CO_2 excretion occurred through the skin, while about 65% of the O_2 uptake was accomplished by the primitive lung.

The price for bimodal gas exchange was a reduction in the external armor or scales, leading to nakedness in many air breathing fish and amphibians where skin became the dominant organ for CO_2 excretion. This arrangement also meant that the primitive lungs or air organs did not have to ventilate continuously to maintain the required CO_2 tension and that the lung could fill intermittently to extract

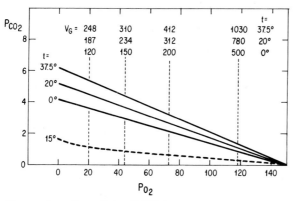

Figure 1.   From Rahn (1966b).

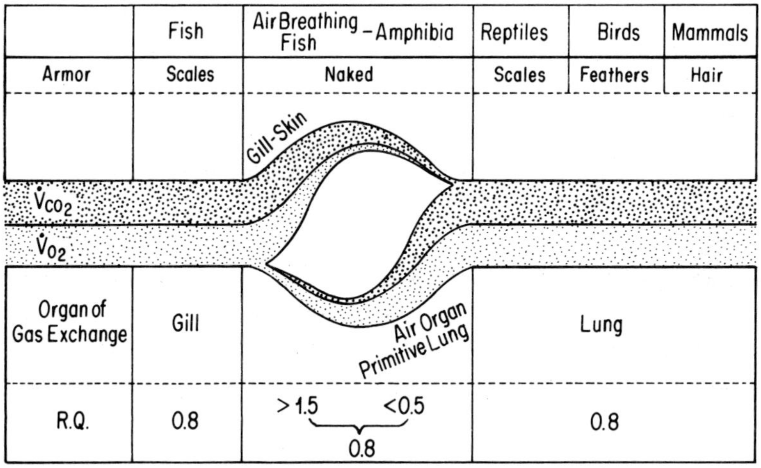

	Fish	Air Breathing Fish – Amphibia		Reptiles	Birds	Mammals
Armor	Scales	Naked		Scales	Feathers	Hair
$\dot{V}_{CO_2}$ $\dot{V}_{O_2}$		Gill-Skin				
Organ of Gas Exchange	Gill	Air Organ Primitive Lung		Lung		
R.Q.	0.8	>1.5 <0.5 0.8		0.8		

Figure 2.   Redrawn from Rahn and Howell (1976).

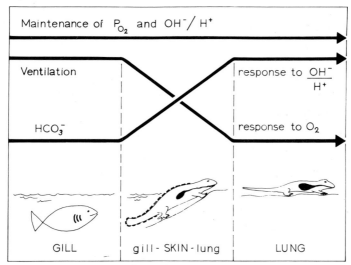

Figure 3.   From Rahn (1966a) and Dejours (1978).

most of the O_2. In other words, it was a type of discontinuous breath-holding
which provided the necessary O_2, while the CO_2 level was regulated primarily by
convective water or air movements across the gills and skin.

Table 1.   Mean values of the O_2 uptake and CO_2 output in lungs and gill-skin
           systems, expressed as percent of the total gas exchange. R, respiratory
           exchange ratio, for each system and Total R (from Rahn and Howell, 1976)

	Lung			Gill-skin			
	CO_2	O_2	R	CO_2	O_2	R	Total R
Air breathing fish (n = 8)	20	67	0.26	80	33	3.5	0.91
Terrestrial amphibians (n = 11)	26	65	0.35	74	35	2.0	0.86

Not until respiratory mechanics of lung ventilation had sufficiently evolved to
take care of all CO_2 excretion as well as O_2 uptake do we encounter the modern
lung as seen in reptiles, birds, and mammals. With this development skin armor
could reappear, which in turn reduced evaporation to the point that reptiles could
escape the moist environment of the air-water interface to become true terrestrial
animals. This skin armor eventually led to feathers and hair and endothermy, which
today is found in 24% of the 52,000 species of vertebrates. In this sense one can
argue that endothermy had to await the development of the modern lung.

EXTERNAL CONSTRAINTS

At 20°C CO_2 in water is 28 times more soluble than O_2, and effluent water from the
gills of a water breather exchanging these gases at an RQ = 1.0 will have O_2 and
CO_2 tensions confined to the 20°C line shown in Figure 3. Any particular gas tension along this line is a function of the gill ventilation, and these are shown
expressed in ml·min^{-1} for 1 ml O_2 (STPD) uptake. The slope of these lines varies
with temperature in distilled or unbuffered water. Sea water is buffered and decreases the slope of this line (Dejours, 1978). In either case, for equivalent O_2
tension, the effluent water and therefore the arterial blood have extremely low
CO_2 tensions compared with air breathing vertebrates where O_2 and CO_2 solubility
in N_2 (air) are essentially the same.

This is shown in Figure 4-A where the gill- and lung areas represent typical
values for gill water and lung air in cold-blooded vertebrates. Thi bimodal gas
exchange of aquatic forms is shown in Figure 4-B. In this case the skin-gill

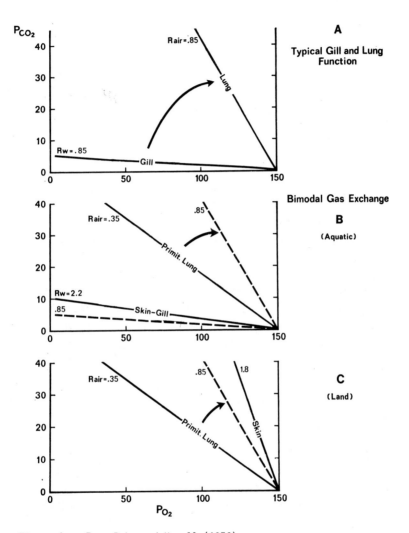

Figure 4.   From Rahn and Howell (1976).

system exchanges water at an exchange ratio of about 2.2, while the primitive lung has an exchange ratio of about 0.35. Figure 4-C shows the typical gas exchange of a terrestrial amphibian where both respiratory organs are exposed to air. The skin exchanges mostly CO_2 at an exchange ratio of about 1.8 and the lung at an exchange ratio of about 0.35.

INTERNAL CONSTRAINTS

Acid-base regulation in cold-blooded vertebrates is best described as a function of body temperature. Figure 5 shows the mean arterial pH of fish based on 14 reports including 9 species taken from the report of Reeves (1977). This slope is essentially identical to that found in air breathing vertebrates. It parallels the change in pH of neutral water and thereby indicates that it is not pH, per se, which is regulated, but rather the ratio of hydroxyl to hydrogen ions, or the OH^-/H^+ ratio, which remains constant over the normal body temperature range and preserves a constant relative alkalinity (Howell et al., 1970).

Transition from water to air breathing, therefore, had to adjust to the external constraints and the mandatory rise in P_{CO_2}, as well as to the internal constraints of maintaining a constant OH^-/H^+ ratio whatever the body temperature. This was achieved by the appropriate increase in plasma bicarbonate. Table 2 provides an

Table 2. Examples of acid-base parameters of water and air breathers (from Dejours, 1981)

	Trout	Bullfrog		Turtle
		Tadpole	Adult	
Temperature °C	20	20	20	20
pH	7.8	7.8	7.9	7.8
P_{CO_2} Torr	2.4	2.0	13	25
$[HCO_3^-]$ meq·L^{-1}	4.6	4.0	32	49

example of a trout and a turtle, at a temperature of 20°C. Note that the pH and therefore the OH^-/H^+ ratio are the same, but the low P_{CO_2} of the trout is balanced by a low HCO_3^- level of 4.6, while the high P_{CO_2} of the turtle is balanced by an HCO_3^- level of 49. Even more interesting is the transition from water to air breathing during the normal development of the bullfrog, where similar changes in P_{CO_2} and $[HCO_3^-]$ are observed, while the pH and OH^-/H^+ ratio remain

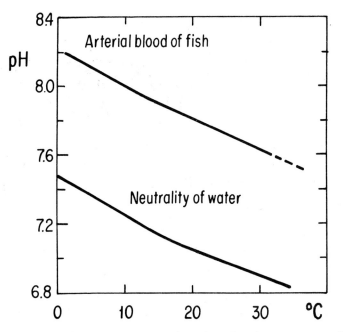

Figure 5.   Arterial pH of fish as a function of body temperature (redrawn from Reeves, 1977).

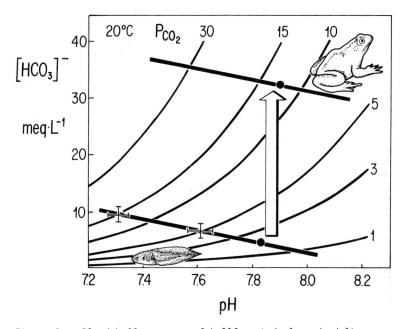

Figure 6.   Blood buffer curves of bullfrog tadpole and adult.

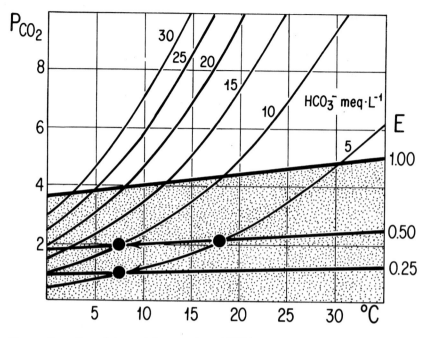

Figure 7.   From Rahn and Baumgardner (1972).

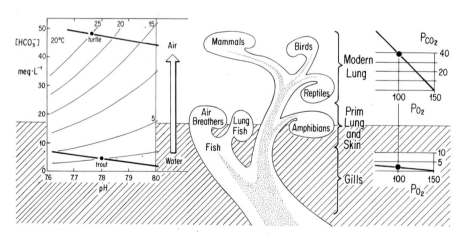

Figure 8.   Overview of transition from water to air breathing. See text.

constant. This is also illustrated in Figure 6 which shows the same data transferred to the $[HCO_3^-]$-pH diagram, including the blood buffer curves.

It is finally of interest to consider the constraints of acid-base regulation of water breathers imposed by changes in body temperature. In Figure 7 is plotted the P_{CO_2} as a function of body temperature, as well as the isopleth of plasma bicarbonate corresponding at each temperature to the blood pH values of fish (Figure 5). The shaded area indicates restraints imposed by the water medium on water breathers. It is in a sense a prison from which they can escape only by breathing air. For example, let's look at a fish at 18°C with a pH = 7.85 and a $[HCO_3^-]$ = 5 meq·L^{-1}. From the right-hand scale one can see that its gills are extracting 50% of O_2 from the water flowing over them. If this fish is now exposed to 8°C, it must double the ventilation to decrease its P_{CO_2} from 2 to 1 Torr if the HCO_3^- concentration and the normal pH and OH^-/H^+ ratio are to be maintained. On the other hand, to maintain the original ventilation at this new temperature the HCO_3^- concentration must nearly double to maintain a normal OH^-/H^+ ratio and would require the necessary ion exchange across the gills. Thus changes in body temperature involve compromises. Readjustment of HCO_3^- concentration is a slow process, requiring ion exchange across the gills. From this point of view water breathers live under a great handicap whenever body temperatures change. On the other hand, when air breathers change body temperature, they change their lung P_{CO_2} easily and quickly without requiring a change of their blood and tissue CO_2 stores (Howell et al., 1970).

OVERVIEW

Figure 8 provides an overview of the transition from water to air breathing. The evolutionary tree traces our ancestors back to water breathing fish, of which some 30,000 species presently inhabit our globe. A few hundred air breathing fish and some 2,600 amphibians remain today at the water-air interface where they are caught with their bimodal gas exchange and their naked, unarmored skin. Some 6,500 reptiles today represent the pioneers which developed the modern lung where respiratory mechanics provided sufficient ventilation to control the CO_2 elimination as well as the O_2 uptake. This event in turn allowed them to rearmor their skin, prevent excessive water loss, and to become the first truly terrestrial animals. The diagram to the right of the tree shows the external constraint that had to be overcome by the change of the respiratory medium from water to air; this was the mandatory increase in P_{CO_2}. The diagram to the left of the tree illustrates the internal constraints in this transition, the increase in blood and tissue bicarbonate while maintaining a constant relative alkalinity at all body temperatures.

REFERENCES

Dejours P (1978) Carbon dioxide in water- and air-breathers. Respir Physiol 33: 121-128.

Dejours P (1981) Principles of Comparative Respiratory Physiology, Elsevier/ North-Holland, Amsterdam pp. 1-265.

Howell BJ, Baumgardner FW, Bondi K, Rahn H (1970) Acid-base balance in cold- blooded vertebrates as a function of body temperature. Am J Physiol 218: 600- 606.

Rahn H (1966a) Aquatic gas exchange: Theory. Respir Physiol 1: 1-12.

Rahn H (1966b) Gas transport from the external environment to the cell. In: de Reuck AVS, Porter R (eds): Development of the lung. Ciba Foundation Sym- posium. J. & A. Churchill Ltd, London; pp. 3-23.

Rahn H, Baumgardner FW (1972) Temperature and acid-base regulation in fish. Respir Physiol 14: 171-182.

Rahn H, Howell BJ (1976) Bimodal gas exchange. In: Hughes GM (ed): Respiration of amphibious vertebrates. Academic Press, London; pp. 271-285.

Reeves RB (1977) The interaction of body temperature and acid-base balance in ectothermic vertebrates. Ann Rev Physiol 39: 559-586.

HOW DO INTERTIDAL INVERTEBRATES BREATHE BOTH WATER AND AIR?

Jean-Paul Truchot

Laboratoire de Neurobiologie et Physiologie Comparées
CNRS UA 1126, Université de Bordeaux I
Place du Docteur B. Peyneau, 33120 Arcachon, France

At low tide, intertidal invertebrates can be either exposed to air or retained in small water bodies with very variable respiratory conditions. Many of these animals have high anaerobic capacities, but a number of morphological, behavioral and physiological adaptations allow some of them to exploit successfully both respiratory media.

INTRODUCTION

Ebb and flow of the tide result in periodic emersion and submersion of a small coastal belt called the intertidal zone. The midpart of this zone remains uncovered for up to 10 hours twice a day in a semi-diurnal tide regime. Increasing periods of air exposure are thus experienced by organisms living higher on the shore and it is their ability to cope more or less with these extreme conditions which explains the typical zonation pattern of species observed on most shores. Animal life in the intertidal zone includes a relatively small number of specialized species which are rarely found elsewhere. These animals must basically solve two main problems at low tide : i) avoid excessive dessication if exposed to air ; ii) maintain a minimum level of energy production, either aerobically or anaerobically. Dessication may be partially prevented in exploiting a variety of suitable habitats such as crevices, algal shelters, sediment burrows, and even residual waters. But, in any case, intertidal animals have to face very adverse respiratory conditions at low tide, either in confined water or in air. Some rely upon anaerobic metabolism, but many can use atmospheric oxygen. In fact, most of the physiological effects resulting from the different properties of water and air as respiratory media can be observed in intertidal animals.

THE ATMOSPHERE AS AN OXYGEN SOURCE : MORPHOLOGICAL AND BEHAVIORAL ADAPTATIONS

Most intertidal invertebrates are basically designed for water-breathing, i.e. they have gills or a gas- and water-permeable integument as the respiratory organ. Under air exposure however, evaporative water loss can be reduced by various means, for example by closure of the shell in bivalves or of the operculum in many gastropods and in barnacles. Yet, isolation from the aerial medium is never perfectly tight and many intertidal animals are known to use atmospheric oxygen

Comparative Physiology: Life in Water and on Land. P. Dejours, L. Bolis, C.R. Taylor, E.R. Weibel (eds.) Fidia Research Series, IX-Liviana Press, Padova © 1987

when exposed. In fact, the ability to breathe air is variously developed as shown by selected examples of the ratios $\dot{M}_{O_2}(air)/\dot{M}_{O_2}(water)$ presented in Table I. Whatever the animal group, it is commonly observed that the higher the level of the shore where the animal thrives, the higher its reliance upon atmospheric oxygen when exposed. Some species living in the upper intertidal zone may even show oxygen consumption rates higher in air than in water. Moreover, humidity conditions as well as the degree of dessication the animal has experienced much influence the extent of the use of atmospheric oxygen. Let us examine in more detail some examples of the morphological and behavioral adaptations allowing these forms to breathe air while avoiding excessive water loss.

TABLE 1. Some air to water ratios of oxygen consumption in intertidal animals ([*])

Species	Temperature (°C) (Air and water)	$\dot{M}_{O_2}(air)/\dot{M}_{O_2}(water)$
Pollicipes polymerus (1) (gooseneck barnacle)	10	5.0
Monodonta turbinata (2) (trochid snail)	20	1.18
Carcinus maenas (3) (shore crab)	15	1.18
Gibbula rarilineata (2) (trochid snail)	20	0.76
Modiolus demissus (4) (ribbed mussel)	20-23	0.66
Cardium edule (5) (common cockle)	10	0.28
Cryptochiton stelleri (6) (giant cradle)	17	0.54

([*]) Species listed from high to low shore position in the intertidal zone.
(1) Petersen et al. (1974) ; (2) Houlihan and Innes (1982) ; (3) Taylor and Butler (1978) ; (4) Booth and Mangum (1978) ; (5) Widdows et al. (1979) ; (6) Petersen and Johansen (1973).

Barnacles are typical inhabitants of the upper intertidal zone on rocky shores. When exposed, some individuals exhibit a small, diamond-shaped opening between the four opercular valves, called the micropylar aperture or pneumostome (Barnes et al, 1963 ; Grainger and Newell, 1965). In fact, upon emersion, these barnacles expel water from and take gas into the mantle cavity. Then, the pneumostome opens and closes periodically, promoting air renewal mainly by diffusion but also probably by convective pumping movements of the cirri, which have been observed a short time after opening or before closure. Several observations indicate that this behavior is closely regulated in order to allow aerial respiration while avoiding excessive water loss (Grainger and Newell, 1965). In a barnacle population, the frequency of open pneumostomes is directly related to atmospheric humidity and inversely related to the time of exposure in given humidity conditions. Hence, in the field, more barnacles are open at lower than at higher shore levels and the percent open pneumostomes is lower the greater the evaporative weight loss. Analysis of air bubbles expelled at reimmersion also shows that oxygen depletion is more marked for barnacles exposed in dry air than in a wet atmosphere (Grainger and Newell, 1965). Interestingly, the pneumostome tends to open more frequently in hypoxic gas (Barnes et al., 1963), demonstrating the respiratory importance of this behavior. Air breathing can sustain normal aerobic metabolism during usual emersion times in the field since no lactate accumulation takes place for several hours air exposure. However, lactate is produced after longer periods when the animals have to resort to anaerobic glycolysis to avoid excessive dehydration.

In contrast with subtidal species, most intertidal bivalves are able to retain water for a long time in their pallial cavity when exposed to air (see for example Coleman and Trueman, 1971). They, however, periodically separate their valves, exposing a thin, water-coated, mantle membrane to the atmosphere, a behavior known as "shell gaping", which allows some degree of air-breathing (Lent, 1968). Aerial O_2 consumption is very variable among intertidal bivalves and these animals commonly accumulate an oxygen debt during air exposure (Moon and Pritchard, 1970 ; Bayne et al., 1976 ; Widdows et al., 1979). Again, gaping appears more marked in a wet than in a dry atmosphere. Evaporative water loss cools the mantle fluid to a temperature lower than ambient in exposed Modiolus demissus (Lent, 1968), thus improving temperature tolerance during air exposure. But, measurements of mantle fluid P_{O_2} have also stressed the respiratory significance of the gaping behavior. Whereas P_{O_2} values decrease to 0 in valve-clamped mussels, they reach steady state levels between 15 and 40 Torr in various free-gaping bivalves during exposure to air (Lent, 1968 ; Moon and Pritchard, 1970 ; Bayne et al., 1976), indicating that O_2 diffuses across the shell gape. The accessibility of the tissues to gases in gaping bivalves is also clear from data showing that various metabolites are labelled in cockles Cardium edule and mussels Mytilus edulis exposed to an atmosphere containing [14]CO_2 (Ahmad and Chaplin, 1977). Gaping as an adaptation for aerial oxygen acquisition has however been questioned because the well-known high anaerobic capacity of most bivalves may not necessitate the maintenance of aerobic metabolism during air exposure. It has even been suggested that gaping may be important in enabling escape of CO_2 produced by buffering of acidic endproducts by the calcareous shell (de Zwaan and Wijsman, 1976 ; see also Pamatmat, 1984).

Many intertidal gastropods are also known to use aerial oxygen when exposed. Some forms living in the upper part of the littoral zone have even lost their ctenidial gill and they use their vascularized mantle cavity as a lung. The ability of gill-bearing forms to breathe air correlates with an increased vascularization of the mantle skirt (see Houlihan and Innes, 1982). Recent studies have also shown that this ability depends strongly on the retention of fluid within the mantle cavity at low tide. This may allow the gill to remain functional in avoiding collapse, which is borne out by P_{O_2} values amounting 50-70 Torr in pallial fluid of exposed snails (Houlihan, 1979 ; Houlihan et al., 1981 ; Houlihan and Innes, 1982). The exact mechanism by which oxygen is transferred from ambient air to the submerged gill remains, however, to be elucidated.

Lamellar gills of crabs appear at first glance ill-suited for aerial respiration because the effective exchange area could be expected to be reduced in air by either collapsing or water retention between the gill leaflets. Yet, high levels of aerial oxygen consumption have been reported for typical intertidal crabs such as Carcinus or Pachygrapsus (Taylor and Butler, 1978 ; Houlihan and Innes, 1984). Most likely, water is drained away from the branchial cavity in these animals which continue to ventilate their gills when in air (Taylor and Butler, 1978). Extensive collapse of the gills may be avoided thanks to a particularly developed marginal canal ensuring better rigidity of the lamellae (Drach, 1930), a mechanical reinforcement which culminates in the form of extensive cuticular ridges or tufts maintaining the gill platelets apart in true terrestrial crabs (Cameron, 1981). By contrast, other intertidal or semi-terrestrial crabs such as Cardisoma carnifex (Wood and Randall, 1981) retain some water in their gill chamber when emerged. Most of these crabs ventilate their gills with air, but some representatives of the families Grapsidae and Ocypodidae have been reported to recirculate gill water externally over the carapace to aerate it.

GAS EXCHANGE IN WATER AND IN AIR

Although many intertidal invertebrates can maintain variable levels of oxygen uptake in air, their gas exchange performance appears often more or less impeded. In fact, most of them become internally hypoxic when exposed, as shown by decreased blood P_{O_2} values (Table 2). An exception may be found in the gooseneck barnacle, Pollicipes polymerus, which consumes more oxygen in air than in water (Table 1) and shows accordingly an increase of hemolymph P_{O_2} (Petersen et al., 1974). The best way to understand the mechanisms enabling intertidal animals to breathe both water and air is to compare the responses to air exposure in nearby intertidal and subtidal species. Few studies except in crabs allow such comparisons to be done (Fig. 1).

The emersed intertidal shore crab, Carcinus maenas, actively ventilates its gill chambers, but the branchial extraction coefficient of oxygen from air is very low ($\leqslant 2$ % vs 30-50 % in water). Oxygen uptake is maintained at the same level as in

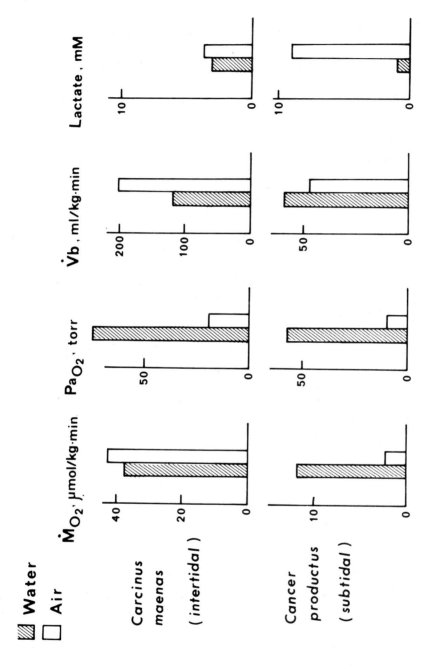

Fig. 1. Comparison of respiratory variables in an intertidal and in a subtidal crab species, breathing in air and in water (from Taylor and Butler, 1978 ; and DeFur and McMahon, 1984a).

TABLE 2. Oxygen partial pressures in arterial blood of intertidal animals
breathing in well aerated water or in air.

Species	Temperature (°C) (water and air)	Pa_{O_2} (Torr) in water	Exposure time	Pa_{O_2} (Torr) in air
Carcinus maenas (1)	15	74.9	2-3 h	18.8
Cancer productus (2)	10	59	4 h	21
Mytilus edulis (3)	12	44	4 h	25
Modiolus demissus (4)	20-23	43	6 h	29
Cryptochiton stelleri (5)	10	85	40 min	29
Pollicipes polymerus (6)	10	42	4 h	50

(1) Taylor and Butler (1978) ; (2) DeFur and McMahon (1984a) ; (3) Jokumsen and
Fyhn (1982) ; (4) Booth and Mangum (1978) ; (5) Petersen and Johansen (1973) ;
(6) Petersen et al. (1974)

water (Taylor and Butler, 1978 ; Table 1), despite a marked fall of Pa_{O_2} (Table 2)
indicating that the capacity of the gills to take up oxygen in air is limited, most
probably by diffusion. This, however, does not result in a commensurate decrease
of postbranchial hemolymph O_2 concentration because the oxygen saturation of the
respiratory pigment hemocyanin remains high. In contrast with most other aquatic
animals, emersion of Carcinus in air does not induce bradycardia and cardiac
output is actually increased (Taylor and Butler, 1978), enabling transport of
enough oxygen to the tissues despite internal hypoxia. In the less-well adapted
subtidal species, Cancer productus, gill ventilation is also maintained in air,
but gill collapse not only causes diffusion limitation but also probably restricts
hemolymph flow, resulting in a reduced cardiac output and in a strongly decreased
oxygen uptake during air exposure (DeFur and McMahon, 1984a). Accordingly, Cancer
productus exhibits increased lactate levels when emerged in air, whereas no
lactate accumulation takes place in Carcinus. The comparison of the responses of
these two crabs in fact indicates that successful exploitation of the aerial
medium at low tide by basically water-breathing animals requires an effective
functioning of all the components of the respiratory system. This requirement is
certainly not met in all intertidal forms which often exhibit reduced oxygen
uptake in air and either must depress their overall metabolism or have to resort
to anaerobic pathways.

Another important aspect of gas exchange in intertidal animals concerns CO_2
excretion and acid-base balance. Because carbon dioxide is much more soluble than
oxygen in water whilst both gases have the same capacitance coefficients in air,
blood P_{CO_2} levels are uniformly lower in water breathers than in air breathers
(Dejours, 1978). Yet, the pH values are found approximately the same at similar

body temperature because air breathers maintain a commensurately higher extra-cellular bicarbonate concentration and are thus in a state of compensated hypercapnic acidosis compared to their water-breathing ancestors. Transition from water- to air-breathing during amphibian metamorphosis also results in an increase of blood P_{CO_2} with only minor changes of pH (Erasmus et al., 1970/71 ; Just et al., 1973).

Air exposure of intertidal animals similarly induces CO_2 retention and acidosis. This has been found in all forms tested to date : crabs (Truchot, 1975a ; Taylor and Butler, 1978 ; DeFur and McMahon, 1984b) ; bivalves (Jokumsen and Fyhn, 1982); cirripeds (Petersen et al., 1974). However, in addition of the purely physical effects stated above, several other mechanisms could contribute to the increase of CO_2 and H^+ levels in air-exposed intertidal animals : severe limitation of gas exchange in air ; buildup of acidic endproducts from anaerobic metabolism, for example. In view of these additional processes which obviously could promote CO_2 accumulation, it is somewhat surprising that internal P_{CO_2} levels increase only moderately upon emersion and never reach values typically found in air breathers. At least in crabs as Carcinus, this may simply result from notably high ventilation rates in air. In fact, CO_2 output may not be ventilation-limited, but only diffusion-limited. Residual water in gill chambers or mantle cavities may also constitute an additional sink for CO_2 excretion (Wood and Randall, 1981).

RESPIRATORY PROBLEMS IN TIDEPOOLS

At low tide, some intertidal forms may avoid air exposure in taking refuge in residual water bodies on the shore. If the risk of dessication is suppressed in such tidepools, much respiratory problems remain because very dense plant and animal populations often result in particularly extreme breathing conditions in these habitats. Dissolved oxygen accumulates during the day when photosynthesis is active but may be completely depleted at night, and CO_2 and pH vary accordingly (see Truchot and Duhamel-Jouve, 1980). Also such waters could warm up to 30-32 °C on summer days or be completely frozen in winter.

Among tidepool inhabitants, there are sessile forms which must cope with these extremes at low tide, and mobile animals which can eventually escape into air when respiratory conditions can no more sustain normal aerobic metabolism. Elaborate behavioral and physiological responses have evolved in some intertidal animals, enabling them to exploit equally well both the aerial and aquatic respiratory media when retained at low tide in rockpools. One of the best studied example is again the shore crab, Carcinus maenas. In shallow hypoxic water, this animal exhibits a behavior known as the "emersion response", first described by Bohn (1897) and thoroughly studied in recent years (Taylor and

Butler, 1973 ; Taylor et al., 1973 ; Wheatly and Taylor, 1979). The crab raises
the anterior border of its cephalothorax above the water level and reverses
respiratory pumping causing air to bubble through the water retained in the gill
chamber and to be expelled by the normally inhalent openings at the base of the
walking legs. The threshold P_{O_2} below which this behavior is observed is well
defined and depends on the temperature and the salinity, two factors influencing
the metabolic level. Namely, the emersion response is triggered at higher P_{O_2}
values when the temperature is raised and the salinity lowered (fig. 2). Gill
water is aerated in the process and its P_{O_2} goes well higher than ambient during
the emersion response (fig. 2). As a consequence, postbranchial hemolymph P_{O_2} and
C_{O_2} are also increased, allowing transport of more oxygen to the tissues. This is
favored by an immediate and sustained tachycardia during the emersion response,
which most probably reflects an increased cardiac output.

In warm water at about 28 °C, when the threshold P_{O_2} for emersion went near 150
Torr (fig. 2), the shore crab exhibits a so-called "emigration behavior",
spontaneously leaving normoxic water to migrate into air (Taylor and Wheatly,
1979). Rather than being a means to avoid heat stress, this response is most
probably of respiratory significance, since I never observed it in hyperoxic
natural conditions, even at water temperature up to 31-32 °C.

When submerged at low tide in rockpools, intertidal animals such as the shore
crab can maintain a nearly constant level of aerobic metabolism despite large
variations of oxygen availability in water. They hyperventilate their gills at
night in hypoxic water and, conversely, hypoventilate during the day in hyperoxic
conditions (Jouve-Duhamel and Truchot, 1983). Laboratory experiments in which the
water oxygenation was changed at constant values for other factors have shown that
these strong ventilatory responses much influence blood P_{CO_2} levels and acid-base
balance, resulting in an hypocapnic alkalosis in hypoxia and in an hypercapnic
acidosis in hyperoxia (Truchot, 1975b). These simplified experimental conditions
do not reflect the complexity of the natural situation however, since many other
factors, particularly carbon dioxide and temperature, concomitantly vary in
rockpools at low tide. Recent experiments using an artificially reconstituted
tidepool (Truchot, 1986) have shown that there are surprisingly few acid-base
disturbances in Carcinus exposed to ambient changes mimicking those occurring
at low tide in rockpools. This mainly results from counteractive influences of
oxygen and carbon dioxide variations which are always closely linked in such
habitats.

To conclude, it has long been appreciated that intertidal animals have impressive
capacities to cope with a very variable and adverse environment, and particularly
to breathe in air and in water, two very contrasted respiratory media. This

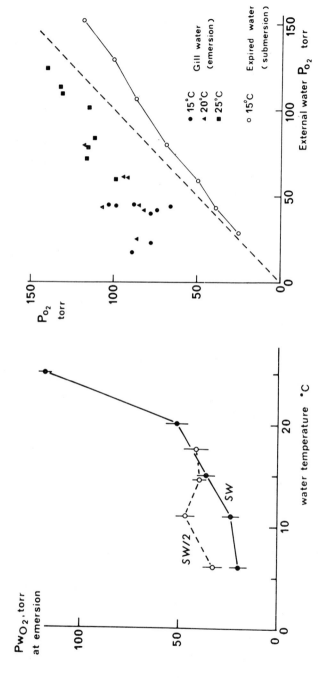

Fig. 2. Left : threshold water P_{O_2} triggering the emersion response in the shore crab Carcinus maenas, as a function of water temperature and at two salinities. Right : P_{O_2} values in gill water of Carcinus maenas during the emersion response at three temperatures (closed symbols) and in water expired by submerged crabs at various ambient hypoxic levels (open symbols) (from Wheatly and Taylor, 1979)

ability probably results from a complex array of relatively specialized
morphological, behavioral and physiological adaptations, which should be
appreciated together in an integrated way and in experimental conditions as close
as possible as those prevailing in the natural setting.

REFERENCES

Ahmad TA, Chaplin AE (1977) The intermediary metabolism of Mytilus edulis (L.)
and Cerastoderma edule (L.) during exposure to the atmosphere. Biochem Soc
Trans 5 : 1320-1323.

Barnes HD, Finlayson DM, Piatigorsky J (1963) The effect of dessication on the
behavior, survival and general metabolism of three common cirripedes. J Anim
Ecol 32 : 233-252.

Bayne BL, Bayne CJ, Carefoot TC, Thompson RJ (1976) The physiological ecology
of Mytilus californianus Conrad. 2. Adaptations to low oxygen tension and air
exposure. Oecologia 22 : 229-250.

Bohn G (1897) Sur le renversement du courant respiratoire chez les Décapodes.
C r hebd Séanc Acad Sci Paris 125 : 539-542.

Booth CE, Mangum CP (1978) Oxygen uptake and transport in the lamellibranch
mollusc Modiolus demissus. Physiol Zool 51 : 27-32.

Cameron JN (1981) Brief introduction to the land crabs of the Paluau Islands :
stages in the transition to air breathing. J Exp Zool 218 : 1-5.

Coleman N, Trueman ER (1971) The effect of aerial exposure on the activity of the
mussels Mytilus edulis (L.) and Modiolus modiolus (L.). J Exp Mar Biol Ecol
7 : 295-304.

DeFur PL, McMahon BR (1984a) Physiological compensation to short-term air
exposure in red rock crabs, Cancer productus Randall, from littoral and sub-
littoral habitats. I. Oxygen uptake and transport. Physiol Zool 57 : 137-150.

DeFur PL, McMahon BR (1984b) Physiological compensation to short-term air
exposure in red rock crabs Cancer productus Randall, from littoral and sub-
littoral habitats. II. Acid-base balance. Physiol Zool 57 : 151-160.

Dejours P (1978) Carbon dioxide in water- and in air-breathers. Respir Physiol
33 : 121-128.

Drach P (1930) Etudes sur le système branchial des Crustacés Décapodes. Arch
Anat Micr Paris 26 : 83-133.

Erasmus B de W, Howell BJ, Rahn H (1970/71) Ontogeny of acid-base balance in the
bullfrog and chicken. Respir Physiol 11 : 46-53.

Grainger F, Newell GE (1965) Aerial respiration in Balanus balanoides. J Mar Biol
Ass UK 45 : 469-479.

Houlihan DF (1979) Respiration in air and water of three mangrove snails. J Exp
Mar Biol Ecol 41 : 143-161.

Houlihan DF, Innes AJ (1982) Respiration in air and water of four mediterranean
trochids. J Exp Mar Biol Ecol 57 : 35-54.

Houlihan DF, Innes AJ (1984) The cost of walking in crabs : aerial and aquatic
oxygen consumption during activity of two species of intertidal crab. Comp
Biochem Physiol 77A : 325-334.

Houlihan DF, Innes AJ, Dey DG (1981) The influence of mantle cavity fluid on the
aerial oxygen consumption of some intertidal gastropods. J Exp Mar Biol Ecol
49 : 57-68.

Jokumsen A, Fyhn HJ (1982) The influence of aerial exposure upon respiratory and osmotic properties of haemolymph from two intertidal mussels, Mytilus edulis L. and Modiolus modiolus L. J Exp Mar Biol Ecol 61 : 189-203.

Jouve-Duhamel A, Truchot JP (1983) Ventilation in the shore crab Carcinus maenas (L.) as a function of ambient oxygen and carbon dioxide : field and laboratory studies. J Exp Mar Biol Ecol 70 : 281-296.

Just JJ, Gatz RN, Crawford EC (1973) Changes in respiratory function during metamorphosis in the bullfrog, Rana catesbeiana. Respir Physiol 17 : 276-282.

Lent CM (1968) Air-gaping by the ribbed mussel, Modiolus demissus (Dillwyn) : effects and adaptive significance. Biol Bull 134 : 60-73.

Moon TW, Pritchard AW (1970) Metabolic adaptation in vertically-separated populations of Mytilus californianus Conrad. J Exp Mar Biol Ecol 5 : 35-46.

Pamatmat MM (1984) Metabolic heat flow patterns in the intertidal mussel Ischadium (=Modiolus=Geukensia) demissum demissum during aerial and underwater respiration. Int Rev Ges Hydrobiol 69 : 263-275.

Petersen JA, Fyhn HJ, Johansen K (1974) Ecophysiological studies of an intertidal crustacean, Pollicipes polymerus (Cirripedia, Lepadomorpha) : aquatic and aerial respiration. J Exp Biol 61 : 309-320.

Petersen JA, Johansen K (1973) Gas exchange in the giant cradle Cryptochiton stelleri (Middendorff). J Exp Mar Biol Ecol 12 : 27-43.

Taylor EW, Butler PJ (1973) The behaviour and physiological responses of the shore crab Carcinus maenas during changes in environmental oxygen tensions. Neth J Sea Res 7 : 496-505.

Taylor EW, Butler PJ, Sherlock PJ (1973) The respiratory and cardiovascular changes associated with the emersion response of Carcinus maenas (L.) during environmental hypoxia. J Comp Physiol 86 : 95-115.

Taylor EW, Butler PJ (1978) Aquatic and aerial respiration in the shore crab, Carcinus maenas (L.), acclimated to 15 °C. J Comp Physiol B 127 : 315-323.

Taylor EW, Wheatly MG (1979) The behaviour and respiratory physiology of the shore crab, at moderately high temperatures. J Comp Physiol B 130 : 309-316.

Truchot JP (1975a) Blood acid-base changes during emersion and reimmersion of the intertidal crab, Carcinus maenas (L.). Respir Physiol 23 : 351-360.

Truchot JP (1975b) Changements de l'état acide-base du sang en fonction de l'oxygénation de l'eau chez Carcinus maenas. J Physiol Paris 70 : 583-592.

Truchot JP (1986) Changes in the acid-base state of the shore crab Carcinus maenas exposed to simulated tidepool conditions. Biol Bull 170 : 506-518.

Truchot JP, Duhamel-Jouve A (1980) Oxygen and carbon dioxide in the marine intertidal environment : diurnal and tidal changes in rockpools. Respir Physiol 39 : 241-254.

Wheatly MG, Taylor EW (1979) Oxygen levels, acid-base status and heart rate during emersion of the shore crab into air. J Comp Physiol B 132 : 305-311.

Widdows MG, Bayne BL, Livingstone DR, Newell RIE, Donkin P (1979) Physiological and biochemical responses of bivalves molluscs to exposure to air. Comp Biochem Physiol 62A : 301-308.

Wood CM, Randall DJ (1981) Oxygen and carbon dioxide exchange during exercise in the land crab Cardisoma carnifex. J Exp Zool 218 : 7-22.

Zwaan A de, Wijsman TCM (1976) Anaerobic metabolism in bivalvia (Mollusca). Characteristics of anaerobic metabolism. Comp Biochem Physiol 54B : 313-324.

HOW DO AMPHIBIANS BREATHE BOTH WATER AND AIR?

Donald C. Jackson

Department of Physiology and Biophysics
Brown University
Providence, Rhode Island, U.S.A.

As a process common to both water and air, cutaneous respiration is the characteristic exchange mechanism in amphibians.
Its importance is typically greater in small animals, under
resting conditions, and at low temperature. Although
circumstantial evidence for active control is persuasive,
conclusive experimental evidence is lacking for effective,
homeostatically-relevant control of this mode. The skin
serves as a low cost alternative to the metabolically costly
pulmonary pump, and as an exclusive exchanger under low
demand metabolic states and in certain highly adapted
salamanders. Because of the dominance of pulmonary control,
the search for active cutaneous control should focus on
amphibians satisfying all exchange requirements through the
skin.

INTRODUCTION

In terms of their respiratory gas exchange, amphibians are a remarkably diverse
group, ranging from strictly aquatic species with gills to highly terrestrial
species with primary reliance on lungs for exchange. Between these extremes are
amphibians that rely to varying extents on the three primary gas exchange organs
of this group: gills, lungs, and skin.

The shift from an aquatic to a terrestrial existence ultimately involves a change-
over from branchial to pulmonary exchange, but the common transitional feature
linking these habitats and characterizing the amphibians as a group is a dependence on the integument as a gas exchange surface. Structural adaptations permit
effective use of the skin as a respiratory exchanger in these animals, but these
adaptations cannot select specifically for only the exchange of gases, so

Comparative Physiology: Life in Water and on Land. P. Dejours, L. Bolis, C.R. Taylor, E.R. Weibel
(eds.) Fidia Research Series, IX-Liviana Press, Padova © 1987

amphibians also utilize their skin in important ways for the transfer of water
and various solutes.

Amphibians, as their name attests, are a group associated with the water-air
interface. With few exceptions, all begin their lives in water and spend their
entire lives in or near water. For the purposes of this discussion, the amphibians of interest are those that live their adult lives at the margins of the
interface (gilled or highly terrestrial species will not be considered). Species
that live at the interface have permeable skin surfaces and may or may not possess
lungs. The lunged species are called bimodal breathers because their respiratory
exchange changes from strictly skin breathing when fully submerged, to aquatic
skin exchange and lung ventilation of air when at the surface, to gas exchange
with the air by both the skin and the lungs when out of the water. Because skin
exchange is the common mode in each medium, my focus will be on this aspect, particularly with respect to the currently controversial topic of control. For
further reading, a comprehensive review on the subject of cutaneous exchange has
recently been published (Feder and Burggren, 1985).

ADAPTATIONS OF AMPHIBIAN SKIN FOR GAS EXCHANGE

Compared to other vertebrates, amphibians possess skin that is highly adapted as
a respiratory organ. The diffusion path is relatively short, sometimes as little
as 12μ (Czopek, 1965). In most cases, the capillaries penetrate only to the
outer part of the dermis so that the epidermis constitutes the diffusion barrier,
but capillaries reportedly reach into the epidermal layer in some species
(Guimond and Hutchison, 1973).

The diffusion barrier also includes any mucous coating on the skin and the
unstirred layer of respiratory medium (water or air) that is in contact with the
skin. The importance of the unstirred layer in water has recently been demonstrated by Burggren and Feder (1985) on the bullfrog, Rana catesbeiana. By
alternately stirring and not stirring the water bathing the frogs, they could
elicit an increase and decrease, respectively, in cutaneous O_2 uptake, presumably
by an effect on the overall diffusion path length. Due to the high diffusivity of
molecules in the gas phase, boundary layer problems are probably less important
in air. This is supported by the observation of Gatz et al. (1974) that mixed
arterial (cardiac) PCO_2 in the plethodontid (lungless) salamander, Desmognathus
fuscus is about 6 torr. This is far lower than predicted if a significant exchange limitation existed in the gas phase (Howell and Rahn, 1976) and suggests
that the 'ventilation' of the skin surface in these salamanders approaches the

infinite level predicted by the gas exchange model of Piiper and Scheid (1975). According to this model, the gas phase would offer no resistance to exchange as the PO_2 at the skin surface would be equal to ambient PO_2.

IMPORTANCE OF CUTANEOUS GAS EXCHANGE

1. Exclusively skin-breathing amphibians. The importance and indeed the necessity of cutaneous exchange cannot be questioned in the plethodontid, or lungless, salamanders. Except for a minor contribution by the buccopharyngeal cavity, these animals exchange all their respiratory gases across the skin (Whitford and Hutchison, 1965). This total reliance on the skin has not proven to be a handicap because the plethodonts are a highly successful, widely distributed group that comprises the majority of urodelan species. Although typically found in cool, moist habitats, either in or near running water, their range extends into the tropics (Porter, 1972), and their rate of O_2 consumption can increase with temperature (Whitford and Hutchison, 1965) and exercise (Feder, 1978). As anyone who has tried to capture one can attest, they can be quite active and elusive.

Many urodeles possess lungs, but in some instances these are sac-like structures with relatively small exchange area and are ineffective for gas exchange. One such species, the large aquatic hellbender, Cryptobranchus alleganiensis, can reach 74 cm in length and close to a kg in body mass (Nickerson and Mays, 1973), yet respires almost entirely through its skin (Guimond and Hutchison, 1973).

2. Bimodal breathers. The importance of the skin in the gas exchange of bimodal amphibians is indisputable. It frequently accounts for most of the CO_2 loss and a smaller but still significant fraction of the O_2 uptake. The relative contributions of the skin and lungs to gas exchange do vary, however, and depend on a variety of characteristics of the animal and its environment. It is beyond the scope of this paper to discuss these in any detail, but the reader is once again referred to the excellent and comprehensive review by Feder and Burggren (1985) for a more thorough treatment. It is the thesis of the present writer, however, in contrast to the authors of the review, that skin gas exchange is relatively stereotyped in nature; i.e., its performance is not subject in acute situations to any substantial adaptive responses, and that its main functions are to relieve the burden of the energetically more costly pulmonary mechanism (Gans, 1970) and to permit prolonged periods of aerobic submergence, such as during hibernation.

Before addressing the topic of control directly, I will briefly consider some of the factors that affect the distribution of gas exchange between the lungs and the skin. Structural differences between species in epidermal thickness and capillary density (Czopek, 1965; Saint-Aubain, 1982), or overall surface area (Hutchison et al., 1968) determine the diffusing capacity of the skin and help account for the wide range of aquatic respiratory abilities observed. Some of these structural characteristics can adaptively change in animals chronically exposed to hypoxic conditions (Burggren and Mwalukoma, 1983). In animals of sim-ilar structural type, cutaneous VO_2 increases in importance as size diminishes. For example, Ultsch (1973) found that critical PO_2 (the ambient PO_2 below which $\dot{V}O_2$ is no longer regulated) in the salamander, Siren lacertina, decreased as a function of body size. The explanation, which assumes that body surface area is proportional to skin diffusing capacity, is based on the observation that surface area is a function of (body mass)$^{0.67}$, whereas VO_2 increases with (body mass)$^{0.75}$. Therefore, $\dot{V}O_2$ rises with a steeper slope than skin exchange capacity. Lung surface area of amphibians, on the other hand, varies with (body mass)$^{0.98}$ (Tenney and Tenney, 1970), and can accommodate the increased demand. As body size decreases, the converging functions of $\dot{V}O_2$ and skin exchange capacity may meet, and at this size and below, all resting O_2 uptake can occur through the skin.

Exchange distribution also is affected by changes in metabolic rate. An increase in rate, whether due to higher body temperature (Hutchison et al., 1968; Whitford and Hutchison, 1965) or activity (Gottlieb and Jackson, 1976) reduces the rela-tive contribution of skin exchange. Conversely, at low temperature and reduced activity, $\dot{V}O_2$ may fall to a level that can be supplied completely by cutaneous uptake.

CONTROL OF GAS EXCHANGE

The above considerations of gas exchange distribution lead naturally to the topic of respiratory control. If an amphibian must increase its rate of gas exchange, by which mechanism will it accomplish this - by the skin, by the lung, or by both? Do these mechanisms serve to maintain a normal state of blood gas and acid-base homeostasis? The premise stated in the previous section, that the diffusion capacity of the skin is constant, leads inevitably to the conclusion that the skin exchange is not the control site. Is this a valid conclusion?

In his classic study of the subject, Krogh (1904) asserted that skin O_2 exchange in the frog is passive. His use of the term 'passive', however, must be under-stood in the context in which it was used. At that time, Krogh and others believed that O_2 uptake by the lung occurred by active secretion, so by saying that O_2 uptake was passive in the skin he meant that it occurred by simple diffusion. A current school of thought also suggests that cutaneous gas exchange is passive, but a different meaning is intended now. Passive in this sense means that the exchange is free-running with little or no adaptive response to changing requirements. We now know that passive diffusion underlies the movement of respiratory gases at all exchange sites, but the diffusive flux can potentially be accelerated by active mechanisms, such as capillary recruitment, increased blood perfusion, and increased ventilation.

1. Pulmonary control. There is convincing evidence that bimodal amphibians exert active control over their pulmonary gas exchange. As described earlier, bullfrogs meet changing metabolic requirements by adjusting pulmonary gas exchange (Whitford and Hutchison, 1965), and by so doing regulate blood acid-base balance (Gottlieb and Jackson, 1976; Mackenzie and Jackson, 1978). Ranids also respond to chemostimulation with increased ventilation. Hypoxic and hypercapnic breathing mixtures caused hyperventilation in Rana esculenta (Smyth, 1939) and elevated water PCO_2 induced a ventilatory response that maintained arterial PCO_2 (Jackson and Braun, 1979). These representative examples indicate that lunged amphibians, like other air-breathing vertebrates, alter pulmonary ventilation to match their gas exchange requirements.

2. Cutaneous control. Active control of skin respiration is much less obvious and has proven difficult to establish experimentally. Many of the same experiments that provide clear evidence for pulmonary control fail to reveal evidence for cutaneous control. For example, variations in skin CO_2 loss in bullfrogs, whether induced by spontaneous activity (Gottlieb and Jackson, 1976), by submergence in hyperoxic water (Gottlieb and Jackson, 1976), by temperature change (Mackenzie and Jackson, 1978), or by CO_2 breathing (Jackson and Braun, 1979), could all be accounted for by parallel variations in blood PCO_2. No change in the diffusion capacity of the skin for CO_2 was apparent in any of these conditions. In addition, the giant salamander, Cryptobranchus alleganiensis, that depends almost entirely on skin respiration, suffered a severe and prolonged respiratory acidosis when forcibly exercised, in contrast to the effective respiratory control exhibited by the toad, Bufo marinus, a species with an effective pulmonary exchanger (Boutilier et al., 1980).

These data support the experimental and theoretical conclusions of Gatz et al. (1975) that gas exchange in a plethodontid salamander is largely limited by the relatively fixed diffusion capacity of its skin and is relatively insensitive to variations in blood flow. This conclusion represented a serious challenge to the functional significance for respiratory exchange of the well-documented vascular responsiveness in amphibian skin (e.g., Poczopko, 1957). However, recent data and re-interpretations have challenged the general applicability of this analysis. Burggren and Moalli (1984) observed apparent CO_2 retention in bullfrogs moved from water to air that was correlated with a reduction in skin blood flow. They suggested that de-recruitment of cutaneous capillaries had occurred, perhaps related to a need for water conservation, and that this reduced the effective surface area for skin exchange. The analysis of Gatz et al. (1975) was based on a single capillary model (constant area) with variable flow.

Recent studies by Malvin and Hlastala (1986 a and b) provide convincing evidence that skin vascular adjustments can affect skin gas exchange, independent of changes in partial pressure gradient across the skin. Using mass spectrometry, they measured the loss of CO_2, Freon, and Halothane from a small encapsulated area on the abdominal surface of frogs, Rana pipiens, equilibrated with the gases. Because gas efflux had no effect on partial pressures within the animal, change in flux could be attributed to variations in blood flow or capillary recruitment. They found pronounced decreases in gas effluxes in response to lowered capsule PO_2 or elevated PO_2 on the general body surface (Malvin and Hlastala, 1986a) and to increased lung PO_2 (Malvin and Hlastala, 1986b). Extrapolation of these observations to the whole skin surface and to the overall control of skin gas exchange is impossible, however, because of the dual blood supply to the skin of anurans (Moalli et al., 1980). The ventral abdominal skin surface is supplied predominantly by systemically derived vessels that increase flow (in the bullfrog) during enforced apnea. In contrast, skin flow originating from the pulmocutaneous artery falls during apnea and, as a consequence, overall flow to the skin is relatively unaffected. Application of Malvin and Hlastala's technique to other skin regions on the frog and to amphibians, such as salamanders, with a single cutaneous blood supply, would be a useful next step.

None of the studies cited demonstrate cutaneous control that is effective for regulating general respiratory homeostasis. Perhaps bimodal amphibians are the wrong animals to study because an effective available pulmonary mode may preclude the necessity for cutaneous control. Better subjects for study may be amphibians that are limited to skin respiration, either the lungless species or lunged species under conditions where ventilation does not occur. The exercise study on Cryptobranchus (Boutilier et al., 1980) is such a study, but the skin control of this animal, if it exists, may not have the capacity to deal with the exchange load associated with exercise. Oxygen uptake of quiescent amphibians breathing exclusively with their skin has been measured as a function of ambient PO_2, but the results are inconsistent. Bentley (1975) observed a linear decrease in $\dot{V}O_2$ of submerged salamanders, Amphiuma means, with declining water PO_2, whereas Beckenbach (1975), studying several species of plethodontid salamanders, and Pinder (1985), studying submerged, paralyzed bullfrogs at $5^{\circ}C$, reported regulation of $\dot{V}O_2$ to PO_2 values well below ambient. Pinder attributed the regulatory behavior to capillary recruitment, although an alternative interpretation is that at the low metabolic rates prevailing in his experiment no diffusion limitation was apparent until the ambient PO_2 could no longer saturate the end-capillary blood. In view of the high affinity of blood for O_2 at the low temperature, this critical PO_2 may be well below ambient. Such an interpretation would require no vascular changes. In the study by Ultsch (1973) of Siren, the largest individuals, like Bentley's Amphiuma, were O_2 conformers at all PO_2 values below ambient. In smaller Siren, however, critical PO_2 was lower and a zone of O_2 regulation was present. The unanswered question in this and in the studies by Beckenbach and Pinder is: what is happening to skin diffusion capacity in the regulatory zone? Is it high and constant down to the critical PO_2, or is the diffusing capacity increasing with progressive hypoxia by vasodilatation and capillary recruitment until a maximum is reached at the critical PO_2?

The ability of lungless salamanders to increase O_2 uptake during exercise may provide convincing evidence for control. In a recent as yet unpublished study by M. E. Feder (personal communication), O_2 consumption was observed to increase more than 10-fold during treadmill exercise by plethodontid salamanders. This result is hard to explain without a major circulatory contribution and an increase in skin diffusing capacity. An intriguing aspect of these data is the O_2 uptake and delivery by the blood. Because the skin circulation in these salamanders is in parallel with the muscles (and other tissues), and because venous blood draining these beds presumably mixes in the return path to the heart, both the skin and the skeletal muscles are supplied by arterial blood of the same composition. At the skin, in the exercise state observed, O_2 is loaded into the blood from the arterial saturation point at 10 times the resting rate, while in the active muscles, O_2 is unloaded from the very same saturation point at 10 times the resting rate. An aerobic scope of this magnitude in an animal with only skin respiration and inefficient circulatory design is surely a remarkable finding.

ACID-BASE REGULATION

As noted above, a variety of experimental approaches have concluded that regulation of blood PCO_2 is dependent on pulmonary ventilation in bimodal amphibians.

Effective control of skin exchange may be lacking under conditions in which gas flux is high, and instead restricted to animals or conditions in which the skin is the exclusive respiratory surface. The demonstration that a plethodontid salamander, for example, could regulate its blood PCO_2 and pH independent of variations in metabolic CO_2 production would be convincing evidence for control of skin gas exchange.

The regulation of blood pH in amphibians conforms to the alphastat pattern, in which blood pH varies with body temperature with a slope of about 0.015 U/^{o}C. Indeed, Reeves' landmark paper (1972) introducing the concept of alphastat control was based on studies of the bullfrog. Subsequent studies on this and other amphibian species have confirmed the relationship (Mackenzie and Jackson, 1978; Withers, 1978; Burggren and Wood, 1981). Interestingly, the hellbender, Cryptobranchus, exhibited alphastat control between 5 and 25°C whether submerged or with access to air, so that in this predominantly skin-breathing salamander, pulmonary exchange is not required (Moalli et al., 1981). This apparent cutaneous control, however, can be explained by proportionate temperature-dependent changes in PCO_2 and in CO_2 production (increased gradient and increased flux) with constant skin diffusing capacity. In this instance, the passive exchange behavior of skin works to the animal's advantage by providing alphastat control without the cost of an active physiological mechanism. For pH control to occur in the face of metabolic change at constant temperature, on the other hand, an active mechanism is required, but thus far has not been documented in a strictly skin-breathing amphibian.

REFERENCES

Beckenbach AT (1975) Influence of body size and temperature on the critical oxygen tension of some plethodontid salamanders. Physiol Zool 48: 338-347.

Bentley PJ (1975) Cutaneous respiration in the congo eel Amphiuma means (Amphibia: Urodela). Comp Biochem Physiol 50A 121-124.

Boutilier RG, McDonald DG, Toews DP (1980) The effects of enforced activity on ventilation, circulation and blood acid-base balance in the aquatic gill-less urodele, Cryptobranchus alleganiensis; a comparison with the semi-terrestrial anuran, Bufo marinus. J exp Biol 84: 289-302.

Burggren WW, Feder ME (1986) Effect of experimental ventilation of the skin on cutaneous gas exchange in the bullfrog. J exp Biol 121: 445-450.

Burggren W, Moalli R (1984) "Active" regulation of cutaneous gas exchange by capillary recruitment in amphibians: experimental evidence and a revised model for skin respiration. Resp Physiol 55: 379-392.

Burggren W, Mwalukoma A (1983) Respiration during chronic hypoxia and hyperoxia in larval and adult bullfrogs (Rana catesbeiana). 1. morphological responses of lungs, skin and gills. J exp Biol 105: 191-203.

Burggren WW, Wood SC (1981) Respiration and acid-base balance in the salamander Ambystoma tigrinum: influence of temperature acclimation and metamorphosis. J comp Physiol 144: 241-246.

Czopek J (1965) Quantitative studies on the morphology of respiratory surfaces in amphibians. Acta Anat 62: 296-323.

Feder ME (1978) Effect of temperature on post-activity oxygen consumption in lunged and lungless salamanders. J exp Zool 206:179-190.

Feder ME, Burggren WW (1985) Cutaneous gas exchange in vertebrates: design, patterns, control and implications. Biol Rev 60: 1-45.

Gans C (1970) Respiration in early tetrapods - the frog is a red herring. Evolution 24: 723-734.

Gatz RN, Crawford EC, Piiper J (1974) Respiratory properties of the blood of a lungless and gill-less salamander, Desmognathus fuscus. Resp Physiol 20: 33-41.

Gatz RN, Crawford EC, Piiper J (1975) Kinetics of inert gas equilibration in an exclusively skin-breathing salamander, Desmognathus fuscus. Resp Physiol 24: 15-29.

Gottlieb G, Jackson DC (1976) Importance of pulmonary ventilation in respiratory control in the bullfrog. Am J Physiol 230: 608-613.

Guimond RW, Hutchison VH (1973) Aquatic respiration: an unusual strategy in the hellbender Cryptobranchus alleganiensis alleganiensis (Daudin). Science 182: 1263-1265.

Howell BJ, Rahn H (1976) Regulation of acid-base balance in reptiles. In: Gans C, Dawson WR (eds): Biology of the Reptilia, vol 5, Physiology A. Academic Press, London; pp. 335-363.

Hutchison VH, Whitford WG, Kohl M (1968) Relation of body size and surface area to gas exchange in anurans. Physiol Zool 41: 65-85.

Jackson DC, Braun BA (1979) Respiratory control in bullfrogs: cutaneous versus pulmonary response to selective CO_2 exposure. J Comp Physiol 129: 339-342.

Krogh A (1904) On the cutaneous and pulmonary respiration of the frog. Skand Arch Physiol 15: 328-419.

MacKenzie JA, Jackson DC (1978) The effect of temperature on cutaneous CO_2 loss and conductance in the bullfrog. Respir Physiol 32: 313-323.

Malvin GM, Hlastala MP (1986a) Regulation of cutaneous gas exchange by environmental O_2 and CO_2 in the frog. Respir Physiol, in press.

Malvin GM, Hlastala MP (1986b) Control of cutaneous blood flow and gas exchange by intrapulmonary oxygen content in the frog. Fed Proc 45: 758.

Moalli R, Meyers RS, Jackson DC, Millard RW (1980) Skin circulation of the frog,
    Rana catesbeiana: distribution and dynamics. Respir Physiol 40: 137-148.
Moalli R, Meyers RS, Ultsch GR, Jackson DC (1981) Acid-base balance and
    temperature in a predominantly skin-breathing salamander, Cryptobranchus
    alleganiensis. Respir Physiol 43: 1-11.
Nickerson MA, Mays CE (1973) The hellbenders: North American "giant salamanders",
    Milwaukee Public Museum, Milwaukee pp. 54-55.
Piiper J, Scheid P (1975) Gas transport efficacy of gill, lungs and skin: theory
    and experimental data. Respir Physiol 23: 209-221.
Pinder AW (1985) Respiratory physiology of the frogs Rana catesbeiana and Rana
    pipiens: influences of hypoxia and temperature. Ph.D. diss, Univ of
    Massachusetts, Amherst.
Poczopko P (1957) Further investigations on the cutaneous vasomotor reflexes in
    the edible frog in connexion with the problem of regulation of the cutaneous
    respiration in frogs. Zool Poloniae 8: 161-175.
Porter KR (1972) Herpetology, W.B. Saunders, Philadelphia, pp. 110-111.
Reeves RB (1972) An imidazole alphastat hypothesis for vertebrate acid-base
    regulation: tissue carbon dioxide content and body temperature in bullfrogs.
    Respir Physiol 14: 219-236.
Saint-Aubain ML de (1982) The morphology of amphibian skin vascularization
    before and after metamorphosis. Zoomorphology 100: 55-63.
Smyth DH (1939) The central and reflex control of respiration in the frog.
    J Physiol, London 95: 305-327.
Tenney SM, Tenney JB (1970) Quantitative morphology of cold-blood lungs:
    Amphibia and Reptilia. Respir Physiol 9: 197-215.
Ultsch GR (1973) A theoretical and experimental investigation of the relation-
    ships between metabolic rate, body size, and oxygen exchange capacity. Respir
    Physiol 18: 143-160.
Whitford WG, Hutchison VH (1965) Gas exchange in salamanders. Physiol Zool 38:
    228-242.
Withers PC (1978) Acid-base regulation as a function of body temperature in
    ectothermic toads, a heliothermic lizard, and a heterothermic mammal. J.
    therm Biol 3: 163-171.

OXYGEN TRANSPORT IN AQUATIC AND TERRESTRIAL VERTEBRATES

Stephen C. Wood

Department of Physiology
School of Medicine
The University of New Mexico
Albuquerque, New Mexico, U.S.A.

Many aspects of the external milieu impact upon the respiratory properties of blood. Water is often polluted or otherwise deficient in oxygen, making adaptations to hypoxia necessary. Air breathing ectotherms must adapt to hypoxia from an internal mechanism –intracardiac shunt. The principle of symmorphosis applies to red cells in that their number and qualities are sufficient for the required O_2 transport.

INTRODUCTION

The central theme of this symposium is the contrast between aquatic and terrestrial life. The central theme of this chapter should then be the impact of aquatic versus terrestrial life on the respiratory properties of blood. However, there is no direct impact. There are important indirect impacts of living in water vs. air on the respiratory properties of blood. These are temperature, oxygen availability, diving, cost of locomotion, acid–base regulation, cardiovascular function, and other topics presented in this volume.

First, the reason there is no direct impact of breathing medium on hemoglobin function is that the oxygen affinity of hemoglobin relates only to oxygen partial pressure (PO_2) and not to O_2 content of the breathing medium. Consequently, there is no inherent reason to expect a difference in respiratory properties of blood between aquatic and terrestrial animals. However, the transition from aquatic to terrestrial life is generally conceived, both in the phylogenetic and ontogenetic context, as being accompanied by large and adaptive changes in oxygen affinity of blood. Breathing air instead of water does have significant impact on the work of breathing, the ventilation required per unit O_2 uptake, and acid–base regulation (Rahn (1966)). But the oxygen affinity of blood has no intrinsic relationship to breathing air instead of water. Indeed, a more basic question than the impact of breathing medium on red cell function is the importance of red cells at all.

Comparative Physiology: Life in Water and on Land. P. Dejours, L. Bolis, C.R. Taylor, E.R. Weibel (eds.) Fidia Research Series, IX-Liviana Press, Padova © 1987

ARE RED BLOOD CELLS IMPORTANT FOR ECTOTHERMS?

Many ectotherms are anemic relative to homeotherms. Some species have
significant amounts of nonfunctional (met) hemoglobin. Some species can survive
the complete loss of their red cells (Flores and Frieden (1968)). Furthermore,
the oxygen affinity of blood is tremendously variable among different species,
often with no convincing correlation with breathing medium or natural history.
These facts suggest that the role of the red cell in metabolic homeostasis is
less important in ectotherms than in homeotherms.

The control of red cell mass and function in ectotherms does appear to be
relatively poor compared with homeotherms. What may account for this? One
possibility is that the importance of the red cell is reduced due to low O_2
demand, which is lower by orders of magnitude. However, I think that the true
relationship between red cell function and O_2 demand provides a good example of
the concept of symmorphosis (Taylor and Weibel, (1981)); i.e., the principle
that structural elements of a physiological system are regulated to satisfy (but
not exceed) the requirements; in this case O_2 transport. It seems reasonable
that red cell mass, oxygen affinity, and other components of the O_2 transport
system (e.g., diffusing capacity and cardiac output) have evolved to be well
matched to the metabolic machinery and O_2 demand and not to the environment per
se.

OXYGEN DISSOCIATION CURVE AND THE ENVIRONMENT

The tradition of correlating O_2 affinity of blood with the environment dates
back to Krogh and Leitch (1919). Dominant environmental factors in vertebrate
evolution were the availability of water, O_2 concentration, and temperature.
These remain dominant factors in the ontogeny as well as the daily life of
modern ectotherms. Consequently these factors have been studied often and much
has been made about the adaptability of red cells to the environment. However,
valid conclusions about the appropriateness of a given O_2 dissociation curve
must be based on more than just the environmental conditions and a
physico-chemical description of the O_2 dissociation curve. The best and ultimate
test of the physiological significance of a particular dissociation curve or
shift of that curve is the effect on arterial PO_2 and mixed venous PO_2. However,
studies that meet these criteria and can apply the dissociation curve to the
Fick principle, $VO_2 = Q \times (CaO_2 - CvO_2)$, are rare.

Other factors complicate a physiological assessment of the oxygen dissociation

curves of ectotherms as well as homeotherms. One factor is the tremendous
reserve for other components of the O_2 transport system, particularly cardiac
output and flow distribution to compensate for altered O_2 uptake. Consequently,
for most tissues the position of the dissociation curve is relatively
unimportant. Also, chronic changes in O_2 affinity of human blood due to mutant
Hb are compensated for by changes in Hb concentration (Parer (1970)). As a
result, blood P_{50} values ranging from 12 to 70 Torr can provide roughly the same
O_2 delivery (CaO_2 $-CvO_2$) at similar mixed venous PO_2 values and cause no
problems in otherwise healthy individuals (Parer (1970)). However, vital organs
(heart, brain, liver) have O_2-consuming processes that are much more dependent
on a higher PO_2 and therefore on the affinity of blood for O_2 (Woodson (1979)).

The position of the O_2 dissociation curve is much more labile in ectotherms than
homeotherms due to the large daily fluctuations of temperature and pH. This
further complicates efforts to assess the significance of the curve position.
Among amphibians the range of blood P_{50} is similar to the range reported for
human Hb mutants, 12 to 70 Torr. Similar large ranges in P_{50} exist for fish and
reptiles. Although the trend for correlations between P_{50} and environmental
variables is present, the accessory data needed to permit generalizations are
still too few.

Adaptations to Hypoxia

The role of O_2 affinity in hypoxia adaptation remains controversial. The debated
point is whether a right shift or a left shift of the O_2 dissociation curve is
better suited to life at high altitude. Both ontogenetic and interspecific
differences in the response to hypoxia have fostered this debate. Earlier
studies on human sojourners to altitude showed that both 2,3-DPG levels and P_{50}
are increased (Lenfant et al. (1968)). In contrast, mammals in utero and
mammals native to high altitude (e.g., llama) have bloods with higher O_2
affinity than adults or their sea-level counterparts (Bartels (1964; 1970)).

It is now fairly clear that the important factor to consider is the degree of
hypoxia (Turek and Kreuzer (1972); Aberman (1977)). A right-shifted dissociation
curve is advantageous (unloading is increased more than loading is reduced) up
to an altitude of ~ 4,300 m (Shappell and Lenfant (1974)). There is convincing
experimental evidence that a left-shifted dissociation curve is adaptive during
more severe hypoxia; i.e., only those animals with a (pharmacologically)
left-shifted dissociation curve survive (Eaton et al. (1974)). When the hypoxia
is due to internal problems (e.g., intracardiac shunt) different principles

apply (see below) and a right-shifted curve is always adaptive, unless it is so far shifted that O_2 loading in the normoxic gas exchanger is impaired (Wood (1984)).

Aquatic ectotherms, even at sea level, often encounter levels of hypoxia equivalent to those found at altitudes > 4,300 m. Furthermore, water PO_2 may have daily cycles (due to photosynthesis and other factors) with values ranging from 10 to 300 Torr (Zaaijer and Wolvekamp (1958)). Fishes living in certain midwater pelagic regions of the ocean encounter similar large ranges of water PO_2, from ~ 1 Torr at a 300-m depth to 150 Torr during their daily migration to the surface (Douglas et al. (1976)). Longer-term changes in O_2 levels are experienced by those species that migrate. More dramatic change, in terms of total O_2 available, occurs during the transition from aquatic to aerial respiration with amphibian metamorphosis and in facultative air-breathing fishes (see next section).

The comparative data on red cell function during hypoxia are quite consistent. Except for the fishes that inhabit the O_2-minimum zones of the ocean, the consistent finding in fish exposed to hypoxia is a left-shifted dissociation curve due to decreased red cell organic phosphates (reviewed by Weber (1982)). The major compensation in fishes in the O_2-minimum zone, like that of some amphibians, seems to be a reduced metabolic rate (Douglas et al. (1976)).

Another mechanism for hypoxia compensation is more rapid and potent than the hematological response. Available only to ectotherms, this is behavioral thermoregulation to a lower than normal preferred body temperature. This was predicted based on model calculations (Wood (1984)) and has been documented in lizards (Hicks and Wood (1985)) and amphibians (Wood et al.(1985)). By reducing their body temperature during hypoxia, an ectotherm has 3 immediate benefits: (1) lowered O_2 demand by roughly 11 % per $^{\circ}$ C; (2) left shifted O_2 dissociation curve providing increased arterial saturation; (3) decreased or removed hypoxic ventilatory drive (Wood (1984)).

Transition From Water Breathing to Air Breathing

The transition from water breathing to air breathing in vertebrate evolution has two modern parallels: interspecific and intraspecific or ontogenetic. There are pronounced interspecific differences in the dependence on air breathing among lungfish (Lenfant et al. (1966)) and amphibians (Lenfant and Johansen (1967)). These studies and others (see Johansen and Weber (1976)) have shown a clear correlation between blood O_2 affinity and dependence on air breathing. As first suggested by Krogh and Leitch (1919), the O_2 dissociation curve is progressively right shifted as the dependence on air breathing increases. Such interspecific comparisons may be interesting and provocative, but their utility as a basis for physiological or evolutionary concepts is limited by genetic diversity as well as allometric and morphological variables. These problems are partially solved by studying the transition from aquatic to aerial respiration that occurs within a species, i.e., the ontogeny of O_2 transport.

Developmental changes in Hb function have been studied in all classes of vertebrates. With rare exception the characteristic fetal-maternal shift of placental mammals has been found to accompany the development of other vertebrates (Bartels (1970)). In many species the increase in P_{50} that occurs with birth, hatching, or metamorphosis is associated with the synthesis of a new Hb. However, that is not necessarily the cause of the change in O_2 affinity. For human Hb, the increase in P_{50} is due to a greater sensitivity of adult Hb than fetal Hb to 2,3-DPG (Bauer (1974)). For amphibians, the increase in P_{50} at metamorphosis (or birth in the case of viviparous species) is due to an increase in the amount of red cell ATP (Toews and MacIntyre (1977); Wood (1971)). These studies support the generalization based on interspecific differences; i.e., the right shift of the O_2 dissociation curve during development is invariably associated with a switch to (or increased dependence on) air breathing. The lower O_2 affinity is argued to be adaptive to the 20-to 50-fold increase in O_2 availability as well as the increased cost of locomotion on land (Johansen and Weber (1976); Wood and Lenfant (1979)).

The traditional argument for the significance of a right-shifted O_2 dissociation curve is an increase in O_2 delivery or tissue O_2 extraction ((CaO_2 $-CvO_2$)/CaO_2)). An alternative means of increasing O_2 extraction is to keep O_2 affinity constant but increase the sigmoidicity of the curve. This is happens in Ambystoma tigrinum. With growth from 2 g to 120 g, P_{50} doesn't change but the Hill coefficient increases from ~ 2 to ~ 3 (Burggren et al., unpublished).

ORGANISMIC ADAPTATIONS

Oxygen Uptake and Metabolic Machinery

The resting metabolic rate compiled for numerous species shows a 30:1 ratio of homeotherms to ectotherms (Hemmingsen (1960)). In some cases it is greater. A 100-g bird at rest expends ~ 10 times as many calories as a 100-g lizard at the same body temperature (Bartholomew (1972)). This is equivalent to an O_2 uptake that is 48-fold higher. The total drop in PO_2 from air to cells is probably about the same in these animals. Consequently the O_2 conductance (VO_2/dPO_2) must be ~ 30 times higher in birds. The increased demand this places on the O_2 transport system of homeotherms is reflected in all components: lung structure, complete separation of systemic and pulmonary circulation, increased perfusion pressures, cardiac performance, and red cell properties. The hematologic correlates of homeothermy, following the principle of symmorphosis, include increased O_2 capacity, increased buffer capacity, and carefully regulated O_2 affinity.

What mechanisms account for the significant increase in the "cost of living" for birds and mammals? After almost a century of considerable speculation, the answers are only now beginning to emerge. One recent study shows that at least some of the extra heat in homeotherms comes from an increase in mitochondrial activity and relative organ size (Else and Hulbert (1981)). Another recent argument suggests that the relative leakiness of cell membranes in endotherms requires a higher turnover of ATP thus, a higher cost of living and inherent sensitivity to hypoxia (Hochachka (1986)). The argument that the lower cost of living reduces the importance of the red cell for O_2 delivery in ectotherms, if valid at all, applies only to animals at rest, the condition under which the metabolic rates previously discussed were measured. However, the role of the red cell during activity is of greater physiological interest. Much less is known about the cost of active living and how ectotherms compare with homeotherms. Clearly there are tremendous behavioral differences in activity levels among both ectotherms and homeotherms. However, calculations by Hemmingsen (1960) of maximum, sustainable metabolic rate suggested that the ratio of homeotherms to poikilotherms approached unity. A recent review of data for maximum, burst metabolic rate also suggested little difference between homeotherms and at least some poikilotherms (Prothero (1979)). Data for maximum O_2 uptake in animals ranging in size from a 31-mg blowfly to a 100-ton blue whale fell on a single regression line with a slope of 0.75. Perhaps the evolution of homeothermy did not increase the maximum metabolic capacity but simply boosted the resting metabolic rate to a higher fraction of the maximum.

Circulatory Shunts

These absence of cardiovascular shunts in mammals and birds greatly enhances O_2 transport potential. Both components of systemic oxygen transport (cardiac output and arterial $[O_2]$) are generally higher. In addition, higher systemic pressures prevail permitting the capacity for redistribution of blood flow.

Among lower vertebrates only varanid lizards possess similar circulatory attributes (Johansen (1979)). Central vascular or intracardiac shunts affect O_2 transport in ectotherms in both obvious and subtle ways. Clearly a right-to-left shunt will lower arterial saturation and, unless total flow increases. lower the maximum O_2 deliverable to tissues (i.e., cardiac output x CaO_2). However, the effect of a shunt on arterial PO_2 is not widely appreciated. The normal relationship of dependent to independent variables is reversed and PO_2 becomes a dependent variable of O_2 content. In this situation, the oxygen affinity of blood becomes more important in determining arterial PO_2 than gas exchange. In fact, arterial PO_2 can exceed PO_2 in the lungs, if the affinity decreases dramatically; e.g., with profound acidosis (Wood (1984)). This in vivo phenomenon is analogous to the mixing method, an in vitro technique used to measure the P_{50} of blood (Haab et al. (1960)). With this method, known ratios of saturated and desaturated blood are mixed anaerobically to produce 50% (or any other percent) saturation. The PO_2 of this mixture is then measured, providing one point on an O_2 dissociation curve. Any factor that right-shifts the dissociation curve will, for a given saturation, increase the measured PO_2. The same principle applies when shunting occurs in normal (e.g., amphibians and reptiles) and abnormal (e.g., cyanotic heart disease) circulations. Rossoff et al. (1980) used a two-compartment model to describe this phenomenon. The mixing of bloods from the perfect compartment and the shunt compartment determines the O_2 content of the systemic arterial blood. In this model the shunt compartment also includes any blood that does not equilibrate with lung PO_2 for reasons other than anatomical shunt, i.e. ventilation-perfusion mismatch or impaired diffusion.

A key point is that the saturation of blood in the perfect compartment is a function of PO_2 and O_2 affinity;
$$SpO_2 = f(PpO_2, P_{50})$$
where SpO_2 is O_2 saturation in perfect blood and PpO_2 is O_2 pressure in perfect blood. The PpO_2 is equal to the lung PO_2 and is therefore a function of inspired PO_2, lung PCO_2, and the respiratory exchange ratio. The O_2 content of this blood (CpO_2) is a function of saturation and Hb concentration: $CpO_2 = f(SpO_2, Hb)$
In systemic arterial blood, the O_2 content is a function of the shunt fraction (Q_S/Q_T) and the O_2 contents of mixed venous and perfect blood; consequently
$$CaO_2 = (Q_S/Q_T)(CvO_2) + (Q_{NS}/Q_T)(CpO_2)$$
where CaO_2 is O_2 content, Q_S/Q_T is the non-shunt fraction, and CvO_2 is mixed venous O_2 content. Saturation of systemic arterial blood is a function of CaO_2 and Hb concentration. Systemic PaO_2 becomes a dependent variable of arterial

saturation of O_2 (SaO_2) and P_{50};

$$PaO_2 = f(SaO_2, P_{50})$$

Thus, for a given saturation, PaO_2 will be higher in blood with lower O_2 affinity. Models of gas exchange in mammals predict this phenomenon (higher PaO_2 with lower O_2 affinity) in physiological shunts (V/Q mismatch) and anatomical shunts (Turek and Kreuzer (1972); Rossoff et al. (1980)). Experimental data have verified this model in dogs with physiological or artificial shunts (Frans et al. (1979); Wood (1982)), in amphibians and reptiles with normal intracardiac shunts (Wood (1982)), and in humans with congenital heart disease (Berman et al. (1986)).

The altitude model of hypoxemia is fundamentally different from the above shunt model. When lung PO_2 is reduced, the degree of hypoxia determines whether a right or left shift of the dissociation curve is adaptive. However, when a shunt is present, a right shift of the curve is generally adaptive within physiological limits of P_{50}; i.e., both PaO_2 and PvO_2 are increased. For ectotherms, a typical factor inducing a right shift of the dissociation curve is elevated body temperature. This also increases O_2 uptake. If tissue O_2 uptake is diffusion limited, as suggested by Robin (1982), it would undoubtedly be facilitated by the higher capillary PO_2. It seems reasonable that this aspect of temperature $-PO_2$ relationship could be a component of the metabolic Q_{10}. One situation in which a right-shifted curve and high body temperature may be deleterious is during environmental hypoxia. If the curve is right shifted due to a relatively high preferred body temperature, there is little reserve for hypoxia in the upper, flat portion of the curve. Therefore, during hypoxia, blood leaving the pulmonary compartment may not be fully saturated, further reducing systemic arterial saturation. The behavioral compensation for hypoxia (voluntary hypothermia) also alleviates this potential problem by left-shifting the dissociation curve and increasing O_2 loading in the lungs.

ACKNOWLEDGEMENT

Supported by NSF Grant PCM 8300472 and NATO Grant RG.86/0021.

REFERENCES

Abermann A (1977) Crossover PO_2, a measure of the variable effect of increased P_{50} on mixed venous PO_2. Am Rev Resp Dis 115:173-175.

Bartels H (1964) Comparative physiology of oxygen transport in mammals, Lancet

2:601-604

Bartels H (1970) Prenatal Respiration, North-Holland, Amsterdam p .50.

Bartholomew GA (1972) Body temperature and energy metabolism. In: Gordon MS (ed): Animal Physiology: Physiology: Principles and Adaptations (2nd ed), Macmillan, New York.

Bauer C (1974) On the respiratory function of hemoglobin. Rev Physiol Biochem Pharmacol 70:1-31.

Berman W Jr, Wood SC, Yabek S, Dillon F, Fripp R, Burstein R Systemic oxygen transport in congenital heart disease. Circulation, In Press.

Douglas EL, Friedl WA, Pickwell GV (1976) Fishes in oxygen-minimum zones: blood oxygenation characteristics. Science 191:957-959.

Eaton JW, Skelton TD, Berger E (1974) Survival at extreme altitude: protective effect of increased hemoglobin-oxygen affinity. Science 183:743-744.

Else PL, Hulbert AJ (1981) Comparison of the "mammal machine" and the "reptile machine": energy production. Am J Physiol 240 (Regulatory Integrative Comp Physiol 9):R3-R9.

Flores G, Frieden E (1968) Induction and survival of hemoglobin-less and erythrocyte-less tadpoles and young bullfrogs. Science 159:101-103.

Frans A, Turek Z, Yokota H, Kreuzer F (1979) Effect of variations in blood hydrogen ion connection on pulmonary gas exchange of artificially ventilated dogs. Pfluegers Arch 380:35-39.

Haab PE, Piiper J, Rahn H (1960) Simple method for rapid determination of an O_2 dissociation curve of the blood. J Appl Physiol 15: 1148-1149.

Hemmingsen AM (1960) Energy metabolism as related to body size and respiratory surfaces and its evolution. Rep Steno Mem Hosp Nord Insulinlab 9:1-110.

Hicks JW, Wood SC (1985) Temperature regulation in lizards: Effects of hypoxia. Am J Physiol 248:R595-600.

Hochachka P (1986) Defense strategies against hypoxia and hypothermia. Science 231:234-241.

Johansen K (1979) Cardiovascular support of metabolic functions in vertebrates. In: Wood SC, Lenfant C (eds): Lung Biology in Health and Disease. Evolution of Respiratory Processes: A Comparative Approach Dekker, New York; pp. 107-192.

Johansen K, Weber RE (1976) On the adaptability of hemoglobin function to environmental conditions. In: Spencer Davis P, Sunderlund N (eds): Perspectives in Experimental Biology, Zoology, and Botany. Pergamon, New York; pp. 219-234.

Krogh A, Leitch I (1919) The respiratory functions of the blood in fishes. J Physiol Lond 52:288-297.

Lenfant C, Johansen K (1967) Respiratory adaptations in selected amphibians.

Respir Physiol 2:247–260.

Lenfant C, Johansen K, Grigg GC (1966) Respiratory properties of blood and pattern of gas exchange in the lungfish Neoceratodus forsteri. Respir Physiol 2:1–21.

Lenfant, C, Torrance J, English E, Finch CA, Reynafarje C, Ramos C, Faura J (1968) Effect of altitude on oxygen finding by hemoglobin and on organic phosphate levels. J Clin Invest 47:2652–2656.

Parer JT (1970) Oxygen transport in human subjects with hemoglobin variants having oxygen affinity. Respir Physiol 9:43–49.

Prothero JW (1979) Maximal oxygen consumption in various animals and plants. Comp Biochem Physiol A Comp Physiol 64:463–466.

Rahn H (1966) Evolution of the gas transport system in vertebrates. Proc R Soc Med 59:493–494.

Robin ED (1982) Tissue O_2 utilization. In: Loeppky JA, Riedesel ML (eds): Oxygen transport to human tissues. Elsevier/North Holland, New York; pp. 179–186.

Rossoff L, Zeldin R, Hew E, Aberman A (1980) Changes in blood P50. Effects on oxygen delivery when arterial hypoxemia is due to shunting. Chest 77:142–146.

Shappell SD, Lenfant C (1974) Physiological role of the oxyhemoglobin dissociation curve. In: Surgenor D (ed): The Red Blood Cell (2nd ed). Academic Press, New York; pp. 842–873.

Taylor, CR, Weibel, ER (1981) Design of the mammalian respiratory system. I. Problem and strategy. Respir Physiol 44:1–10.

Toews D, MacIntyre D (1977) Blood respiratory properties of a viviparous amphibian. Nature Lond 266:464–465.

Turek Z, Kreuzer F (1972) Effects of shifts of the O2 dissociation curve upon alveolar–arterial O2 gradients in computer models of the lung with ventilation–perfusion mismatching. Respir Physiol 45:133–139.

Weber RE (1982) Intraspecific adaptation of hemoglobin function in fish to oxygen availability. In: Addink ADF, Spronk N (eds): Exogenous and endogenous influences on metabolic and neural control. Pergamon Press, Oxford; pp. 87–102.

Wood SC (1971) Effect of metamorphosis on blood respiratory properties and erythrocyte adenosine triphosphate level of the salamander. Dicamptodon ensatus. Respir Physiol 12:53–65.

Wood SC (1982) Effect of O2 affinity on arterial PO2 in animals with central vascular shunts. J Appl Physiol 53:1360–1364.

Wood SC (1984) Cardiovascular shunts and oxygen transport in lower vertebrates. Am J Physiol 247 (Regulatory Integrative Comp Physiol 16):R3–R14.

Wood SC, Lenfant C (1979) Oxygen transport and oxygen delivery. In: Wood SC, Lenfant C (eds): Lung Biology in Health and Disease. Evolution of Respiratory Processes: A Comparative Approach, Marcel Dekker, New York; pp. 193–223.

Wood SC, Dupre' RS, Hicks JW (1985). Voluntary hypothermia in hypoxic animals. Acta Physiol Scand 124:46.

Woodson RD (1979) Physiological significance of oxygen dissociation curve shifts. Crit Care Med 7: 368–373.

Zaaijer JP, Wolvekamp P (1958) Some experiments on the hemoglobin oxygen affinity in the blood of the ramshorn (<u>Planorbis</u> <u>corneus</u> <u>L.</u>). Acta Physiol Pharmacol Neerl 7: 56–77.

RESPIRATORY REGULATIONS IN WATER AND AIR BREATHERS: PHYSIOLOGICAL CONTRASTS

Pierre Dejours

Laboratoire d'Etude des Régulations Physiologiques
associé à l'Université Louis Pasteur,
Centre National de la
Recherche Scientifique
23 rue Becquerel, 67087 Strasbourg, France

Water breathers are mainly sensitive to changes of the oxygenation of their milieu; they hypoventilate in hyperoxia and become hypercapnic; they hyperventilate in hypoxia and become hypocapnic. They are relatively insensitive to changes in ambient carbon dioxide.

Air breathers, living in a milieu which for a given O_2 tension has a much higher O_2 concentration than water, breathe much less than water breathers and are relatively hypercapnic. Air breathers are sensitive to changes of both O_2 and CO_2 ambient pressures, except in extreme hypoxia during which they hyperventilate irrespective of arterial hypocapnia.

All ventilatory reactions to environmental O_2 and CO_2 conditions are viewed as having as priority a proper oxygenation of the body. In case of conflict between a proper oxygenation and a proper CO_2 clearance, oxygenation has precedence; the CO_2 clearance is subordinate.

Since animal life requires the uptake of oxygen and the clearance of carbon dioxide, how do organisms react to variations of the oxygen and carbon dioxide characteristics of the milieu? It is out of question to consider all possible cases, rest, exercise, temperature-induced variations of oxygen consumption, etc., in all zoological groups for which at least some information is available. I will limit the discussion to what might be considered as essentials.

RESPONSES TO VARIATIONS OF AMBIENT OXYGENATION

A water is normoxic when it is in equilibrium with air at sea level, that is at an O_2 partial pressure of about 150 Torr.

Water breathers. The table shows the qualitative variations of ventilation induced

Comparative Physiology: Life in Water and on Land. P. Dejours, L. Bolis, C.R. Taylor, E.R. Weibel (eds.) Fidia Research Series, IX-Liviana Press, Padova © 1987

Table. Action of hypoxia, of hyperoxia and of hypercapnia in normoxia on ventilation, $\dot{V}$, in seawater and freshwater animals. For a more complete list of references, see Wilkes et al. (1981) and Toulmond and Tchernigovtzeff (1984).

Ref.	Species	Method	Hypoxia	Hyperoxia	Hypercapnia in normoxia
1	ANNELIDS Arenicola marina	expired water collection	increases $\dot{V}$	decreases $\dot{V}$	
2	CRUSTACEANS Carcinus maenas	"	increases $\dot{V}$		small or no increase of $\dot{V}$
3	Astacus lepto-dactylus	"	increases $\dot{V}$	decreases $\dot{V}$	variable increase of $\dot{V}$
4	MOLLUSKS Anondonta cygnea	visual observ.	increases $\dot{V}$	decreases $\dot{V}$	
5	FISHES Anguilla vulg.	expired water collection	increases $\dot{V}$		increase $\dot{V}$ slightly
	Salmo shasta	"	"		
6	Catostomus com.	Saunders' method	increases $\dot{V}$		"
	Ictalurus neb.	"	"		"
	Cyprinus carpio	"	"		"
7	Cyprinus carpio	opercular mechanogram	increases $\dot{V}$	decreases $\dot{V}$	"
8	several Cyprinidae	modified Saunders' method	increases $\dot{V}$	decreases $\dot{V}$	"
	Salmo trutta	"	increases $\dot{V}$	decreases $\dot{V}$	increases $\dot{V}$
9	several marine fishes	opercular water sampling		decreases $\dot{V}$	
10	AMPHIBIANS tadpole Rana catesbeiana	expired water collection	increases $\dot{V}$	decreases $\dot{V}$	

1 Toulmond and Tchernigovtzeff (1984)

2 Jouve-Duhamel and Truchot (1983)

3 Massabuau et al. (1984)

4 Koch and Hers (1943)

5 Van Dam (1938)

6 Saunders (1962)

7 Peyraud and Serfaty (1964)

8 Dejours (1973)

9 Dejours, Toulmond and Truchot (1977)

10 West and Burggren (1982)

by aquatic hypoxia or hyperoxia in some selected water breathers belonging to a few phyla. It is well known that hypoxia increases ventilation in all animals studied. The ventilatory depression induced by hyperoxia (Dejours, 1973) is less well documented; but all observations show it to be a general phenomenon (see Toulmond and Tchernigovtzeff, 1984). Note that aquatic hyperoxia is a natural phenomenon.

Figure 1 gives examples of the changes of specific ventilation, $\dot{V}w/\dot{M}_{O_2}$, in three groups of water breathers as a function of the concentration of O_2 in water, $C_{I_{O_2}}$; the top of Figure 2 shows another example in the crayfish. At constant $\dot{M}_{O_2}$, any change in $\dot{V}w$ is coupled with a reciprocal change of $(P_{I_{O_2}}-P_{E_{O_2}})$ and of $(P_{E_{CO_2}}-P_{I_{CO_2}})$. Although the relation between $(P_{E_{CO_2}}-P_{I_{CO_2}})$ and body fluid P_{CO_2}, $e.g.$ arterial Pa_{CO_2}, is complex, particularly in aquatic animals, any increase of $(P_{E_{CO_2}}-P_{I_{CO_2}})$ leads to a hypercapnia in body fluids.

When ventilation is depressed by hyperoxia, the differences $(P_{I_{O_2}}-P_{E_{O_2}})$ and $(P_{E_{CO_2}}-P_{I_{CO_2}})$ do actually increase, and a hypercapnia and an acidosis are observed. The reverse effect, hypocapnia and hypocapnic alkalosis, is observed in animals whose ventilation is enhanced by hypoxia. Figure 2 is an example of the action of the change in ambient P_{O_2} on $\dot{V}/\dot{M}_{O_2}$ and on P_{CO_2}, pH and $[HCO_3^-]$ of the prebranchial blood of the crayfish. I do not know of any exception to these modifications in specific ventilation and acid-base balance brought about by variations of the ambient oxygenation. It is sometimes stated that hypoxic hyperventilation does not lead to hypocapnia, because the arterial P_{CO_2} value is already very low, but when the ambient conditions are well controlled and blood carefully sampled and analyzed, a hypocapnic alkalosis is always observed (see Truchot, 1987).

Animals differ, however, in their capacity to compensate hyperoxia-induced hypercapnic acidosis and hypoxia-induced hypocapnic alkalosis. Figure 2 shows that in the crayfish *Astacus leptodactylus* 24 hours after exposure to a change of O_2 tension, the acidosis of hyperoxia and the alkalosis of hypoxia persist, and the respiratory acid base balance (ABB) changes are only slightly compensated metabolically, even after several weeks of hyperoxia-induced hypercapnia (Dejours and Beekenkamp, 1977). In other species, as in the shore crab (Truchot, 1975), the deviations of the blood ABB due to environmental changes of oxygenation are completely compensated. Truchot (1987) gives an extensive review of the literature on this problem in various mollusks, crustaceans and fishes.

Air breathers. The variations of ventilation as a function of ambient P_{O_2} has been extensively studied in vertebrates (see summary by Bouverot, 1985). Prolonged

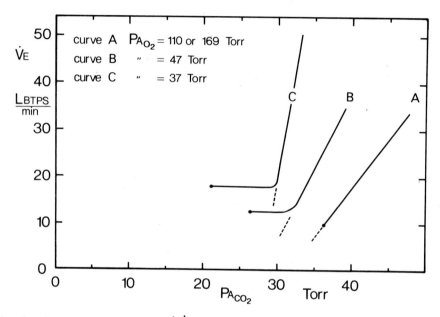

Fig. 1. Specific ventilation, $\dot{V}/\dot{M}_{O_2}$, as a function of the inspired O_2 concentration, $C_{I_{O_2}}$, in water- and air breathers. The horizontal bars under the abscissa scale indicate the baronormoxic range for water breathers (left side) and for air breathers (right side). The width of the bars take into account the pressure range of 140-160 Torr, the variation of the capacitance coefficient and of the water vapor effect with temperature. The intercepts of the vertical dashed line at abscissa $C_{I_{O_2}} = 1.2$ mmol·L^{-1} with the water breathers' and air breathers' lines are marked by asterisks. The insert shows some characteristics of these intercepts. E designates the O_2 extraction coefficient.

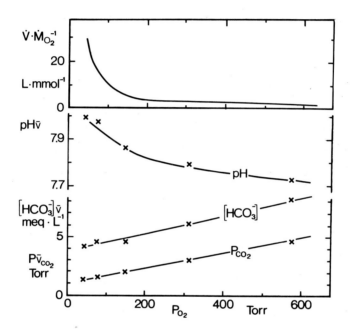

Fig. 2. Specific ventilation and pH, [HCO$_3^-$] and P$_{CO_2}$ of prebranchial (mixed venous) hemolymph in 17 crayfish *Astacus leptodactylus* at various water oxygen
tensions. Ambient temperature was 13°C, and the water pH and P$_{CO_2}$ were regulated
at 8.40 and 0.8 Torr (from Sinha and Dejours, 1980).

hyperoxia, let us say 300-400 Torr, leads to no change in ventilation in humans, but causes a prolonged fall in dogs, rats and chickens. All normoxic air-breathing vertebrates differ from normoxic water breathers, vertebrates or not, by the fact that the hyperoxia-induced depression of ventilation is much less pronounced in air breathers than in water breathers.

In air breathers, hypoxia entails an increase of ventilation which may be extremely high in marked hypoxia as shown by Figure 1. Obviously such an increase of ventilation leads to a hypocapnia and an alkalosis. If the altitude is not too high, the hypocapnic alkalosis may be compensated by a decrease in $[HCO_3^-]$, but at very high altitude, as humans on the upper slopes of Mount Everest, there occurs a marked alkalosis as predicted by Dejours (1979, 1981) and actually observed by West et al. during the 1981 expedition to Everest (see West, 1983). The lowest value of alveolar P_{CO_2} recorded there was 7.5 Torr.

Comparison of air breathers to water breathers. Figure 1 shows the specific ventilation in water- and air breathers respectively. The differences between air- and water breathers are striking. One may see that for each group in baronormoxia, indicated by horizontal bars below the abscissa, the specific ventilation is much higher in water breathers than in air breathers. At the same value of CI_{O_2} (1.2 $mmol \cdot L^{-1}$), the theoretical points (asterisks) on the water breathers' and air breathers' lines differ markedly: the specific ventilation is much higher in the air breather than in the water breather. It is presumably so high in the air breathers because the oxygen in air is there at a very low partial pressure, a pressure head which has to diminish at each step of the respiratory system from the air down to the mitochondria. It may mean that the O_2 partial pressure is more important than the actual O_2 concentration in driving the air breathers' ventilation.

For poikilothermic animals Shelton et al. (1986) observed that the reptiles may breathe intermittently because air is very rich in oxygen. As a consequence broad variations of blood P_{O_2}, P_{CO_2} and pH exist in this group. However, ventilation is a continuous process in water breathing fish because the relatively scarcity of oxygen in this medium and in birds and mammals because of their high oxygen demand.

RESPONSES TO VARIATIONS OF AMBIENT CARBON DIOXIDE

Water breathers. The table shows that ambient hypercapnia does not greatly increase the ventilation of water breathers. When one wants to study the possible action of ambient hypercapnia, the CO_2-enrichement of the water to be inhaled should be limited to a few torrs of CO_2, an increase which may occur in natural conditions. A marked hypercapnia will not teach one much about the control of breathing, since it may very well depress the ventilation by inducing a narcotizing acidosis, an effect for which it is sometimes used purposely.

However, hypercapnia induces some hyperventilation in several normoxic water breathers. Massabuau et al. (1984) have suggested that animals whose ventilation responds to hypercapnia are only those with a relatively high energy metabolism, but whether this is a general rule is unknown. However, we have seen that hyperoxia always results in hypoventialtion and hypercapnia. I have argued that this hyperoxia-induced hypercapnia implies that the hyperoxic animal is insensitive to hypercapnia, and most probably to ambient hypercapnia (Dejours, 1973). Indeed, hyperoxia has been seen to depress considerably or entirely the ventilatory responsiveness to hypercapnia. It is the case in various Cyprinadae, the trout (Dejours, 1973), the crayfish *Astacus leptodactylus* (Massabuau et al., 1984) and the shore crab (Jouve-Duhamel and Truchot, 1983). That hypercapnia has no effect on the ventilation of hyperoxic water breathers may seem rationally obvious, but it had to be tested experimentally.

Air breathers in mild hypoxia, normoxia or hyperoxia, inhaling CO_2-enriched mixtures increase their ventilation, which moderates the pulmonary and arterial hypercapnia, as initially reported by Haldane and Priestley (1905). However, there are three circumstances in which the ventilatory responsiveness to CO_2 is small or absent.

(1) In fossorial mammals and birds, hypercapnia-induced hyperventilation is not as high as in open air mammals (see Boggs, Kilgore and Birchard, 1984).

(2) In diving mammals and birds, the ventilatory response to carbon dioxide inhalation is less marked than in non-diving animals (see *e.g.* for the harbor seal Robin et al., 1963).

(3) Extreme hypoxia. Nielsen and Smith (1951) studied two human subjects at various levels of ambient oxygenation. A very low oxygen ambient pressure (alveolar P_{O_2} at either 47 or 37 Torr) entails a very important hyperventilation with a marked hypocapnia (Figure 3), a phenomenon which has been widely confirmed (*e.g.* Velasquez, 1959). The originality of the paper by Nielsen and Smith is that they studied the ventilatory response to the inhalation of hypercapnic mixtures in these hypoxic conditions. Figure 3 shows the results obtained in one subject. In the normoxic or slightly hyperoxic subject (Fig. 3, curve A), hypercapnia induced hyperventilation. In marked hypoxia (curve B, PA_{O_2} = 47 Torr; curve C, PA_{O_2} =

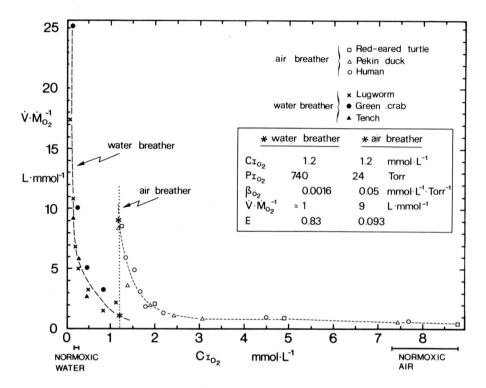

Fig. 3. Ventilatory flow rate, $\dot{V}_E$, as a function of alveolar P_{CO_2}, Pa_{CO_2}, in one human subject breathing air or a slightly hyperoxic gas (curve A), or a hypoxic mixtures with Pa_{O_2} equal to 47 Torr (curve B) or 37 Torr (curve C). Point at the lower left of each curve: the subject breathed a gas mixture with a negligible amount of CO_2. To obtain the solid part of each curve the subject was given a CO_2-enriched mixture to breathe. In the hypoxic conditions B and C, a very marked increase of ventilation occurred only above a certain Pa_{CO_2} threshold (redrawn from Nielsen and Smith, 1951).

37 Torr), there was a P_{CO_2} threshold value, 30 to 32 Torr, above which a hyperventilatory response to CO_2 was seen. In the case of curve C, the P_{ACO_2} value of the subject breathign 7.9% O_2 in N_2 at sea level was 21 Torr. When the subject was given CO_2-containing mixtures to breathe, the pulmonary ventilation remained unchanged until P_{ACO_2} was about 30 Torr; above that value the ventilation increased steeply. In short, hypoxic subjects are insensitive to variations of blood P_{CO_2} (and pH) below a certain P_{CO_2} level. The slope of the ventilatory CO_2 response curve, an index of ventilatory CO_2 sensitivity, was the steeper the greater the hypoxia. The two stimuli, hypercapnia and hypoxia, interact positively.

One may look at the insensitivity of very hypoxic subjects to variations of P_{CO_2} (below a certain CO_2 threshold) in a teleonomic way. Obviously if the subjects were still sensitive to hypocapnia below 30 Torr, they would breathe less, would become still more hypoxic, and their life would be endangered. The data on the CO_2 sensitivity of very hypoxic subjects are scarce, presumably because such a degree of hypoxia may not be harmless. The effect has, however, been confirmed in goats (Lahiri et al., 1971).

Figure 3 concerning one human subject in normoxia (A) shows the ventilatory response as a function of P_{ACO_2} is as a straight line; but more often the response to hypercapnia is curvilinear. The slope, the index of CO_2 responsiveness, is rather low near the normocapnic value and increases progressively to become approximately linear for marked hypercapnia (Anthonisen, Bartlett and Tenney, 1965; Dejours et al., 1965; Forster et al., 1982).

It is common to believe that carbon dioxide is the most important stimulant of external respiration. This erroneous conviction comes from the study of mammals, of man in particular, at sea level, in normoxic conditions. But breathing the modern atmosphere at sea level is the result of a long evolution. Comparative physiological studies show that oxygenation is the primary goal of respiration and that carbon dioxide clearance is subordinate to it. In case of conflict between a proper oxygenation and a proper carbon dioxide clearance, it it sht first condition which has precedence.

COST OF BREATHING
IN WATER- AND AIR-BREATHERS

Peter Scheid

Institut für Physiologie
Ruhr-Universität
D 4630 Bochum
F.R.G.

Due to the differences in the physical properties of air and water and to structural and functional differences of the respiratory apparatus, work and cost of breathing are different between air-breathers and water-breathers. In air-breathers, elastic and flow-resistive forces are of relevance, while flow-resistive and inertial forces appear to determine work of breathing in water-breathers. In man, even at high ventilation, O_2 cost constitutes only a small fraction of the total metabolism, but this may be larger in fish, although reliable data are not available.

With the evolution of specialized gas exchange surfaces for the transfer of O_2 and CO_2, the animals had to be equipped with a special respiratory apparatus for the steady renewal of the respired medium. Due to the great variety in structure of the respiratory gas exchangers and to the substantial differences in the physical properties of the two respiratory media - water and air - (Dejours, 1981) fundamental differences might be expected between animals in respect of the cost of breathing, and of its potential role in limiting the level of activity of the whole animal.

I restrict this brief review to selected representatives of two classes of vertebrates, mammals, particularly man, for air-breathers, and fish for water-breathers. Discussion of other vertebrate classes and of invertebrates should certainly be a fruitful task, however, very little concern has yet been devoted to it. A number of excellent reviews have dealt with the subject of the work and cost of breathing in man (Otis, 1954, 1964; Campbell, 1958; Campbell et al., 1970; Roussos, 1985) and in fish (Ballintijn, 1972; Alexander, 1967; Shelton, 1970; Holeton, 1980); they should be consulted for detailed references.

I. DEFINITIONS AND PROBLEMS

When a human subject performs work the rate of **work output**, $\dot{W}_{out}$, can easily be measured, and the corresponding O_2 uptake, can be obtained simultaneously by subtracting from the total O_2 uptake, $\dot{M}$, the steady

Comparative Physiology: Life in Water and on Land. P. Dejours, L. Bolis, C.R. Taylor, E.R. Weibel (eds.) Fidia Research Series, IX-Liviana Press, Padova © 1987

state level during zero external work. This O_2 cost can be converted to the rate of **energy input**, $\dot{W}_{in}$, and hence the **efficiency** of performing external work by the skeletal muscles is calculated as $\eta = \dot{W}_{out}/\dot{W}_{in}$. Typical values for efficiency in skeletal muscle range between 20 and 25%.

Prerequisite for this energetic balance between input and output and the calculation of efficiency is that all external work is recorded and that the extra O_2 consumption measured is indeed used to perform this work, e.g. by aerobically working skeletal muscle. In breathing, however, the external work, e.g. imparted on gas or water, is typically a small fraction of the total, and internal work, performed in moving the respiratory apparatus itself, is larger; and the O_2 cost of breathing constitutes a small fraction of the total metabolism only, and is thus very hard to measure with accuracy. Hence, literature values for work, cost and efficiency of breathing span an enormous range, both in man and fish, and it is extremely difficult to separate methodological from biological reasons for this variation.

II. AIR BREATHERS : MAN

A. Work of breathing

Work is performed against three main types of force:

(1) Elastic (including gravitational) forces developed by the lung and the thorax when inflated above the resting volume; elastic recoil of the lung is determined both by elastic fibers with their geometric arrangement and by surface forces acting on the liquid/gas interface.

(2) Flow-resistive forces developed when moving air through the airways and in non-elastic tissue movement.

(3) Inertial forces developed in air and tissue acceleration.

(4) Distorting forces developed when the chest geometry is changed without volume change.

Work of breathing (per breath) is typically measured by recording simultaneously the pleural (or esophageal) pressure, P_{pl}, and the changes in lung volume, ΔV_L, measured from gas flux at the airways' opening

$$W_{resp} = \int P_{pl} \cdot dV_L \qquad \qquad \dots (1)$$

This approach neglects work performed against inertial forces, which are regarded small because of the low density of air and the low acceleration of tissues; against distorting forces, which are pronounced only during high ventilation; against elastic forces of the thorax and non-elastic forces of tissue movement.

Analysis of the pressure-volume relationships reveals the following:

(1) During quiet breathing, expiration is passive in that the elastic energy stored in the lung during inspiration suffices for expiratory

air movement, and inspiratory muscles perform negative work (pliome-
tric contraction). From the energetic point of view a mode of breathing where inspiration extends above and expiration below the resting
level would be less costly since less elastic energy is wasted. Birds
appear to expire below the resting level, and it would be interesting
to analyze the energetic consequences.

(2) Work of breathing can particularly easily be estimated by recording
P_{Pl} and ΔV_L in heavy breathing, when expiration is active and no
negative work is performed by the inspiratory muscles.

Estimates of the rate of work (or power) of breathing, $\dot{W}_{resp}$, vary
widely; a reasonable estimate during quiet breathing in man is 0.12 cal
per L of ventilation. Agreement exists, however, that $\dot{W}_{resp}$ increases
more than linearly with ventilation ($\dot{V}$), according to a relationship

$$\dot{W}_{resp} = k_1 \cdot \dot{V}^2 + k_2 \cdot \dot{V}^3 \qquad \qquad ... (2)$$

Equation (2) has been deduced by Otis et al. (1950) for the situation
of no negative work performed during expiration, and the first term on
the right-hand side has been interpreted to correspond to laminar, the
second term to turbulent flow condition. The results of Margaria et al.
(1960) would thus suggest turbulent work during moderately high ventilation to account for more than 50% of the total work of breathing. This is
in conflict, however, with the results of Dean and Visscher (1941) who
compared air breathing with breathing a He/O_2 mixture, in which the
viscosity is somewhat higher and the density substantially lower than in
air. The results suggest that turbulent work is insignificant during
normal breathing and becomes important only with constricted airways.
Thus, eq.(2) should probably be used in descriptive terms only. When
breathing SF_6/O_2 mixtures, the density is increased more than fourfold
and gas flow is essentially turbulent throughout the bronchial tree; in
this case, work of breathing is indeed significantly elevated (Glauser et
al., 1967).

B. O_2 cost of breathing

O_2 consumption of the respiratory muscles, $\dot{M}_{O_2,resp}$, cannot directly be
measured; it is usually estimated from the difference in total O_2 uptake,
$\dot{M}_{O_2}$, measured at rest and at various levels of hyperpnea, which are
induced by addition of dead space or by CO_2 inhalation or are effected
voluntarily, while the subject remains at rest.

At rest, values for respiratory O_2 uptake per unit ventilation,
$\dot{M}_{O_2,resp}/\dot{V}$, of 0.2 to 1 ml per L ventilation have thus been obtained;
hence, $\dot{M}_{O_2,resp}$ is only 1-2% of the resting $\dot{M}_{O_2}$. With increased $\dot{V}$,
$\dot{M}_{O_2,resp}$ has been found to increase in a similar way as does $\dot{W}_{resp}$, but
there is a considerable scatter both between subjects in one study and
between averages of different studies (Otis, 1964; Roussos, 1985). At
intermediate levels of $\dot{V}$, between 20 and 100 $L \cdot min^{-1}$, $\dot{M}_{O_2,resp}/\dot{V}$ ranges

from 0.5 to 3.5 ml·L^{-1}, and at the highest levels of $\dot{V}$, 100–240 L·min^{-1}, it reaches values of above 8 ml·L^{-1}, corresponding to absolute levels of $\dot{M}_{O_2,resp}$ up to 1700 ml·min^{-1}. The increase in $\dot{M}_{O_2,resp}$ with increasing $\dot{V}$ reflects not only the increased work of breathing but also the recruitment of muscle groups in the arms and trunk, which are not respiratory.

The variability in results could have several reasons. Depending on the mode by which hyperventilation is induced – dead space, inspired CO_2, voluntary hyperpnea – different muscle groups could be involved, and this is particularly true when the subjects are studied in different posture. In addition, voluntary hyperventilation causes hypocapnia with increased airway resistance; furthermore, the high ventilatory levels, above 100 L·min^{-1}, can only be attained for short periods of time, and steady state becomes a problem. Bartlett et al. (1958) found different relations between $\dot{M}_{O_2,resp}$ and $\dot{V}$ depending on the respiratory rate. At any $\dot{V}$ there appears to be an optimal frequency for minimal work, and the subjects tend to choose this frequency, unless it is fixed by the experimental protocol. Thus, differences in frequency can account for part of the variability as well.

From measurements of O_2 consumption and blood flow in dog diaphragm, Roussos (1985) has estimated that maximum O_2 consumption of the entire respiratory muscle mass in a 70 kg man could be as high as 1500–1800 ml·min^{-1}, matching the highest experimental estimates. This O_2 consumption would require a minimum blood flow of 8.5 L·min^{-1}. To pump blood at this high rate, the heart requires an extra amount of O_2, estimated at 100–150 ml·min^{-1}, which is included in $\dot{M}_{O_2,resp}$.

C. Efficiency of the respiratory apparatus

Efficiency is the ratio in the respiratory machine of work output and energy input. As discussed above, work output is probably underestimated; but what about energy input? Increased O_2 consumption with increased ventilation does not only reflect miometric, i.e. positive, work of respiratory muscles. In addition, negative work is performed; antagonistic muscle groups are innervated leading to thorax distortion without external work; non-respiratory muscle groups are recruited; cardiac muscle requires more O_2. Thus, the total O_2 consumption of the respiratory machine, reflecting energy input, is more than that of the miometrically contracting, synergistic respiratory muscles, and the efficiency is expected to be low and dependent on ventilation. It thus makes little sense to compare values for respiratory efficency with those of functionally isolated skeletal muscle, as this efficiency concerns the total respiratory apparatus rather than the respiratory muscles alone.

The experimental values of efficiency range from a few to 25%, largely depending on how the increase in work of breathing is obtained (Otis, 1964). For example, the efficiency for a given work of breathing of 25 cal·min^{-1} was 5% at isocapnic hyperventilation, 1.7% during breathing against a low, and less than 1% when breathing against a high resistance (Roussos, 1985). The highest reported values of efficiency, 20-25%, are those of Milic-Emili and Petit (1960), obtained in recumbent subjects with minimal contribution by postural muscles.

Aside from differences in the mode of breathing, the fractional contribution of muscle fiber types will also affect the efficiency (Roussos, 1985). Fast-twitch glycolytic fibers with less O_2 consumption are recruited during high ventilation, yielding higher efficiency from measurement of O_2 uptake.

D. Limitation of ventilation

Several factors could theoretically limit ventilation, among them the mass of respiratory muscles. With increasing exercise a point is eventually reached where the O_2 cost for a further increment in ventilation exceeds the O_2 gained by this increment (Otis 1954; 1964). This critical level of $\dot{V}$ has been calculated between 120 and 170 L·min^{-1} (Margaria et al., 1960; Shepard, 1966), corresponding to the highest ventilation observed during maximum exercise and to the highest ventilation that normal subjects can sustain for 3 - 4 min (Campbell et al., 1970).

III. WATER BREATHERS : FISH

A. Physical properties of water vs. air

Several physical characteristics appear to render water much less suited as a respiratory medium than air (cf. Dejours, 1976). O_2 solubility of water is some 30 times less than that of air, and declines with increasing temperature, when the metabolic demand of most fish increases. Viscosity and density of water are about 60 and 800 times that in air.

B. Work of breathing

The respiratory system in fish acts as a double pump, consisting of the buccal force pump separated from paired opercular suction pumps by a continuous mesh of the gills. Buccal and opercular valves ensure unidirectional water flow from the mouth towards the opercula, both during inspiration and expiration. Much of the knowledge about this system was derived from pressure recordings in the buccal and opercular cavities (Shelton, 1970; Ballintijn, 1972) and from water velocity recordings therein (Holeton and Jones, 1975).

In principle the same set of forces acts in fish as in mammals against which work must be performed by the respiratory muscles to create gill water flow, but the relative significance is different due firstly to the submersion in the dense medium, secondly to the breathing of the dense, viscous medium, and thirdly to the structural and functional differences in the breathing apparatus (Hughes, 1984; Piiper and Scheid, 1975, 1984). Gravitational forces are absent in a neutrally buoyant fish; any movement of the respiratory system imparts, however, energy on the surrounding dense and viscous medium. This holds for water-dwelling air-breathers as well, and Hong et al. (1969) found the work of breathing in subjects submerged to the neck significantly elevated, mainly due to an increase in elastic work. Water-breathing, on the other hand, eliminates surface forces and thus reduces elastic work. Flow resistive and inertial forces are expected to be much higher in water-breathers than in air-breathers because of the higher viscosity and density of water compared with air.

There is no direct equivalent of the mammalian pleural pressure in fish, and the estimation of work of breathing from pressure-volume recordings is not possible in the flow-through system of the fish. Thus, attempts at estimating the external work of the respiratory pump have concentrated on the resistive work of the respiratory water flow, neglecting elastic forces as well as forces to move tissues. Any comprehensive analysis of flow-resistive work in the system is complex due to the unsteady gill water flow and to an ill-defined, spatially and temporally varying gill geometry. Results are thus even wider open to critique than in mammals.

Alexander (1967) has estimated the work for gill irrigation at 0.03 cal per liter of water, about one fourth the estimate of work of breathing in man at rest. Because of the low O_2 solubility in water, however, the work needed to move one unit of O_2 would be sizeably larger in the resting fish than in man. This estimate of Alexander has been criticized as an oversimplification (Ballintijn, 1972), since it assumes differential pressure and gill sieve resistance to remain constant throughout the respiratory cycle. In fact, differential pressure is far from constant and nothing is known about fluctuations in gill sieve resistance.

Furthermore, simultaneous recordings of pressure and flow at the buccal entrance and opercular exit of water as well as deep in the buccal cavity in the carp reveal phase differences between the fluctuating pressure and flow signals, indicating that inertial forces cannot be neglected (Holeton and Jones, 1975). These measurements also suggest that the kinetic energy imparted by the respiratory muscles onto the water may be a significant fraction of the total work. As pointed out by Holeton and Jones (1975) 'any adequate analysis of the hydrodynamics of fish ventilation can only be made when the data on water velocities and cross-sectional areas are obtained in addition to hydrostatic pressures'. Such measurements are still lacking.

C. O_2 cost and efficiency of breathing

The same problems in measuring O_2 cost of breathing and interpreting the results exist for fish as for man. In addition, fish like most animals cannot be made to hyperventilate spontaneously, and any method of increasing breathing bears the risk of changing other factors that affect the O_2 consumptions. Thus estimates of O_2 cost obtained from manipulation of ventilation by means of pharmacological or chemical intervention have rightly been criticized (Cameron and Cech, 1970; Holeton, 1980; Ballintijn, 1972). They are largely responsible for the wide range in results, 0.5% to 70% of the total metabolism.

Some estimates in the literature of O_2 cost of breathing have been obtained by estimating work of breathing from pressure recordings and assuming a value for the efficiency of the respiratory apparatus (Shelton, 1970; Davis and Randall, 1973; Alexander, 1967).

Estimates of the efficiency of breathing are not more reliable than the components from which this parameter is calculated, cost and work of breathing.

D. Gill ventilation and locomotion : Ram ventilation

The basic requirement to enable fish to extract water across the gill epithelia is that the respiratory surfaces be irrigated. So far we have assumed that the fish dwells in still water, and that the ventilatory water flow is produced by the action of the respiratory pump. However, many fish irrigate their gills by simply opening their mouth when swimming or remaining stationary in flowing water, and this mode has been termed ram ventilation (Muir and Kendall, 1968).

Ram ventilation has been observed in a number of near-shore, midwater and pelagic fish whose habit is to cruise. These fish convert from pump to ram ventilation when reaching a critical velocity relative to the water (Roberts, 1975; Steffensen and Lomholt, 1983).

Ram ventilation must be regarded as an extreme example for the potential interaction between locomotor and respiratory activity, in which the locomotor muscles perform the entire work of breathing. In contrast, during pump ventilation locomotor energy is obtained from the respiratory muscles when expired water is expelled from the opercular cavity. With ram ventilation, the muscle mass available for respiration is enormous and will not be limiting for O_2 uptake, as has been suggested by Cameron and Cech (1970).

The work of breathing can be expected to be smaller with ram ventilation since much of the inertial work is eliminated with the steady flow of water. Freadman (1979) has measured metabolism in the striped bass (Morone saxatilis) and the bluefish (Pomatomus saltatrix) over a range of swimming speeds and found indeed substantial metabolic savings on the transition from cyclic pump to ram ventilation. Steffensen (1985) found a saving of 10% of the metabolism when the trout switched from pump to ram ventilation at a speed of about 0.7 body lengths per sec. This value constitutes a lower estimate of the total cost of breathing since in ram ventilation work of steady gill irrigation must still be performed, in this case by locomotor mucles, and their O_2 cost is included in the total metabolic budget.

Particularly interesting are studies in the sharksucker, Echeneis naucrates, who attaches to a shark and performs ram ventilation on the expense of the locomotor activity of the host, the only respiratory energy expenditure being spent by the sucking muscles (Steffensen and Lomholt, 1983; Steffensen, 1985). From differences in pump and host-powered ram ventilation , Steffensen and Lomholt have estimated the cost of pump breathing in this animal at 3.7 to 5.7% of the total metabolism.

Somewhat similar to ram ventilation is an experimental set-up in which water is made to flow at a continuous stream over the gills by virtue of hydrostatic pressures applied between separated head and body compartments (Cameron and Cech, 1970). Davis and Randall (1973) have thus estimated the work of irrigating the gills of trout at about 0.003 cal per liter of water which is one order of magnitude below the estimate of Alexander (1967). Jones and Schwarzfeld (1974) have forced ventilation at different speeds in the trout whereby the animal shifted from active to (forced) ram ventilation; as a result, metabolism dropped about 10%. The efficiency calculated from estimates of the work of irrigation was very low, less than 2%.

IV.   CONCLUSIONS

Work of breathing in animals is difficult to assess and estimates are uncertain, particularly in fish during pump ventilation. The O_2 measured as cost of breathing is not only needed for respiratory but also for other skeletal muscles and the heart. Experimental estimates span a wide range, in man as in fish. The variability reflects partly physiological differences due to differences in the mode of breathing, partly experimental difficulties. In man O_2 cost of breathing does not appear to limit the level of ventilation. In fish, inertial forces (aside from dead space problems) would render a dead-end, tidal gas exchanger impossible, and even the flow-through gill system poses serious problems when being pump-ventilated; ram ventilation appears to be an advantageous type of breathing since flow is made steady and since locomotor muscles are enticed to perform respiratory work.

REFERENCES

Alexander R (1967) Functional Design in Fishes. London, Hutchinson University Library.

Ballintijn CM (1972) Efficiency, mechanics and motor control of fish respiration. Respir Physiol 14: 125-141

Bartlett, RG Jr, Brubach HF, Specht H (1958) Oxygen cost of breathing. J Appl Physiol 12: 413-424.

Cameron JN, Cech JJ (1970) Notes on the energy cost of gill ventilation in teleosts. Comp Biochem Physiol 34: 447-455.

Campbell, EJM (1958) The Respiratory Muscles and the Mechanics of Breathing, Year Book Publ, Chicago.

Campbell EJM, Agostoni E, Newsom Davis J (1970) The Respiratory Muscles. Mechanics and Neural Control. Lloyd-Luke Ltd, London.

Davis JC, Randall DJ (1973) Gill irrigation and pressure relationships in rainbow trout (_Salmo gairdneri_) J Fish Res Board Can 30: 99-104.

Dean RB, Visscher MB (1941) The kinetics of lung ventilation. An evaluation of the viscous and elastic resistance to lung ventilation with particular reference to the effects of turbulence and the therapeutic use of helium. Am J Physiol 134: 450-468.

Dejours P (1976) Water versus air as the respiratory media. In: Hughes GM (ed): Respiration of Amphibious Vertebrates. Academic Press, London, New York, San Francisco; pp 1-15.

Dejours P (1981) Principles of Comparative Respiratory Physiology, 2nd ed. Elsevier/North-Holland, Amsterdam, New York, Oxford.

Freadman MA (1979) Swimming energetics of striped bass (_Morone saxatilis_) and bluefish (_Pomatomus saltatrix_): Gill ventilation and swimming metabolism. J exp Biol 83: 217-230

Glauser SC, Glauser EM, Rusy BF (1967) Gas density and the work of breathing. Respir Physiol 2: 344-350

Holeton GF, Jones DR (1975) Water flow dynamics in the respiratory tract of the carp (_Cyprinus carpio_ L.). J exp Biol 63: 537-549.

Holeton GF (1980) Oxygen as an environmental factor of fishes. In: Ali MA (ed): Environmental Physiology of Fishes. Plenum Press, New York and London; 7-32.

Hong SK, Cerretelli P, Cruz JC, Rahn H (1969) Mechanics of respiration during submersion in water. J Appl Physiol 27: 535-538.

Hughes GM (1984) General anatomy of the gills. In: Hoar WS and Randall DJ (eds): Fish Physiology, Vol XA. Academic Press, New York, San Francisco, London; pp. 1-72.

Jones DR, Schwarzfeld (1974) The oxygen cost to the metabolism and efficiency of breathing in trout (_Salmo gairdneri_). Respir Physiol 21:241-254.

Margaria R, Milic-Emili G, Petit JM and Cavagna G (1960) Mechanical work of breathing during muscular exercise. J Appl Physiol 15: 354-358.

Milic-Emili G, Petit JM (1960) Mechanical efficiency of breathing. J Appl Physiol 15: 359-362.

Muir BS, Kendall JI (1968) Structural modifications in the gills of tunas and some other oceanic fishes. Copeia 2: 388-398.

Otis AB, Fenn WO, Rahn H (1950) Mechanics of breathing in man. J App Physiol 2: 592-607.

Otis AB (1954) The work of breathing. Physiol Rev 34: 449-458.

Otis AB (1964) The work of breathing. In: Fenn WO and Rahn H (eds): Handbook of Physiology , Section 3: Respiration, Vol. I. Am Physiol Soc, Washington DC; pp. 463-476.

Piiper J, Scheid P (1975) Gas transport efficacy of gills, lungs and skin: theory and experimental data. Respir Physiol 23: 209-221.

Piiper J, Scheid P (1984) Model analysis of gas transfer in fish gills. In: Hoar WS and Randall DJ (eds): Fish Physiology, Vol. XA. Academic Press, New York, San Francisco, London; pp. 229-262.

Roberts JL (1975). Active branchial and ram gill ventilation in fishes. Biol Bull 148: 85-105.

Roussos C (1985) Energetics. In: Roussos C, Macklem PTM (eds): The Thorax, Part A. Marcel Dekker, New York, Basel; pp. 437-492.

Shelton G (1970) The regulation of breathing. In: Hoar WS and Randall DJ (eds): Fish Physiology, Vol.IV. Academic Press, New York, London; 293-359.

Shephard RJ (1966) The oxygen cost of breathing during vigorous exercise. Quart J exp Physiol 51:336-350.

Steffensen JF (1985) The transition between branchial pumping and ram ventilation in fishes: Energetic consequences and dependence on water oxygen tension. J exp Biol 114: 141-150.

Steffensen JF, Lomholt JP (1983) Energetic cost of active branchial ventilation in the sharksucker Echeneis naucrates. J exp Biol 103: 185-192.

CARDIOVASCULAR IMPLICATIONS OF THE TRANSITION FROM AQUATIC TO AERIAL RESPIRATION

Fred N. White and James W. Hicks

Physiological Research Laboratory
Scripps Institution of Oceanography
University of California, San Diego
La Jolla, California 92037, U.S.A.

Analysis of cardiopulmonary control of gas exchange among reptiles suggests that the transition from water to air breathing was accompanied by a shift of emphasis from control of O_2 uptake and ion transport for acid-base balance via cardiobranchial mechanisms in fishes to close control of CO_2 and acid-base balance by the cardiopulmonary system in lung breathers. O_2 transport appears to be a more passive process determined by molecular features of hemoglobin.

INTRODUCTION

The nature of the modifications of the cardiovascular system which accompanied the transition from water to air-breathing is obscured by the fact that these events occurred in the dim geological past. We are left with contemporary forms which are not representative of key stem groups.

All contemporary air-breathing fishes and amphibians utilize extrapulmonary avenues of gas exchange, especially for CO_2 excretion (Rahn and Howell, 1976). Romer (1972) contended that the presence of a well-developed rib cage and dermal scales in the Seymouriamorphic, reptile-like amphibians of the Permian indicates near or complete dependence on the lung for total gas exchange. He concluded that contemporary amphibians are, in respiratory habits, not primitive but specialized and degenerate.

Amphibians probably originated from Sarcopterygian stock via Crossopterygians. Present-day lung fishes represent a side branch of the line leading to amphibians and thus may offer the only extant clues to those early transitions which led to the adoption of lung-breathing by primitive amphibians. While Protopterus and Lepidosiren exhibit high potential for meeting O_2 requirements across the

Comparative Physiology: Life in Water and on Land. P. Dejours, L. Bolis, C.R. Taylor, E.R. Weibel (eds.) Fidia Research Series, IX-Liviana Press, Padova © 1987

lungs, the gill-skin avenue accounts for over 50% of CO_2 exchange (Rahn and Howell, 1976).

Among extant lower vertebrates, complete dependence on the lung for total gas exchange is only evident in reptiles in cases where the skin has been modified to reduce water loss or to provide protection by armor.

SIMILARITIES: LUNG FISH, AMPHIBIANS, REPTILES

In considering the cardiorespiratory transitions from water to air breathing, it may be useful to search for features among fishes, amphibians, and reptiles which are common. Here we ask: What is preserved?

Coupling of heart rate to the phase of ventilation accomplishes matching of maximal gill blood and water flows in the Port Jackson shark (Satchell, 1971) and in teleost fishes during hypoxia (Randall and Daxboeck, 1984). In Propterus, lung-breathing amphibians, and most reptiles examined, the breathing phase of the intermittent respiratory cycle is accompanied by elevated heart rate and increased pulmonary blood flow relative to breath-holding (Johansen, 1972, 1984; Shelton and Boutilier, 1982; White, 1976). Among all of these groups, there persists a vagal cholinergic vasoconstrictor innervation of the gas exchanger. The function is not clearly established in fishes. In lung fish, the pulmonary arterial vasoconstrictor area originates near the junction of the ductus arteriosus (Fishman et al., 1984). The situation is similar in amphibians and non-crocodilian reptiles, the constrictor area originating immediately distal to the ligamentum arteriosum (remnant of the ductus). This feature, shared by lung fish, lung-breathing amphibians, and reptiles, may be considered a homologue which has disappeared in endotherms. Among ectotherms, pulmonary blood flow and distribution of cardiac output appears to be modulated, during the phases of breathing, by this cholinergic control of pulmonary vascular resistance. In both amphibians and intermittently breathing reptiles, the absence of a complete ventricular septum is associated with right-to-left intracardiac shunting during breath-holding when pulmonary vascular resistance is elevated. It is also likely that the control of fluid balance across the pulmonary exchange surface is guarded by the presence of a variable pulmonary resistance within the pulmonary arterial system.

An additional similarity among amphibians and reptiles is the maintenance of low respiratory exchange ratio (Re) of the lung gas during breath-holding. This has been attributed to cutaneous CO_2 exchange in amphibians (Shelton and Boutilier,

1982). However, in the turtle, Pseudemys scripta, low Re of breath-hold is
followed by recovery of the expected RQ during the brief ventilatory periods
which occupy, on average, about 15% of the breathing cycle. This was referred to
as cyclic CO_2 exchange (Ackerman and White, 1979). As pointed out by Rahn and
Howell (1976): "It is the handling of carbon dioxide by the lung system that
appears to have been the greatest obstacle in the transition from water-breathing
to the modern lung, which is first encountered in our present-day reptiles."

Cardiopulmonary Gas Exchange and Temperature, a "Steady State" Analysis:
Emergence from water to land is associated with the rapidly changing thermal
conditions which impact both metabolism and blood gas transport properties. The
relationship of convective flows of blood and gas at different body temperatures
has been addressed for the turtle (White et al., 1985) utilizing the analytical
framework elaborated by Dejours (1981), and placed in a formal model by Piiper
and Scheid (1977). The analysis requires knowledge of both O_2 (Maginnis et
al., 1980) and CO_2 (Weinstein et al., 1986) dissociation curves, O_2 and CO_2
exchange, lung perfusion and ventilation, partial pressure differences for
expired and inspired gases, and blood gases for pulmonary arterial and venous
blood. Key to the analysis is the extension of the Fick equation to the general
form: $\dot{M}gas = \dot{V}medium \cdot \beta gas \cdot Pgas$. The product of a convective flow and the
effective capacitance coefficient (β) is defined as a conductance (G). Thus, for
the perfusive transport of CO_2 in blood: $Gperf = \dot{V}b \cdot \beta$. The widely used $\dot{V}/\dot{Q}$
concept of mammalian gas exchange is replaced by Gvent/Gperf for both O_2 and
CO_2 transport in order to account for the large differences in β values for
O_2 and CO_2 in blood. Gvent/Gperf values may then be related to the relative
resistances (Δp values) attributed to the major processes of cardiopulmonary gas
exchange. A necessary assumption of the analysis is that steady state conditions
exist (not strictly true).

The air convection requirement, as reported for a number of reptiles (Jackson,
1978), is inverse to body temperature, BT (Figure 1A). This pattern is
responsible for a direct relationship between BT and arterial P_{CO_2}. Blood
convection requirement, on the other hand, is maintained at a relatively stable
level, resulting in an inverse $\dot{V}_E/\dot{V}_b$ (Figure 1B). It appears that the higher
BT range is associated with underventilation and overperfusion relative to lower
temperatures. When placed in the context of Gvent/Gperf, a quite different
perspective emerges (Figure 2). At all BT's, Gvent/Gperf for CO_2 remains low
and stable, while the Δp values for each process remain essentially fixed.
Gvent/Gperf for O_2 varies inversely with BT while Δp's are variable. This
finding for Gvent/Gperf (O_2) indicates an under-ventilation/over-perfusion

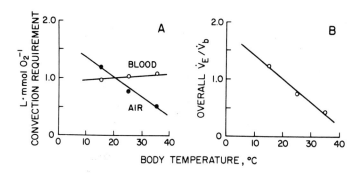

Figure 1.A. Blood and air convection requirements of *Pseudemys scripta* as functions of body temperature. B. Ventilation-perfusion ratios shift inversely with body temperature.

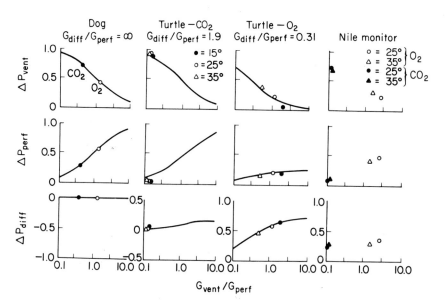

Figure 2. Relative resistances (p's) for the processes of ventilation, perfusion, and diffusion as functions of ventilation-perfusion conductance ratios. Experimental points are shown in relationship to theoretical lines derived from equations in Piiper and Scheid (1977). Data on Nile monitor from Hicks, Ishimatsu and Heisler (personal observations).

trend at higher BT's and is a partial explanation for a significant decline in arterial O_2 saturation levels at temperatures above 30°C. Recent observations on the Nile monitor indicate similar trends in Gvent/Gperf for both gases. Low Gvent/Gperf for CO_2, relative to O_2, is accounted for solely by the higher β values for CO_2 in blood. This analysis suggests that O_2 transport is a more passive process than for CO_2, the latter being subject to regulation at constant content across a broad temperature range. Additional indications of the primacy of CO_2 control are the direct and linear relationships between Gperf CO_2 and total ventilation at various temperatures. Further, the relationship between air convection requirement and BT stands in direct linear relation to blood β for CO_2. Elevation in Gvent/Gperf, above 30°C, in the service of correcting falling blood O_2 saturation values would result in alkalosis and reduction in CO_2 stores. This was not observed. Behavioral selection of temperatures below 30°C are characteristic of this species. Insofar as the nature of the transition from water to air-breathing can be inferred from this study, full reliance on lung-breathing was accomplished by a shift in emphasis from cardiobranchial control of O_2 transport in fishes (Randall and Daxboeck, 1984; Dejours, 1981) to cardiopulmonary control of CO_2 in air-breathers. Acid-base regulation has likewise shifted from active transport mechanisms in gills to greater reliance on stable Gvent/Gperf for CO_2, a mechanism which allows rapid adjustment to body temperature alterations.

The Unsteady State: Periodic breathing in the turtle is accompanied by steady depletion of O_2 from the lung and low CO_2 influx during breath-holding (Figure 3). Heart rate declines and pulmonary vascular resistance rises, under cholinergic vasoconstrictor control, resulting in a large right-to-left intracardiac shunt. These trends are reversed during breathing. At first thought, this pattern of diverting venous blood to the arterial circuit would appear to be counter-productive to O_2 transport. However, 85% of the O_2 stores in the lungs may be utilized during breath-holding at 25°C, sufficient to sustain aerobic metabolism at resting levels for over 40 min (Ackerman and White, 1979).

From knowledge of lung gas composition, and estimates of pulmonary flow and partial pressure gradients between lung gas and pulmonary arterial and venous blood and an estimated right-left shunt of 70%, points for systemic and pulmonary arterial (mixed venous) and pulmonary venous blood may be related to the CO_2 dissociation curve (Figure 4A) at early, mid, and late times of a 21 min breath-hold (White, 1985). During early breath-hold, the expected positive, mixed venous to pulmonary venous PCO_2 gradient is seen. At reduced Vb,

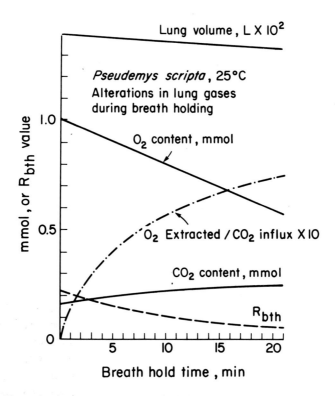

Figure 3. Alterations in lung gas composition during breath-holding in Pseudemys scripta at 25°C. Data from White (1985).

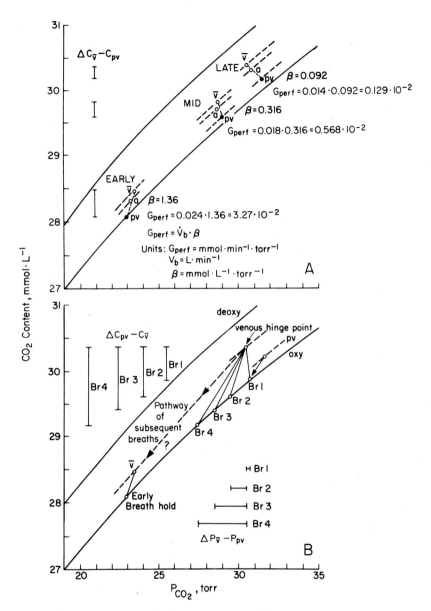

Figure 4.A. Pulmonary venous (pv), mixed venous (v) = pulmonary arterial, and systemic arterial (a) points are referenced to the CO_2 dissociation curve of Pseudemys scripta during a breath-hold of 21 min at 25°C. B. Probable course of CO_2 difference between pv and v blood during a 4-breath sequence following a 21 min breath-hold.

the blood O_2 extraction progressively increases as steeper portions of the O_2 curve are encountered (elevated βO_2). The result is time-dependent elevation in Haldane effect. When Haldane effect (He) exceeds the respiratory exchange ratio, β rotates in a counter-clockwise direction from the pv point ($-\beta$) and $-(Pv-Ppv)$ CO_2 develops, as shown theoretically in the analysis of Meyer et al. (1976) for the avian parabronchus. This is illustrated for the turtle lung at mid and late stages of the breath-hold of Figure 4A. The establishment of $-(Pv-Ppv)$, in the presence of right-to-left shunt, creates $-(Pa-Ppv)$ which becomes increasingly negative as He increases and Re falls with breath-hold time. Low Re, by slowing the decline in pH in pulmonary blood, has the effect of minimizing right Bohr shift and thus facilitating access to lung O_2 stores.

At the terminus of a 21 min breath-hold, it is estimated that of the CO_2 produced by metabolism, 17.8% is found in the lungs, 26.4% in blood, and 55.79% is buffered in tissues (White, 1985). While the CO_2 content stored in blood exceeds that of the lung, it is of interest that the PCO_2 of the arterial and mixed venous blood becomes progressively lower than that of the lung gas, as noted above. This trend is strongly influenced by the Haldane effect. This effect is associated with a decrease in systemic to venous β values and is expected to prolong the time to reach the CO_2 mediated threshold for central neural stimulation of ventilation (breaking point of mammalian physiology). It is perhaps one explanation of the ability to maintain aerobic breath-holds of more than 40 min in freely diving animals at 25°C. Utilizing a model analysis based on the data of White (1985), and using principles found in Farhi and Rahn (1955, 1960) and in Cherniack et al. (1974), we have calculated the time to reach an arterial PCO_2 of 8 torr above that at the start of breath-hold. We have assumed that this increment in PCO_2 will stimulate ventilation via central chemoreceptors. A threshold in this magnitude may be deduced from Jackson et al. (1974) and observations of Levin, Hicks and White (personal observations) on CO_2 stimulation of ventilation via lung gas perfusion at subthreshold to threshold levels of CO_2. Figure 5 contrasts the time to threshold under conditions of differing He effect, shunt magnitude and tissue buffering capacity. Normal tissue buffering capacity, based on White (1985), is estimated at 0.04 mmol CO_2 per torr rise in $PvCO_2$ per kg of tissue. A He of -0.28 mmol $CO_2 \cdot$ mmol O_2^{-1} was reported by Weinstein et al. (1986). R-L shunt of approximately 70% was observed by White et al. (1985) during breath-holding. The analysis emphasizes a small positive effect of shunt level, per se, on time to threshold. However, shunt interacts to magnify He with breath-hold time (Figure 4A). Both the magnitude of He and tissue buffer capacity are positively correlated with the threshold time. This combination of

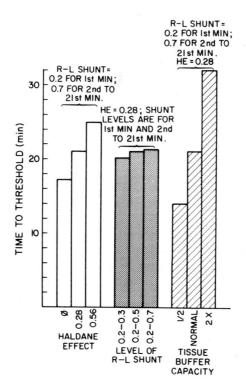

Figure 5. Calculated effects of varying Haldane effect, level of right–left shunt and tissue buffering capacity on the time to ventilatory threshold. See text for further explanation.

factors, tissue buffering, He, and shunt all combine to reduce the rate of rise in arterial P_{CO_2}, an effect which may prolong breath-hold time.

R-L shunt has the effect of magnifying He and elevating CO_2 content in both arterial and venous blood (Figure 4A). The CO_2 stored in blood during breath-holding is thus augmented. At the onset of ventilation, tissue stores are unavailable for pulmonary transport until the circulation time has elapsed. Circulation time, estimated at 40 sec, is a period sufficient for about 4 breaths. This suggests that while the P_{CO_2} of lung gas is progressively lowered by ventilation, the venous point will remain unaffected in the early phase of breathing, representing a "hinge point" from which progressively larger extraction of CO_2 by the lung proceeds prior to the appearance at the lung, of CO_2 previously stored in tissue (Figure 4B). The result is a more efficient accessibility to CO_2 stores in the early phase of ventilation. Efficient gas exchange during this phase is also conserving of energy costs of ventilation, which has been estimated at about 16% of $\dot{M}_{O_2}$ in resting turtles at 25°C.

CONCLUSION

Contemporary species which may serve as indicators of the cardiovascular transition from water to air-breathing (lung-fish, amphibians, reptiles) exhibit similarities: (1) cholinergic constrictor innervation of the pulmonary artery; (2) coupling of heart rate to phase of the respiratory cycle; (3) right-to-left intracardiac shunting during breath-holding; and (4) maintenance of low respiratory exchange ratio in lung gas during breath-holding. Only in reptiles does complete dependence on lungs for gas exchange occur. The turtle, as temperature is altered, exhibits close regulation of ventilation-perfusion conductance ratio for CO_2, a feature which maintains constant CO_2 content of body fluids and acid-base regulation. At temperatures between 30 and 35°C, O_2 saturation of hemoglobin falls sharply while CO_2 regulation persists. Cardiopulmonary regulation of CO_2 in lung breathers offers rapid adjustments in acid-base regulation as temperature changes. This stands in contrast to cardiobranchial control of O_2 uptake in fishes where active transport of ions is the primary mechanism of acid-base control. During the unsteady state of breath-holding, interactions among intracardiac shunts, Haldane and Bohr effects and buffering properties of tissues and blood result in compartmentalization of CO_2 stores, primarily to tissue and blood, blunting of right Bohr shift at the lung level, and prolongation of the time to ventilatory stimulation. Right-left shunt, through its influence on Haldane effect, elevates CO_2 stores in blood, allowing more effective off-loading of CO_2 during the early phase of ventilation. Natural selection appears to have favored mechanisms in lung breathers which actively regulate CO_2 balance, while allowing effective O_2 consumption through the molecular behavior of hemoglobin within thermal limits which are controlled by behavioral thermoregulation.

Acknowledgements. Supported by National Institutes of Health Grant HL 17731 and National Science Foundation Grant DCB 8504238. James W. Hicks is a recipient of an NIH Postdoctoral Training Fellowship.

REFERENCES

Ackerman RA, White FN (1979) Cyclic carbon dioxide exchange in the turtle, Pseudemys scripta. Physiol Zool 52: 378-389.

Cherniack NS, Longobardo GS, Fishman AP (1974) The behavior of carbon dioxide stores of the body during unsteady states. In: Nahas G, Schaefer K (eds): Carbon dioxide and metabolic regulations. Springer-Verlag, New York; pp. 324-338.

Dejours P (1981) Principles of comparative respiratory physiology, Elsevier/North-Holland, Amsterdam pp. 110-115.

Farhi LE, Rahn H (1955) Gas stores of the body and the unsteady state. J Appl Physiol 7: 472-484.

Farhi LE, Rahn H (1960) Dynamics of changes in carbon dioxide stores. Anesthesiology 21: 604-614.

Fishman AP, DeLaney RG, Laurent P, Szidon JP (1984) Blood shunting in lungfish and humans. In: Johansen K, Burggren W (eds): Cardiovascular shunts. Munksgaard, Copenhagen; pp. 88-95.

Jackson DC, Palmer SE, Meadow WL (1974) The effects of temperature and carbon dioxide breathing on ventilation and acid-base status of turtles. Respir Physiol 20: 131-146.

Jackson DC (1978) Respiratory control in air-breathing ectotherms. In: Davies D, Barnes C (eds): Regulation of ventilation and gas exchange. Academic Press, New York; pp. 93-130.

Johansen K (1972) Heart and circulation in gill, skin, and lung breathing. Respir Physiol 14: 193-210.

Johansen K (1984) A phylogenetic overview of cardiovascular shunts. In: Johansen K, Burggren W (eds): Cardiovascular shunts. Munksgaard, Copenhagen; pp. 17-32.

Maginnis LA, Sang YK, Reeves RB (1980) Oxygen equilibria of ectotherm blood containing multiple hemoglobins. Respir Physiol 42: 329-343.

Meyer M, Worth H, Scheid P (1976) Gas-blood CO_2 equilibrium in parabronchial lungs of birds. J Appl Physiol 41: 302-309.

Piiper J, Scheid P (1977) Comparative physiology of respiration: functional analysis of gas exchange organs in vertebrates. In: Widdicombe JG (ed): International review of physiology, respiration physiology 2, Vol. 14. University Park Press, Baltimore; pp. 219-253.

Rahn H, Howell BJ (1976) Bimodal gas exchange. In Hughes GM (ed): Respiration in amphibious vertebrates. Academic Press, New York; pp. 271-285.

Randall D, Daxboeck C (1984) Oxygen and carbon dioxide transfer across fish gills. In: Hoar WS, Randall DJ (eds): Fish physiology, Vol. 10 (Gills) pt. A. Academic Press, New York: pp. 263-314.

Romer AS (1972) Skin breathing--primary or secondary? Respir Physiol 14 (nos. 1 and 2): 183-192.

Satchell GH (1971) Circulation in fishes, Cambridge University Press, London.

Shelton G, Boutilier RG (1982) Apnoea in amphibians and reptiles. In: Butler PJ (ed): Control and coordination of respiration and circulation. Cambridge University Press, London; pp. 245-273.

Weinstein Y, Ackerman RA, White FN (1986) Influence of temperature on the CO_2 dissociation curve of the turtle, Pseudemys scripta. Respir Physiol 63: 53-63.

White FN (1976) Circulation. In: Gans C, Dawson W (eds): Biology of the reptilia, Vol. 5. Academic Press, New York; pp. 275-334.

White FN (1985) Role of intracardiac shunts in pulmonary gas exchange in chelonian reptiles. In: Johansen K, Burggren W (eds): Cardiovascular shunts, Alfred Benzon Symposium 21. Munksgaard, Copenhagen; pp. 296-305.

White FN, Hicks JW, Ishimatsu A (1985) Intracardiac shunts during intermittent ventilation in unrestrained turtles (personal observations).

Part II

WATER, IONIC EXCHANGES, OSMOREGULATION IN THE INTACT ORGANISM

OSMOREGULATION IN AMPHIBIANS

Vaughan H. Shoemaker

Department of Biology
University of California
Riverside, California, U.S.A.

Adult amphibians live in a wide variety of habitats that present
an array of osmoregulatory challenges. The solutions to these
problems are analyzed in terms of rates of influx and efflux of
water, sodium and nitrogen partitioned between skin, kidney,
bladder and gut.

INTRODUCTION

Among vertebrates, the transition between aquatic and terrestrial life is
nowhere more prevalent than in the Amphibia. Almost all begin life as
water-breathing larvae, and air breathing adults occupy an impressive array of
habitats. Some are virtually confined to fresh water, some live in saline
waters, many move freely between land and water, and some are independent of
standing water for long periods. Of the latter, most require moist or protected
habitats, but a few can face direct exposure to hot semi-arid environments.
This diversity, combined with the generally low resistance of amphibian skin to
osmotic and evaporative water fluxes, has aroused curiosity about osmoregulation
in this group and led investigators to search for physiological differences
between aquatic and terrestrial forms. Reviews on various aspects this topic
are available (Deyrup, 1964; Bentley, 1966, 1971; Warburg, 1972; McClanahan,
1975; Shoemaker and Nagy, 1977; Alvarado, 1979; Balinsky, 1970, 1981).

Fluxes of water, electrolytes and nitrogen comprise the major components of the
osmoregulatory process. Osmoregulation requires that in- and effluxes of water
and solutes balance over some appropriate time scale, but strict homeostasis is
not required. In fact, amphibians are quite tolerant of fairly wide and
long-term deviations from "normal" water content and solute concentration, which
are important in their ability to exploit a variety of environments. In what
follows I will sketch a semi-quatitative picture of the major components of the
osmoregulatory process in aquatic, semi-terrestrial, and terrestrial amphibian
adults, and point out differences which appear to be adaptive for a particular
life-style. This requires drawing data from numerous studies, and taking some
liberties in an attempt to present a general picture.

Comparative Physiology: Life in Water and on Land. P. Dejours, L. Bolis, C.R. Taylor, E.R. Weibel
(eds.) Fidia Research Series, IX-Liviana Press, Padova © 1987

AQUATIC EXISTENCE

Amphibians in fresh water all take up large volumes of water by osmosis, and counteract this influx by producing an equivalent volume of urine. Electrolyte losses in the urine, and diffusive losses across the skin, are balanced in fasted animals by active cutaneous uptake. Considerable effort has been devoted to interspecific comparisons to determine whether primarily aquatic amphibians cope more effectively with the freshwater environment than their more terrestrial counterparts. Although such comparisons are complicated by differences in body size, taxon, experimental technique and uncertainties concerning the ecology of some species, several trends are apparent.

Schmid (1965) found that survival time in distilled water of some north temperate anurans was greatest in more aquatic forms (Rana > Hyla > Bufo). Using isolated skin preparations he found that osmotic fluxes in these species were inversely correlated with survival time. A similar result was obtained by Mullen and Alvarado (1976) whose study included a primarily aquatic genus (Ascaphus) and was conducted using intact animals. Surface specific rates of uptake were almost four times higher in Bufo than in Ascaphus, and the difference in the permeability of the skin (correcting for the osmotic gradient) was three-fold. Compilations of values from numerous studies confirm this trend in both anurans and urodeles (Bentley, 1971, Alvarado, 1979). The aquatic newt Notopthalmus viridescens, which returns to water after a terrestral eft phase, has a very high permeabilty to water (Brown and Brown, 1977).

Sodium fluxes of fasting amphibians generally parallel osmotic influxes, with more aquatic species exchanging considerably smaller amounts of sodium (Greenwald, 1972; Mullen and Alvarado, 1976). Rates of active sodium uptake depend on the sodium concentration in the medium and exhibit Michaelis-Menton kinetics. The higher unidirectional uptake rates in the more terrestrial species reflect a higher capacity or Vmax of the system, whereas transport in aquatic species is characterized by a higher affinity (low Km). Both aquatic and semi-terrestrial species show an increased Vmax and decreased sodium efflux when depleted of sodium (Greenwald, 1972). Presumably the high affinity of aquatic species is adaptive in very dilute waters, but actual rates of uptake at low environmental concentrations (ca. 0.1 mM) are similar in aquatic and terrestrial species (see Mullen and Alvarado, 1976). Both aquatic and semi-terrestrial species remain in sodium balance in dilute fresh water despite large differences in sodium influxes (see Greenwald, 1972). It is not clear whether the greater sodium effluxes of terrestrial species result from urinary or from cutaneous losses. Greenwald (1971) concluded for Rana pipiens that 90% of sodium efflux occured via the kidney and calculated urine sodium concentration to be >10 meq/l. Data of Mullen and Alvarado indicate that the primary avenue for sodium efflux is diffusion across the skin, and that urinary sodium fluxes account for only 10-20% of the total. They measured urinary sodium concentrations of about 1 meq/l which are consistent with this conclusion. Sodium resorption by the urinary bladder may be important in establishing these low concentrations (Middler et al, 1968). Notopthalmus newts show the highest sodium uptake rates known (ca. 60 µeq/g day; Wittig and Brown, 1977), of which 70% is lost via the skin and the remainder via the urine at a concentration of 10 meq/l. Available data thus suggest that the skins of aquatic amphibians are less leaky to both sodium and water, and both factors contribute to their lower rates of sodium flux.

Aquatic adult amphibians excrete nitrogen as a variable mixture of urea and ammonia (see Balinsky, 1970). Most appear to be capable of excreting nitrogen primarily as urea when water turnover is low, but a few may not be able to do so (Shoemaker and McClanahan, 1980). Schmid (1968) measured urinary nitrogen concentrations in a variety of anurans in the field and found urea was the predominant nitrogen waste in aquatic as well as semi-terrestrial species, contributing 80-98% of the total and ranging from 30-80 mM. Apparently either urea or ammonia can be used effectively by aquatic amphibians, and reasons for interspecific differences in excretory patterns are presently obscure.

A general picture of water, sodium and nitrogen fluxes in a freshwater amphibian is shown in Figure 1. Nitrogenous wastes must be excreted at appreciable concentrations, despite high rates of water influx and urine production. Dietary sodium input is small compared to unidirectional cutaneous uptake, but may be significant when compared to net uptake across the skin.

Although there are no truly marine amphibians, many species of anurans and urodeles can tolerate saline or brackish waters to varying degrees. This subject has recently been reviewed by Balinsky (1981). For adults, survival in saline media depends on accumulation of solutes so that the body fluids remain slightly hyperosmotic to the environment. Elevations of sodium, chloride and urea generally account for the increased osmolality of the extracellular fluids. Within cells, potassium and free amino acids also increase significantly. The ability to increase urea levels rapidly appears to be characteristic of the most tolerant species. Urea accumulates at least in part because rates of urine production are low due to the reduced osmotic uptake of water, but changes in levels of substrates and enzymes associated with the urea cycle have also been implicated. Most studies of salinity acclimation have used fasted animals, and in view of the high rates of urea production in fed animals it would be interesting to see whether feeding speeds the acclimatory response. Feeding apparently does not affect the final concentrations of urea attained in acclimated animals, but may increase salinity tolerance (Romspert and

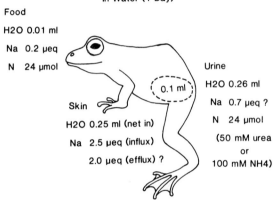

AQUATIC

In Water (1 Day)

Food

H2O 0.01 ml

Na 0.2 µeq

N   24 µmol

Skin

H2O 0.25 ml (net in)

Na 2.5 µeq (influx)

2.0 µeq (efflux) ?

0.1 ml

Urine

H2O 0.26 ml

Na 0.7 µeq ?

N   24 µmol

(50 mM urea

or

100 mM NH4)

Figure 1.

Typical fluxes of water, sodium and nitrogen for an aquatic amphibian living and feeding in fresh water (ca. 1 meq Na/l). All values shown are per gram of body mass (hydrated, bladder empty). Estimates of cutaneous water and sodium influxes are from Alvarado (1979), Mullen and Alvarado (1976), and Greenwald (1972). Partitioning of sodium efflux between skin and urine is uncertain. Values shown are for urinary sodium concentrations of 2-3 mM (0.7 µeq Na ÷ 0.26 ml H2O). Dietary input was estimated from values of Shoemaker and McClanahan (1975) for mealworms (Tenebrio) with the water content adjusted upward to match values more typical of adult insects. Approximate pool sizes for water and sodium in amphibians are 0.8 ml/g and 50-65 µeq/g respectively.

McClanahan, 1981).

SEMI-TERRESTRIAL EXISTENCE

The majority of adult amphibians divide their time in some fashion between land and water. When on land they are highly vulnerable to dehydration. For many species the skin provides no resistance to the evaporation of water (Spotila and Berman, 1976). Water losses on the order of 0.5 ml/g day are typical for medium sized amphibians sitting quietly in slow moving dry air at 25°C. This rate would be greatly increased at higher temperatures, wind velocities or levels of activity, and much reduced at high humidities (see Tracy, 1976). The integument of arboreal frogs apparently provides measurable resistance to evaporation (Wygoda, 1984), but rates of loss are still high (about half those of "typical" frogs). Thus for most amphibians on land, control of evaporative water loss is primarily a matter of behavior and depends on the availability of humid and protected microhabitats. The greatest gap in our understanding of the water economies of free living amphibians derives from this fact. Amphibians are both nocturnal and secretive, and little is known of their whereabouts for most of the time. Field and laboratory studies addressing this question (Dole, 1967; Claussen, 1973; Van Berkum et al, 1982) indicate that both semi-terrestial and arboreal species not only seek high humidities but also utilize moist substrates which permit absorption of water. However, measurement of water fluxes in amphibians on land remains a formidable task.

We can use our knowledge of the physiological characteristics of semi-terrestrial amphibians to make some predictions concerning their degree of independence of water while they are active and feeding. All amphibians apparently have the ability of stop urine production when water is unavailable. Terrestrial forms tend to have large bladders which they fill with dilute urine when water is available, and from which they can resorb water to replace evaporative losses. They are tolerant to the accumulation of urea in the body fluids and to losses of body water in excess of bladder reserves. When dehydrated, osmotic uptake across the skin is very rapid so that body water is rapidly restored (generally within one hour) when water is available. These adaptive characteristics have long been appreciated in anurans (Bentley, 1966) and they may well be equally important in urodeles (Brown and Brown, 1980).

Either of two factors may limit the time a semi-terrestrial amphibian spends away from water. Exposure to even moderately desiccating conditions will make the time very short. For example, a typical 20 g frog exposed to moving air at 25°C and 40% RH loses 5% of its body mass/hr (Wygoda, 1984). Bladder reserves of 30% body mass would last 6 hr, and after 12 hr the animal would be dehydrated to near its lethal limit. For a 2 g animal the total survival time would be

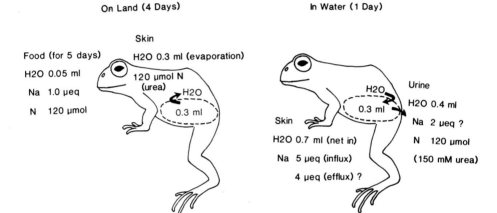

Figure 2.

Estimated fluxes of water, sodium and nitrogen for a semi-terrestrial amphibian feeding on land (right), and returning to water to rehydrate and excrete excess solutes (left). The animal is assumed to be in a protected habitat so that evaporation does not exceed bladder reserves during the period on land. Units and sources for other fluxes are as in Figure 1. In the example shown, urea concentration in the body fluids would increase by 75 mM (120 μmol N/g ÷ 2 mol N/mol urea ÷ 0.8 ml/g) during the 4 day period on land.

about 6 hr, and for a 200 g animal, 24 hr. In the absence of high rates of evaporative loss, the partitioning of time between land and water would be dictated by considerations of nitrogen excretion. An amphibian that forages on land (Figure 2), would accumulate a considerable amount of nitrogen waste (urea) over four days. If the animal returned to water with an empty bladder (but otherwise normally hydrated), it would have to spend about a day in water to flush out the urea stored in its body fluids and to refill the bladder before embarking on another foray. The implication of this is that a feeding ureotelic amphibian would have to spend about 20% of its time in water even if evaporative water losses were minimized. The other solution is to find a sufficiently moist substrate so that water uptake exceeds evaporation permitting continued production of urine (see Dole, 1967). Assuming urinary urea concentration to be 100 mM, a urine flow of 0.12 ml/g day would be required. This is similar to the average daily urine production estimated for an amphibian shuttling between land and water (Fig 2) and, surprisingly, is about half of the rate of urine production of a fully aquatic amphibian. This suggests that the higher osmotic permeability of terrestrial amphibians is beneficial in permitting them to flush out urea and replenish bladder stores rapidly. Moreover, it may not be feasible for aquatic amphibians to reduce rates of osmotic uptake much below those observed if they are to excrete dietary nitrogen as urea and/or ammonia.

TERRESTRIAL EXISTENCE

Amphibians that inhabit semi-arid environments typically aestivate below ground and do not feed for long periods during the dry season. Many of these form cocoons of dead epidermis which greatly restrict water transfer (see Lee and Mercer, 1967; Loveridge and Withers, 1981; McClanahan et al, 1983), but some rely only on the protection provided by the soil. If the soil is dry burrowed amphibians cannot osmoregulate. However, changes in body water content and solute concentration are slow enough to permit them to wait out the dry season, and accumulation of urea can be beneficial in fossorial species that do not form cocoons (Delson and Whitford, 1973; McClanahan, 1975).

Some arboreal species are exceptional in having physiological characteristics that allow them to remain abroad under dry conditions. Climbing frogs representing several genera have skins that are highly resistant to evaporation (see Withers et al, 1982). Tree frogs of the genera Chiromantis and Phyllomedusa are the most striking examples known to date. They not only lose little water by evaporation, but also use uric acid as the primary vehicle for nitrogen excretion (Loveridge, 1970; Shoemaker et al, 1972). These adaptations combine to provide for an osmoregulatory pattern that differs dramatically from those of other amphibians (Figure 3). Even when exposed to dry air, water input

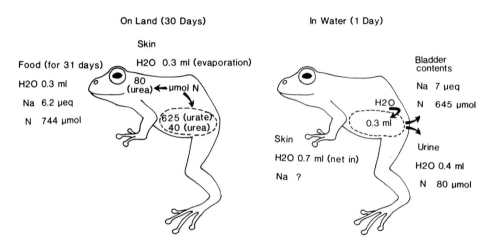

Figure 3.

Fluxes of water, sodium and nitrogen in arboreal frogs adapted to terrestrial existence by virtue of high cutaneous resistence to evaporative water loss and use of uric acid for excretion of nitrogen. Even under dry conditions (left), the animal may offset evaporation with water in the food and most of the nitrogen wastes are deposited in the bladder. When water is available (right) the bladder is emptied immediately. Although not shown, bladder reserves can also be used on land as in Fig 2. Units are as in Figure 1. Fluxes shown are based on Drewes et al (1977), and Shoemaker and McClanahan (1975). Fecal losses are neglected because they comprise only a small portion of the total.

from the food is of the same order as evaporative loss. The kidneys continue to function and most of the nitrogen waste is deposited in the urinary bladder as precipitated uric acid where it is held until water is available. Accumulation of urea in the body fluids is slow and similar to the rate seen in fasting ureotelic frogs (Jones, 1980). Large urinary bladders provide a water reserve (Shoemaker and Bickler, 1979).

When water is available, Phyllomedusa and Chiromantis require little time to restore balance. The bladder is quickly voided when the frog enters water. This eliminates 85% of the nitrogen input from a month of feeding, as well as most of the electrolytes acquired in the food. A substantial fraction of the sodium and potassium stored in the bladder is coprecipitated with the urate. One day in water is more than adequate to refill the bladder and excrete the modest amounts of urea and salts stored in the body fluids while on land. It now appears that Phyllomedusa sauvagei need not seek standing water or a wet substrate to gain water. They have recently been shown to drink rainwater striking the head, taking in water through the mouth at a rate of about 0.2 ml/g hr while a smaller amount (ca. 0.05 ml/g hr) enters via the skin (McClanahan and Shoemaker, 1986). Thus a rain of a few hours duration could provide the same water influx as for one day spent in water.

LITERATURE CITED

Alvarado RH (1979) Amphibians. In: Malioy GMO (ed): Comparative Physiology of Osmoregulation in Animals. Academic Press, London; pp. 261-303.

Balinsky JB (1970) Nitrogen metabolism in amphibians. In: Campbell JW (ed): Comparative Biochemistry of Nitrogen Metabolism. Vol 2. Academic Press, New York; pp. 519-637.

Balinsky JB (1981) Adaptation of nitrogen metabolism to hyperosmotic environment in amphibia. J Exp Zool 215: 335-350.

Bentley PJ (1966) Adaptations of Amphibia to arid environments. Science 152: 619-623.

Bentley PJ (1971) The Amphibia. In: Endocrines and Osmoregulation. Springer-Verlag, New York; pp. 161-197.

Brown PS, Brown SC (1977) Water balance responses to dehydration and neurohypophysial peptides in the salamander, Notophthalmus viridescens. Gen Comp Endocrinol 31: 189-201.

Brown SC, Brown PS (1980) Water balance in the California newt, Taricha torosa. Am J Physiol 238: R113-118.

Claussen DL (1973) The water relations of the tailed frog, Ascaphus truei, and the Pacific treefrog, Hyla regilla. Comp Biochem Physiol 44A: 155-171.

Delson J, Whitford WG (1973) Adaptation of the tiger salamander, Ambystoma tigrinum, to arid habitats. Comp Biochem Physiol 46A: 631-638.

Deyrup IJ (1964) Water balance and kidney. In: Moore J (ed): Physiology of Amphibia. Vol 1. Academic Press, New York; pp. 251-328.

Dole JW (1967) The role of substrate moisture and dew in the water economy of leopard frogs, Rana pipiens. Copeia 1967: 141-149.

Drews RC, Hillman SS, Putnam RW, Sokol OM (1977) Water, nitrogen and ion balance in the African treefrog Chiromantis petersi Boulenger (Anura: Rhacophoridae), with comments on the structure of the integument. J Comp Physiol 116:257-267.

Greenwald L (1971) Sodium balance in the leopard frog (Rana pipiens). Physiol Zool 44: 149-161.

Greenwald L (1972) Sodium balance in amphibians from different habitats.

Physiol Zool 45: 229-237.

Jones RM (1980) Metabolic consequences of accelerated urea synthesis during seasonal dormancy of spadefoot toads, Scaphiopus couchi and Scaphiopus multiplicatus. J Exp Zool 212: 255-267.

Lee AK, Mercer EH (1967) Cocoon surrounding desert-dwelling frogs. Science 157: 87-88.

Loveridge JP (1970) Observations on nitrogenous excretion and water relations of Chiromantis xerampelina (Amphibia, Anura). Arnoldia 5: 1-6.

Loveridge JP, Withers PC (1981) Metabolism and water balance of active and cocooned African bullfrogs Pyxicephalus adspersus. Physiol Zool 54: 203-214.

McClanahan LL (1975) Nitrogen excretion in arid-adapted amphibians. In: Hadley NF (ed): Environmental Physiology of Desert Organisms. Dawden, Hutchison, and Ross Publ, Stroudsberg. pp. 106-116.

McClanahan LL, Ruibal R, Shoemaker VH (1983) Rate of cocoon formation and its physiological correlates in a ceratophryd frog. Physiol Zool 56: 430-435.

McClanahan LL, Shoemaker VH (1987) Behavior and thermal relations of the arboreal frog Phyllomedusa sauvagei. National Geographic Research 3:in press.

Middler SA, Kleeman CR, Edwards E (1968) The role of the urinary bladder in salt and water metabolism of the toad, Bufo marinus. Comp Biochem Physiol 26: 57-68.

Mullen TL, Alvarado RH (1976) Osmotic and ionic regulation in amphibians. Physiol Zool 49: 11-23

Romspert AP, McClanahan LL (1981) Osmoregulation of the terrestrial salamander, Ambystoma tigrinum, in hypersaline media. Copeia 1981: 400-405.

Schmid WD (1965) Some aspects of the water economies of nine species of amphibians. Ecology 46: 261-269.

Schmid WD (1968) Natural variations in nitrogen excretion of amphibians from different habitats. Ecology 49: 180-185.

Shoemaker VH, Balding D, Ruibal R, McClanahan LL, Jr. (1972) Uricotelism and low evaporative water loss in a South American frog. Science 175: 1018-1020.

Shoemaker VH, Bickler PE (1979) Kidney and bladder function in a uricotelic treefrog (Phyllomedusa sauvagei). J Comp Physiol 133: 211-218.

Shoemaker VH, McClanahan LL, Jr. (1975) Evaporative water loss, nitrogen excretion and osmoregulation in phyllomedusine frogs. J Comp Physiol 100: 331-345.

Shoemaker VH, McClanahan LL (1980) Nitrogen excretion and water balance in amphibians of Borneo. Copeia 1980: 446-451.

Shoemaker VH, Nagy KA (1977) Osmoregulation in amphibians and reptiles. Annu Rev Physiol 39: 449-471.

Spotila JR, Berman EN (1976) Determination of skin resistance and the role of the skin in controlling water loss in amphibians and reptiles. Comp Biochem Physiol 55A: 407-411.

Tracy, CR (1976) A model of the dynamic exchanges of water and energy between a terrestrial amphibian and its environment. Ecol Monographs 46: 293-326.

Van Berkum F, Pough FH, Stewart MM, Brussard, PF (1982) Altitudinal and interspecific differences in the rehydration abilities of Puerto Rican frogs (Eleutherodactylus). Physiol Zool 55: 130-136.

Warburg MR (1972) Water economy and thermal balance of Israeli and Australian amphibia from xeric habitats. Symp Zool Soc Lond 31: 79-111.

Withers PC, Hillman SS, Drewes RC, Sokol OM (1982) Water loss and nitrogen excretion in sharp-nosed reed frogs (Hyperolius nasutus: Anura, Hyperollidae). J Exp Biol 97: 335-343.

Wittig KP, Brown SC (1977) Sodium balance in the newt, Notophthalmus viridescens. Comp Biochem Physiol 58A: 49-52.

Wygoda WL (1984) Low cutaneous evaporative water loss in arboreal frogs. Physiol Zool 57: 329-337.

ACID-BASE BALANCE AND NITROGENOUS WASTE EXCRETION IN FISHES: THE AQUATIC TO AMPHIBIOUS TRANSITION

David H. Evans

Department of Zoology, University of Florida
Gainesville, FL 32611, USA
Mt. Desert Island Biological Laboratory
Salsbury Cove, ME 04672, USA

Because of limited buffering capacity, and the constraints of branchial gas exchange and renal function, fishes respond to acid-base disturbances by movements of either acid or base equivalents across the gills and cellular membranes to maintain extra- and intracellular pH. However, amphibious fishes may be forced to compensate only intracellular pH because of limited access to necessary ions in the environment. Preliminary data indicate that some amphibious fish species may have some limited abilities to expel respiratory CO_2 and switch from ammono- to ureotelism.

The generalized terrestrial vertebrate maintains acid-base balance by a combination of extracellular and intracellular buffers, pulmonary ventilation to control blood P_{CO_2} and various solute transport steps in renal tubules (such as Na^+/NH_4^+, Na^+/H^+, and Cl^-/HCO_3^- exchanges) which can excrete acid or base equivalents (ABE; e.g., Davenport, 1974; Valtin, 1983; Koeppen et al., 1985). Nitrogenous wastes are excreted via the kidney, predominantly in the form of urea by the mammals and uric acid by the reptiles and birds. These pathways for acid-base balance, and the actual nitrogenous wastes involved, result from aerial gas exchange, as well as selective pressures favoring morphological and physiological changes to promote water conservation in the terrestrial environment. The evolutionary shift from ammonia excretion in primarily aquatic vertebrates (fishes and some amphibia) to urea excretion and, subsequently, uric acid excretion is a classic scenario in the history of the vertebrates, despite well-known exceptions such as uricotelic amphibians and ammonotelic crocodilians (Schmidt-Nielsen, 1979). It is only in the past 15 years that comparative physiologists have begun to appreciate that pathways for excretion of ABE also have changed during the evolution of terrestrial vertebrates from aquatic ancestors. The intent of this review is to outline briefly the early evolution of this changing pattern of acid-base and nitrogen regulation in the fishes (including some amphibious forms) and also to

Comparative Physiology: Life in Water and on Land. P. Dejours, L. Bolis, C.R. Taylor, E.R. Weibel (eds.) Fidia Research Series, IX-Liviana Press, Padova © 1987

demonstrate that the mechanisms of epithelial transport of relevant solutes
(e.g., ionic exchanges) have been conserved. Whenever possible, the most
recent reviews and chapters are referenced, rather than original publications,
in order to facilitate entry into an extensive literature.

Acid-base balance and nitrogen excretion by fishes has been extensively
reviewed recently (e.g., Cameron, 1978; Heisler, 1984a; Evans, 1986). One
problem that fishes face is that their total body buffering capacity is far
below that of terrestrial vertebrates because of relatively low blood volumes,
hematocrits, and bicarbonate and non-bicarbonate buffer concentrations
(Heisler, 1984a). This low, body-buffer capacity means that physico-chemical
compensation for acid-base disturbances plays a relatively smaller role in
fishes than it does in mammals. The constraints of the morphology and
physiology of aquatic gas exchange dictate that, theoretically, respiratory
compensations for acid-base disturbances in fishes may be relatively minor
(Rahn, 1966). The aquatic medium has a lower oxygen content than air (e.g.,
Dejours, 1981) which means that water breathers must move substantial volumes
of medium past the gas exchanger in order to maintain blood P_{O_2}. Correlated
with this large ventilation rate, and the counter-current nature of the
irrigation and perfusion of the fish gill (e.g., Randall and Daxboeck, 1984),
the resting blood P_{CO_2} of fishes is generally less than 5 torr (Heisler,
1984a). Thus, fishes are relatively hyperventilated with respect to
elimination of CO_2 and it is unlikely that increases in ventilation in response
to acidosis could produce substantial changes in blood P_{CO_2} and, hence, pH. In
addition, reductions in ventilation in response to alkalosis may compromise
oxygen uptake. Experimental measurements of ventilatory responses to at least
acidosis indicate little (e.g., Dejours, 1973) or only transient (e.g., Randall
et al., 1976) changes in irrigation of the gills.

Thus, because the buffering capacity and respiratory control mechanisms
are relatively weak in fishes, solute transport mechanisms are the dominant
physiological control sites for acid-base regulation. Despite the presence of
glomeruli (except for a few marine species), proximal tubules, and distal
tubules (except in most marine species) in the fish kidney (Evans, 1979), its
role in acid-base regulation is rather marginal. In a relatively large number
of acid-base stress studies it has been shown that the kidneys of both marine
and freshwater species do respond to altered internal pHs (produced by
hypercapnia, acid/base infusion, exercise, hypoxia, or hyperoxia) with altered
excretion of ABE, but the fractional role in total compensation is less than 1%
in marine species and generally less than 25% in freshwater species (Heisler,
1984a).

The dominant site of baseline or experimentally stimulated excretion of ABE is the gills. Heisler et al. (1976) showed that hypercapnic, larger spotted dogfish (Scyliorhinus stellaris) compensated the resulting acidosis by excreting acid equivalents, or influxing base equivalents across the branchial epithelium in order to increase blood HCO_3^- concentrations from approximately 8 mM to 20 mM, and thereby nearly re-establish control pH of the blood within 10 hours after the initiation of hypercapnia. Interestingly, the excretion of ammmonia did not change during the experiment. More recently, Toews et al. (1983) and Claiborne and Heisler (1984) have observed, at least qualitatively, the same responses in hypercapnic, marine eels (Conger conger) and freshwater carp (Cyprinus carpio), respectively, except that ammonia excretion from the carp increased under hypercapnic conditions. In addition, the time-course of compensation was also under 10 hours in the conger eel, but over 46 hours in the carp. This difference in speed of compensation may be related to the ionic content of the respective media (see below). Other stressful conditions (e.g., temperature changes, exercise, hypoxia, injected acid or base loads, acidification of the external medium, and hyperoxia) produce similar branchial responses in terms of net acid secretion (e.g., Heisler, 1984a; Evans, 1986). Interestingly, while acidotic freshwater fish species appear to excrete some acid equivalents via a stimulated ammonia efflux, acidotic marine species do not normally display any change in ammonia efflux (Heisler, 1984a).

In addition to possessing epithelial mechanisms which can control blood pH via changes in net ABE fluxes across the gills, it is clear that fishes also modulate intracellular and extracellular pHs in the face of various environmental stresses via transfer of ABE across the cellular membranes of muscle cells (e.g., Heisler, 1984b). In fact, Cameron (1980) calculated that branchial extrusion mechanisms could only account for some 20-30% of the compensation of a mineral acid load infused into the channel catfish, Ictalurus punctatus, the remainder being buffered presumably intracellularly. Cameron and Kormanik (1982) extended this work and found that 32% of an infused base load and 52% of an infused acid load were buffered intracellularly in the catfish. It is clear that further data are needed to determine the relative importance of transepithelial vs. transcellular compensatory mechanisms in maintaining consistent extra- and intracellular pH in the face of acid-base disturbances.

Reeves (1977) has proposed that, in poikilotherms, blood P_{CO_2} is regulated in order to maintain consistent fractional dissociation of imidazole rings on peptides such as histidine, the "alphastat hypothesis". However, various authors (e.g., Cameron, 1984; Heisler, 1984a) suggest that the trans-epithelial

and trans–membrane transfers of ABE in order to compensate pH changes, as well
as the generally observed changes in fish blood or intracellular pH seen with
temperature changes, are not consistent with this model.

The actual mechanisms for these transepithelial ABE transfers are
generally thought to be via ionic antiports such as Na^+/H^+ and Cl^-/HCO_3^-
exchanges (see Evans, 1986), but very few experiments have examined this point
directly. Evans (1982) has provided rather direct evidence for Na^+/H^+ exchange
by showing that acid extrusion by hypercapnic or mineral acid loaded Gulf
toadfish (Opsanus beta) and dogfish shark pups (Squalus acanthias) is dependent
upon external Na^+. Interestingly, these data also demonstrate that, despite
the fact that such an exchange is inappropriate for Na^+ balance, marine fishes
also extrude unwanted acid equivalents via branchial Na^+/H^+ exchange. More
recent studies have shown that the hagfish (Myxine glutinosa) displays
Na^+-sensitive acid excretion and Cl^--sensitive base excretion, indicating that
these branchial ionic exchange systems arose for acid-base regulation before
the vertebrates entered freshwater, and were subsequently utilized for ionic
regulation in these dilute salinities (Evans, 1984). Nevertheless, more data
are necessary before we can be certain that fish utilize these antiport systems
for branchial acid-base regulation. No data have been gathered to indicate
that similar antiports function in the extracellular to intracellular
compartment transfers, but data from a variety of other tissues (e.g., Boron,
1983) indicate that they probably play a major role in fish cellular pH
regulation as well. It is, of course, an interesting example of evolutionary
conservatism that these mechanisms of cellular pH regulation have been retained
in tissues, such as branchial and renal epithelia, for regulation of
extracellular pH. In light of recent re-interpretation of some fundamental
concepts of acid-base balance (e.g., Stewart, 1983), it is appropriate to add
the proviso that it may be more correct to speak in terms of net movements of
Na^+ vs. Cl^- across membranes or epithelia affecting the "Strong Ion Difference"
and, hence, the pH of a solution. In any event, the same antiports (e.g.,
Na^+/H^+, Na^+/NH_4^+, and Cl^-/HCO_3^- exchanges) would probably be involved.

The gill is also the site of excretion of ammonia from fishes (e.g.,
Evans, 1982). The mechanisms for trans-branchial ammonia extrusion have
recently been outlined by Evans and Cameron (1986), and consist of non-ionic
diffusion of NH_3, ionic diffusion of NH_4^+, as well as basolateral (serosal) and
apical (mucosal) Na^+/NH_4^+ exchange. The relative roles of these transport
pathways are currently under investigation and preliminary data indicate that,
at least in O. beta, Na^+/NH_4^+ exchanges may account for some 50% of the
extrusion, with the non-ionic diffusion predominating over ionic diffusion in
the remainder (Evans, More, and Robbins, unpublished).

Some species of fishes are capable of breathing air, usually because of patent connections to the swim bladder or specialized morphological extensions of the branchial, buccal, or opercular epithelium. Air-breathing fishes are generally found to have reduced gill surface areas and perfusion patterns presumably secondary to a decreased emphasis on branchial oxygen uptake, and to avoid loss of oxygen into the often hypoxic water (Randall et al., 1981). Bimodal breathing is common in these species, with CO_2 excretion into the water continuing through the branchial epithelium, parallel with oxygen uptake from air (e.g., Rahn and Howell, 1976). There does seem to be some limitation on CO_2 excretion under these conditions, however, since air-breathing fishes often display higher blood P_{CO_2} (and compensatory blood HCO_3^- concentrations to maintain similar pH) than water-breathing species (e.g., Randall et al., 1981). It is clear, however, that severe limitation on branchial water contact (either by reduced gill surface area or by reduction in water ventilation of gills) will jeopardize the animal's ability to regulate its acid-base status, not to mention its ability to excrete ammonia, as long as the gills remain the dominant site for these processes. Heisler (1982) has presented the only relatively complete study of acid-base balance of a fish under these circumstances. When the tropical, fresh-water teleost Synbranchus marmoratus (which does not possess any specialized aerial gas-exchange epithelium was stimulated to switch to exclusive air-breathing (because of aquatic hypoxia) blood P_{O_2} did not change, but blood P_{CO_2} rose from approximately 6 torr to approximately 25 torr, while blood HCO_3^- concentration increased only transiently above the normal 25 mM, and then returned to control levels within five hours. Blood pH therefore fell some 0.6 units, and remained uncompensated. However, intracellular (heart and white muscle) pH remained nearly constant by a combination of intracellular non-bicarbonate buffering, cellular uptake of base equivalents produced in the extracellular spaces by non-bicarbonate buffering, and some uptake of base equivalents from the medium. Thus, intracellular pH was maintained at the expense of extracellular pH. This is energetically efficient since Heisler (1982) calculated that it would have taken 5 times as much base to reach a similar degree of compensation in the extracellular fluids because of the extremely high HCO_3^- concentration in the blood of Synbranchus. This degree of transepithelial transfer of ABE is presumably not possible because of the limited contact time between the medium and the gills, as well as the relatively small pool of either HCO_3^- or Na^+ in the local, soft-water medium (assuming transepithelial transfers would be either Na^+/H^+ or reversed Cl^-/HCO_3^- exchanges). Importantly, urine acid efflux accounted for only some 3% of the compensation, so stimulated renal extrusion of acid was apparently not available as a compensatory pathway.

Thus, it is clear that the morphological limitations in Synbranchus dictate that the respiratory acidosis produced by aerial gas exchange is compensated only intracellularly by addition of base equivalents. This is because the limited access to water at the gill epithelium (or skin) precludes substantial transfers of acid or base equivalents for extracellular pH compensation. However, it appears that other species, with more developed aerial gas exchange surfaces may be able to avoid the acidosis produced by incomplete aerial or aquatic CO_2 excretion. For instance, Burggren (1979) demonstrated that the blue gourami (Trichogaster trichopterus) is able to adjust the ventilation of its specialized aerial gas exchange surface (labyrinth organ derived from the epibranchial region of the first and second gill arches) to meet the needs of both O_2 uptake and CO_2 excretion, in the face of hypoxia or hypercapnia in either the gaseous or water phase. Of special interest is the finding that the aerial ventilation rate of the labyrinth increased five-fold when the water was made hypercapnic. Unfortunately, we have no blood or intracellular pH data for these experiments, so it is unclear if pH equilibrium was maintained under these conditions, but apparently the aerial gas exchanger is utilized to excrete CO_2 when the gill CO_2 excretion is compromised in this species.

Compromise of gill function in air-breathing or amphibious fishes may also affect their ability to excrete nitrogen in the form of ammonia. Heisler (1982) did find that ammonia efflux from Synbranchus declined during air breathing, presumably because branchial extrusion mechanisms were reduced secondary to the limited contact with environmental water. It is not known if this decline was produced by decreased nitrogen metabolism or a shift to another form of nitrogeneous end-product. Measurement of blood and urine ammonia and urea concentrations would have been very instructive since various studies (e.g., Gordon et al., 1978) have found that some intertidal, marine species increased their rate of urea efflux vs. ammonia efflux (as well as the sum of both products) when deprived of water for 12-26 hours and then returned to sea water. Importantly, these species excrete more than 50% of their nitrogen waste as urea even when in sea water. Unfortunately, fluxes were monitored from whole animals so that conclusions about renal vs. extra-renal pathways cannot be reached. A more recent study (Davenport and Sayer, 1986) indicates that the intertidal amphibious teleost, Blennius pholis, excretes only some 19% of its nitrogen waste as urea and does not alter either the total rate or the proportion due to urea when emersed in air, and then returned to sea water. It is clear that more data are needed on the mechanisms of acid-base regulation and nitrogen excretion by amphibious fishes, both during their aquatic phases and after initiation of aerial gas exchange.

In summary, current data indicate that fishes normally compensate for pH disturbances by predominantly transepithelial and transcellular movements of ABE rather than by ventilatory or renal changes. However, limited access to water in some amphibious species prompts intracellular compensation at the expense of extracellular pH. Fish normally excrete ammmonia across the gills by both diffusion and Na^+/NH_4^+ exchange, and apparently rates are stimulated by acidosis only in freshwater species. Semiterrestriality in some species is associated with an increased excretion of urea relative to ammonia.

The National Science Foundation, most recently PCM 8302621, has supported the author's research discussed in this review.

REFERENCES

Boron WF (1983) Transport of H^+ and of ionic weak acids and bases. J. Membrane Biol. 72: 1-16.

Burggren WW (1979) Bimodal gas exchange during variation in environmental oxygen and carbon dioxide in the air breathing fish Trichogaster trichopterus. J. Exp. Biol. 82: 197-213.

Cameron JN (1978) Regulation of blood pH in teleost fish. Resp. Physiol. 33: 129-144.

Cameron JN (1980) Body fluid pools, kidney function, and acid-base regulation in the freshwater catfish Ictalurus punctatus. J. Exp. Biol. 86: 171-185.

Cameron JN (1984) Acid-base status of fish at different temperatures. Am. J. Physiol. 246: R452-R459.

Cameron JN, Kormanik GA (1982) The acid-base responses of gills and kidneys to infused acid and base loads in the channel catfish, Ictalurus punctatus. J. Exp. Biol. 99: 143-160.

Claiborne JB, Heisler N (1984) Acid-base regulation in the carp (Cyprinus carpio) during and after exposure to environmental hypercapnia. J. Exp. Biol. 108: 25-43.

Davenport HW (1974) The ABC of Acid-Base Chemistry, University of Chicago Press, Chicago, 124 pgs.

Davenport J, Sayer MDJ (1986) Ammonia and urea excretion in the amphibious teleost Blennius pholis (L.) in sea-water and in air. Comp. Biochem. Physiol. 84A: 189-194.

Dejours P (1973) Problems of control of breathing in fishes. In: Bolis L, Schmidt-Nielsen K, Maddrell SHP (eds.): Comparative Physiology, North-Holland Publishing, Amsterdam: pp. 117-133.

Dejours P (1981) Principles of Comparative Respiratory Physiology,
Elsevier/North-Holland, Amsterdam: pp. 23-35.

Evans DH (1979) Fish. In: Maloiy GMO (ed.): Comparative Physiology of
Osmoregulation in Animals, Academic Press, London: pp. 305-390.

Evans DH (1982) Mechanism of acid extrusion by two marine fishes: The teleost,
Opsanus beta, and the elasmobranch, Squalus acanthias. J. Exp. Biol. 97: 289-
299.

Evans DH (1984) Gill Na^+/H^+ and Cl^-/HCO_3^- exchange systems evolved before the
vertebrates entered fresh water. J. Exp. Biol. 113: 465-469.

Evans DH (1986) The role of branchial and dermal epithelia in acid-base
regulation in aquatic vertebrates. In: Heisler N (ed.): Acid-base Regulation
in Animals, Elsevier, Amsterdam: pp. 401-444.

Evans DH, Cameron JN (1986) Gill ammonia transport. J. Exp. Zool., in press.

Gordon MS, Ng WW, Yip AY (1978) Aspects of the physiology of terrestrial life
in amphibious fishes. III. The Chinese mudskipper Periophthalmus
cantonensis. J. Exp. Biol. 72: 57-75.

Heisler N (1982) Intracellular and extracellular acid-base regulation in the
tropical fresh-water teleost fish Synbranchus marmoratus in response to the
transition from water breathing to air breathing. J. Exp. Biol. 99: 9-28.

Heisler N (1984a) Acid-base regulation in fishes. In: Hoar WS, Randall DJ
(eds.): Fish Physiology, Vol. XA, Academic Press, Orlando: pp. 315-401.

Heisler N (1984b) Role of ion transfer processes in acid-base regulation with
temperature changes in fish. Am. J. Physiol. 246: R441-R451.

Heisler N, Weitz H, Weitz AM (1976) Hypercapnia and resultant bicarbonate
transfer processes in an elasmobranch fish. Bull. Sur. Physiolpathol. Resp.
12: 77-85.

Koeppen B, Giebisch G, Malnic G (1985) Mechanism and regulation of renal
tubular acidification. In: Seldin DW, Giebisch G (eds.): The Kidney:
Physiology and Pathophysiology, Raven Press, New York: pp. 1491-1525.

Rahn H (1966) Aquatic gas exchange: Theory. Resp. Physiol. 1: 1-12.

Rahn H, Howell BJ (1976) Bimodal gas exchange. In: Hughes GM (ed.):
Respiration of Amphibious Vertebrates, Academic Press, London: pp. 271-285.

Randall DJ, Daxboeck C (1984) Oxygen and carbon dioxide transfer across fish
gills. In: Hoar WS, Randall DJ (eds.): Fish Physiology, Vol. XA, Academic
Press, Orlando: pp. 263-314.

Randall DJ, Heisler N, Drees F (1976) Ventilatory response to hypercapnia in the larger spotted dogfish Scyliorhinus stellaris. Am. J. Physiol. 230: 590–594.

Randall DJ, Burggren WW, Farrell AP, Haswell MS (1981) The Evolution of Air Breathing in Vertebrates, Cambridge University Press, Cambridge: pp. 29–36.

Reeves RB (1977) The interaction of body temperature an acid-base balance in ectothermic vertebrates. Ann. Rev. Physiol. 39: 559–586.

Schmidt-Nielsen K (1979) Animal Physiology, Cambridge University Press, Cambridge: pp. 362–376.

Stewart PA (1983) Modern quantitative acid-base chemistry. Can. J. Physiol. Pharmac. 61: 1442–1461.

Toews DP, Holeton GF, Heisler N (1983) Regulation of the acid-base status during environmental hypercapnia in the marine teleost fish Conger conger. J. Exp. Biol. 107: 9–20.

Valtin H (1983) Renal Function, Little Brown & Co., Boston: pp. 195–248.

WATER FLUX SCALING

Kenneth A. Nagy and Charles C. Peterson

Laboratory of Biomedical and Environmental Sciences and
Department of Biology, University of California,
Los Angeles, California, U.S.A.

Allometric analyses of daily rates of water flux, measured
using tritiated or deuterated water, show strong correlations
among several taxa of captive vertebrates and invertebrates,
and free-ranging vertebrates. Water-breathing animals have
much higher rates of water flux than air-breathers, and endothermic air-breathers have higher water fluxes than air-
breathing ectotherms. Desert-dwelling eutherians, birds and
reptiles have lower water fluxes than nondesert species.
"Water use effectiveness" values (ratios of ml water flux per
day to kJ energy metabolized per day, measured in the field
with doubly labeled water) reveal that, among air-breathers,
the taxonomic differences in daily water fluxes apparently
result mainly from differences in daily energy turnover. However,
within taxa, desert species have comparatively low water use
effectiveness values.

INTRODUCTION

Water is an essential ingredient of animal life. The availability of water in
the environment differs dramatically between aquatic habitats as compared with
terrestrial habitats. Deserts are the most arid, and are defined as deserts on
the basis of water availability. One of the major challenges faced by animal
life during the colonization of land has been that of maintaining water balance.
As animals invaded progressively drier habitats, they had to find ways either to
obtain water from their environment, to reduce their water requirements by
restricting water losses, to tolerate temporary water imbalances until water
became available, or to achieve some combination of these feats. The degree to
which various animals have achieved a reduced dependence on water availability is
reflected in the amount of water they process through their bodies each day.
Daily rates of water flux have been measured by means of isotope tracers in
hundreds of animal species. We have collected all available data on water flux,

Comparative Physiology: Life in Water and on Land. P. Dejours, L. Bolis, C.R. Taylor, E.R. Weibel
(eds.) Fidia Research Series, IX-Liviana Press, Padova © 1987

and are analyzing these data by least-squares regressions of log water flux rate
on log body mass (allometry). This paper summarizes some of our preliminary
results and interpretations.

METHODS

Measurements of water flux through intact animals were obtained from the literature. All measurements were made using deuterated or tritiated water, which
traces the movement of water molecules between organism and environment. In
water-breathing animals, water flux consists primarily of diffusion across
respiratory surfaces, with water intake due to feeding and drinking accounting
for a relatively small portion of total flux. In air-breathing animals, diffusion
of water (mostly as vapor) across respiratory surface and integument is much
lower, with ingestion accounting for most water intake. In habitats having low
humidities, such as deserts, water influx consists nearly entirely of water consumed as food and drink, along with water produced in vivo via energy metabolism.

Scaling of water flux was assessed by allometric analysis for animals in the
following taxonomic groups: eutherian mammals, marsupial mammals, birds,
reptiles, amphibians, fishes, arthropods (separated into those that breathe air
and those that breathe water), mollusks and polychaete annelid worms. One of our
goals is to evaluate the water relations of animals living undisturbed in their
natural habitats. Thus, we further separated the data into the two categories of
1) studies done in the laboratory or in outdoor enclosures on captive wild
animals, or on domestic animals under all conditions, and 2) studies done in the
field. Where permitted by available data, field measurements were further divided
into those done on desert animals as opposed to nondesert animals. Analysis of
covariance (ANCOVA) was used to evaluate the statistical significance of
differences between groups.

SCALING IN CAPTIVE AND DOMESTIC ANIMALS

The regression lines for these data sets tend to fall into three groups (Fig. 1).
Water-breathing animals (mollusks, fish, water-breathing arthropods) have the
highest daily water fluxes, air-breathing endothermic vertebrates (eutherian and
marsupial mammals, birds) have water flux rates that are one to two orders of
magnitude lower, and air-breathing ectothermic animals (reptiles and air-breathing
arthropods) have the lowest flux rates. The data sets for annelids and amphibians
are too small to analyze allometrically, but polychaete annelids group near
mollusks. Small amphibians, all of which are aquatic larvae (tadpoles), have

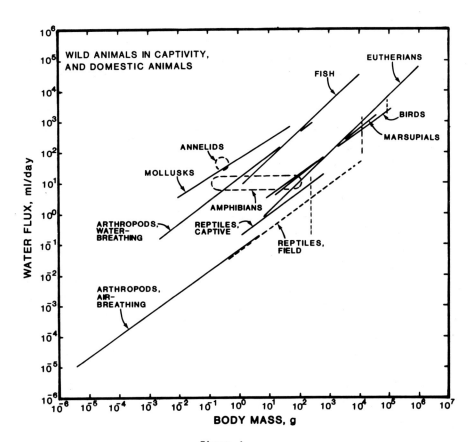

Figure 1

Allometry of water flux in captive wild animals and domestic
animals. Axes are logarithmic, and dashed vertical lines indicate
ranges within species (from left: ground squirrel, coyote, human)

water fluxes similar to those of fish, while the larger, terrestrial adult amphibians have flux rates near those of endotherms. This hierarchy of water flux is expected in view of the large diffusional exchange of water across respiratory surfaces of water-breathing animals, and the much higher energy and food requirements of terrestrial endotherms, compared to terrestrial ectotherms.

The variation of data points around each regression line is quite large. This reflects the wide range of conditions and stresses to which researchers have exposed experimental animals, and illustrates the plasticity of water flux rate in individual animals. Within a wide range, one can generate almost any water flux rate one desires in captive animals. The dashed vertical lines in Figure 1 illustrate the variation possible within species, the most extreme of which is represented by a ground squirrel that was normally active in a cold environment at one extreme, and was hibernating at the other extreme.

The regression line for captive reptiles is close to the lower end of the eutherian line. This is primarily due to the preponderance of aquatic and amphibious species (sea snakes, fresh-water snakes, turtles) in that data set. A more representative line is the one for free-living reptiles (dashed line in Fig. 1), which includes xeric as well as mesic and hygric species. Water flux in free-living reptiles scales the same as it does in air-breathing arthropods (ANCOVA, slopes and intercepts do not differ significantly). The reptile taxon is the only one in which water fluxes in the field are lower than those in captive animals. Water fluxes in free-ranging eutherians, marsupials and birds are higher than those in captive animals.

SCALING IN FREE-LIVING ANIMALS

Few field studies have been done on water flux in water-breathing animals. This is understandable because their high fluxes dictate very short measurement periods (minutes or hours), which render field studies difficult and of dubious value. Among terrestrial vertebrates, the regression lines for free-ranging eutherians, marsupial and birds cluster together, and are much higher than the line for reptiles (Fig. 2). The relatively steep slope of the eutherian line is partly due to a preponderance of desert species among the data available for small eutherians.

ANCOVA comparisons of desert and nondesert species within taxa show significant differences in water flux scaling within eutherians, birds and reptiles. (Only two species of desert-dwelling marsupials have been studied, and these data do not yield a significant allometric regression.) Among reptiles and birds, the

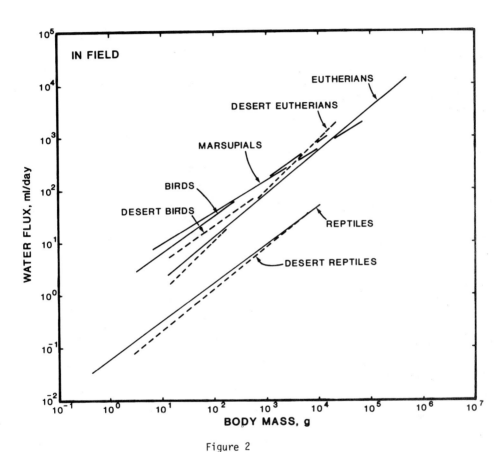

Figure 2

Allometry of water flux in free-living wild animals. Solid lines represent regressions for all species, dashed lines show desert species only (lines for nondesert species not shown)

regression lines for desert species have similar slopes but significantly lower
intercepts than the lines for nondesert species, indicating that desert reptiles
and birds have lower water flux rates than do similar-sized nondesert species.
(The regression lines for nondesert animals are not shown in Fig. 2.) The
regression for desert eutherians has a significantly greater slope than that for
nondesert eutherians, indicating that small desert eutherians have lower water
fluxes than do their nondesert counterparts. The only large desert eutherian for
which field water flux measurements are available, is the collared peccary
(Dicotyles tajacu; Zervanos and Day 1977), which had very high water fluxes while
eating succulent cactus. Thus, most desert vertebrates have lower water fluxes
than nondesert vertebrates of similar size.

WATER USE EFFECTIVENESS

The results summarized in Figures 1 and 2 indicate that 1) air-breathing animals
have low water fluxes compared with animals that breathe water, 2) among air
breathers, ectotherms have lower water fluxes than endotherms, and desert
eutherians, birds and reptiles have lower water fluxes than their nondesert
counterparts. Do these differences reflect physiological adaptations which
conserve water in progressively drier habitats? We suspect that the difference
between water- and air-breathers is primarily due to morphological properties
(internal lungs and tracheae vs. gills) associated with effective respiratory gas
exchange in air vs. water, rather than physiological water balance adaptations,
and that lower water fluxes in air-breathers are a necessary consequence of the
designs of these respiratory systems, along with the environmental property of
lower water availability in air.

The differences between endothermic and ectothermic air-breathers, and between
desert and nondesert species, may reflect physiological water conservation
adaptations, but these differences could also be mere consequences of the well-
known differences in rates of energy metabolism between these groups. Reptiles
and terrestrial arthropods have much lower metabolic rates than do endotherms,
and desert animals usually have lower metabolic rates than related nondesert
animals (Bartholomew 1982, Schmidt-Nielsen 1983). Animals with low rates of
living, as reflected by low metabolic rates, would be expected to have low water
flux rates as well.

This hypothesis can be tested by comparing metabolic-rate-specific water fluxes
among groups of animals. The ratio of ml water flux per kJ energy metabolized,
which we term "water use effectiveness" (WUE in parallel with a similar ratio

used by plant physiologists), has the advantage of simultaneously normalizing whole-animal water fluxes for differences in energy metabolism due to body mass, taxon and habitat. WUE values indicate the amount of water used per unit of energy processed and they can be analogized to the amount of motor oil used per unit gasoline consumed by motor vehicles. Because water flux in a given animal may vary widely while metabolic rate varies little (as in a heat-stressed dog with ad lib. drinking water versus the same dog without food or water), it is important when making comparisons between species to use only WUE values for animals that are in a steady-state (as indicated by maintenance of constant body mass). WUE values for free-living animals may be readily calculated from doubly labeled water results, by dividing daily water flux rates (measured by tritium or deuterium) by daily rates of carbon dioxide production (measured as the difference between hydrogen and oxygen isotope washout rates) after conversion to energy equivalents in kilojoules. Doubly labeled water measures both water flux and energy flux simultaneously in an individual animal (Lifson and McClintock 1966, Nagy 1980).

Doubly labeled water does not provide accurate measurements of CO_2 production by water-breathing animals because water fluxes are so high relative to CO_2 fluxes that washout rates of the hydrogen isotope and the oxygen isotope are virtually identical. To estimate WUE values for water-breathers, we used metabolic rate values from Prosser (1973) for selected species in which water flux rates have also been measured.

Not surprisingly, water-breathing animals have water use effectiveness values that are more than three orders of magnitude greater than WUEs of air-breathing animals (Fig. 3). Among free-living air-breathing animals, WUEs for endotherms are similar, and ectotherms (reptiles and arthropods) use as much or more water per unit of energy metabolized as do endotherms. Thus, the large difference in whole-organism water flux between endotherms and air-breathing ectotherms is completely accounted for by the difference in energy metabolism. The suggestion that reptiles are better-adapted to arid habitats than are endotherms, because reptiles have comparatively low water fluxes (Fig. 2), appears to be an over-simplification.

Desert-dwelling eutherians, birds and reptiles do have lower WUE values in the field than their nondesert relatives (Fig. 3). The comparatively low WUEs of desert animals suggests that low metabolic rates alone do not account for their low water flux rates. This indicates that desert animals employ, in their natural habitats, at least some of the physiological, behavioral and morpho-

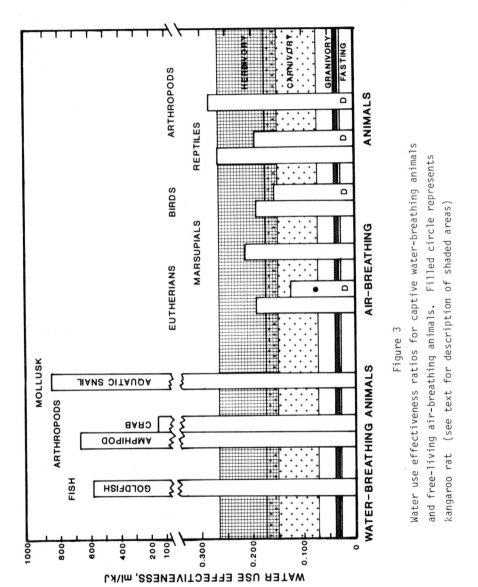

Figure 3

Water use effectiveness ratios for captive water-breathing animals
and free-living air-breathing animals. Filled circle represents
kangaroo rat (see text for description of shaded areas)

logical mechanisms for water conservation that have been abundantly documented
in the laboratory.

A theoretical framework for evaluating WUE values of terrestrial animals may be
derived from knowledge of the chemical composition and digestibilities of various
diets. If an animal consumes just enough food to achieve energy balance, and is
able to maintain water balance on the preformed and metabolic (oxidation) water
provided by that diet without drinking extra water, then the WUE for that animal
is equivalent to the ratio of water yield to energy yield of its diet. We
calculated WUE values for some typical diets, such as vegetation (water contents
of 62 and 72% of fresh mass), animal matter (lipid-rich mealworms and lean whole
vertebrates), and seeds (preformed water contents of 0 and 10% of fresh mass),
using water and energy equivalents in Nagy (1983), as well as for a starving
animal metabolizing body fat alone, and these are shown as shaded areas in
Fig. 3.

Kangaroo rats (Dipodomys merriami), which are able to survive quite well in
captivity with a diet of air-dried seeds (ca. 10% preformed water; Schmidt-
Nielsen and Schmidt-Nielsen 1951), have an average WUE value while living in
their natural habitat (calculated from Mullen 1971) that is twice as high as
expected if they consumed dry seeds alone (filled circle in Fig. 3). Free-
ranging kangaroo rats apparently do not make continuous use of their remarkable
capacity for conserving water. One of the valuable aspects of the WUE ratio is
that it reflects what animals actually do in nature, and such knowledge often
elicits questions and hypotheses for subsequent study. For example, the kangaroo
rat WUE shows that these animals are able to get more water than expected. Do
they consume arthropods or vegetation in addition to seeds, or do they store
seeds in humid places and consume them after they have taken up more water
hygroscopically?

FUTURE RESEARCH

This review uncovered a paucity of information about water flux rate of free-
living snakes, amphibians (the xeric-dwelling Chiromantis and Phyllomedusa are
especially interesting), desert marsupials, and desert and nondesert air-
breathing arthropods. Labeled water studies of these animals would be very
useful. A year-round study of any desert animal would provide valuable insights
into seasonal aspects of water balance, which are poorly understood at present.

ACKNOWLEDGEMENTS

We thank Mark Gruchacz, Brian Henen and Amy Roberts for their help with various aspects of this review, which was supported by U.S. Department of Energy Contract DE-AC03-76-SF00012.

REFERENCES

Bartholomew GA (1982) Energy metabolism. IN: Gordon MS (ed): Animal physiology, Fourth edition. MacMillen Publ. Co., New York; pp 46-93.

Lifson N, McClintock R (1966) Theory of use of the turnover rates of body water for measuring energy and material balance. J Theoret Biol 12:46-74.

Mullen RK (1971) Energy metabolism and body water turnover rates of two species of free-living kangaroo rats, Dipodomys merriami and Dipodomys microps. Comp Biochem Physiol 39A:379-390.

Nagy, KA (1980) CO_2 production in animals: analysis of potential errors in the doubly labeled water method. Am J Physiol 238:R466-R473.

Nagy KA (1983) The doubly labeled water ($^3HH^{18}O$) method: a guide to its use. Univ of Calif Los Angeles Publ No 12-1417.

Prosser CL (1973) Comparative animal physiology, Third edition, WB Saunders Co., Philadelphia pp 182-185.

Schmidt-Nielsen K (1983) Animal physiology, Third edition, Cambridge Univ Press, Cambridge pp 177-222.

Schmidt-Nielsen B, Schmidt-Nielsen K (1951) A complete account of water metabolism in kangaroo rats and an experimental verification. J Cell Comp Physiol 38:165-182.

Zervanos SM, Day GI (1977) Water and Energy requirements of captive and free-living collard peccaries. J Wildl Manage 41:527-532.

WATER DEPLETION AND RAPID REHYDRATION IN THE HOT AND DRY TERRESTRIAL ENVIRONMENT

Amiram Shkolnik and Itzhak Choshniak

Department of Zoology, The George S. Wise Faculty of Life Sciences
Tel Aviv University, Ramat Aviv, Israel

Desert antelopes feed selectively on plants that in addition to energy contain enough water to meet the animals' water requirements. Livestock depend on drinking surface water. Goats and sheep, raised in the desert for economic production, require three times more water than their kindred in the wild. Nevertheless, in order to meet their energy requirements they graze far away from the sparsely distributed water sources. Bedouin goats are watered only once every 2-4 days. When given access to water they gulp it in volumes that often exceed 40% of their body mass. The water voluminously imbibed is stored in the goat's rumen and, by being gradually released into the permeable parts of the animals gut, helps maintain the osmotic stability of the body proper. All body fluid compartments are maintained equally well hydrated during the prolonged grazing period. The goat's kidney responds to the drinking by reducing its blood perfusion and consequently the urinary water loss that is likely to increase in fully hydrated animals. The capacity to store water was also described in sheep but does not seem to be shared by desert ruminants in the wild.

Water balance: Aquatic vs. terrestrial animals.

Dehydration poses a serious threat to life on land. Nevertheless, in certain hyperosmotic aquatic habitats, the drain on the body water of an animal may far exceed that of terrestrial organisms.

In seepages that discharge brine of high salinity into the Dead Sea, schools of Cyprinodont fish, Aphanius dispar, are often found. The salinity of the water surrounding this fish ranges from near that of fresh-water to several times that of sea-water. Aphanius fish have often been observed swimming in the estuary zone of the Dead Sea proper, where the water salinity ranges from 2000 - 5000 mosm/kg H_2O. When maintaining these fish in a saline solution at twice the concentration of sea-water, Skadhauge and Lotan (1974) measured in them a daily drinking rate that exceeded 50% of the fish body mass.

Comparative Physiology: Life in Water and on Land. P. Dejours, L. Bolis, C.R. Taylor, E.R. Weibel (eds.) Fidia Research Series, IX-Liviana Press, Padova © 1987

The Judean and Negev deserts, dry, sun-scorched wildernesses, border the Dead Sea. Camels, sheep, goats and gazelles roam these deserts, constantly exposed to the harsh climate. Nubian ibexes and spiny mice inhabit the barren escarpment only a few metres away from the water where the cyprinodont fish swarm. The daily water turnover rates that we have measured for these animals, in their natural habitat, range between 6% of the animal's body mass (camels, Dorcas gazelles) and 14% (Bedouin goats). In desert rodents, daily water turnover rates ranging between 5-10% of their body mass were recorded.

In both the hypoosmoregulating fish and in the terrestrial mammal, water loss to the environment is a continuous process that never ceases. The fish also replenishes its water loss continuously; the ingestion of the medium in which the fish is submerged never ceases, and its rate matches that at which the water is lost to the environment. On land, however, mammals drink intermittently and in the arid terrestrial environment the interval between drinking bouts is often prolonged. In the desert, even during the extreme heat of the summer, this interval between drinking bouts may be extended over several days. During this prolonged period of water deprivation, mammals may become severely dehydrated.

The capacity to withstand a prolonged water deprivation period lends the desert mammals an ecological freedom to move far away from the water sources that are widely distanced in that area; it determines the animal's chance of survival and often even of thriving in an area where animals dependent on frequent drinking will perish within a few days. The ability of a mammal to withstand dehydration has long been described as an important adaptation to life in the desert (Schmidt-Nielsen et al, 1958; Schmidt-Nielsen, 1964).

Dependence on drinking : goats vs. gazelles

Since 1950, when Schmidt-Neilsen et al. worked out the water balance sheet for the kangaroo rat, we are no longer puzzled by the way that desert rodents manage to balance their water economy without drinking. Desert ruminants are, however, too large to avoid the harsh climate by exploiting protective shelters in the same way as rodents. Indeed, herds of Bedouin goats and small groups of Dorcas gazelles roam the barren deserts of the Middle East during the heat of the day, often at several days walking distance from any water source. How then to these mammals manage to meet their water requirements?

We penned Bedouin goats and Dorcas gazelles in the desert during summer. The animals, all of a similar body mass (18-20 kg), were fed dry alfalfa hay and were watered ad lib once daily. It was found that the goats that were of a body size similar to that of the gazelles consumed, however, three times more water than the gazelles. In agreement with their high water requirement, the overall

Table 1: Body water and water relations:Bedouin goats vs. Dorcas gazelles

	HTO space (% body mass)	Evans Blue space (% body mass)	Daily drinking ml/kg	WTO ml/kg.day	Water loss during 4 days (kg)	Drinking capacity liters
Goat	78.8	6.5	116	141	7	8
Gazelle	67.3	4.4	37	58	2.3	2.5

From Shkolnik et al., 1980 and Maltz and Shkolnik, 1984

water turnover rate of the goats also exceeded three times that of the gazelles. Nevertheless, after three days of water deprivation the goats still maintained their appetites while the gazelles stopped eating altogether. On the fourth day following the last bout of drinking, the body mass of the goats amounted to 70% the initial value, but the animals continued to eagerly consume the dry roughage. Body mass of the gazelles on day 4 amounted to 84% of the initial value, but the animals were in such poor condition that the experiment had to be terminated in order not to jeopardize their lives.

At the termination of the four day water deprivation period, the animals were given free access to water and all regained their initial mass. The goats, however, gulped down a volume of water exceeding 40% of their dehydrated body mass while the gazelles consumed only 15% of their dehydrated body mass.

The limited capacity of Dorcas gazelles to withstand dehydration in face of the extremely frugal water economy in which they exist, surprised us. The enigma was resolved when we realised that gazelles ranging free in their natural habitat may never drink. Field ecologists have reported that during the dry season gazelles move to the arid wadi beds where they browse selectively on three plant species, all containing appreciable amounts of water. The Acacia leaf is most favoured by these gazelles. Knowing the energy requirement of a gazelle from our studies with the penned gazelles, we calculated that this requirement can be met by consuming 1200 g of Acacia leaves per day. The water (performed and oxidative) that a gazelle obtains in this way exceeds the rate of water turnover measured by us in the gazelles penned in the desert. Taylor (1968a) found African antelopes also capable of balancing their water expenditure from the water they gained from their food. Oryx and gazelles survive without drinking even when grazing on dry grass, provided they forage at night when air humidity is high. The dry but hygroscopic plant material absorbs enough moisture from the air to provide all the water required by these antelopes (Taylor, 1968b).

Table 2.Water relations of a Dorcas gazelle grazing on leaves of Acacia tortilis
in the Negev desert of Israel at the end of the summer.

Water turnover rate ml/day	Water gain while consuming 5800 kJ/day in 1250 g Acacia ml/day
800	Pref. = 750
	Met. = 112
	862

The Bedouin goats, as well as the cattle and sheep of the arid zones, have
been selected throughout the ages to be productive in the extreme terrestrial
environment. The water economy of these domesticated animals is by no means as
frugal as that of their wild kindred. Stocked by their owners at high densities,
they cannot afford to be selective feeders. They consume all edible plant
material, irrespective of its dryness. In order to meet their energy requirement
on the meagre desert pasture, they forage far away from the water sources and
drink only once every few days. Nevertheless, even following four days of
continuous grazing in mid summer in the water depleted desert, the body water
content and the plasma volume of the Bedouin goats are still maintained within a
range considered normal for mammals (Fig. 1).

Following their copious drinking, goats contain water in excess of the
volume usually held by other mammals. It is only this quantity of water that is
utilized during their prolonged grazing period away from a water source. Unlike
the gazelle, and the "camels that drink to restore water content rather than in
anticipation of further needs" (Schmidt-Neilsen, 1964), goats store water as if
they do anticipate the prolonged grazing trips ahead of them. Having stored water
in excess, there is no further need to the goats to guard their plasma fluid
level, any more than they guard the level of other body fluid compartments.
Their water loss is indeed equally shared by all compartments.

Gazelles lack the protection of a shepherd or owner to tend them. Storing
water would render the wild antelopes highly vulnerable to predators. Having
economized their water expenditure, they are independent of drinking. However,
when their water intake level is low, the maintenance of their plasma fluid is as
essential to them as it is to the camel (Schmidt-Nielsen, 1964; Siebert and
MacFarlane, 1975) and the other animals that lack the capacity to store water
(Denny and Dawson, 1975; Horowitz, 1984; Zurovsky et al., 1984).

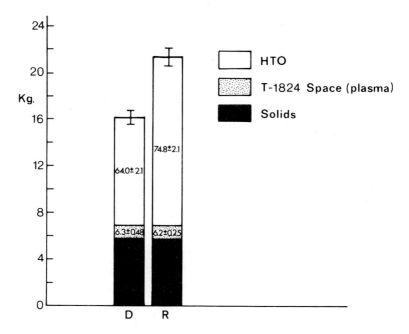

Fig. 1: Body mass, total body water and plasma volume (in kg and as % of body
         mass - assuming body fluid density = 1) in Bedouin goats grazing in the
         desert. D = 4 days following drinking; R = 12 h following drinking (Mean ±
         SD, n = 10).

Table3. <u>Body mass, body solids, total water and plasma volume in dehydrated</u>
         <u>Bedouin goats, Marwari sheep and Dorcas gazelles</u>
              (in % value of the water repleted animal)

	Body mass	Body solids	HTO space	Evans Blue space
goat	71	100	64	64
sheep[*]	80	96	69	69
gazelle	89	109	69	84

----

[*]From Puroit et al. (1972)

Storing water

Unlike the wild antelopes, livestock are dependent on a water supply. They may withstand severe dehydration but ultimately they rely on drinking surface water in order to replenish the water they lose to the environment. Camels, donkeys, sheep and goats, the traditional livestock of the Bedouin in the desert, often graze at several days walking distance from any watering site. Upon arrival at a water-hole they rapidly rehydrate by ingesting, in one drinking bout, a volume of water likely to cause fatal hemolysis in most non-desert mammals (Bianka, 1970; Siebert and MacFarlane, 1975). No ill effect, however, was observed to follow the voluminous drinking of the Bedouin livestock, even when the volume of water thus imbibed exceeded 40% of the goat's dehydrated body mass. Although there has been much fascination regarding the capacity of livestock indigenous to the desert to rapidly rehydrate, the subject has been far less researched than the capacity of these animals to withstand dehydration.

Hecker et al. (1964) in Australia, drew attention to the role played by the rumen in a sheep in serving as a water reservoir that helps maintain the osmotic stability of the body proper. In the Bedouin goat we demonstrated that the water so copiously gulped down by the animal is first retained in its rumen (Choshniak et al., 1984). This compartment basically specialising in maintaining fermentation, is more spacious in desert livestock than in their non-desert kindred and may thus serve also for water storage (Shkolnik and Choshniak, 1984). The rumen can store the imbibed water for an extended period, releasing it only gradually into the other body fluid compartments. During the first hour following drinking by the Bedouin goat, only 1.5% of the total fluid recorded in the rumen at that time flowed out into the permeable sections of the gut posterior to the rumen. A gradual increase in the outflow of the ruminal fluid was later recorded. Nevertheless, as the osmolality in the rumen continuously increased, it posed no threat to the osmotic homeostasis of the body. Plasma osmolality, as well as Na concentration, in the rehydrated goat dropped gradually and slowly. At the same time, a gradual drop in the plasma protein concentration was also recorded, pointing to a continuous volume expansion. The slow changes in the humoral parameters are compatible with the role assigned to the rumen as a water storer.

Kidney function following rehydration

It was highly noticeable that no increase in the urine flow succeeded the voluminous drinking by the goats. As if to help conserve the water ingested, immediately following the drinking the urine flow dropped to rates 50% lower than those recorded in the severely dehydrated animal. The pre-drinking level was barely regained even 4 hours after drinking. A similar drop in urine flow was
Water depletion and rapid rehydration

Table 4. Rumen fluid in Bedouin goats before and after drinking: Volume, osmolality and outflow

Rumen fluid	Before drinking	Time following drinking (h)			
		0-1	1-3	3-5	5-7
Volume[a](ml)	1738	5277	5091	4585	4260
Osmolality[b] (mOsm/kg)	328	91	111	122	131
Outflow of rumen fluid (ml/h)		74	211	240	

From Choshniak et al. (1984).

[a]Volumes extrapolated to the beginning of each time interval.
[b]Average of samples taken at 20 min intervals.

also reported for rehydrated Merino sheep in Australia (Blair West et al., 1972, 1979), for camels and to some extent even in cattle (Siebert and Macfarlane, 1975). In the Bedouin goat as well as in the Merino sheep and the camel, the drop in urine flow was not only more marked than in the cattle but also lasted much longer. A negative ratio of free water clearance to osmotic clearance was maintained in the goat according to the calculation depicted in Table 7 during the 4 hour measuring period that followed the drinking. Such a ratio also points to a substantial contribution of the kidney to the conservation of water in the rehydrated goat.

Table 5. Blood plasma osmolality, Na and protein concentrations in goats dehydrated to 25%-30% of their initial body mass and following drinking to satiety

| | Before drinking | Time following drinking (h)[*] | | | |
		0-1	1-2	2-3	3-4
osmolality (mOsm/kg)	349	343	329	328	320
Na[+](mM)	154.7	156.5	146.2	145.4	145.2
Protein %	100	104	97	94	89

[*]Average of samples taken every 20 minutes.   From Choshniak et al. (1986).

Table6. Urine flow and urine osmotic and Na concentrations in goats dehydrated to 25-30% of their body mass and following drinking to satiety

| | Before drinking | Time following drinking (h)[*] | | | |
		0-1	1-2	2-3	3-4
Flow (ml/min)	0.17	0.09	0.13	0.15	0.22
Osmolality (mOsm/kg)	1771	1643	1518	1518	1515
Na[+] (mM)	80.3	71.6	60.0	21.2	37.7

[*]Average from samples taken at 20 minute intervals.   From Choshniak et al. (1984)

In addition to its contribution to water conservation, the kidney in the rehydrated goat also helps to conserve solutes. Na excretion in the rehydrated goats drops as does osmotic clearance. The drop in the rate of Na excretion, however, considerably exceeds that recorded for osmotic clearance. In the wake of the outstandingly high drinking capacity of the Bedouin goat, the conservation of NaCl by the kidney is of obviously great significance; water can be lost and then regained rapidly. If sodium is retained during dehydration, even to hypernatremic levels, it will be diluted back to normal following the rapid rehydration.

Rapid rehydration and the modulation of kidney function

The changes in endocrine activity that were recorded in the Bedouin goat following its drinking (Wittenberg et al, 1986) were similar to those reported for the rehydrated sheep (Blair West et al., 1972, 1979). Rehydration by

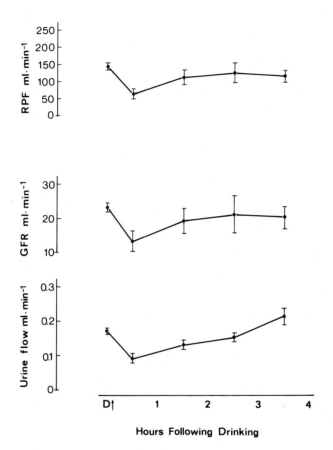

Fig. 2: Renal plasma flow, glomerular filtration rate and urine flow in Bedouin
goats dehydrated to 25-30% of their initial body mass (D) and following
drinking to satiety (- drinking time).

Table 7. Free water clearance, osmotic clearance, Na excretion and filtration fraction in goats dehydrated to 25-30% their initial body mass and following drinking to satiety

| | Before drinking | Time following drinking (h)[*] | | | |
		0-1	1-2	2-3	3-4
CH_2O (ml/min)	-.81	-.55	-.60	-.66	-.74
Cosm (ml/min)	1.14	.75	.74	.93	.87
CH_2O/Cosm	-.71	-.73	-.81	-.71	-.81
FF %	16	20	17	18	18
Na excretion (mM/min)	13.7	6.4	7.8	7.7	8.3

[*]Average from samples taken at 20 minute intervals. From Choshniak et al. (1984)

increasing the animal's total body water, as well as its plasma volume, might be expected to augment the effective renal plasma flow (ERPF) and the GFR and ultimately lead to an increase in the urine flow. In the newly rehydrated Bedouin goat, the 50% drop in the urine flow is emphasized by a similar drop in the GFR and the ERPF (fig.2). It was suggested that a factor participating in modulating the perfusion of the kidney, plays a significant role in bringing about the changes that were recorded in the kidney function.

References

Bianca W (1970) Effects of dehydration, rehydration and overhydration on the blood and urine of oxen. Brit Veterinary J 126:121-132.

Blair West JR, Brook AH, Simpson PA (1972) Renin responses to water restriction and rehydration. J Physiol 226:1536.

Blair West JR, Brook AH, Gibson A, Morris M, Pullan TP (1979) Renin anatidiuretic hormone and kidney in water restriction and water rehydration. J Physiol 294:181-193.

Choshniak I, Wittenberg C, Rosenfeld J, Shkolnik A (1984). Rapid rehydration adn kidney function in the black Bedouin goat. Physiol Zool 57(5):573-579.

Choshniak I, Mittenberg C, Shaham, D (1986) Rehydrating Bedouin goats with saline: Rumen and kidney function. (in preparation).

Denny MJS, Dawson TJ (1975) Effects of dehydration on body water distribtion in desert kangaroos. Am J Physiol 229:251-254.

Hecker JF, Budtz-Olsen DE, Ostwald M (1964) The rumen as a water store in sheep. Australian J Agr Res 15:961-968.

Horowitz M (1984) Thermal dehydration and plasma volume regulation: Mechanism and control. In: Hales JRS (ed): Thermal Physiology. Raven Press, N.Y.

Maltz E, Shkolnik A (1984) Lactating strategies of desert ruminants: the Bedouin goat, ibex and desert gazelle. Symp Zool London 51:193-213.

Puroit GR, Ghosh PK, Taneja GC (1972) Water metabolism in desert sheep. Aus J agric Res 23:685-691.

Schmidt-Nielsen K (1958) Animals and arid conditions: Physiological aspects of productivity and management. In: White GF (ed):The future of arid lands. Washington DC. Am Assoc Adv Sci 43:368-382.

Schmidt-Nielsen K (1952) A complete account of the water metabolism in kangaroo rats and an experimental verification. J Cell Comp Physiol 38:165-182.

Schmidt-Nielsen K (1964) Desert animals: physiological problems of heat and water. Clarendon, Oxford.

Shkolnik A, Maltz E, Choshniak I (1980) The role of the ruminant's digestive tract as a water reservoir. In: Ruckebush Y, Thiven P: Digestive Physiology and metabolism in ruminants. MTP Press; pp.731-741.

Shkolnik A, Choshniak I (1984) Physiological responses and productivity of goats. In: Yousef MK (ed): Stress physiology in livestock. CRC; pp.39-57.

Skadhauge E, Lotan R (1974) Drinking rate and oxygen consumption in the euryhaline teleost Aphanius dispar in water of high salinity. J exp Biol 60:547-556.

Siebert BD, Macfarlane WV (1975) Dehydration in desert cattle and camels. Physiol Zool 48:36-48.

Taylor CR (1968a) The minimum water requirements of some East African bovids. Symp Zool Soc London 21:195-206.

Taylor CR (1968b) Hygroscopic food: A source of water for desert antelopes. Nature; London 219:181.

Wittenberg C, Choshniak I, Shkolnik A, Thurau K, Rosenfeld J (1986) Effect of dehydration and rapid rehydration on renal function and on plasma renin and aldosteron levels in the black Bedouin goat. Pflügers Archiv 406:405-408.

Zurovsky Y, Ovadia M, Shkolnik A (1984) Conservation of blood plasma fluid in hamadryas baboons after thermal dehydration. J Appl Physiol 57:768-771.

Part III

**NUTRIENT ABSORPTION:
AQUATIC VS TERRESTRIAL**

NUTRIENT TRANSPORT
ACROSS THE INTEGUMENT OF INVERTEBRATES

Stephen H. Wright

Department of Physiology
University of Arizona
Tucson, AZ 85724, U.S.A.

The capacity to accumulate dissolved organic material (DOM) from sea water directly into cells of the integument is a ubiquitous feature of "soft-bodied" marine invertebrates, though similar phenomena seem less wide-spread in freshwater species. These processes are saturable, show structural specificity, and are capable of accumulating material from nanomolar external concentrations. Rates of net amino acid uptake into adult and larval molluscs, from ambient levels representative of the environment of these animals, suggest that integumental transport of DOM can play a significant role in animal nutrition. The regulation and mechanism of epidermal transport pathways, as well as the distribution of DOM in the environment, will be major areas of interest in the future.

INTRODUCTION

The ability of aquatic organisms to accumulate dissolved organic material (DOM) has been the source of speculation and study since the turn of the century (see Jørgensen, 1976, for an excellent review of the early work in this area). The "modern" era of interest in the role of DOM in animal nutrition was stimulated by the observations made in the early 1960's by Stephens and his coworkers (see Stephens, 1972). Using radiolabeled substrates they showed that the capacity to accumulate DOM from sea water via carrier-mediated processes was a common property of the integument of "soft-bodied" marine invertebrates. In fact, only marine arthropods appear to lack this facility, though evidence for integumental DOM transport by freshwater invertebrates remains incomplete (Stephens, 1972).

In the 1970's interest turned to whether the uptake of radiolabeled DOM was indicative of a net accumulation of substrate; i.e., does the efflux of endogenous, non-labeled material exceed the influx of labeled substrate? Stephens (1975) made use of the then recently introduced reagent, fluorescamine, to demonstrate a net uptake of naturally occurring primary amines (PA's) into marine polychaetes. Shortly thereafter Stephens and I (Wright and Stephens, 1977) showed that the net accumulation of PA's into marine mussels was closely approximated by the rate of influx of radiolabeled amino acid. Subsequent studies using a variety of different

Comparative Physiology: Life in Water and on Land. P. Dejours, L. Bolis, C.R. Taylor, E.R. Weibel (eds.) Fidia Research Series, IX-Liviana Press, Padova © 1987

animals consistently showed that the "epidermal" transport processes of the general body wall can effect a net accumulation of substrates from micromolar concentrations in surrounding sea water. However, observations that some PA(s) was lost from animals (Bayne and Scullard, 1977; Wright and Stephens, 1978) still left the "net flux" question in doubt.

Over the last several years questions concerning the net flux of DOM, particularly amino acids, into marine invertebrates have been largely resolved. Extremely sensitive chromatographic procedures now permit the quantitative and qualitative analysis of nanomolar levels of amino acid. Using such techniques several laboratories have documented the ability of animals such as marine mussels and echinoderms to accumulate dissolved free amino acids (DFAA's) from external concentrations of less than 10 nanomoles per liter, well below the lower limit of concentrations found in natural waters.

There can be no doubt of the general ability of marine invertebrates to accumulate DOM from external solution. Today's issues are concerned with the quantitative role of such processes in animal physiology. In the present review I will consider the current status of DOM uptake, particularly the uptake of DFAA, in the nutrition of bivalve molluscs. The interested reader is also directed to recent reviews on epidermal transport in polychaetes (Gomme, 1984) and coelenterates (Schlichter, 1984).

AMOUNT AND NATURE OF DOM IN THE OCEAN

DOM represents an enormous pool of reduced carbon. However, the volume of the ocean is such that DOM concentrations are quite low; dissolved organic carbon is on the order of 0.5-2 mg/l (Williams, 1975). Less than 10% of this has been identified chemically. DFAA's average approximately 0.5 to 2% of the total, but the existence of sensitive chemical assays for amino acids, combined with their importance in invertebrate metabolism, has made the uptake of DFAA's the focus of interest in this area.

The concentration of amino acids in marine waters is extremely dependent upon the environment in question. Recent reports of DFAA levels in near shore waters range from as high as 1-3 μM (e.g., Siebers and Winkler, 1984) to a low of between 10 and 100 nM (Henrichs and Williams, 1985). This range quite likely represents the actual variability of this parameter in the marine environment. Despite their low concentration, DFAA's nevertheless represent a nutrient source in near shore waters of the same general size as the pool of phytoplankton available to filter-feeders (Parsons, 1975).

NUTRITIONAL ROLE OF INTEGUMENTAL DOM TRANSPORT IN MARINE BIVALVES

In the 1960's and 1970's a number of studies documented both the presence and general characteristics of epidermal DOM transport in marine invertebrates (reviewed by Jørgensen, 1976, and Stewart, 1979), including a number of reports on

DFAA uptake in bivalve molluscs (Wright, 1982). The potential impact of organic solute exchange across epidermal surfaces is particularly apparent in the case of marine bivalves. The filter-feeding habit exposes these animals to large volumes of substrate-containing media. However, the same sea water serves as a "sink" into which endogenous material can be lost. Therefore, bivalve epidermal tissues offer excellent model systems for the study of both the role and mechanism of integumental uptake.

Péquignat (1973) first reported that the gill is the primary site of accumulation of glycine and glucose in a bivalve, i.e., the mussel, Mytilus edulis. Isolated gill tissue subsequently became the routine experimental system for the study of epidermal transport in bivalves, though it is worth noting that other epidermal tissues, such as the mantle, can be responsible for at least 20-30% of whole animal uptake (Jørgensen, 1983; Wright and Secomb, 1984). Unfortunately, alterations in normal ciliary activity which arise upon isolation of the gill result in the formation of "unstirred layers" above the transporting surfaces. Consequently, the apparent Michaelis constants (i.e., medium concentrations resulting in half-maximal uptake; K_t's) of transport processes in the isolated gill are larger than the "effective" values noted in intact, actively pumping animals (Wright and Stephens, 1978). Nevertheless, the information from studies with gill tissue, when combined with data from whole animal studies, has resulted in a relatively detailed picture of both the mechanism of epidermal transport, and the potential role of these processes in the physiology of several species of bivalve.

Kinetics of Transport in Bivalves - Amino acid uptake in gill tissue from several species of mussel, clam, and oyster, is saturable and described by Michaelis-Menten kinetics (see Stewart, 1979). There are several, separate transport pathways with specificities that favor the transport of distinct structural classes of amino acid. Neutral (zwitterionic), α-amino acids appear to share one or possibly two distinct pathways (Stewart, 1979; Wright, 1985), while separate pathways exist for i) anionic, α-amino and ii) neutral, β-amino substrates (Wright, 1985). Maximal rates of uptake (J_{max}'s) into gills are large compared to those reported for other epidermal systems, probably stemming from the favorable surface-to-volume ratio of gills compared to, for example, the body wall of an annelid worm. Typical values for J_{max} range from 7 to 35 μmol/(g wet gill wt-hr) (Wright, 1985, and unpublished observations). For comparative purposes, this can be expressed per unit of "nominal" area of the gill (i.e., without the influence of apical microvilli): 4 to 20 x 10^{-12} mol/(cm^2-sec). On a "mammalian" standard, such values would be representative of "low capacity" transport processes; e.g., the area specific J_{max} for proline transport in mouse intestine is about 9 x 10^{-9} mol/(cm^2-sec) (Karasov, personal communication). However, as emphasized below, epidermal transporters have a "high affinity" for substrate (i.e., comparatively low values for K_t). Given the low metabolic rates of bivalves, the result is a process with a potential for acquiring substrate at rates which are of nutritional significance to these animals.

The Michaelis constant provides a measure of the apparent affinity of a transporter for substrate, and therefore is indicative of the extent to which transport capacity can effectively be realized. Studies with intact animals (Jørgensen, 1983; Wright and Stephens, 1978), and recent work with isolated gill tissue in which the influence of unstirred layers was reduced by in vitro activation of lateral cilia (Wright and Secomb, 1984; Wright, 1985) have found that the apparent K_t's for amino acid uptake into the working gill are on the order of 2-9 μM. The "true" K_t's of these transporters are expected to be even lower, ranging from 0.2 to 5 μM (Gomme, 1982; Wright and Secomb, 1984). This phenomenon stems from the influence on kinetic measurements of the unidirectional, laminar flow of water between gill filaments; gradients of substrate concentration form as a result of the transport activity of the "upstream" transport sites (Wright and Secomb, 1984). However, as stressed by Jørgensen (1983), the apparent K_t measured in the intact animal is the parameter that applies to estimates of the ability of epidermal uptake to accumulate nutrients from the surrounding water.

Influx vs. Efflux: Net Amino Acid Uptake in Bivalves - Though it was evident from studies measuring net amino acid flux by monitoring clearance of "fluorescamine positive material" (FPM) that epidermal uptake could result in a net accumulation of substrate, it was also apparent that FPM was lost from mussels to sea water (e.g., Wright and Stephens, 1978). There was speculation that such a loss could represent a significant caloric loss from mussels (Bayne and Scullard, 1977). More recent work involving the direct chromatographic determination of amino acids in sea water has demonstrated, however, that the FPM lost from mussels in these earlier studies was probably ammonia, and epidermal uptake processes have been shown to produce a net accumulation of amino acids from external concentrations of 10 nanomoles per liter and less (e.g., Manahan et al., 1983). The evidence now suggests that a net accumulation of substrate can be expected to occur from any concentration of DFAA that marine mussels are likely to encounter in the environment.

Nutritional Potential of DOM Uptake by Bivalves - The obvious approach for a direct assessment of the role of epidermal uptake in bivalve nutrition would be to monitor growth of animals in an environment free of particulate foodstuffs, but with added supplements of DOM at concentrations representing a reasonable range of those measured in the environment. Unfortunately, such an approach has not been possible in studies with adult organisms because of complications arising from the inevitable presence of contaminating microorganisms. Recently, progress has been made using direct methods to assess the role of DOM in larval nutrition, as I will discuss below.

Estimates of the nutritional potential of DOM in adult invertebrates have been primarily based on comparisons of i) the calculated rate of net input of DOM through epidermal pathways, to ii) estimates of the need of animals for substrate to support measured metabolism. The most widely used measure of metabolism has been oxygen consumption (Q_{O_2}), and recent evidence indicates that during active ventilation Q_{O_2} does reflect total energy metabolism in bivalves as determined by direct calorimetry (Hammen, 1983). Q_{O_2} does not reflect the nutrient needs required to support growth or reproductive costs. Nevertheless, if a source of nutrients can provide a significant fraction of an organism's requirement for oxidizeable substrate, it should be considered in any assessment of the animal's energy budget. Conversely, if the input can only provide a few per cent of the animals oxidative needs, it is not likely to play an important role in supporting whole animal nutrition.

The most detailed case concerning the nutritional potential of DFAA uptake by a bivalve can be made for Mytilus. The kinetics of uptake are well-established (e.g., Wright and Stephens, 1978), and the oxidative requirements of these animals are well-understood (Bayne et al., 1976). For the present purpose, a J_{max} of approximately 9 μmol/(g dry body wt-hr) can be used; this represents an average value based upon the kinetics of transport of different structural classes of amino acid (Wright, 1985; corrected for uptake at 15°C), and the qualitative profile of DFAA in sea water. It is instructive to consider this uptake capacity in terms of the delivery of oxidizeable substrate. The complete oxidation of an amino acid requires approximately 1 ml O_2 per 10 μmol of substrate (assuming an average mol. wt. of 100). Thus, at J_{max} epidermal pathways can deliver DFAA's capable of sustaining a Q_{O_2} of 0.9 ml O_2/(g-hr). I stress that this calculation neither implies nor necessitates that all accumulated substrates are funneled into oxidative pathways. It is merely a means to gauge the potential of the epidermal input in terms that offer a general perspective of the process.

The realization of this nutritional potential is, of course, dependent upon availability of substrate in the environment. Figure 1 shows the relationship between the rate of delivery of oxidizeable substrate and environmental substrate concentration. The stippled area depicts the relevant range of total DFAA measured in near shore waters, while the horizontal dashed lines represent the J_{max} and Q_{O_2} of M. californianus (Bayne et al., 1976). Depending upon the use of "standard" or routine" values for Q_{O_2}, net DFAA uptake from a 1 μM concentration can account for between 33 and 77% of the animals needs for oxidizable substrate. Quantitatively similar results are obtained using measured rates of uptake and metabolism in M. edulis (Jørgensen, 1983, and Bayne et al., 1976, respectively). A 1 μM concentration does represent the upper end of that likely to be encountered by Mytilus.

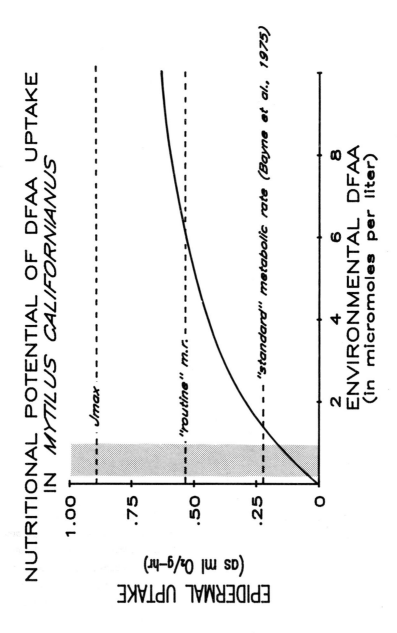

However, exposure to 200 to 700 nM levels of DFAA, well-within the measured limits for near-shore waters, would still support from 8 to 55% of measured O₂ consumption. Though the degree of "significance" such an input represents is subjective, a process capable of sustaining 10 to 70% of an animal's oxidative needs should be considered when developing an energy budget. Furthermore, it is worth stressing that the epidermal input represents an opportunistic approach to the accumulation of foodstuffs. When exposed to DFAA net uptake into mussels will occur at rates which can be significant to whole animal nutrition; when DFAA levels are low, as discussed below, epidermal transporters still play an important role is maintaining intracellular levels of these substrates.

I should emphasize that it is not clear to what extent material accumulated by the gill is translocated to other tissues in the animal. There is speculation that Mytilus gill may be metabolically isolated from the rest of animal (Jørgensen et al., 1986), and there is evidence that the translocation of amino acids from the gill may be very slow (Jørgensen, 1983). If accumulated DFAA's are restricted to the gill, then their contribution to the metabolic needs of this organ may be extreme; Jørgensen (1982) has made such calculations and found that uptake can account for >100% of the gill's Q_{O_2}. However, he suggested that this was an overestimate of the nutritional impact that uptake has in the in situ situation because laboratory measurements of uptake do not reflect a "down regulation" of transport that may occur in nature as a result of the prolonged exposure of organisms to moderate levels of substrate. When mussels are exposed to 2 mM concentrations of amino acid in the laboratory for 24 hrs, there is an apparent down regulation of uptake from micromolar concentrations (Jørgensen, 1983). The meaning of this observation is unclear, though, in light of recent field measurements of amino acid uptake in M. edulis (Siebers and Winkler, 1984) showing that net clearance of naturally occurring DFAA is essentially constant during exposures of at least one week to sea water. Nevertheless, the regulation of epidermal transport processes, as well as the fate of accumulated substrates in animal tissues, remains an important question for future study.

Nutritional Potential of DFAA Transport in Bivalves: Recycling of Endogenous Substrates - Because the concentration of amino acid in gill tissue is so high (>100 μmol/g wet wt), passive losses of these compounds were once thought to represent a significant caloric drain on these animals (e.g., Bayne and Scullard, 1977). Recent studies, however, suggest that rates of amino acid loss to sea water are very low (e.g., Wright and Secomb, 1986). Gomme (1981) has suggested that epidermal transport processes may play a significant role in recovering endogenous substrates lost from surface cells through passive processes. This hypothesis has been supported and extended through the introduction of mathematical models of transport in Mytilus gill which take into account the relationship between gill

structure, fluid movement through the gill, and the kinetics of epidermal transporters in this organ. Calculations using these models indicate that the uptake
capacity of the gill can result in a recovery of between 30% (Wright and Secomb,
1984) and 90% (Gomme, 1982) of the endogenous DFAA lost to the passing water
stream. If the routine loss of taurine, for example, were left unchecked, it
would result in a ≈5% loss of the epidermal pool of this compound per day (Wright
and Secomb, 1986). This would be a large loss of both amino nitrogen and reduced
carbon. By sparing a significant fraction of this loss, epidermal transport i)
sustains the intracellular free amino acid pool and thereby influences the nutritional status of the animal; and ii) maintains cellular levels of osmolytes
involved in cell volume regulation.

Epidermal Transport in Larval Bivalves - Shortly after fertilization, glycine accumulation is activated in Crassostrea gigas larvae (Manahan, 1983a). The velum
becomes the principal site of glycine uptake in pediveliger larvae of C. gigas,
Ostrea edulis, and M. edulis, though following settlement, uptake into the newly
developed gill buds becomes apparent (Manahan and Crisp, 1983). Expressed on a
weight basis, J_{max}'s for uptake are approximately 10-times those observed in adult
bivalves, presumably reflecting differences in surface-to-volume-ratios (Manahan,
1983b). Furthermore, rates of uptake into larvae exceed those occurring into
natural population of marine bacteria (based on an equal weight of material;
Manahan and Richardson, 1983), suggesting that larvae can effectively compete with
microorganisms for the "standing-crop" of DFAA.

Indirect estimates of the effect DFAA on larval nutrition are also complicated
by our poor understanding of the effective concentration of these compounds in
surface water. DFAA levels in the off-shore planktonic habitat are usually
reported to be 20 to 200 nM. However, there is evidence that DOM in the water
column has a very "patchy" distribution, with great spatial variation in concentration being the norm. Smith (1986) found that the concentration of dissolved
organic nitrogen in samples collected almost simultaneously varied by one
thousand-fold over a distance of <2 m, leading to the possibility that free
swimming larvae may be exposed, transiently or perhaps routinely, to higher
substrate levels than "bulk water" samples would suggest. In any event, rates of
amino acid uptake from concentrations that have been measured in the water column
are such that epidermal uptake represents a significant potential source of supplemental nutrition during early stages of bivalve growth and development.

Of particular interest is the recent observation that dissolved substrates can
influence growth of a larval bivalve in axenic culture. Langdon (1983) reported
that shell growth in C. gigas larvae was significantly greater in cultures reared
in the presence of a mixture of DOM compared to control, starved cultures. The
use of axenic bivalve larvae offers an excellent opportunity to make direct,

rather than indirect, assessments of the effect of DOM on bivalve nutrition, and should be an effective approach for future studies in this area.

CONCLUSION

Epidermal transport of DOM is a ubiquitous feature of the integument of soft-bodied marine invertebrates. Rates of net uptake from environmentally relevant concentrations are such that "epidermal inputs" should be considered in the energy budget of many species. In the case of marine bivalves, the process represents an "opportunistic" approach to the acquisition of foodstuffs: when exposed to DFAA, net uptake will occur at rates of nutritional significance; when DFAA levels are low, epidermal transport will still play a key role in sparing the loss of intracellular solutes. Future research will focus on such questions as: i) is epidermal uptake regulated; ii) what is the metabolic fate of accumulated material; iii) what is the role of uptake in the growth and development of organisms (with a particular emphasis on the use of recently developed axenic cultures of invertebrate larvae); iv) what is the distribution of DOM in the specific habitat of organisms under study; and, finally, v) what is the mechanism(s) of this process?

Acknowledgement - Studies from my laboratory were supported by National Science Foundation grants PCM82-16745 and DCB85-17769.

REFERENCES

Bayne BL, Scullard C (1977) Rate of nitrogen excretion by species of Mytilus (Bivalvia: Mollusca). J Mar Biol Assoc UK 57: 355-369.

Bayne BL, Thompson RJ, Widdows J (1976) Physiology: I. In: Bayne, B (ed): Marine Mussels: Their Ecology and Physiology. Cambridge University Press, Cambridge; pp. 121-206.

Gomme J (1981) Recycling of D-glucose in collagenous cuticle: a means of nutrient conservation? J Membrane Biol 62: 47-52.

Gomme J (1982) Laminar water flow, amino acid absorption, and amino acid recycling in the mussel gill. Am Zool 22: 989.

Gomme J (1984) Annelida: permeability and epidermal transport. In: Bereiter, J, Matoltsy, AG, Richards, KS (eds): Biology of the Integument, Vol 1, Invertebrates. Springer-Verlag, New York; pp. 323-367.

Hammen, CS (1983) Direct calorimetry of marine invertebrates entering anoxic states. J Exp Zool 228: 397-403.

Henrichs SM, Williams PM (1985) Dissolved and particulate amino acids and carbohydrates in the sea surface microlayer. Mar Chem 17: 141-163.

Jørgensen CB (1976) August Putter, August Krogh, and modern ideas on the use of dissolved organic matter in aquatic environments. Biol Rev 51: 291-328.

Jørgensen CB (1982) Uptake of dissolved amino acids from natural sea water in the mussel Mytilus edulis. Ophelia 21: 215-221.

Jørgensen CB (1983) Patterns of uptake of dissolved amino acids in mussels (Mytilus edulis). Mar Biol 73: 177–182.

Jørgensen CB, Møhlenberg F, Sten-Knudsen O (1986) Nature of relation between ventilation and oxygen consumption in filter feeders. Mar Ecol Prog Ser 29: 73–88.

Langdon CJ (1983) Growth studies with bacteria-free oyster (Crassostrea gigas) larvae fed on semi-defined artificial diets. Biol Bull 164: 227–235.

Manahan DT (1983a) The uptake of dissolved glycine following fertilization of oyster eggs, Crassostrea gigas (Thunberg). J Exp Mar Biol Ecol 68: 53–58.

Manahan DT (1983b) The uptake and metabolism of dissolved amino acids by bivalve larvae. Biol Bull 164: 236–250.

Manahan DT, Crisp DJ (1983) Autoradiographic studies on the uptake of dissolved amino acids from sea water by bivalve larvae. J Mar Biol Assoc UK 63: 673–682.

Manahan DT, Richardson K (1983) Competition studies on the uptake of dissolved organic nutrients by bivalve larvae (Mytilus edulis) and marine bacteria. Marine Biol 75: 241–247.

Manahan DT, Wright SH, Stephens GC (1983) Simultaneous determination of uptake of 16 amino acids by a marine bivalve. Am J Physiol 244: R832–R838

Parsons TR (1975) Particulate organic carbon in the sea. In: Riley, JP, Skirrow, G (eds): Chemical Oceanogr. Vol 2. Academic Press, New York; pp. 301–363.

Péquignat E (1973) A kinetic and autoradiographic study of the direct assimilation of amino acids and glucose by organs of the mussel Mytilus edulis. Mar Biol 19: 227–244.

Schlichter D (1984) Cnidaria: Permeability, epidermal transport and related phenomena. In: Bereiter-Hahn, J, Matoltsy, AG, Richards, KS (eds): Biology of the Integument Vol 1. Invertebrates. Springer-Verlag, New York; pp. 79–95.

Siebers D, Winkler A (1984) Amino-acid uptake by mussels, Mytilus edulis, from natural sea water in a flow-through system. Helogolander wiss Meeresunters 38: 189–199.

Smith DF (1986) Small-scale spatial heterogeneity in dissolved nutrient concentrations. Limnol Oceanogr 31: 167–171.

Stephens GC (1972) Amino acid accumulation and assimilation in marine organisms. In: Campbell, JW, Goldstein, L (eds): Nitrogen Metabolism and the Environment. Academic Press, New York; pp. 155–184.

Stephens GC (1975) Uptake of naturally occurring primary amines by marine annelids. Biol Bull 149: 397–407.

Stewart MG (1979) Absorption of dissolved organic nutrients by marine invertebrates. Oceanogr Mar Biol Ann Rev 17: 163–192.

Williams PJ le B (1975) Biological and chemical aspects of dissolved organic material in sea water. In: Riley, JP, Skirrow, G (eds): Chemical Oceanogr. Vol 2. Academic Press, New York; pp. 301-363.

Wright SH (1982) A nutritional role for amino acid transport in filter-feeding marine invertebrates. Am Zool 22: 621-634

Wright SH (1985) Multiple pathways for amino acid transport in Mytilus gill. J Comp Physiol B 156: 259-267.

Wright SH, Secomb TW (1984) Epidermal taurine transport in marine mussels. Am J Physiol 247: R346-R355.

Wright SH, Secomb TW (1986) Epithelial amino acid transport in marine mussels: role in net exchange of taurine between gills and sea water. J Exp Biol 121: 251-270.

Wright SH, Stephens GC (1977) Characteristics of influx and net flux of amino acids in Mytilus californianus. Biol Bull 152: 295-310.

Wright SH, Stephens GC (1978) Removal of amino acid during a single passage of water across the gill of marine mussels. J Exp Zool 205: 337-352.

NUTRIENT TRANSPORT BY INVERTEBRATE GASTROINTESTINAL ORGANS AND THEIR DIVERTICULA

Gregory A. Ahearn

Department of Zoology, 2538 The Mall,
University of Hawaii at Manoa, Honolulu,
Hawaii 96822, U.S.A.

Intestines of Echinodermata, Mollusca, and Arthropoda exhibit
carrier-mediated transport processes for amino acids and sugars
which display ion cotransport, energy-dependency, sensitivity to
phloridzin, and competitive inhibitory effects of structurally
similar substances. Digestive diverticula in each group
structurally appear to be major sites of nutrient absorption.
Crustacean hepatopancreatic brush border membrane vesicles
exhibit at least 3 distinct carrier mechanisms for amino acids
and sugars which are sensitive to pH and respond to the
transmembrane Na gradient or membrane potential.

I. Introduction

The gastrointestinal tracts of multicellular organisms are often complex
and variable in structure and exhibit a variety of physiological roles which aid
in the maintenance of stable internal homeostatic conditions as well as participate in the regulation of normal growth and development. Because of medical
implications the most detailed studies of gastrointestinal physiology have been
conducted on mammals. However, over 95% of living metazoan organisms are not
mammals, but are invertebrates which also exhibit highly complex gastrointestinal
organs. It is therefore appropriate to examine, in similar detail, the physiological properties of gut function in other animal groups, for only in this way can
a complete understanding of the processes involved in gastrointestinal physiology
be adequately assessed and their possible adaptive significance ascertained.
This review summarizes information concerning organic solute transport by the
gastrointestinal tracts of the three most anatomically complex invertebrate phyla:
Echinodermata, Mollusca, and Arthropoda. Mechanisms of nutrient transport by
absorptive organs of each animal group are characterized in general terms and
recent information from the author's laboratory about sugar and amino acid
absorption in crustacean gut are discussed in detail.

II. Echinodermata.

Segments of intestine from the sea urchins, Paracentrotus lividus and

Comparative Physiology: Life in Water and on Land. P. Dejours, L. Bolis, C.R. Taylor, E.R. Weibel
(eds.) Fidia Research Series, IX-Liviana Press, Padova © 1987

Echinus esculentus, displayed active transport of glucose, while segments of
stomach wall exhibited both active glucose and galactose transport; fructose was
not actively transported by either gut region (Bamford and James, 1972; Bamford
et al., 1972). Intestinal carrier transport of glucose in Paracentrotus displayed
an apparent binding constant (K_t) of 0.4 mM, while that of galactose was 0.2 mM.
The esophagus, stomach, and intestine of Echinus all demonstrated active L-alanine
transport with the kinetic constants for transapical transfer being related to
expected substrate availability in each gut region (James and Bamford, 1974).
Furthermore, the stomach preparation of this species displayed active transport of
L-alanine, D-alanine, L-leucine, and L-lysine, while L-aspartate was not actively
accumulated (Bamford and James, 1972).

       Isolated digestive glands (pyloric diverticula) of asteroids transport
sugars and a variety of amino acids by mechanisms which are inhibited by the metabolic poison sodium iodoacetate (Ferguson, 1964, 1968). Most of the transport
studies of this complex organ were performed on intact pieces of tissue incubated
in a saline medium containing radiolabelled organic solute. Therefore, uptake of
isotope reflected activities of epithelial basolateral membranes and was analogous
to movements of substrate from coelomic fluid to diverticulum tissues in living
animals. These studies showed that the basolateral surface of starfish digestive
glands was capable of accumulating fifteen different amino acids from the bathing
coelomic fluid. Inhibitory and stimulatory effects of one amino acid on the transport of another suggested that more than one transport mechanism for this group of
compounds existed on the basolateral epithelial pole of this tissue. No studies
have been conducted on apical or transepithelial transport of nutrients in echino-
derm digestive diverticula so, at present, the role of this large organ in sugar
and amino acid absorption is still unclear.

III. Mollusca.

       The intestine appears to be a major site of sugar and amino acid trans-
port among the Polyplacophora (chitons) and the Gastropoda (snails). Lawrence and
Lawrence (1967) demonstrated active transport of D-glucose and 3-o-methyl-D-glucose
in the chiton, Cryptochiton stelleri, which was inhibited by anaerobic conditions.
D-mannose and D-fructose were not actively absorbed by the intestine in this
species. In another study of the same animal, glycine, L-alanine, DL-proline,
AIBA, and DL-lysine were actively transported by lower intestinal segments, while
L-glutamic acid was not (Greer and Lawrence, 1967). As with sugar transport,
anaerobic incubation significantly depressed the extent of amino acid absorption.
Carrier-mediated sugar absorption was characterized for the intestine of the
terrestrial snail Cryptomphalus hortensis, although in this species active solute
transport was sodium-dependent and insensitive to oxygen deprivation (Barber et al.,
1975a,b). The intestine of the marine gastropod, Aplysia californica, actively

transports a variety of sugars and amino acids in conjunction with sodium from
luminal solution to blood (Gerencser, 1981, 1985). In this species the net flow
of organic solutes and sodium is electrically coupled with that of chloride so
that nutrient and salt absorption occur simultaneously.

Many workers regard the hepatopancreas or the caecum as the primary
sites of nutrient absorption in bivalves (Yonge, 1926) chitons (Fretter, 1937),
gastropods (van Weel, 1961) and cephalopods (Bidder, 1957; Boucaud-Camou , et al.,
1976). Much of the evidence for this contention is based on cytology and histochemistry, but several workers have used tracer substances to follow the course
of dietary components through the gastrointestinal tract. Bidder (1957) used
suspensions of carmine dye to demonstrate the absorptive function of the Octopus
liver (i.e., hepatopancreas); van Weel (1961) showed that ingested iron saccharate
accumulated in the epithelium of the snail (Achatina fulica) midgut gland (i.e.,
hepatopancreas), but not in the intestinal epithelium, suggesting a greater role
of the former in absorption; 14C-glycine injected into a living crab before consumption by Octopus vulgaris appeared most rapidly and accumulated to the greatest
extent in cells of the cephalopod caecum, pancreas, and liver, while virtually
none of the isotopic label was recovered from intestinal cells, leading workers
to suggest a minimal role of the latter organ in nutrient absorption in this
animal (Boucaud-Camou et al., 1976).

IV. Arthropoda.

A.   Insecta.

Early studies of nutrient transport in insects indicated that sugars and
amino acids were absorbed in both the tubular midgut and its anterior diverticula
(caeca), with the latter structures transferring the bulk of the solute load to
the blood (Treherne, 1958, 1959). These studies showed that both types of compounds were passively absorbed from lumen to blood down concentration gradients
established by the net flow of ions and water in the same direction. For sugars
diffusion was facilitated by the rapid conversion of monosaccharides, such as
glucose, into the disaccharide trehalose in the blood, a process which maintained
a steep concentration gradient across the gut wall.

More recently, isolated intestinal sheets of lepidopteran larvae (Hyalo-
phora cecropia), clamped in modified Ussing chambers, demonstrated active net
transmural α-aminoisobutyric acid transport from lumen to blood (Nedergaard, 1972).
This active amino acid transport was inhibited by lack of oxygen, was independent
of the active transport of potassium by the tissue, and was highly sensitive to
transmural electrical potential difference.

In the last 6 years a variety of potassium-dependent amino acid transport mechanisms have been disclosed for the brush border membrane of the midgut
epithelium in herbivorous lepidopteran larvae (Hanozet et al., 1980, 1984). The

brush border vesicle preparation from this insect (Philosamia cynthia) also
demonstrated the capacity for significant D-alanine transport and its inhibition
by the L isomer (Hanozet et al., 1984). Furthermore, the D isomer was shown to
exhibit active transmural transport across intact intestinal tissue and its net
flux was strongly reduced by the presence of L-alanine. Whereas all L-amino acids
examined in this species displayed K cotransport, D-alanine transport appeared to
illustrate a strong Na-dependency. The phytophagous diet of this animal is high
in potassium and low in sodium and may explain the occurrence of these unique
cotransport systems.

B.   Crustacea

    1.   Nutrient influxes across intestinal brush border membrane.

        The intestinal apical membrane of the marine shrimp, Penaeus marginatus,
exhibited high affinity, carrier-mediated, sodium-dependent transport of glycine
(Ahearn, 1974). Proline and alanine were competitive and non-competitive inhibitors of glycine influx, respectively. Later studies on this same animal (Ahearn,
1976) examined in detail the role of Na as cosubstrate with glycine in mucosal
membrane transport.

        In the freshwater prawn, Macrobrachium rosenbergii, intestinal 3H-L-
lysine transport across the mucosal epithelial border occurred through two
parallel carrier-mediated processes, a high-affinity system displaying Micheal is-
Menten kinetics and a low-affinity, non-saturable mechanism that had a transfer
rate which was a linear function of luminal lysine concentration (Brick and
Ahearn, 1978). The high affinity process was Na-dependent and inhibited by both
arginine and iodoacetate. The low affinity mechanism displayed homoexchange
diffusion and was Na-independent, inhibited by arginine, and unaffected by iodoacetate. 3H-L-alanine influx across the apical membrane in the perfused freshwater prawn intestine was also Na-dependent (Wyban et al., 1980). 3H-D-Glucose
transport across the intestinal apical membrane of the freshwater prawn displayed
Na-dependent, Micheal is-Menten kinetics with no discernible diffusion component
(Ahearn and Maginniss, 1977). Phloridzin was a potent competitive inhibitor of
glucose influx (K_i = 3.6 x 10^{-3}M), galactose was a weak inhibitor (mechanism
undetermined), and fructose had no evident effect on glucose uptake.

    2.   Transmural transport of organic solutes by crustacean intestine.

        Transmural transport of both 3H-D-glucose and 3H-L-alanine (1 mM) across
the intestine of the freshwater prawn were linear functions of time in both directions after a short 10-min lag period. A significant net flux of labelled sugar
toward the blood was observed, but was only 10% or less of that for sugars and
amino acids in vertebrate intestine (Ahearn and Maginniss, 1977). Unidirectional
transmural fluxes of 3H-L-alanine across prawn intestine were not significantly
different from one another, suggesting the absence of net amino acid movement

across the tissue (Wyban et al., 1980).

3.   Brush border transport of sugars and amino acids by crustacean
     hepatopancreas.

It has been suggested for some time, as a result of histological and
ultrastructural studies, that the hepatopancreas is the major site of nutrient
absorption in crustaceans, with the tubular intestine playing a minor role in many
species (Yonge, 1924; van Weel, 1974; Gibson and Barker, 1979). Until recently,
there have been no direct measurements of solute transfer across either the
hepatopancreatic brush border membrane or the epithelial cell layer to clarify
organ function. In order to directly define apical and basolateral nutrient
transfer properties of crustacean hepatopancreatic cells, membrane vesiculation
techniques developed by mammalian gastrointestinal and renal physiologists were
applied to this invertebrate organ.

a.   Sugar transport by hepatopancreatic BBMV.

Isolated and purified brush-border membrane vesicles (BBMV) were
prepared from Atlantic lobster (Homarus americanus) hepatopancreas (Ahearn et al.,
1985) using a modification of methods developed for mammalian epithelia by Kessler
et al. (1978). Transport studies using these BBMV were conducted using the
Millipore filtration technique of Hopfer et al. (1973). Incubation of BBMV in a
medium containing 150 mM NaCl and 0.1 mM 3H-D-glucose resulted in rapid uptake of
the sugar into the vesicles with a maximal accumulation occurring by 1 min,
followed by a slow reduction in sugar content to equilibrium at 60 min (Fig. 1).
Maximal transient glucose uptake (overshoot) by these vesicles was generally 3 to
4 times the equilibrium values. The overshoot phenomenon of glucose accumulation
was not observed when incubation medium either contained 150 mM KCl, 300 mM
mannitol, or 150 mM NaCl + 0.5 mM phloridzin. Apparent glucose influx rates,
estimated at 15 sec incubations were: 27.6 (NaCl) and 3.9 (KCl) pmol/mg protein/15
sec, suggesting a seven-fold greater entry rate in sodium medium. These results
indicate a clear Na-dependency of glucose uptake by BBMV and its significant
inhibition by the drug, phloridzin.

Gastric contents of crustaceans are acidic (Gibson and Barker, 1979) and
are periodically flushed through hepatopancreatic tubules, where they come into
contact with apical surfaces of absorptive cells. A series of experiments was
conducted to determine the effect of luminal pH on glucose transport by hepatopancreatic BBMV. Vesicles were loaded with mannitol adjusted to pH 6.0, 7.4, and
8.0 and were incubated in media of the same pH containing NaCl and 0.1 mM 3H-D-
glucose. At each pH an Na-dependent glucose uptake overshoot at 1 min incubation
occurred, but the initial rate of uptake at 15 sec and the extent of overshoot
were markedly greater at pH 6.0 than at either of the other two pH conditions
(Ahearn et al., 1985).

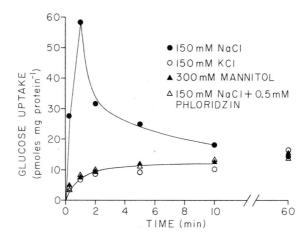

Figure 1.   Time course of 0.1 mM 3H–D–glucose uptake by lobster hepatopancreatic BBMV loaded with 300 mM mannitol at pH 7.4 and incubated in media of various compositions at the same external pH.   After Ahearn et al. (1985).

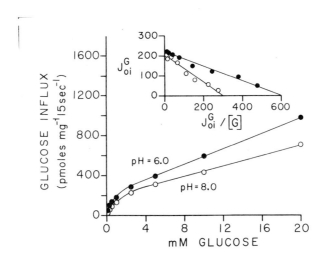

Figure 2.   Influence of bilateral pH conditions on concentration dependence of glucose influx (J_{oi})(15 sec exposures).   Vesicles were loaded with 300 mM mannitol at each pH and were exposed to identical external pH conditions in a 150 mM NaCl solution containing respective 3H–D–glucose concentrations (0.1 to 2.0 mM). Inset is an Eadie-Hofstee plot of calculated carrier influxes after subtracting non-saturable entry.   Half saturation constants and maximal transfer velocities were obtained from the slopes and vertical intercepts, respectively.   After Ahearn et al. (1985).

To ascertain how pH exerted its effect on glucose transport, 15 sec sugar influx as a function of external sugar concentration was assessed at two pH extremes, pH 6.0 and 8.0. As shown in Fig. 2, glucose influx was a curvilinear function of glucose concentration at each pH and was greater at every external sugar concentration at pH 6.0 than at pH 8.0. Glucose influx for both conditions was described as the sum of at least two independent processes operating simultaneously: 1) a Micheal̄is-Menten carrier mechanism illustrating saturation kinetics, and 2) a linear entry system having a rate that was proportional to the external glucose concentration. These two processes acting together at each pH follow the equation:

$$J_{oi} = ((J_M \, [\text{Glu}]) \, / \, (K_t + [\text{Glu}])) + P \, [\text{Glu}] \qquad (1)$$

where J_{oi} is total glucose influx in pmol/mg protein/15 sec; J_M is apparent maximal carrier-mediated influx; K_t is apparent glucose concentration resulting in 1/2 J_M; [Glu] is external glucose concentration, and P is the rate constant of the linear entry process, which can operationally be defined as an apparent diffusional permeability coefficient. Apparent diffusional sugar influx was subtracted from total glucose entry at each external concentration providing an estimate of the carrier transport component at each pH. These calculated carrier-mediated glucose influx values were graphed in Eadie-Hofstee plots in the inset of Fig. 2 to yield estimates of the apparent transport constants K_t and J_M. Results indicated that an alteration in pH from 8.0 to 6.0 stimulated glucose influx because it significantly enhanced apparent sugar binding affinity (reduced K_t) and increased apparent membrane diffusional permeability (P). Maximal glucose influx was virtually unaffected by pH (Ahearn et al., 1985).

   b. Amino acid transport by hepatopancreatic BBMV.

Figure 3 illustrates the results of a series of experiments designed to test the effect of a single transmembrane ion gradient on 0.05 mM 3H-L-alanine uptake by lobster BBMV. Individual cation gradients (H, Na, or K) had no influence on apparent vesicular alanine influxes (15 sec uptake) or the extent of amino acid accumulation against a concentration gradient. In contrast, a transmembrane Cl gradient, directed inward, in the presence of an acidic external environment led to a four-fold increase in initial alanine uptake rate and a transient amino acid uptake overshoot where vesicles contained approximately 2.5 times the equilibrium alanine content. Similar results were obtained with lysine transport across lobster BBMV (Ahearn and Clay, 1986a).

In contrast to the Na-insensitivity of alanine and lysine transport, transmembrane transfer of 0.05 mM 3H-L-glutamate strongly responded to an inwardly-directed NaCl gradient at acidic pH (Fig. 4). At neutral pH conditions, a similar NaCl gradient was ineffective as a stimulant of glutamate transfer (Ahearn and Clay, 1986b).

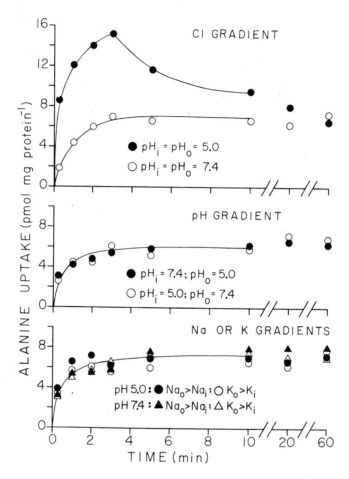

Figure 3. Effects of a single transmembrane ion gradient on the time course of 0.05 mM 3H–L–alanine uptake by lobster hepatopancreatic BBMV. In upper panel vesicles were loaded with 100 mM Na–gluconate at either pH 5.0 or 7.4 and were incubated in 100 mM NaCl at identical external pH. In the central panel vesicles were loaded with, and incubated in, 100 mM NaCl, and had proton gradients directed inward or outward. In the lower panel vesicles were loaded with 100 mM choline chloride at either pH 5.0 or 7.4 and were incubated in 100 mM NaCl or KCl at identical external pH. After Ahearn et al. (1986).

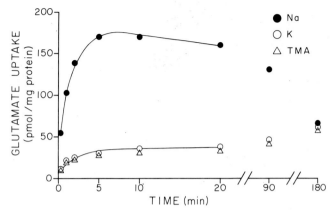

Figure 4. Effects of transmembrane ion gradients on the time course of 0.05 mM 3H-L-glutamate uptake by lobster hepatopancreatic BBMV. Vesicles were loaded with 200 mM mannitol at pH 5.0 and were incubated in media at the same pH containing either 100 mM NaCl, KCl, or TMA-Cl and the labelled amino acid. Each point is the mean of three replicates. After Ahearn and Clay (1986b).

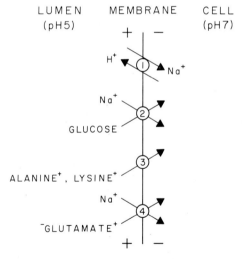

Figure 5. Model for nutrient transport by crustacean hepatopancreatic epithelial brush border membrane. At least 3 distinct carrier proteins for sugars and amino acids reside on this membrane as well as an Na/H antiport mechanism for cation exchange and luminal acidification (Ahearn, 1986). A fifth distinct transport process for Na-independent transfer of amino acids typified by L-leucine may also be present at this location. Direction of arrows indicate proposed mode of transfer for each solute, that is, movement against or down a concentration gradient. Details of this model are discussed in the text. After Ahearn (1986).

Both the transports of L-alanine and L-lysine were electrogenic in the absence of cation gradients at acidic external pH and were stimulated by a trans-membrane electrical potential established either by varying external anions with different rates of membrane diffusion, or in response to a valinomycin-induced K diffusion potential directed-outward (electrically negative vesicular interior) (Ahearn et al., 1986; Ahearn and Clay, 1986a). At acidic pH, L-glutamate trans-port by vesicles loaded with K and incubated in a medium containing Na and the ionophore valinomycin appeared to be electrically silent and was unaffected by the presence of an outwardly-directed K diffusion potential (Ahearn and Clay, 1986b).

pH had a strong effect on the BBMV transports of the three amino acids alanine, lysine, and glutamate (Ahearn et al., 1986; Ahearn and Clay, 1986a,b). In all three cases a reduction in pH from 7.0 to 4.0 significantly increased both initial rates of transport (influx) as well as the extent of amino acid accumula-tion within vesicles against a concentration gradient. It was suggested that a drop in pH protonated amino acid carboxyl groups converting alanine and lysine into Na-insensitive, cationic species responsive to transmembrane electrical potential, and Na-sensitive glutamate into a neutral zwitterion which employed the transmembrane Na gradient for transfer (Ahearn, 1986).

A current model of lobster hepatopancreatic brush border membrane and its distinct sugar and amino acid transporters and Na/H antiporter is shown in Fig. 5 (Ahearn, 1986). At least two, possibly three, distinct brush border amino acid carrier processes were characterized for lobster BBMV on the basis of cis inhibition and trans-stimulation experiments. External L-lysine strongly inhibit-ed the uptake of 3H-L-alanine by BBMV and when the former amino acid was added to the interior of these vesicles the transfer rate of labelled alanine was markedly stimulated (Ahearn et al., 1986), suggesting that both amino acids may share a common mode of carrier transfer. In another study measuring 3H-L-lysine influx into lobster BBMV, both external alanine and arginine proved to be strong competi-tive inhibitors of labelled amino acid transfer, while L-leucine was a non-competitive inhibitor of lysine entry (Ahearn and Clay, 1986a). These results suggested that two different Na-independent, carrier-mediated transport mechanisms may reside on hepatopancreatic BBMV, one serving alanine and lysine and the other used by amino acids typified by L-leucine. In addition, a third distinct amino acid carrier process which exhibits Na-dependent transport of 3H-L-glutamate also appeared to occur in these membranes (Ahearn and Clay, 1986b).

V. Acknowledgements

This work was funded by NSF grant no. DCB85-11272 which is gratefully acknowledged.

VI. References.

Ahearn, G. A. (1974) Kinetic characteristics of glycine transport by the isolated midgut of the marine shrimp, Penaeus marginatus. J. Exp. Biol. 61: 677–696.

Ahearn, G. A. (1976) Co-transport of glycine and sodium across the mucosal border of the midgut epithelium in the marine shrimp, Penaeus marginatus. J. Physiol. 258: 499–520.

Ahearn, G. A. (1986) Nutrient transport by the crustacean gastrointestinal tract: recent advances with vesicle techniques. Biol. Rev. (submitted for publication).

Ahearn, G. A. and Clay, L. P. (1986a) Membrane-potential-sensitive, Na-independent lysine transport by lobster hepatopancreatic brush border membrane vesicles. J. Exp. Biol. (submitted for publication).

Ahearn, G. A. and Clay, L. P. (1986b) Na-dependent glutamate transport by lobster hepatopancreatic brush border membrane vesicles. J. Exp. Biol. (submitted for publication).

Ahearn, G. A., Grover, M. L., and Dunn, R. E. (1985) Glucose transport by lobster hepatopancreatic brush border membrane vesicles. Am. J. Physiol. 248: R133–R141.

Ahearn, G. A., Grover, M. L., and Dunn, R. E. (1986) Effects of Na+, H+, and Cl- on alanine transport by lobster hepatopancreatic brush border membrane vesicles. J. Comp. Physiol. B, 156 (in press).

Ahearn, G. A. and Maginniss, L. A. (1977) Kinetics of glucose transport by the perfused mid-gut of the freshwater prawn, Macrobrachium rosenbergii. J. Physiol. (Lond.) 271: 319–336.

Bamford, D. R. and James, D. (1972) An in vitro study of amino acid and sugar absorption in the gut of Echinus esculentus. Comp. Biochem. Physiol. 42A: 579–590.

Bamford, D. R., West, B., and Jeal, F. (1972) An in vitro study of monosaccharide absorption in echinoid gut. Comp. Biochem. Physiol. 42A: 491–500.

Barber, A., Jordana, R., and Ponz, F. (1975a) Effect of anaerobiosis, dinitrophenol and fluoride on the active intestinal transport of galactose in snail. Rev. Esp. Fisiol. 31: 125–130.

Barber, A., Jordana, R., and Ponz, F. (1975b) Sodium dependence of intestinal active transport of sugars in snail (Cryptomphalus hortensis Muller). Rev. Esp. Fisiol. 31: 271–276.

Bidder, A. M. (1957) Evidence for an absorptive function of the "liver" of Octopus vulgaris Lam. Pubbl. Staz. zool. Napoli 29: 139–150.

Boucaud-Camou, E., Boucher-Rodoni, R., and Mangold, K. (1976) Digestive absorption in Octopus vulgaris (Cephalopoda: Octopoda). J. Zool. Lond. 179:261–271.

Brick, R. W. and Ahearn, G. A. (1978) Lysine transport across the mucosal
    border of the perfused midgut in the freshwater prawn, Macrobrachium
    rosenbergii. J. comp. Physiol. 124: 169–179.

Ferguson, J. C. (1964) Nutrient transport in starfish. II. Uptake of nutrients
    by isolated organs. Biol. Bull. 126: 391–406.

Ferguson, J. C. (1968) Transport of amino acids by starfish digestive glands.
    Comp. Biochem. Physiol. 24: 921–931.

Fretter, V. (1937) The structure and function of the alimentary canal of some
    species of polyplacophora (Mollusca). Trans. R. Soc. Edinb. 59: 119–163.

Gerencser, G. A. (1981) Effects of amino acids on chloride transport in Aplysia
    intestine. Am. J. Physiol. 240: R61–R69.

Gerencser, G. A. (1985) Transport across the invertebrate intestine, IN: Gilles,
    R., Gilles-Baillien, M. (eds.): Transport processes, iono- and osmoregula-
    tion. Springer-Verlag, Berlin, pp. 251–264.

Gibson, R. and Barker, P. L. (1979) The decapod hepatopancreas. Oceanog. Mar.
    Biol. 17:285–346.

Greer, M. L. and Lawrence, A. L. (1967) The active transport of selected amino
    acids across the gut of the chiton (Cryptochiton stelleri). 1. Mapping
    determinations and effects of anaerobic conditions. Comp. Biochem. Physiol.
    22: 665–674.

Hanozet, G. M., Giordana, B., and Sacchi, V. F. (1980) K-dependent phenylalanine
    uptake in membrane vesicles isolated from the midgut of Philosomia cynthia
    larvae. Biochim. Biophys. Acta 596: 481–486.

Hanozet, G. M., Giordana, B., Parenti, P., and Guerritore, A. (1984) L- and
    D-alanine transport in brush border membrane vesicles from Lepidopteran mid-
    gut: evidence for two transport systems. J. Membrane Biol. 81: 233–240.

Hopfer, U., Nelson, K., Perrotto, J., and Isselbacher, K. J. (1973) Glucose
    transport in isolated brush border membrane from rat small intestine. J.
    Biol. Chem. 248: 25–32.

James, D. W. and Bamford, D. R. (1974) Regional variation in alanine absorption
    in the gut of Echinus esculentus. Comp. Biochem. Physiol. 49A: 101–113.

Kessler, M., Acuto, O., Storelli, C., Murer, H., Muller, H., and Semenza, G.
    (1978) A modified procedure for the rapid preparation of efficiently trans-
    porting vesicles from small intestinal brush border membranes. Their use in
    investigating some properties of D-glucose and choline transport systems.
    Biochim. Biophys. Acta 506: 136–154.

Lawrence, A. L. and Lawrence, D. C. (1967) Sugar absorption in the intestine of
    the chiton, Cryptochiton stelleri. Comp. Biochem. Physiol. 22: 341–357.

Nedergaard, S. (1972) Active transport of α-aminoisobutyric acid by the isolated
    midgut of Hyalophora cecropia. J. Exp. Biol. 56: 167–172.

Treherne, J. E. (1958) The absorption of glucose from the alimentary canal of the locust Schistocerca gregaria (Forsk.). J. Exp. Biol. 35: 297–306.

Treherne, J. E. (1959) Amino acid absorption in the locust (Schistocerca gregaria Forsk.). J. Exp. Biol. 36: 533–545.

van Weel, P. B. (1961) The comparative physiology of digestion in molluscs. Am. Zoologist. 1: 245–252.

van Weel, P. B. (1974) Hepatopancreas? Comp. Biochem. Physiol. 47A: 1–9.

Wyban, J. A., Ahearn, G. A., and Maginniss, L. A. (1980) Effects of organic solutes on transmural PD and Na transport in the intestine of freshwater prawns. Am. J. Physiol. 239: C11–C17.

Yonge, C. M. (1924) Studies on the comparative physiology of digestion. II. The mechanism of feeding, digestion, and assimilation in Nephrops norvegicus. J. Exp. Biol. 1: 343–390.

Yonge, C. M. (1926) The digestive diverticula in the lamellibranchs. Trans. R. Soc. Edinb. 56: 703–718.

NUTRIENT REQUIREMENTS AND THE DESIGN AND FUNCTION OF GUTS IN FISH, REPTILES, AND MAMMALS

William H. Karasov

Department of Wildlife Ecology
University of Wisconsin
Madison, Wisconsin 53706, U.S.A.

Abstract. What changes in gut design and nutrient absorption are associated with differences in daily nutrient intake between fish, reptiles, and mammals? To answer this question I compared glucose and proline uptake rates in 23 species from the three classes using uniform methodology. The uptake rate for the entire gut for these nutrients in mammals was 13 times higher than in fish and four times higher than in reptiles. The main basis for faster absorption in the mammals and reptiles was that the intestine operates at a higher temperature (both mammals and reptiles) and that the area of the intestine is greater (mammals only).

Introduction. This paper is concerned with the design and function of the intestine in its absorptive role in relation to the nutrient requirements of vertebrates. A major difference between fish and terrestrial vertebrates is that the latter tend to have higher daily energy requirements, due possibly to low transport costs in fish, higher body temperatures in ectothermic reptiles and to endothermy in the case of mammals and birds. Predictive relationships for daily energy requirements in kJ/d as a function of body mass in grams, for example, are $0.14 \ g^{0.81}$ in fish at a temperature of 15 °C (NRC 1981), $0.22 \ g^{0.80}$ in free-living iguanid lizards and $7.4 \ g^{0.67}$ in rodents (Nagy 1982). Higher metabolic rates must be fueled by more food. The higher food intake rates of mammals, as compared with reptiles, are possible because transit times of individual meals are about ten times faster in the mammals (Karasov et al. 1986a). Are the higher food intake rates and faster transit times met by a higher absorptive capability of the intestine? We will see that the answer to this question in most cases is yes, and we will attempt to assess quantitatively the basis for this within and between taxa.

A priori we can list several ways in which the intestine could achieve a higher absorptive capability and thereby make possible increased feeding rates: (1)

Comparative Physiology: Life in Water and on Land. P. Dejours, L. Bolis, C.R. Taylor, E.R. Weibel (eds.) Fidia Research Series, IX-Liviana Press, Padova © 1987

higher passive permeability to nutrients; (2) higher affinity carriers; (3)
more active transport sites; (4) greater intestinal surface area at the
macroscopic and microscopic level; (5) the advantage of operating the
intestine at high temperatures all day (as in most mammals and birds) or part
of the day (as in reptiles, amphibians, and some fish) rather than at low
temperatures (as in most fish). Passive permeability to nutrients appears to
be similar in ectothermic reptiles and endothermic mammals (Karasov et
al. 1985a), but there are too few data for fish for comparison and so we will
omit detailed consideration of the first possibility. Because it is also
difficult for technical reasons to establish unambiguously differences between
species in carrier affinity (Karasov et al. 1985b), we will focus on the
latter three factors affecting absorptive capability. A sixth factor which
affects the extent of absorption is the retention time of digesta in the gut
and we shall consider this near the end of the paper.

For simplicity, let us view the apportioning of intestinal nutrient uptake
capacity between physiological, environmental, and anatomic causes as follows
(Karasov et al. 1985a):

$$J_{s,T} = (J_T/X)(X) \qquad\qquad (1)$$

where $J_{s,T}$ is the summed uptake rate over the entire length of the small
intestine at a given temperature T, J_T/X is the uptake rate per unit intestine
(e.g. cm length or cm^2 nominal area) at the temperature, and X is an anatomic
measure of the small intestine. $J_{s,T}$ can be conveniently measured using the
everted sleeve method (Karasov and Diamond 1983) by measuring solute uptake
per cm at saturating concentrations at several positions along the small
intestine and then interpolating uptake rates linearly between successive
positions and summing over the length of the small intestine (Karasov et
al. 1983). Because we are attempting to account for most of the transport
activity in the gut, and because fish and terrestrial vertebrates differ in
their gastrointestinal tract morphology, it is necessary to consider first
where transport occurs in the gut.

Where transport occurs. Intestines of fish are generally a single unsectioned
tube and have transport activity all along the length (Buddington and Diamond
1987a,b). In many terrestrial vertebrates there is a distinct hind gut which
is separated from the small intestine by a valve and which functions in the
conservation of electrolytes and water. The existence of carrier-mediated
uptake of sugars and amino acids in this region is disputed. Should the large

intestine be included in the quantitation of $J_{s,t}$ in terrestrial vertebrates?
Uptake rates in the large intestine have been found to be less than 10% of
those in the small intestine (Ilundain and Naftalin 1981, Karasov et
al. 1985a). In desert iguanas, chuckwallas, and desert woodrats the large
intestine constitutes a third to a half of the total surface area of the gut
but its contribution to $J_{s,T}$ is less than 7% for both glucose and proline
(Karasov et al. 1985a). Thus, the large intestine can be excluded from
quantitation of $J_{s,T}$ because its contribution appears to be negligible.

A modification of the digestive tract which is unique to fish is the presence
of blind diverticula located in the proximal gut adjacent to the pyloric
sphincter. These pyloric caeca are present among certain families of fish,
sometimes within only certain species within a family (Buddington and Diamond
1987a). Although the caeca of some fish species constitute a significant
portion of the post-gastric surface area, their function in digestion and
absorption has been uncertain. Buddington and Diamond (1987a) quantitated the
caecal contribution relative to that of the rest of the intestine in four fish
species in which the caeca contributed 16-70% of the total post-gastric area
of the gut. They found that for glucose and proline the caeca contribute
about the same percentage to uptake as to total gut area. Hence, the caeca
are an adaptation for increasing gut area and in some fish species may be the
most important region of nutrient absorption.

Having identified the regions of the digestive tract where most transport
activity occurs in fish, reptiles, and mammals, we shall now analyze the
patterns of variation in intestinal nutrient uptake as they relate to
metabolic requirements.

<u>Relationships between intestinal summed uptake, body size, and taxa</u>. If
higher metabolic rates and hence feeding rates are met by a higher absorptive
capability of the intestine, then we should find that endothermic mammals have
higher summed uptake rates than ectothermic reptiles and fish, and that within
each group summed uptake rate increases with increasing body mass. Summed
uptake rates for the entire gut (including pyloric caeca when present) in
11 fish species are plotted in Fig. 1 along with summed uptake rates for the
small intestine in nine mammal species and three reptile species. The
measurements on fish were performed at 20 °C, all others at 37 °C, though I
will take into consideration below the effect of temperature. The clearest
illustration of how summed uptake varies among these groups can be made by
adding together for each species the summed uptake rates for glucose and

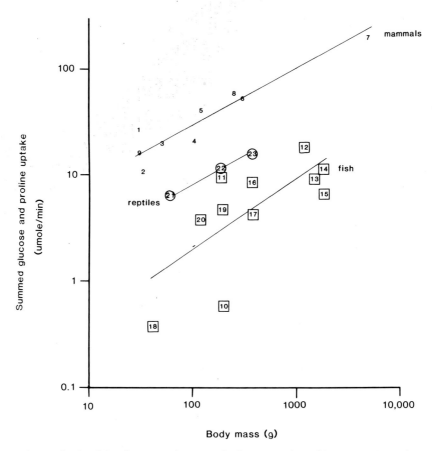

<u>Figure 1</u>. Relationships between the sum of glucose and proline summed uptake
rates and body mass in fish, reptiles, and mammals. For each species summed
uptake for both solutes is at either 25 or 50 mM. The relationships do not
differ significantly in slope. When fitted to a common slope (0.57) the
calculated proportionality coefficients are: fish, 0.17; reptiles, 0.58;
mammals, 2.15. Species designations and sources are as follows: mammals
(Karasov et al. 1985a,b unless otherwise indicated) - 1, mouse; 2, kangaroo
rat; 3, <u>Artibeus</u> fruit bat; 4, hamster; 5, desert woodrat; 6, Belding's ground
squirrel; 7, green monkey; 8, lab rat (Karasov and Debnam 1986); 9, vole
(Karasov and Meyer, unpublished data); fish (Buddington and Diamond 1987a,b
unless otherwise indicated) - 10, opaleye (Karasov et al. 1985b); 11, common
carp; 12, channel catfish; 13, white sturgeon; 14, striped bass; 15, cod; 16,
rainbow trout; 17, largemouth bass; 18, striped bass; 19, grass carp; 20,
tilapia; reptiles (Karasov et al. 1985a) - 21, desert iguana; 22, chuckwalla;
23, box turtle

proline. We might consider this an index of the total absorptive capability of the intestine for sugars and amino acids. The physiological rationale for this is that glucose transport, and possibly amino acid transport too, varies with dietary composition (see the Chapter by JM Diamond in this volume) and so analysis of the transport of only one of these nutrients might introduce a bias when comparing vertebrates with different dietary habits.

Within each taxa summed uptake rate is an increasing function of body mass. The relations are usefully described by allometric equations of the general form $Y = aX^b$ and are linear when plotted on logarithmic coordinates. The small differences in slope are not statistically significant (analysis of covariance, Dunn and Clark 1974) and so data for each taxa are fitted to the common slope of 0.57. When the resultant proportionality coefficients (intercept at unity) are compared, mammals significantly exceed both reptiles and fish, by four times and thirteen times respectively ($p<0.001$). Reptiles significantly exceed fish by about three times ($p<.05$). What is the basis for differences in summed uptake within and between taxa within the context of equation 1?

Physiological, anatomical, and environmental components of intestinal nutrient uptake. To determine whether differences in tissue specific uptake rates (J/X) exist, we can compare uptake rates per cm^2 nominal surface area (Table 1). Comparing uptakes per cm length intestine would be complicated by the fact that luminal diameter, and hence amount of intestine per cm, increases with increasing body size. A complication in making any comparison at the tissue level is that in most species sugar and amino acid transport varies with position (c.f. Karasov et al. 1985a, Buddington and Diamond 1987a). To make our comparisons here, we will use the uptake rates at the position where uptake is highest, usually the proximal or mid intestine.

Plots of uptake per cm^2 as a function of body size (not shown) indicated that there was no dependence of tissue specific transport activity on body size for either glucose or proline transport for any of the taxa over the following body mass ranges: mammals (30-5000 g), reptiles (71-384 g), fish (43-1871 g). Thus, within each taxa the dependence of summed uptake rate on body size (Fig. 1) is probably not due to changes in uptake rate at the tissue level.

Table 1 compares across the three classes of vertebrates glucose and proline uptake rates normalized to cm^2 nominal area. Uptake rates of reptiles do not differ significantly from those of mammals for either glucose or proline.

Uptake rates for proline by fish intestine, however, are significantly lower than for mammal or reptile intestine ($p < .005$ and $p < .01$, respectively, by the t-test). Within fish, glucose uptake rates were significantly higher in the herbivorous/omnivorous species than in the carnivores ($p < .05$), but still below those in herbivorous/omnivorous mammals and reptiles ($p < .001$ in both cases). Notice that the factorial difference between mammals and fish in uptake per cm^2 (ca. 3-5X) does not account for the thirteen-fold difference in summed uptake between mammals and fish. Uptake in fish was measured at 20 $^{\circ}C$, and so Table 1 also compares glucose uptake in the three classes at more similar temperatures. This comparison suggests only a modest difference, if any, between these groups.

Table 1. Comparison of total L-proline uptake and carrier-mediated D-glucose uptake at the tissue level in vertebrates

TAXA	FISH		REPTILES	MAMMALS
Diet	carnivores	herbivore/ omnivore	herbivore/ omnivore	herbivore/ omnivore
PROLINE (nmoles/min, cm^2)				
number of species	5	5	3	8
range	21-266	50-166	158-343	137-500
mean $\pm$ S.E.M.	117 $\pm$ 41	86 $\pm$ 24	271 $\pm$ 57	276 $\pm$ 48
GLUCOSE (nmoles/min, cm^2)				
number of species	5	5	3	8
range	6-30	20-99	276-526	129-489
mean $\pm$ S.E.M.	16 $\pm$ 4	62 $\pm$ 16	357 $\pm$ 85	300 $\pm$ 40
GLUCOSE, AT LOW TEMPERATURE (nmoles/min, cm^2)				
number of species	5	5	2	2
mean $\pm$ S.E.M.	16 $\pm$ 4	62 $\pm$ 16	100 $\pm$ 31	80 $\pm$ 17
temperature ($^{\circ}C$)	20	20	25	20-25
Q_{10} (between 20-37$^{\circ}C$)	-	-	3.4, 2.0	3.4

Uptake rates are at saturating concentrations (i.e. 25-50 mM). Incubation temperatures were 20$^{\circ}C$ for fish and 37$^{\circ}C$ for all others, unless otherwise indicated. Tissues were taken from the region where transport activity was greatest (i.e. the proximal or mid intestine). Sources for data were: Buddington and Diamond 1987a,b; Karasov et al. 1985a,b; Karasov and Debnam 1986).

If there are only modest differences in uptake per cm^2 (i.e. J/X) between these groups, and no differences in uptake per cm^2 among different-sized species within each group, presumably the differences in summed uptake can be attributed to differences in surface area (i.e. X in eq. 1). In order to explore quantitatively this possibility, I summarized in Fig. 2 estimates of nominal surface area (i.e. that of the corresponding smooth tube) in species of varying sizes from the three vertebrate classes. For the mammals and reptiles I drew on several sources (listed in the figure legend) in which measurements were usually made on intestine opened and flattened out. For the fish I took the values reported by Buddington and Diamond (1987a,b), though in their studies they estimated gut diameter differently; as the diameter of a rod giving a good fit to a sleeve of intestine or caecum. (I have found that this latter method yields consistently higher values of nominal surface area than the former method [1.75 $\pm$ 0.11 times higher in two mammal and two reptile species; unpublished data]).

As was the case for summed uptake rate (Fig. 1), intestinal surface area increases with body mass within each class, differences between classes in the mass exponent (i.e. slope) are negligible, and so data for mammals and fish are fitted to the common slope (0.63 $\pm$ 0.04). When the resultant proportionality coefficients are compared, mammals significantly exceed fish by 2.3 times (p<.001). Considering the differences in measurement methodology mentioned above, the actual difference is probably closer to four times. The two reptile species have small intestinal surface areas similar to gut areas of similar-sized fish.

Thus, at least three differences in intestinal transport between endothermic mammals and ectothermic fish and reptiles contribute to higher summed uptake rates in mammals. The least important difference appears to be differences in uptake rate at the tissue level, when uptakes are measured at the same temperature. It appears that mammals may exceed fish, if at all, by a factor of no more than two times in uptake per cm^2 at 20 $^\circ$C, and that they don't exceed reptiles at all. The next most important difference may be that fish intestine operates at a lower temperature than mammal intestine all day, and reptile intestine usually for part of the day (during inactivity when they don't bask). Considering the almost 20 $^\circ$C difference in temperature between fish (15-20 $^\circ$C) and mammals (ca. 37 $^\circ$C) and a Q_{10} for carrier-mediated uptake of 2-3 (Table 1), the temperature difference could account for a four-fold difference in summed uptake. For a reptile thermoregulating at 37 $^\circ$C during daytime and cooling to 24 $^\circ$C at night, the temperature difference could account for a ca. 1.5X difference in summed uptake (Karasov et al. 1986a). An equally important difference between endotherms and ectotherms appears to be

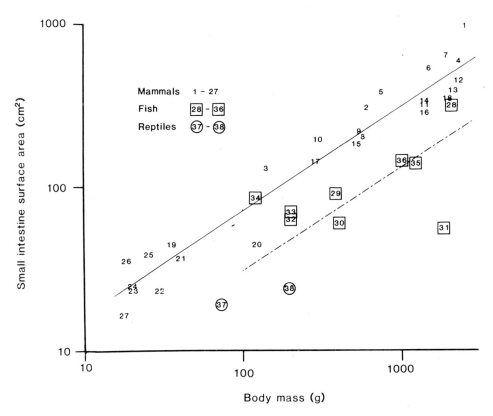

<u>Figure 2</u>. Relationships between nominal surface area of the small intestine
and body mass in fish, reptiles, and mammals. Surface areas for fish include
pyloric caeca when present. Data for each taxa are fitted to a common slope
of 0.63 and the calculated proportionality coefficients are: fish, 1.68;
mammals, 3.94. Species designations and sources are as follows: mammals -
#1-16 domestic and temperate and tropical wild mammal species weighing 43-2100
g from Tables 7 and 8 of Chivers and Hladik (1980); 17, lab rat (Wood 1944);
18, cat (Wood, 1944); 19,20, mouse and woodrat (Karasov et al. 1986a); #21-27,
small mammal species from (Barry, 1976); fish - #28-36 species from
(Buddington and Diamond, 1987a,b); reptiles - 37,38, desert iguana and
chuckwalla (Karasov et al. 1986a).

in amount of intestine. Mammals appear to have about four times more intestinal nominal surface area than fish and reptiles, though this conclusion is based on relatively few data.

The increased metabolic rates associated with increasing body size or associated phylogenetically with the evolution of endothermy involve increased requirements for all nutrients. This increase is met by the evolution of more intestinal tissue, the simplest solution to the problem of absorbing more of everything. The design features of the lung provide an attractive analogy. Increased lung capacity with increasing body size in mammals, or in mammals compared with reptiles, is achieved largely by increasing the surface area of the lung rather than by increasing the gas exchange through each centimeter squared of lung surface (Tenney and Remmers 1963, Tenney and Tenney 1970).

Allometric relations in gastrointestinal tract structure and function. The analysis so far suggests that morphometric analysis is an important area of research in the comparative study of intestinal transport. I have applied allometric scaling procedures in order to understand how morphological parameters such as intestinal surface area are related to nutrient requirements, body size, and so on. (Scaling deals with the structural and functional consequences of changes in size or scale among otherwise similar organisms [Schmidt-Nielsen 1984]). The analysis suggests several areas for future research:

(1) Does the scaling of gut area fully account for the scaling of intestinal nutrient uptake? In this analysis the mass exponents for summed uptake ($M^{0.57}$) and gut nominal surface area ($M^{0.63}$) are close, but the analysis is based on relatively small sample sizes.

(2) Will summed uptake and nominal surface area in amphibians and birds scale in a similar fashion, and will the birds exceed the ectotherms as do the mammals? A recent study of three bird species applying similar techniques to those used here (Karasov et al. 1986b) recorded intestinal glucose and proline uptake rates per cm^2 nominal area which were comparable to those in Table 1. Thus, some simple measurements of nominal surface area in bird species of differing sizes would shed some light on this question.

(3) Why does summed uptake rate not scale with mass raised to the 0.75 power, as does metabolism? One answer might be that nutrient absorption rate does not need to keep pace with metabolism (in contrast to O_2 movement across the

lung which must equal the body tissues' O_2 consumption in order to maintain
aerobic steady state). Instead, many nutrients in a meal are absorbed during
a restricted period of the day (after a meal) and are stored in body tissues
for use throughout the day. The length of that restricted period for
absorption can vary, and so instantaneous maximal nutrient absorption rates
can vary independently of the time-averaged rates over a whole day.

Probably, an important factor to consider is the retention time of food in the
gut. Extraction efficiency can be viewed as follows:

$$\% \text{ absorbed} = \frac{(\text{retention time [min]}) \times (\text{summed uptake rate [moles/min]})}{(\text{quantity of nutrient in gut [moles]})}$$

(eq. 2)

In order for nutrient extraction efficiency to be independent of body size
(which is by no means certain), the following relation should hold:

$$\% \text{ absorbed} \propto M^{1.0} \propto (M^x_{ret} \times M^y_{upt})/M^z_{cap}$$

(eq. 3)

where M^x_{ret} is the manner in which retention time in the digestive tract
scales with body size, M^y_{upt} is the manner in which summed uptake scales, and
M^z_{cap} is the manner in which gut capacity scales. The integration in this
manner of nutrient transport with other aspects of the digestive process
(c.f. Karasov et al. 1986b) would be a challenging but possibly rewarding
exercise.

Acknowledgements. I thank Kate Meurs and Mark Schwalbe for valuable
assistance. Drs. R. K. Buddington and J. M. Diamond graciously shared their
thoughts and their unpublished manuscripts with me. Supported by NSF grant
BSR-8452089, and the Graduate School at the University of Wisconsin.

                                REFERENCES

Barry, RE Jr (1976) Mucosal surface areas and villous morphology of the small
        intestine of small mammals: functional interpretations. J Mammal
        57:273-290.
Buddington RK, Diamond JM (1987a) Pyloric caeca of fish, a "new" absorptive
        organ. Am J Physiol, in press.
Buddington RK, Diamond JM (1987b) Genotypic regulation of intestinal nutrient

transport. J Physiol, in press.

Chivers DJ, Hladik CM (1980) Morphology of the gastrointestinal tract in primates: comparisons with other mammals in relation to diet. J Morph 166:337-386.

Dunn OJ, Clark VA (1974) Applied statistics: analysis of variance and regression, John Wiley and Sons, New York pp.307-335.

Ilundain, A, Naftalin RJ (1981) Na-dependent co-transport of alpha-methyl-D-glucose across the mucosal border of rabbit descending colon. Biochim Biophys Acta 644:316-322.

Karasov WH, Debnam ES (1986) Rapid adaptation of intestinal glucose absorption. Fed Proc 45:537.

Karasov WH, Diamond JM (1983) A simple method for measuring intestinal solute uptake in vitro. J Comp Physiol 152:105-116.

Karasov WH, Pond RS, Solberg DH, Diamond JM (1983) Regulation of proline and glucose transport in mouse intestine by dietary substrate level. Proc Nat Acad USA 80:7674-7677.

Karasov WH, Solberg DH, Diamond JM (1985a) What transport adaptations enable mammals to absorb sugars and amino acids faster than reptiles? Am J Physiol 249:G271-G238.

Karasov WH, Buddington RK, Diamond JM (1985b) Adaptation of intestinal sugar and amino acid transport in vertebrate evolution. In: Gilles R, Gilles-Baillien M (eds): Transport processes, iono- and osmoregulation. Springer-Verlag, Berlin Heidelberg; pp. 227-239.

Karasov WH, Petrossian E, Rosenberg L, Diamond JM (1986a) How do food passage rate and assimilation differ between herbivorous lizards and nonruminant mammals? J Comp Physiol, in press.

Karasov WH, Phan D, Diamond JM, Carpenter FL (1986b) Food passage and intestinal nutrient absorption in hummingbirds. Auk, in press.

Nagy, KA (1982) Energy requirements of free-living iguanid lizards. In: Burghardt FM, Rand AS (eds): Iguanas of the world: their behavior, ecology, and conservation. Noyes Publications, Park Ridge, NJ; pp. 49-59.

National Research Council (1981) Nutrient requirements of coldwater fishes. Natl Acad Sci USA.

Schmidt-Nielsen K (1984) Scaling, why is animal size so important? Cambridge University Press, Cambridge.

Tenney SM, Remmers JE (1963) Comparative quantitative morphology of the mammalian lung: diffusing area. Nature 197:54-56.

Tenney SM, Tenney JB (1970) Quantitative morphology of cold-blooded lungs: amphibia and reptiles. Respir Physiol 9:197-215.

Wood, HO (1944) The surface area of the intestinal mucosa in the rat and cat. J Anat 78:103-105.

INTESTINAL NUTRIENT ABSORPTION IN HERBIVORES AND CARNIVORES

Jared M. Diamond and Randal K. Buddington

Department of Physiology
University of California Medical School
Los Angeles, California 90024
U.S.A.

Herbivores differ from carnivores in intestinal adaptations
for absorbing sugars and amino acids. Among vertebrate
species eating their natural diets, the ratio of amino acid to
sugar transport is highest for carnivores, lower for
omnivores, and lowest for herbivores. These differences arise
in two ways: from reversible adaptation to diet, by induction
and repression of transport proteins, with a delay of hours or
days; and constitutive genetic differences. Species that
experience developmental changes in diet exhibit development
changes in intestinal nutrient transport, those changes being
"hard-wired" in the one case studied so far.

INTRODUCTION

Plants have been widespread for over 350 million years, but it took about 70
million years more before insects began to chew on their leaves (Southwood
1985). Among vertebrates even today, most fish, amphibia, and reptiles
arecarnivores, and herbivory becomes frequent only among birds and mammals.
These facts testify to the difficulties that animals faced in acquiring the
ability to digest plants. Compared to animal matter, plant matter is high in
carbohydrate, low in fat and protein, and poses particular problems of
digestibility associated with fibrous carbohydrates such as cellulose and
lignin.

Two of the digestive adaptations of vertebrate herbivores are well known: a
tendency towards a longer gut than carnivores, and the ability to ferment plant
fiber in specialized fermentation chambers, of which the rumen is the best

Comparative Physiology: Life in Water and on Land. P. Dejours, L. Bolis, C.R. Taylor, E.R. Weibel
(eds.) Fidia Research Series, IX-Liviana Press, Padova © 1987

known and most efficient. These differences between the guts of herbivores and
carnivores are visible, but there must also have been differences at the level
of cell membrane transport mechansims for absorbing nutrients. What
adaptations of these transport systems were required for carnivores to evolve
into herbivores?

MEASUREMENT OF INTESTINAL TRANSPORT

Vertebrate intestines do not absorb ingested complex carbohydrates and proteins
intact. Instead, carbohydrate polymers and protein polymers are first split by
hydrolytic enzymes into smaller units. Intestinal absorption of sugars appears
to be entirely in the form of monosaccharides, while proteins are absorbed in
the form of small peptides as well as of amino acids. Absorption from the
lumen of the intestine is carried out by specialized transport proteins in the
cell membranes of the intestinal mucosal epithelium. Most of these
transporters are active (i.e., they can absorb solutes against concentration
gradients), and most of them are Na^+-dependent. In most vertebrates studied
there is one transporter for glucose and galactose, a separate one for
fructose, and separate transporters for neutral, basic and acidic amino acids
and for imino acids (e.g., proline).

For interspecies comparisons my colleagues and I chose to measure the
intestinal transport of glucose, the most important sugar, and of proline, a
non-essential AA absorbed by a "private" transport protein of its own. We
measure absorption in vitro by a cylindrical segment of intestine that has been
turned inside out and secured over a solid glass rod so that the absorbing
epithelial cells face outside (Karasov and Diamond 1983). After an incubation
of a few minutes carried out under conditions of very rapid stirring, in
solutions to which radioactively labelled nutrients have been added, we measure
the amount of radiolabel taken up into the intestinal mucosa. Thus, our
measurement is of nutrient uptake from the intestinal lumen into the
enterocyte; a separate set of transporters conveys nutrients from the
enterocyte into the blood stream.

How should one express one's measurements of nutrient uptake? Should uptakes
be normalized to the size of the animal, or to the weight or area or length of
the intestine? In studies of the physiological significance of nutrient
transport for the whole animal, the most relevant measure is the intestine's

"total uptake capacity": i.e., the rate of nutrient uptake, integrated over the whole length of the intestine, with nutrient present at a high concentration that saturates the transporter. The ratio of proline transport to glucose transport (P/G ratio) is a convenient index of intestinal adaptations for absorbing protein or carbohydrate.

We have compared 32 vertebrate species from all five higher classes, each species eating its natural diet or a diet simulating the natural one (Karasov, Solberg, and Diamond 1985; Karasov, Buddington, and Diamond 1985; Karasov and Diamond 1985; Diamond, Karasov, Phan, and Carpenter 1986; Karasov, Phan, Diamond, and Carpenter 1986; Buddington and Diamond 1986; Buddington, Chen, and Diamond 1986). For example, among three bird species that we studied, the P/G ratio is highest (5.3) in shrikes, carnivores whose diet is high in protein and very low in carbohydrate; intermediate (1.05) in chickens, omnivores that take in both carbohydrate and protein; and lowest (0.19) in hummingbirds, nectarivores whose diet in effect consists of extremely concentrated sugar solutions. Thus, the intestine of each species is best adapted to absorbing the main type of nutrients in that species' diet.

Fig. 1 summarizes P/G ratios for the 32 species. As in the just-cited example for birds, the ratio is highest in carnivores, intermediate in omnivores, and lowest in herbivores. This pattern is exhibited by all five higher classes of vertebrates. It is still unclear whether there are any differences among species of similar trophic habits from different vertebrate classes.

Most of this variation in P/G ratios arises from species differences in the denominator (glucose transport) rather than the numerator (proline transport). The reason is that all animals require amino acids in order to synthesize their own proteins, but carbohydrates are not nutritionally essential. Only herbivores and omnivores, whose natural diets are high in carbohydrates, have found it worthwhile to make a significant investment in sugar absorption.

PHENOTYPIC ADAPTATIONS TO DIET

For the measurements of Fig. 1, intestines were taken from each species eating its natural diet or a simulated one of similar composition. Thus, Fig. 1 does not indicate whether the P/G ratio is determined proximately or only ultimately by diet.

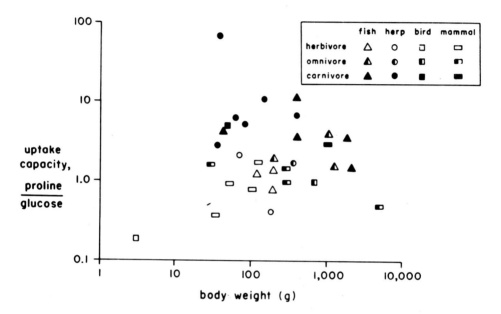

Fig. 1. Relative rates of sugar and amino acid transport in herbivorous, omnivorous, and carnivorous species of all five higher vertebrate classes. Each species was eating its natural diet or else one of similar nutrient composition to the natural diet. The ordinate is the ratio of the intestine's total uptake capacity for proline to its uptake capacity for glucose. This ratio is highest for herbivores, intermediate for omnivores, and lowest for carnivores, and this trend holds true in all five higher vertebrate classes. Herp. = amphibia plus reptiles.

Fig. 2 demonstrates that diet can exert proximate control over intestinal nutrient transporters in mice. The figure depicts results of an experiment in which mice were switched from a no-carbohydrate high-protein ration to an isocaloric high-carbohdyrate medium-protein ration, and then back to the original ration. It is apparent from the figure that sugar transport increases reversibly on the high-carbohdyrate ration, while proline transport increases reversibly on the high-protein ration. Stimulation of each transporter by substrate occurs after a lag of about one day, while down-regulation after removal or lowering of substrate takes several days.

Kinetic analysis shows that these substrate-dependent changes in transporters involve changes in the V_{max} (maximal velocity of transport), not in the K_m (binding constant between transporter and substrate). This suggests that the reversible adaptations of Fig. 2 involve changes in the number of copies of each carrier. For the glucose transporter we were able to test this assumption directly, because the number of copies can be measured by intestinal binding of phlorizin, a competitive specific inhibitor of glucose transport (Ferraris and Diamond 1986). It turned out that the change in transport rate illustrated in Fig. 2 arose entirely from a change in number of copies of the transporter. That is, reversible adaptation to diet involves induction and repression of transport proteins.

Similar tests for reversible adaptation have been made in one carnivorous species, rainbow trout (Buddington 1986), and two omnivorous species, laboratory rat (Scharrer, Wolffram, Raab, Amann, and Agne 1981) and carp (Buddington 1986). Mice are also omnivores. All three omnivores prove capable of reversible regulation of both glucose and proline transport. In contrast, the trout, a strict carnivore, shows no induction of glucose transporters even after six months on a carbohydrate-containing diet. Thus, it may be that only those species whose diets are mixtures of plant and animal foods, and that encounter natural variation in diet composition, are able to regulate their nutrient transporters phenotypically.

GENETIC ADAPTATIONS TO DIET

Since at least omnivores are capable of reversible phenotypic diet-dependent changes in nutrient transporter, are such adaptations the whole explanation for Fig. 1? Because the measurements of Fig. 1 were obtained while each species

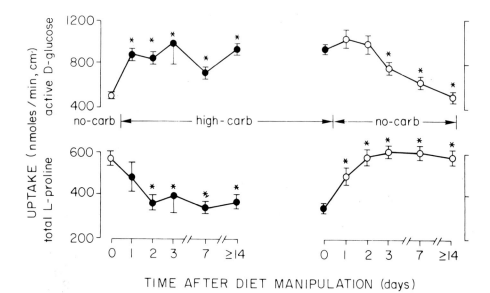

Fig. 2. Reversible phenotypic adaptation of glucose and proline transport to dietary carbohydrate and protein levels, in laboratory mice. At the indicated times mice were switched from a no-carbohydrate, high-protein diet to an isocaloric high-carbohydrate, medium-protein ration and then back again. The ordinate is the rate of active D-glucose uptake (above) or of total L-proline uptake (below) in the proximal jejunum. Note that glucose uptake increases reversibly on the high-carbohydrate ration, while proline uptake increases reversibly on the high-protein (no-carbohydrate) ration.

was eating something like its natural diet, it could be interpreted to mean that there are no fixed genetic differences in intestinal nutrient transporters among carnivores, omnivores, and herbivores. Perhaps, instead, the transporters of each species were just adapting reversibly to the diet that the species was eating at that moment.

In order to assess whether any genetic differences at all contribute to the pattern of Fig. 1, one would like to study a carnivore, an omnivore, and a herbivore while all three were eating the same ration. This experiment is difficult to perform in mammals, because there is no known ration equally acceptable to rabbits and tigers. However, it is possible to maintain fish species of widely differing natural diets in the laboratory on the same artificial ration. Fig. 3 depicts P/G ratios for eight species (two carnivores, two omnivores, three herbivores, and one that switches from carnivory to herbivory as it grows) while eating the same ration. Fig. 3 shows that the carnivores still have relatively the highest P/G ratio, omnivores intermediate values, and herbivores the lowest value. Thus, there are genetic differences among carnivores and herbivores that adapt the former to preferential transport of amino acids, the latter to preferential transport of sugars.

DEVELOPMENTAL CHANGES IN NUTRIENT TRANSPORT

Two of the points in Fig. 3 are half-solid, half-open symbols. These two points refer to the same fish species, the monkey-face prickleback, which underoes a developmental switch from carnivory to herbivory as it grows. The carnivorous small pricklebacks are the ones with the high P/G ratio, while herbivorous large pricklebacks are the ones with the low P/G ratio. Recall, however, that all those pricklebacks in our laboratory were eating the same diet, yet they exhibited size-dependent differences in intestinal nutrient transport. Evidently, the developmental change in nutrient transport is "hard-wired". That is, the switch from carnivory to herbivory is not an immediate external signal that regulates the nutrient transporters. Instead, some internal timing mechanism suppresses proline transport and stimulates glucose transport as the fish approaches the age where under natural circumstances it would switch from carnivory to herbivory.

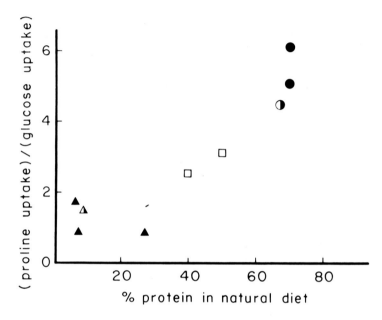

Fig. 3. Ratio of the intestine's total uptake capacity for proline to its uptake capacity for glucose (ordinate), in eight fish species of different natural diets. All species were being maintained in the laboratory on the same artificial diet at the time of the experiments, thereby eliminating proximate effects of species differences in diet. Points for carnivorous species are denoted by solid circles, omnivorous species by open squares, herbivorous species by solid triangles, while the half-solid half-open symbols refer to a species that is carnivorous when young (half-solid half-open circle) and herbivorous when old (half-solid half-open triangle). The abscissa is the protein content of the natural diet. Note that carnivores have relatively higher proline uptake and herbivores have relatively higher glucose uptake, even when proximate effects of differences in diet have been eliminated.

The pattern of change in nutrient transport by the prickleback is clear, and the fact of hard-wiring is also clear. However, the functional significance still admits of two possibilites. One possibility is that, as we have just argued, the nutrient transporters switch in order to adapt the animal to the nutrient content of the natural diet that it will be consuming. However, an alternative interpretation notes that young growing animals, whatever their food source, require more amino acids for growth than do older animals. Thus, the developmental decrease in the P/G ratio for pricklebacks might have evolved because adult pricklebacks no longer need as much amino acid as do young ones. A decisive test between these two theories will require studying a species whose development shift in diet is the reverse of the prickleback's: that is, a shift from herbivory to carnivory. The first hypothesis now predicts that the P/G ratio should increase with age, while the second hypothesis predicts that it should still decrease. The development of herbivorous tadpoles into carnivorous adult frogs provides an ideal test case, and we await the answer with interest.

Buddington RK (1986) Phenotypic adaptation of intestinal nutrient transport to dietary carbohydrate in a carnivore and an omnivore. Submitted to Am J Physiol.

Buddington RK, Chen J, Diamond JM (1986) Genetic and phenotypic adaptations of intestinal nutrient transport to diet. Submitted to J Physiol.

Buddington RK, Diamond JM (1986) Pyloric caeca of fish, a "new" absorptive organ. Am J Physiol, in press.

Diamond JM, Karasov WH, Phan D, Carpenter FL (1986) Digestive physiology is a determinant of foraging bout frequency in hummingbirds. Nature 320: 62-63.

Ferraris RP, Diamond JM (1986) Use of phlorizin binding to demonstrate induction of intestinal glucose transporters. Submitted to J Membrane Biol.

Karasov WH, Buddington RK, Diamond JM (1985) Adaptation of intestinal sugar and amino acid transport in vertebrate evolution. In: Gilles R, Gilles-Baillien MG (eds): Transport processes, iono- and osmoregulation. Springer, Berlin; pp. 227-239.

Karasov WH, Diamond JM (1983) A simple method for measuring intestinal solute uptake in vitro. J Comp Physiol 152: 105-116.

Karasov WH, Diamond JM (1985) Digestive adaptations for fueling the cost of endothermy. Science 228: 202-204.

Karasov WH, Phan D, Diamond JM, Carpenter FL (1986) Food passage and intestinal nutrient absorption in hummingbirds. Auk, in press.

Karasov WH, Solberg DH, Diamond JM (1985) What transport adaptations enable mammals to absorb sugars and amino acids faster than reptiles? Am J Physiol 249: G271-G283.

Scharrer E, Wolffram S, Raab W, Amann B, Agne N (1981) Adaptive changes of amino acid and sugar transport across the brush border of rat jejunum. In: Robinson JWL, Dowling RH, Riecken E-O (eds): Mechanisms of intestinal adaptation. MTP: Press, Lancaster; pp. 123-137.

Southwood TRE (1985) Interactions of plants and animals: patterns and processes. Oikos 44: 5-11.

Part IV

NITROGEN AND SULFUR METABOLISM

ORGANIC OSMOLYTE SYSTEMS: CONVERGENT EVOLUTION IN THE DESIGN OF THE INTRACELLULAR MILIEU

George N. Somero

Marine Biology Research Division
Scripps Institution of Oceanography
University of California, San Diego
La Jolla, CA 92093 U.S.A.

Osmotically-concentrated cells accumulate low-molecular-weight organic solutes (free amino acids, carbohydrates, methylamines and urea) as the major intracellular osmolytes. The basic principles that determine the fitness of osmolytes involve protein-water-solute interactions; the osmolyte systems selected have minimal effects on protein structure. For example, similar osmolyte adaptation patterns are seen in urea-rich marine fishes and in the medulla of the mammalian kidney. In both cases, the perturbing effects of urea are counteracted by the stabilizing effects of methylamines like trimethylamine-N-oxide (TMAO) or glycine betaine.

INTRODUCTION

The phenomenon of convergent evolution has long fascinated biologists. At the anatomical and morphological levels of biological organization one finds case after case of a common type of "solution" adopted to solve a particular type of "problem," be that problem one of vision or locomotion. In this review I will examine a particularly good example of convergent evolution at the biochemical level, one that is characteristic of organisms that span the entire phylogenetic range from prokaryotes to plants to animals. This is the evolution of osmotic solute (osmolyte) systems.

Among virtually all water-stressed organisms only a few types of low-molecular-weight organic solutes, which are nitrogenous compounds or carbohydrates, constitute the bulk of the osmotically active fraction of the intracellular fluids (although usually not of the extracellular fluids). The occurrence of these osmolytes in the intracellular fluids of diverse groups of organisms indicates an independent evolutionary "discovery" of an effective way of dealing with the problems of water stress that has been made numerous times, by diverse organisms.

To analyze the selective processes at work in the evolution of osmolyte systems I will address two principal questions. First, what types of organic osmolytes are accumulated, and by whom? Second, what is the physical-chemical basis for the selection of these particular

Comparative Physiology: Life in Water and on Land. P. Dejours, L. Bolis, C.R. Taylor, E.R. Weibel (eds.) Fidia Research Series, IX-Liviana Press, Padova © 1987

types of osmolytes? This analysis will explain why the osmolytes found in cells have a certain Darwinian fitness, and it will also provide us with some general rules to use in evaluating the potential fitness of newly discovered osmolyte systems such as the one recently described in the cells of the mammalian renal papilla. The discussion of the physical-chemical principles involved in the evolution of the internal milieu will also complement Dr. Clegg's analysis in the following presentation, in which the effects of the "crowded" cytoplasm on the structure of the intracellular water are emphasized.

OSMOLYTE DISTRIBUTION PATTERNS: CONVERGENT EVOLUTION AT WORK

There are two principal circumstances in which high intracellular osmolarity is found. First, water stress imposed by the external environment may lead to the build-up of very high osmolarities in the intra- and extracellular fluids. The environments imposing this stress may be terrestrial habitats where desiccation stress is great, or aquatic habitats where salinity is high. Second, high intracellular osmolarities are found in particular organs involved in transport and excretion. The mammalian kidney is an excellent example, for it must generate a urine that is strongly hyperosmotic to the intracellular fluids of a typical mammalian cell. The mammalian kidney faces a second problem relating to high solute concentrations; one of the major solutes in the kidney is urea, a strong perturbant of proteins that diffuses freely across membranes. In our analysis of osmolyte systems, then, we will need to focus both on total osmotic content and on the effects that specific solutes have on the structures and functions of macromolecules like proteins. To see how these solute-related problems are solved, we must first consider the solute compositions of water-stressed cells.

As the data in Table 1 show, widely different organisms have evolved a common set of adaptations to the problems posed by osmotic stress, whether the stress is due to the desiccating effects of the external milieu or arises from internal conditions as in the mammalian kidney. Except for the strongly halophilic Archaebacteria in which extremely high concentrations (2-3 molal) of potassium ion are accumulated, the cells of osmotically-concentrated prokaryotes and eukaryotes accumulate under water stress the suite of low-molecular-weight organic osmolytes listed in this table.

The organic osmolytes accumulated in these diverse types of cells fall into only three chemical groups: (1) carbohydrates (polyhydric alcohols like glycerol; sugars; disaccharides like trehalose and sucrose, and a variety of other compounds; cf. Borowitzka, 1985). (2) free amino acids and their derivatives (cf. Yancey et al., 1982; Clark, 1985), and (3) urea and methylamines (cf. Yancey, 1985).

TABLE 1. The distribution of osmolyte systems (cf. Yancey et al., 1982;
              Borowitzka, 1985; Somero, 1986).

Carbohydrates (polyhydric alcohols, sugars, disaccharides)

> Cyanobacteria (e.g., glucosylglycerol)
> Fungi (e.g., arabitol)
> Lichens (e.g., mannosidomannitol)
> Unicellular algae (e.g., glycerol, isofloridoside)
> Multicellular algae (e.g., mannitol)
> Vascular plants (e.g., glucose, fructose, sucrose)
> Insects (e.g, glycerol, sorbitol)
> Crustaceans (e.g., trehalose in Artemia salina)
> Amphibians (e.g., glycerol in winter Hyla versicola)
> Mammals (sorbitol and inositol in mammalian kidney medulla)

Amino Acids and Amino Acid Derivatives

> Eubacteria (e.g., glutamic acid, proline, glycine betaine)
> Protozoa (e.g., glycine, alanine, proline)
> Vascular plants (e.g, proline, glycine betaine)
> Marine invertebrates (various amino acids)
> Cyclostomes (various amino acids)
> Amphibians (various amino acids in Bufo marinus)

Urea + Methylamines

> Cartilaginous fishes
> Coelacanth (Latimeria chalumnae)
> Mammals (medulla of kidney)

Inorganic Ions

> Archaebacteria (high K^+ in Halobacterium spp.)

The occurrence of these different groups of osmolytes is clearly not based on the phylogenetic status of the organisms. Amino acid osmolytes are accumulated by halotolerant
eubacteria and most, if not all, marine invertebrates and the hagfishes. Carbohydrate
osmolytes are common in fungi, vascular plants, unicellular algae, and are also utilized by
certain invertebrates and amphibians. Of special relevance for the discussion that follows
are the cases in which the osmotic balance of the cell depends on a mixture of urea and
methylamines like trimethylamine-N-oxide (TMAO) and glycine betaine. This seemingly
odd mixture of the strong protein perturbant, urea, and methylamines is found in the "living
fossil" fish, the coelacanth (Latimeria chalumnae), marine cartilaginous fishes, and, as
recently shown, in the kidney cells of mammals (Balaban and Knepper, 1983; Finley, 1984;
Bagnasco et al., 1986).

These convergent evolutionary trends in the occurrences of the organic osmolytes suggest
an important point: the bases for selecting these osmolytes are certain ubiquitous features
of the chemistry of water-solute-macromolecule interactions that apply in all living
organisms.

SOLUTES EFFECTS--AND NON-EFFECTS--ON PROTEIN STRUCTURE AND FUNCTION

To illustrate the selective factors favoring the accumulation of these organic osmolytes, it is appropriate to consider how they affect or, more to the point, how they fail to affect the structures and functions of the macromolecular constituents of cells.

In categorizing the effects of low-molecular-weight organic osmolytes on macromolecular systems, one can employ a simple dichotomous ranking; the solutes can be viewed either as compatible solutes or as counteracting solute systems. Those solutes that are non-perturbing of protein structure and function, at least at physiological concentrations, are called compatible solutes. The carbohydrate and free amino acid osmolytes belong in this category. A very different phenomenon is found for osmolyte systems containing urea and methylamines. Urea and methylamines both perturb the functional and structural proper-ties of proteins, but they do so in opposite directions and with algebraic additivity. Thus, at the appropriate concentrations ratios (roughly two parts urea to one part methylamines), these are systems of counteracting solutes.

To understand the selective advantages of these solute systems and to lay the groundwork for an understanding of what appears to be an especially complex osmotic system in the medulla of the mammalian kidney, I will consider briefly a few examples of how compatible and counteracting solute systems interact with proteins. We then will appreciate why each of these solute adaptation strategies is an effective way of coping with water stress, and why complex mixtures of compatible and counteracting solutes in the medulla of the mam-malian kidney may be an effective means for generating an osmotically concentrated solution in these strongly water-stressed cells.

COMPATIBLE SOLUTES AND PROTEINS. By definition, compatible solutes lack strong effects on macromolecules, at least at physiologically relevant concentrations of these osmolytes. Compatible solutes can be present at widely varying concentrations in the cell without influencing the functional or structural properties of proteins. Compatible solutes thus are ideal for situations in which water stress varies widely in time, since by employing these non-perturbing osmolytes the cell can raise or lower its osmotic content without disrupting macromolecular activity.

Two examples of solute compatibility with protein function are illustrated in Figure 1. Figure 1A shows how the concentration of glycerol can be varied widely without perturb-ing the activity of the enzyme glucose-6-phosphate dehydrogenase of the halophilic green alga Dunaliella viridis. Only when glycerol concentrations greatly exceed those found in the alga is inhibition noted. In contrast to the effects of glycerol, both NaCl and KCl strongly inhibit the enzyme, showing one reason why inorganic ions are poor candidates for use at high concentrations in most cells (cf. Somero, 1986).

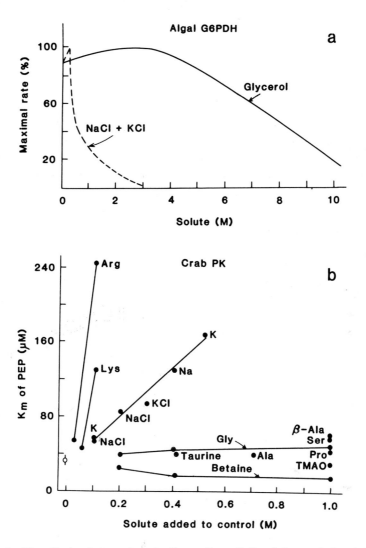

Figure 1. A. The effects of glycerol and salts on the activity of glucose 6 phosphate dehydrogenase of the unicellular alga, Dunaliella viridis (modified after Borowitzka and Brown, 1974). B. The effects of selected free amino acids and salts on the Michaelis-Menten constant (K_m) of phosphoenolpyruvate (PEP) for pyruvate kinase (PK) of the crab Pachygrapsus crassipes (modified after Bowlus and Somero, 1979).

Figure 1B shows the compatibility of certain free amino acids. The enzyme examined was pyruvate kinase (PK) from the marine crab Pachygrapsus crassipes, a species that employs free amino acids as the major intracellular osmotic component. The apparent Michaelis-Menten constant (K_m) of phosphoenolpyruvate (PEP) was not affected by 1M concentrations of the commonly utilized free amino acid osmolytes glycine and proline. NaCl and KCl again are seen to be strongly perturbing of enzymatic activity. Note that not all free amino acids are compatible solutes. The positively charged amino acids arginine and lysine are strongly perturbing of PEP binding to the enzyme. The perturbing effects of arginine and lysine on PK activity, and on the structures and functions of many other proteins (see Bowlus and Somero, 1979; Yancey et al., 1982), reflect another aspect of solute compatibility: most compatible solutes either lack a charge, e.g., the carbohydrate osmolytes, or they bear no net charge, e.g., zwitterionic amino acids like glycine and proline.

Compatibility of certain carbohydrates and free amino acids with protein structure and function has been observed for a number of other proteins from a wide range of species (Yancey et al., 1982). The lack of perturbation by these solutes is not dependent on the species source of the protein, indicating that the effects of the solutes reflect ubiquitous features of protein-solute-water interactions, as discussed later. The stabilizing effects of carbohydrates, especially trehalose, on membrane structure during desiccation stress have also been noted (Crowe et al., 1984), suggesting that compatible solutes may play a number of important roles in water-stressed organisms.

Counteracting Solute Systems. The accumulation of compatible solutes by a diverse array of prokaryotes, plants and animals attests to the selective advantages of using these non-perturbing osmolytes under conditions of water stress. It may seem paradoxical, therefore, that some water-stressed organisms accumulate strongly perturbing organic osmolytes instead of compounds like alanine, proline, glycerol or trehalose. Most puzzling, at least at first inspection, is the use of urea at high concentrations in a number of groups of animals, notably the marine cartilaginous fishes and the coelacanth. Urea is a very strong perturbant of protein structure and function, even at the concentrations found in urea-rich fishes (up to approximately 0.5M; Yancey et al., 1982).

While perturbing of proteins, urea does appear to have one major advantage as an osmolyte. This involves metabolic economy. Unlike free amino acids and carbohydrates, urea is a waste product having no further metabolic function; it cannot be oxidized to yield useful energy, and it cannot be polymerized into large molecular species. The cost of using urea, therefore, is not measured in terms of removing from metabolism a compound with a role in energy metabolism or biosynthesis; rather, the cost entailed is measured in terms of potential disruption of the structures and activities of many, and perhaps most, of the proteins in the cell.

There would appear to be three adaptive strategies for coping with the perturbing effects of urea on proteins. First, evolutionary changes in proteins might render them insensitive to urea, at least to the urea concentrations found in the cells of urea-rich organisms. Second, evolutionary changes in proteins might render these proteins urea-requiring in the sense that the optimal values for the structural and functional properties of the proteins of urea-rich organisms might be found only in the presence of physiological concentrations of urea. Third, physiological processes could be brought into play that lead to the accumulation within the cells of one or more solutes that counteract the effects of urea.

All three of these strategies for adaptation to high concentrations of urea have been observed (reviewed by Yancey, 1985). Some proteins are resistant to urea at concentrations of 0.5M and below. However, at least for enzymatic function, this strategy seems to be of lesser importance than other modes of adaptation to urea. In at least one case, the muscle-type (A_4 or M_4) isozyme of lactate dehydrogenase (LDH), a requirement for urea has been observed. The interactions between LDH and its substrate pyruvate for LDHs of cartilaginous fishes attain the proper values only when physiological concentrations of urea are present (Yancey and Somero, 1978).

For most of the proteins studied to date, the third strategy, which we have designated the counteracting solute strategy, seems most important. For a variety of proteins and a number of protein characteristics, we have observed that the methylamine osmolytes (TMAO, glycine betaine, sarcosine) commonly found in the intracellular fluids of urea-rich fishes have the ability to counteract the perturbing effects of urea. Figure 2 illustrates the counteraction phenomenon in the case of the K_m of adenosine diphosphate (ADP) of PK purified from a marine stingray, Urolophis halleri. Urea is a strong competitive inhibitor, raising the K_m of ADP. TMAO, sarcosine, and glycine betaine lower the K_m of ADP. Thus, methylamines, too, can be viewed as disruptive of enzyme function when present alone, i.e., without urea. When both urea and methylamines are present in the solution at a concentration ratio of approximately two parts of urea plus one part methylamines, no net effect on the K_m of ADP was found. As in the case of compatible solutes, these effects of urea and methylamines are not dependent on the species from which the protein is obtained, suggesting that the counteracting solute strategy may be used whenever stress from high urea concentrations is present (Yancey, 1985; Somero, 1986).

THE MAMMALIAN KIDNEY: COPING WITH OSMOTIC AND UREA STRESS

Until very recently, the solutes that bring the intracellular fluids of the renal papilla into osmotic balance with the extracellular fluids were not known. In 1984, Beck and colleagues first showed unambiguously that the inorganic ion concentrations of cells of the renal papilla of the rat were no higher than those of typical mammalian cells. Thus, it appeared that, as in the case of almost all other water-stressed cells, the cells of the mammalian renal papilla must build-up their osmotic concentration with low molecular weight

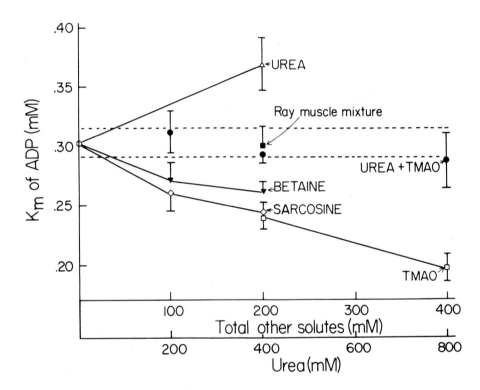

Figure 2. The effects of urea and methylamine solutes on the K_m of ADP of pyruvate kinase of the stingray Urolophis halleri (from Yancey and Somero, 1980). The ray mixture contained 400 mM urea + 65 mM TMAO + 55 mM sarcosine + 50 mM β-alanine + 30 mM glycine betaine.

organic osmolytes.

The nature of these osmolytes has been revealed by the recent studies of Balaban and colleagues (Bagnasco et al., 1986; Balaban and Knepper, 1983) and Finley (1984). Finley (1984)
looked for the occurrence of methylamines in different regions (cortex, outer medulla and
inner medulla) of the kidneys of dogs held with ad libitum access to water. No TMAO
could be detected, but she found substantial concentrations of glycine betaine, which increased from the cortex to the inner medulla. Because no glycine betaine could be detected
in urine, she reasoned that all of the glycine betaine was within the intracellular water.
Estimates of intracellular glycine betaine concentrations in the cells of the inner medulla
averaged approximately 100 mmoles/kg cell water. The glycine betaine concentration in
the inner medulla was roughly one-half of the urea concentration.

Balaban and Knepper (1983) and Bagnasco et al. (1986) reported that glycerylphosphoryl-
choline (GPC) as well as glycine betaine was present at substantial concentrations (up to
21 mmoles/kg wet weight and 56 mmoles/kg wet weight, respectively) in rabbit kidneys
under antidiuresis. The intracellular concentrations were likely much greater than this,
since as in Finley's study, no methylamines could be measured in the urine.

The study of Bagnasco et al. (1986) also showed that sorbitol and inositol were present
at significant levels in the kidney cells of mammals. In antidiuretic rabbits, sorbitol and
inositol concentrations in the inner medulla reached 21 and 11 mmoles/kg wet weight,
respectively. Again, intracellular concentrations must have been markedly higher since
neither carbohydrate osmolyte was detected in the urine.

In the inner medulla of the mammalian kidney, then, it appears that the intracellular fluids
contain sufficient concentrations of urea-counteracting solutes, glycine betaine and,
probably, GPC (see below), to offset perturbation by urea, plus adequate levels of compatible carbohydrate solutes to complete the build-up of intracellular osmolarity to equal
that of the extracellular fluids.

MECHANISMS OF SOLUTE EFFECTS ON PROTEINS

A brief discussion of the proposed mechanisms of solute-protein interactions will indicate
how the solutes exert their effects, and will provide us with a predictive model that will
allow us to make educated guesses about the likely effects of an untested solute on protein
function and structure.

Timasheff and colleagues (cf. Arakawa and Timasheff, 1983, 1985) have provided a
theoretical and empirical foundation for much of our current understanding of how different
low-molecular-weight solutes influence protein structure. They have shown that solutes
that stabilize protein structure are strongly excluded from the organized water that

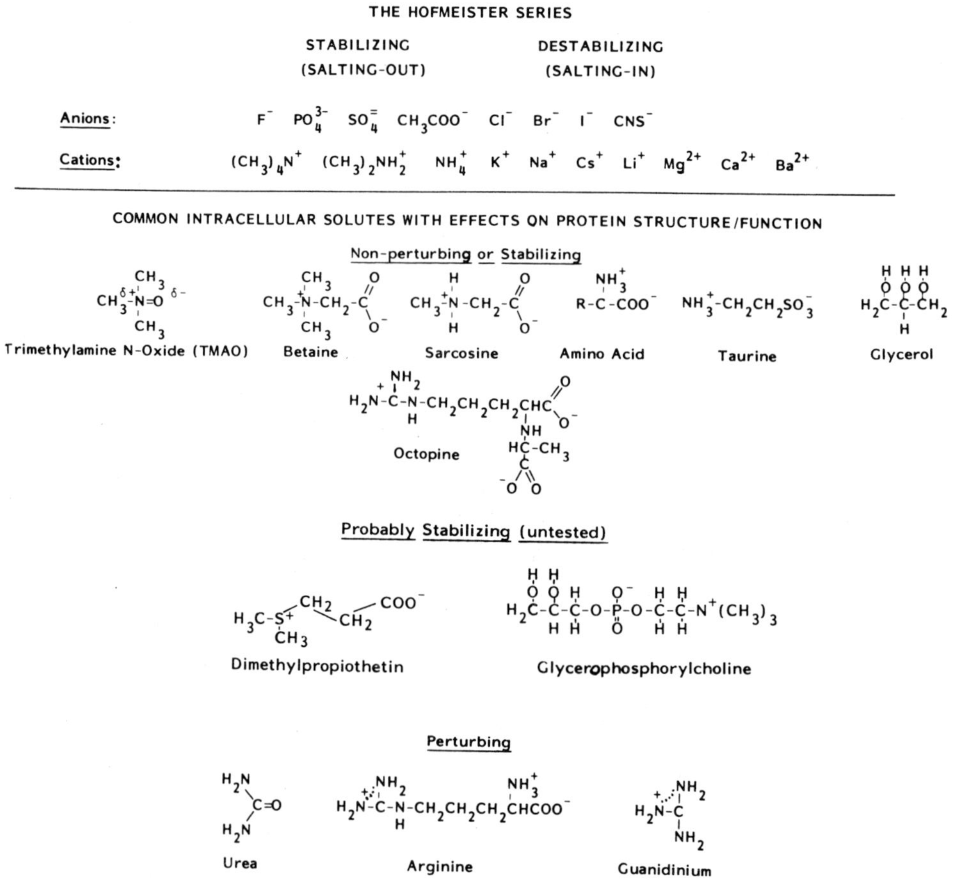

Organic Osmolyte Systems

THE HOFMEISTER SERIES

STABILIZING   DESTABILIZING
(SALTING-OUT)   (SALTING-IN)

Anions: F^- PO_4^{3-} $SO_4^=$ CH_3COO^- Cl^- Br^- I^- CNS^-

Cations: $(CH_3)_4N^+$ $(CH_3)_2NH_2^+$ NH_4^+ K^+ Na^+ Cs^+ Li^+ Mg^{2+} Ca^{2+} Ba^{2+}

COMMON INTRACELLULAR SOLUTES WITH EFFECTS ON PROTEIN STRUCTURE/FUNCTION

Non-perturbing or Stabilizing

Trimethylamine N-Oxide (TMAO)   Betaine   Sarcosine   Amino Acid   Taurine   Glycerol

Octopine

Probably Stabilizing (untested)

Dimethylpropiothetin   Glycerophosphorylcholine

Perturbing

Urea   Arginine   Guanidinium

Figure 3. Solutes having structure–stabilizing or structure perturbing effects on macro-molecules.

surrounds the protein's surface. The structure-stabilizing solutes, e.g., methylamines, must, therefore, partition into the "bulk" phase of the solution.

Thermodynamic considerations show clearly that when structure-stabilizing solutes are present in the solution, the most energetically favorable situation is one where the amount of "bulk" water is maximal, i.e., where the greatest fraction of the total water in the system is available for solvating the structure-stabilizing solutes. This reasoning explains why these solutes can enhance the structural stability of proteins. If proteins unfold in the presence of structure-stabilizing solutes, an increased fraction of the water of the system becomes organized around the protein, and the structure-stabilizing solutes must partition into a smaller fraction of the solution. This creates a very unfavorable entropy situation. Thus, when structure-stabilizing solutes are in a protein solution, the thermodynamically most favorable situation is one where the protein surface area in contact with water is minimal. For this reason protein-stabilizing solutes enhance the stability of the globular structures of proteins and also favor protein-protein interactions in which some of the proteins' surfaces are brought together, with the expulsion of bound water.

Urea and other protein denaturants can penetrate the bound water surrounding proteins and can bind strongly to protein groups, e.g., peptide backbone linkages. Thus, the structure-destabilizing solutes favor an expansion of protein surface area, unlike the stabilizing solutes.

How can one predict what the net effect of a solute on protein structural stability will be? Some clues to the basis of a predictive model are given in Figure 3. Note that the strongest structure-stabilizing solutes contain certain chemical groups, notably methylammonium groups, carboxylate groups, and sulfate groups. Salts containing these groups have long been appreciated as being good reagents for stabilizing and precipitating proteins (e.g., ammonium sulfate, tetramethylammonium ions, etc.). The occurrence of these groups in organic osmolytes was first noted by Clark and Zounes (1977), who proposed that the selection of organic osmolytes was based on their favorable effects on proteins.

One advantage of understanding which chemical groups stabilize protein structure is that we can make educated conjectures about the probable effects of a molecule whose effects on proteins remain to be tested. One such molecule is glycerophosphorylcholine. GPC has the earmarks of a good intracellular osmolyte. Its choline group confers on one tip of the molecule the same appearance as TMAO and glycine betaine (Figure 3). GPC thus should stabilize protein structure and counteract the effects of urea. GPC also contains the compatible glyceryl group in its structure. Finally, GPC bears no net charge, a feature commonly found in compatible solutes. It would seem, then, that GPC is a marvelously fit solute for the tasks an organic osmolyte must serve in the cells of the mammalian renal papilla.

REFERENCES

Arakawa T, Timasheff SN (1983) Preferential interactions of proteins with solvent components in aqueous amino acid solutions. Arch Biochem Biophys 244: 169-177.

Arakawa T, Timasheff SN (1985) The stabilization of proteins by osmolytes. Biophys J 47: 411-414.

Bagnasco S, Balaban R, Fales H, Yang Y-M, Burg M (1986) Identification of intracellular organic osmolytes in renal inner medulla. J Biol Chem, in press.

Balaban R, Knepper MA (1983) Nitrogen-14 nuclear magnetic resonance spectroscopy of mammalian tissues. Am J Physiol 245: C439-C444.

Beck F, Dorge A, Rick R, Thurau K (1984) Intra- and extracellular element concentrations of rat renal papilla in antidiuresis. Kidney Int 25: 397-403.

Borowitzka LJ (1985) Glycerol and other carbohydrate osmotic effectors. In: Gilles R, Gilles-Baillien M (eds): Transport Processes, Iono- and Osmoregulation. Springer-Verlag, Berlin; pp. 437-453.

Borowitzka LJ, Brown AD (1974) The salt relations of marine and halophilic species of the unicellular green alga, Dunaliella: The role of glycerol as a compatible solute. Arch Microbiol 96: 37-52.

Bowlus RD, Somero GN (1978) Solute compatibility with enzyme function and structure: rationales for the selection of osmotic agents and end-products of anaerobic metabolism in marine invertebrates. J Exp Zool 208: 137-152.

Clark ME (1985) The osmotic role of amino acids: discovery and function. In: Gilles R., Gilles-Baillien M (eds): Transport Processes, Iono- and Osmoregulation. Springer-Verlag, Berlin; pp. 412-423.

Clark ME, Zounes M (1977) The effects of selected cell osmolytes on the activity of lactate dehydrogenase from the euryhaline polychaete Nereis succinea. Biol Bull 153: 468-484.

Crowe JH, Crowe LM, Chapman D (1984) Preservation of membranes in anhydrobiotic organisms: the role of trehalose. Science 223: 701-703.

Finley KD (1984) The role of methylamines and free amino acids in protection of mammalian kidney proteins. Masters dissertation, San Diego State University, California.

Somero GN (1986) Protons, osmolytes and the fitness of the internal milieu for protein function. Am J Physiol, in press.

Yancey PH (1985) Organic osmotic effectors in cartilaginous fishes. In: Gilles R, Gilles-Baillien M (eds): Transport Processes, Iono- and Osmoregulation. Springer-Verlag, Berlin; pp. 424-436.

Yancey PH, Clark ME, Hand SC, Bowlus RD, Somero GN (1982) Living with water stress: evolution of osmolyte systems. Science 217: 1214-1222.

Yancey PH, Somero GN (1978) Urea-requiring lactate dehydrogenases of marine elasmobranch fishes. J Comp Physiol 125: 135-141

Yancey PH, Somero GN (1980) Methylamine osmoregulatory solutes of elasmobranch fishes counteract urea inhibition of enzymes. J Exp Zool 212: 205-213.

CELLULAR AND MOLECULAR ADAPTATIONS
TO SEVERE WATER LOSS

James S. Clegg

University of California
Bodega Marine Laboratory
Bodega Bay, California
U.S.A. 94923

Some organisms have the ability to reversibly lose almost
all intracellular water, a condition known as
"anhydrobiosis". Although widespread, we know little about
the underlying mechanisms. This chapter deals chiefly with
the encysted embryos of Artemia, the brine shrimp, and
considers some of the biochemical and biophysical
adaptations that underly anhydrobiosis. Studies on cellular
anhydrobiosis also raise questions about the nature of
their aqueous compartments. Abundant evidence, considered
briefly, indicates that these compartments are not crowded
solutions of macromolecules but are instead dilute,
reflecting a more organized picture of cells than the one
generally adopted.

INTRODUCTION

Adaptations of organisms to arid environments are chiefly those involved with
water conservation. This paper concerns an alternative: organisms which have
adapted to severe water shortage by simply losing virtually all of their water,
but in a fully reversible fashion. This condition, known currently as
"anhydrobiosis", occurs in representatives of many taxa, although apparently
limited to small and often microscopic adults, or other stages of the life
cycle of animals. The long and interesting history of research on
anhydrobiosis has been described in the scholarly review by Keilin (1959) and
more recent work can be found in books edited by Crowe and Clegg (1973 and
1978) and Leopold (1986). These life forms allow us to try to understand

Comparative Physiology: Life in Water and on Land. P. Dejours, L. Bolis, C.R. Taylor, E.R. Weibel
(eds.) Fidia Research Series, IX-Liviana Press, Padova © 1987

adaptations involved with the ability to tolerate dehydration so severe that
even the biological status of the organism is in doubt (Clegg, 1973 and 1986b).
There are other reasons to study anhydrobiotic cells. By examining the
hydration dependence of their ultrastructure and metabolism we may uncover
general relationships between water and the nonaqueous components of living
systems that are obscured in the more ordinary, fully hydrated situation. In
what follows, these two aspects of cellular anhydrobiosis will be considered
briefly.

ARTEMIA: MODEL SYSTEM FOR THE STUDY OF ANHYDROBIOSIS

Artemia, an anostracan crustacean, inhabits saline lakes and solar salt
operations world-wide. It is widely used in commerce and its biology is well
known (see books edited by Bagshaw and Warner, 1979, and three volumes by
Persoone et al. 1980). Bisexual varities of Artemia reproduce either
viviparously or oviparously. In the latter case, encysted dormant embryos at
the gastrula stage are released into the environment which is usually
concentrated sea water. The normal fate of these cysts is desiccation, and
they are capable of repeated hydration/desiccation cycles. It is evident that
the success of this organism is related directly to the desiccation tolerance
of the encysted stage. The cysts are available commercially in kilogram
quantities, but can also be produced in laboratory cultures. Several detailed
accounts of the techniques used to wash, process and experimentally manipulate
these cysts have been published (see Clegg, 1986a). All cysts used in our work
are from the San Francisco Bay area; cysts of other origins may differ in their
specific properties.

When fully hydrated the cysts are spheres of about 210 um diameter, and contain
close to 4000 cells in an inner cell mass. The shell is noncellular and
complex, the outer 6 um of which can be removed by hypochlorite treatment.
That leaves behind only a 2 um layer of a protein-chitin complex called the
embryonic cuticle. The compact inner cell mass has no observable extracellular
space. These features allow us to interpret biochemical and biophysical data
directly in terms of the cells themselves. When fully hydrated, the cysts
contain close to 1.4 gH_2O/g dry mass (g/g), at which the cells contain nearly
1.7 g/g. No desiccation treatment, short of temperatures above 100° C, is
known to kill these cells: Gas-bombardment techniques, at low temperatures
under high vacuum, reduced the water content to <0.007 g/g, without any
measurable effect on cyst viability (Clegg et al.1978). Because of these and

other features, the _Artemia_ cyst provides access to experimental designs that
are not possible with "ordinary cells".

PHYSICAL PROPERTIES OF DRIED AND HYDRATING CYSTS

Figure 1 summarizes results from many previous studies using a variety of
techniques. These results have been interpreted in detail in individual papers
and in a recent summary (Clegg, 1986b) so those efforts will not be repeated.
Here I focus upon one highly unexpected feature of the results from methods
that sample the motions of water. One would suppose that the few water
molecules present in severely desiccated cysts would exhibit greatly reduced
motion because of high energy interactions with nonaqueous components of the
system. Likewise, we would expect to observe an increase in the translation
and rotation of water as the cyst undergoes hydration. Those expectations are
strikingly unfulfilled in _Artemia_ cysts. Thus, nuclear magnetic resonance (Fig
1, NMR) and microwave dielectric measurements (Fig 1, MD) show that as water is
added to dried cysts no corresponding increase occurs in water motion. Indeed,
the opposite is observed at water contents near 0.25g/g. These transitions are
most evident in the case of NMR. The relaxation times, T_1 and T_2, exhibit what
appear to be very sharp transitions, and the self-diffusion coefficient, D,
behaves similarly. Essentially no hysteresis is evident – the same situation
occurs when water is removed from hydrated cysts. Although the generalized
lines for quasi-elastic neutron scattering (Fig 1, QNS) are based on relatively
few data points, this result suggests that most of the water below 0.3g/g does
not behave as though it were the primary hydration layer(s) expected to be
present. The "thermodynamic" measurements (Fig 1, Isotherms, V, DSC) show, as
expected, that marked changes in energetics accompany the uptake of water, no
doubt reflecting the many rearrangements that take place in the nonaqueous
components, as well as the water itself.

What do these findings tell us about the dried cells and their hydration? Why
should the motions of water molecules decrease, when precisely the opposite is
to be expected? We interpret these results in the context of the "water
replacement hypothesis" (WRH), which proposes that trehalose, a non-reducing
disaccharide of glucose, present in large amounts in these cysts, substitutes
for the water of hydration of macromolecules and membranes at low cyst water
contents (<0.3g/g) (Crowe & Clegg, 1978). This water, so released, is believed
to be associated with much smaller solutes, glycerol being one candidate, which
would result in an increase in its motions and account for our results. Thus,

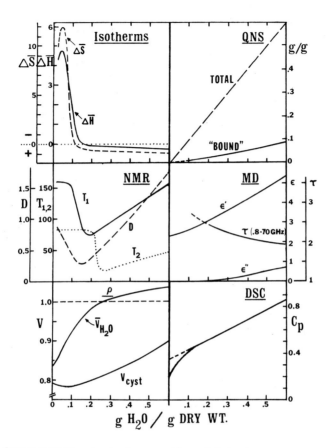

<u>Figure 1.</u> Summary of the physical properties of <u>Artemia</u> cysts at low water contents. <u>Upper left</u>: Differential values for the entropy (S) and enthalph (H) changes accompanying the sorption of water vapor. ΔS is in units of cal/mole degree, ΔH is in units of kcal/mole. <u>Middle left</u>: The self diffusion coefficient (D, in 10^{-6} cm^2/sec) and relaxation times (T_1, T_2, in msec) measured by Nuclear Magnetic Resonance (NMR). <u>Lower left</u>: Cyst specific volume (V, in cm^3/g) calculated from density (ρ) measurements. VH_2O represents estimates for the specific volume of cyst water. The dashed line is the value for liquid water. <u>Upper right</u>: The fraction of water considered to be "bound", compared to the total cyst water, deduced from analysis of quasi-elastic neutron scattering (QNS) spectra. <u>Middle right</u>: Microwave dielectric (MD) measurements of the permittivity (ϵ' and ϵ'' are dimensionless) and relaxation time, τ , in units of picoseconds. <u>Lower right</u>: Heat capacity (Cp) in units of cal/g/degree (1+gH_2O/g dry wt) at 25 C. The dotted line is an extrapolation. (Taken from Clegg, 1986b)

the hypothesis indicates that as the cells lose water they do become
dehydrated, but the damage known to result from the loss of primary hydration
water is mitigated by its replacement with trehalose.

These physical studies were performed in part to test the WRH which was first
advanced in the early 1970s (see Crowe, 1971; Crowe & Clegg, 1973). At that
time, we had only a correlation between the ability of cells to enter
anhydrobiosis, and the accumulation in them of very large amounts of
polyhydroxy compounds such as glycerol and trehalose. Since that time,
extensive studies by John and Lois Crowe, and their students and colleagues,
have provided a firm basis and strong experimental support for the WRH as a
major molecular adaptation in anhydrobiosis. An important finding was their
observation that membranes in anhydrobiotic nematodes did not undergo phase
transitions as do those in nonadapted, ordinary cells (Crowe et al. 1978)
They went on to show that trehalose actually protected sarcoplasmic reticulum
(SR) vesicles against desiccation damage in vitro and their subsequent elegant
work (see J.H. Crowe, 1986; L. Crowe, 1986) clearly established that the
mechanism of dehydration protection in this case involves the interaction of
trehalose with the phospholipid head groups of the membranes. It is
astonishing to note that the Crowe's used lobster muscle as their source of SR
vesicles, suggesting that trehalose may be a "universal" replacer of water at
all membrane surfaces. The message here is that the ability to reversibly lose
cell water may not be as difficult as it might seem. Ostensibly, successful
anhydrobiosis does not require the evolutionary development of unique and
extraordinary macromolecules and organelles whose integrity is maintained,
reversibly, in the absence of water; rather, only that the cells can tolerate
large concentrations of trehalose (and perhaps other substitutes) which we may
suppose are "compatible solutes" (Yancey et al. 1982) as well as water
replacers. No doubt there are other biophysical adaptations that are critical
to anhydrobiosis and we can expect to uncover these with further study.

HYDRATION AND METABOLISM IN ARTEMIA CELLS

What are the cells of Artemia cysts doing as their water contents undergo such
massive fluctuations? Metabolism, as we know it, depends not only upon enzyme
and organelle integrity, but upon the participation of an enveloping aqueous
phase. We have long been interested in the relationships between metabolism
and the aqueous phase properties of cells. Studies that sampled the hydration-
dependence of metabolism in Artemia cysts revealed the presence of three

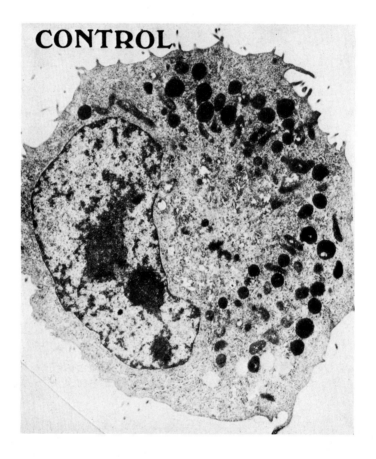

<u>Figure 2.</u> Mouse L-929 cells exposed to physiological conditions (control) and increasing molal concentrations (m) of sorbitol added to control medium. Total magnification is the same in all cases (15,400X). (Modified from Mansell and Clegg, 1983.)

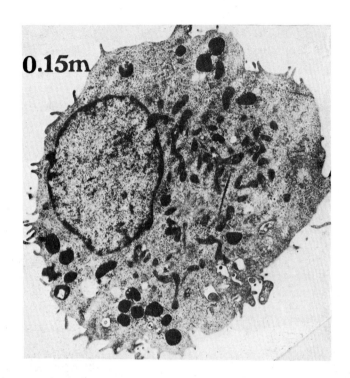

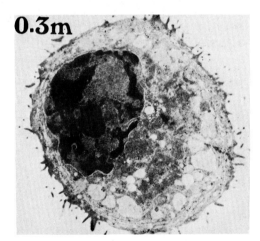

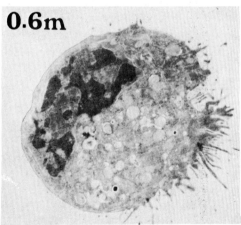

metabolic domains. Between 0 and about 0.3g/g no metabolic activity is
detected, presumably prevented by lack of adequate primary hydration water:
trehalose, while preserving macromolecular and membrane structure, does not
appear to support their functions (Clegg, 1973). Between 0.3 and about 0.65g/g
some metabolism occurs but is limited and restricted to small metabolites.
Above 0.65g/g the cysts exhibit a metabolism characteristic of fully hydrated
cysts; however, the metabolic _rate_ is a strong function of cell water content
up to about 0.85g/g. We suppose this rate dependence is a reflection of
metabolite transport between various intracellular locations and involves such
things as the microviscosity of the transport channels. Once cysts reach water
contents of 0.85g/g, however, their metabolic rates are nearly independent of
further water addition. These impressive features of cyst metabolism reflect a
high degree of organization and coordination in the face of massive changes in
their intracellular aqueous compartments. To explain these results we supposed
that these compartments were probably not the locus of much metabolic activity;
rather, that all or most of the enzymes participating in the intermediary
metabolism we observed were organized into complexes and these, in turn,
loosely associated with cell ultrastructure (Clegg, 1979). An examination of
the literature at that time revealed the existence of scattered supporting
evidence, some of it going back to the 1930s (see Clegg, 1984 and 1986c).
Thus, the model we initially proposed for _Artemia_ cysts was advanced as a
general one.

THE AQUEOUS INTRACELLULAR COMPARTMENTS

It is not widely appreciated that mammalian cells in culture also have
impressive abilities to survive large fluctuations in their volume and water
content (see Mansell and Clegg, 1983; Wheatley et al. 1984a,b). For
example, mouse L-929 cells tolerate easily the osmotic removal of half their
volume and about 70% of their total water. Figure 2 illustrates the
ultrastructure of L cells exposed to hypersmotic conditions generated by the
addition of sorbitol. Cells in 0.3 molal sorbitol supplements or less, resume
their control volumes and water contents within seconds when returned to
physiological conditions and exhibit little irreversible damage even when
exposed to these conditions for an hour at 37° C. In contrast, populations
exposed to higher concentrations of sorbitol (0.6 molal) exhibit appreciable
damage and die if exposure is prolonged. A number of metabolic parameters have
been sampled in L cells at variable water contents and volumes (Mansell &
Clegg, 1983; Clegg & Gordon, 1985). Those studies show that L cells are

remarkably insensitive to sorbitol addition ($\leq$ 0.3 m) and behave much like Artemia cysts do above 0.85g/g.

The aqueous compartments of cells have generally been considered to be crowded solutions, containing high total concentrations of macromolecules, within which much of the metabolic repertoire of cells is presumed to take place. However, that widespread and useful paradigm is not consistent with a growing body of experimental evidence which tells us that a very large fraction of cellular macromolecules are not diffusing freely in three dimensions, and that the aqueous compartments are dilute and not concentrated. That evidence has recently been summarized (see Clegg, 1984; Srivastava & Bernhard, 1986; Welch & Clegg, 1986; Kellermayer et al. 1986). Coming from a variety of independent sources, and different kinds of cells, these results provide compelling reason to abandon the "crowded solution" image of the aqueous compartments in favor of a more organized one, such as that advanced by Porter and colleagues (Porter et al. 1983; Porter, 1986) from high voltage electron microscopy. Our results on L cells are easier to understand on that basis: hyperosmotic exposure causes the removal of water from the dilute aqueous phase surrounding the organized cytomatrix, causing minimal disturbance to the cells because we believe little of consequence seems to be taking place there, at least in terms of metabolism. We suppose that the collapse of the cytomatrix at very high external osmotic pressures results in a prohibitively large reduction in the aqueous compartments, possibly preventing intracellular transport to occur at rates consistent with metabolic and other cellular requirements. Obviously, other factors may be involved.

CONCLUDING COMMENTS

Research on the manipulation of water content in mammalian cells indicates that they are "preadapted" to undergo changes in volume and water content that are never encountered in nature. These abilities do not seem to be adaptations acquired during evolution but reflect instead the basic organization of cells. Viewed in the context of cellular anhydrobiosis, and considering the water replacement hypothesis to be correct, we may tentatively conclude that the ability to reversibly lose virtually all intracellular water is not as formidable a problem as it might appear to be. It will be interesting to determine whether mammalian cells in culture can achieve anhydrobiosis following the microinjection of appropriate amounts of trehalose.

ACKNOWLEDGMENTS

Manuscript costs were provided by a grant from US NSF (PCM 8217311). I am grateful to Diane Cosgrove for skillful manuscript preparation.

REFERENCES

Bagshaw JC, Warner AH (eds) (1979) Biochemistry of _Artemia_ Development, University Microfilms International, Ann Arbor; pp. 1-238.

Clegg JS (1973) Do dried cryptobiotes have a metabolism? In: Crowe JH, Clegg JS (eds): Anhydrobiosis. Dowden, Hutchinson and Ross, Stroudsburg; pp. 141-147.

Clegg JS (1978) Hydration-dependant metabolic transitions and the state of water in _Artemia_ cysts. In: Crowe JH, Clegg JS (eds.): Cell-Associated Water. Academic Press, New York; pp. 363-413.

Clegg JS (1984) Properties and metabolism of the aqueous cytoplasm and its boundaries. Amer J Physiol 246:R133-R151.

Clegg JS (1986a) _Artemia_ cysts as a model for the study of water in biological systesm. In: Packer L (ed): Methods in Enzymology, vol. 127, Membranes: Water, Ions and Biomolecules. Academic Press, New York; in press.

Clegg JS (1986b) The physical properties and metabolic status of _Artemia_ cysts at low water contents: the water replacement hypothesis. In: Leopold C (ed): Membranes, Metabolism and Dry Organisms. Cornell University Press, Ithaca; Chapter 9.

Clegg JS (1986c) On the physical properties and potential roles of intracellular water. In: Welch GR, Clegg JS (eds): Organization of Cell Metabolism. Plenum, New York in press.

Clegg JS, Gordon EP (1985) Respiratory metabolism of L-929 cells at different water contents and volumes. J Cell Physiol 124:299-304.

Clegg JS, Zettlemoyer AC, Hsing HH (1978) On the residual water content of dried but viable cells. Experientia 34:734-735.

Crowe JH (1971) Anhydrobiosis: an unsolved problem. Amer Naturalist 105:563-573.

Crowe JH, Clegg JS (eds) (1973) Anhydrobiosis, Dowden, Hutchinson and Ross, Stroudsburg pp. 1-477.

Crowe JH, Clegg JS (eds) (1978) Dry Biological Systems, Academic Press, New York pp. 1-357.

Crowe JH (1986) Stabilization of membranes and proteins in anhydrobiotic organisms. In: Leopold C (ed): Membranes, Metabolism and Dry Organisms. Cornell University Press, Ithaca in press.

Crowe L (1986) Hydration dependent phase transitions and premeability properties of biomembranes. In: Leopold C (ed): Membranes, Metabolism and Dry Organisms. Cornell University Press, Ithaca in press.

Crowe JH, Lambert DT, Crowe LM (1978) Ultrastructural and freeze-fracture studies on anhydrobiotic nematodes. In: Crowe JH, Clegg JS (eds): Dry Biological Systems. Academic Press, New York pp. 23-50.

Keilin D (1959) The problem of anabiosis or latent life: history and current concept. Proc Roy Soc Lond B 150:149-191.

Kellermayer M, Ludany A, Jobst K, Szucs G, Trombitas K, Hazlewood, CR (1986) Cocompartmentation of proteins and K^+ within the living cell. Proc Natl Acad Sci USA 83:1011-1015.

Mansell JL, Clegg JS (1983) Cellular and molecular consequences of reduced cell water content. Cryobiology 20:591-612.

Leopold C (ed) (1986) Membranes, Metabolism and Dry Organisms, Cornell University Press, Ithaca in press.

Porter KR, Beckerle M, McNivan M (1983) The cytoplasmic matrix. Mod Cell Bio 2:259-302.

Porter KR (1986) Structural organization of the cytomatrix. In: Welch GR Clegg JS (eds): Organization of Cell Metabolism. Plenum, New York in press.

Persoone G, Sorgeloos P, Roels O, Jaspers E (eds) (1980) The Brine Shrimp Artemia vols 1-3. Universal Press, Wettern, Belgium pp.1-1024.

Srivastava DK, Bernhard SA (1986) Enzyme--enzyme interactions and the regulation of metabolic reaction pathways. In: Horecker BL, Stadtman ER (eds): Current Topics in Cellular Regulation vol 29. Academic Press, New York pp. 1-76.

Welch GR, Clegg JS (eds) (1986) Organization of Cell Metabolism, Plenum Press, New York in press.

Wheatley DN, Inglis MS, Clegg JS (1984a) Dehydration of HeLa S-3 cells by osmosis. I kinetics of cellular responses to hypertonic concentrations. Mol Physiol 6:163-182.

Wheatley DN, Inglis MS, Robertson JC (1984b) Dehydration of HeLa S-3 cells by osmosis. II Effects on amino acid pools and protein synthesis. Mol Physiol 6:183-200.

Yancey PH, Clark ME, Hand SC, Bowlus RD, Somero GN (1982) Living with water stress: evolution of osmolyte systems. Science 217:1214-1222.

UPTAKE AND TRANSPORT OF SULFIDE
IN MARINE INVERTEBRATES

James J. Childress

Department of Biological Sciences
University of California
Santa Barbara, California
U.S.A.

Terrestrial organisms have little exposure to reduced sulfur compounds other than those produced within their own digestive tracts. In contrast, reduced sulfur compounds are abundant in many marine habitats, presenting a challenge to organisms due to the toxicity of sulfide and an opportunity because of the energy potentially available from the oxidation of these compounds. Those animals concerned solely with the avoidance of toxicity oxidize sulfide to less toxic sulfur species and do not concentrate sulfide from the environment. Marine invertebrates with sulfur oxidizing prokaryotic chemolithoautotrophic symbionts must not only avoid the toxic effects of sulfide, but must also provide reduced sulfur compounds and oxygen to their symbionts in order to reap the rewards. Some of the elements used for this purpose in different symbioses are: 1)sulfide binding proteins in blood, 2)sulfide binding proteins in symbiont-containing tissues, 3)oxidation of sulfide to less toxic species at the body surface.

INTRODUCTION

One major difference between terrestrial and aquatic environments is the prevalence of reduced inorganic sulfur compounds in aquatic but not terrestrial environments. Many aquatic habitats are characterized by μM concentrations of hydrogen sulfide. Examples include the deep-sea hydrothermal vents, marine hydrocarbon seeps, eel-grass beds, mangrove swamps, areas receiving pulpmill wastes, anoxic basins, deep-sea groundwater seeps, and sewage outfall areas (Cavanaugh, 1985; Felbeck, Somero and Childress, 1983). The sulfide in these habitats ,originates either from geological sources or from the dissimilatory reduction of sulfate by anaerobic bacteria. In at least one of these habitats, animals are found living in sediments containing as much as 10 mM sulfide (Childress and Lowell,1982). In contrast terrestrial animals are virtually never exposed to sulfide except as a product of bacteria living in their oral cavities or guts. The levels in these areas are probably relatively low. For both aquatic and terrestrial organisms some method of detoxifying sulfide is necessary, however, only in the aquatic environment do we find animals capable of obtaining energy and reducing power from the oxidation of sulfide either directly or by bacterial endosymbionts (Cavanaugh, 1985; Felbeck, Somero and Childress, 1983). This paper will consider the challenge of sulfide toxicity in the contrasting cases of those animals which do and those which do not

Comparative Physiology: Life in Water and on Land. P. Dejours, L. Bolis, C.R. Taylor, E.R. Weibel (eds.) Fidia Research Series, IX-Liviana Press, Padova © 1987

possess endosymbiotic chemolithoautotrophic bacteria.

Both those animals which do and those which do not possess endosymbionts face the common problem of avoiding the toxic effects of sulfide on aerobic metabolism. The major toxic effect of hydrogen sulfide is probably as a potent inhibitor of cytochrome-c-oxidase. Hydrogen sulfide, like cyanide, has a very high affinity for this enzyme with concentrations of less than 1 µM being highly inhibitory. In addition, sulfide is a small molecule which appears to be rapidly taken up across respiratory surfaces, therefore animals appear to be unable to exclude it from their bodies (Powell, Crenshaw and Rieger, 1979). In spite of these problems many species of marine animals can tolerate µM levels of sulfide for indefinite periods (Theede, 1973)and some can tolerate even mM concentrations (Vetter, Wells, Kurtsman and Somero, 1986). The common factor in sulfide detoxification in animals is the oxidation of sulfide to less toxic forms. In mammals this activity is found in the liver and kidneys (National Research Council, 1979). This is also the apparent site of oxidation in decapod crustaceans (Vetter, Wells, Kurtsman and Somero, 1986). Arenicola marina apparently carries out this oxidation in its blood by means of a modified hemoglobin produced by contact with sulfide (Patel and Spencer, 1963). Most aquatic animals, however, appear to carry out this oxidation at the body surface (Powell, Crenshaw and Rieger, 1979; Powell and Somero, 1985, 1986). In meiofaunal animals different species oxidize sulfide to different extents (Powell, Crenshaw and Rieger, l980). These workers found elemental sulfur, thiosulfate, sulfite, and sulfate as products of sulfide detoxification in meiofaunal species. All of these are nontoxic or much less toxic than sulfide.

Animals which have endosymbiotic chemolithoautotrophic bacteria have not only the problem of protecting themselves against sulfide toxicity but must also supply the symbionts' metabolic needs, including sulfide or some other reduced sulfur compound. Further, sulfur-oxidizing bacteria themselves generally live in very narrow zones where both sulfide and oxygen are present, but at very low concentrations (Jørgensen, l982; Kelly and Kuenen, l984; Nelson,Revsbech and Jørgensen, 1986). They are often sensitive to elevated concentrations of either sulfide or oxygen. Thus these animals must not only protect their own tissues from toxic effects, they must also protect the endosymbionts from the toxic effects of sulfide and oxygen while at the same time taking these substances from the environment and transporting them to the symbionts in quantity. This problem is similar to the problem of supplying oxygen to root nodule symbionts and to mitochondria (Wittenberg, 1985) with the additional difficulty of sulfide supply. Another supply problem which is unique to these symbioses is that sulfide is rapidly and spontaneously oxidized by oxygen (Fisher and Childress, 1984), a process which could greaatly reduce the availability of sulfide to the symbionts if it were not controlled. We see then that the host's role in the symbiosis is quite complex, going well beyond simply supplying the necessary metabolites to the symbionts. Each of the major animal taxa which has such symbioses appears to have evolved a different set of mechanisms for these purposes. I will describe below the mechanisms used by three very different sulfur based symbioses: the vestimentiferan tubeworm Riftia pachyptila, the vesicomyid clam Calyptogena magnifica and the solemyid clam Solemya reidi. These are the best understood of the sulfur-based symbioses.

Riftia pachyptila

The vestimentiferan tubeworm Riftia pachyptila lives immediately around the deep-sea hydrothermal vents on the Galapagos Rift in the area where the sulfide rich (up to 250 μM), anoxic vent water is mixing with the ambient sulfide-free, oxic water (Hessler and Smithey, 1983). This species has the same body plan as other pogonophorans, that is: the interior of its trunk consists primarily of the trophosome, an organ which contains the bacterial symbionts (Cavanaugh, et al, 1981; Felbeck, 1981; Southward, 1982). This organ is bathed in an abundant coelomic fluid and is highly vascularized (Jones, 1981). The vascular system connects the trophosome to the plume by way of major dorsal and ventral vessels and blood flow appears to be driven by a dorsal heart as well as accessory hearts within the plume. The plume is the organ of gas exchange, having both a large surface area and the circulation of blood very close to its surface (Arp, Childress and Fisher,1985). The trophosome accounts for about 15% of the worms body mass (Childress, Arp and Fisher, l984) and the intracellular bacteria constitute between 15 and 35% of the trophosome (Powell and Somero, l986). The coelomic fluid accounts for 26% of the body mass and the vascular blood is more than 4% (Childress, Arp and Fisher, 1984).

Both the coelomic fluid and the vascular blood contain abundant hemoglobin (Terwilliger, Terwilliger and Schabtach, l980; Arp and Childress, l981). There appear to be two different hemoglobins present which have different molecular weights (~400,000 and ~3,000,000) and different subunit compositions (Terwilliger, Terwilliger and Schabtach, l980; Terwilliger, et al, l985). The fact that the hemoglobins are in very different concentrations in the blood and coelomic fluids indicates that these two compartments are not confluent (Childress, Arp and Fisher, 1984). The blood has a high affinity for oxygen (P_{50}=1-3 torr), a low temperature effect on binding and a slight Bohr effect (Arp and Childress, 1981). The blood oxygen affinity is unaffected by sulfide and there is no spectral indication of either sulfhemoglobin or methemoglobin in the blood of freshly captured worms (Childress, Arp and Fisher, 1984). Other vestimentiferans and pogonophorans may well have much higher affinities for oxygen (Terwilliger, et al, 1985). There appears to be no tissue hemoglobin in R. pachyptila (Wittenberg, 1985). The hemoglobins of R. pachyptila appear to be well adapted as oxygen transport proteins for this symbiosis since the high oxygen affinity would help the animal to obtain oxygen from a sometimes hypoxic environment and provide oxygen to the symbionts at a high concentration but a low partial pressure. The maintenance of low oxygen partial pressure may be important both to retard spontaneous oxidation and to protect the probably microaerophilic symbionts. The insensitivity to sulfide is an obvious benefit for an animal which has high levels of sulfide in its blood.

Freshly captured Riftia pachyptila have blood sulfide concentrations as high as several mM which indicates that this species is concentrating sulfide by about an order of magnitude from its environment (Childress, Arp and Fisher, 1984). This concentrating ability appears to be a result of the binding of sulfide by the blood (Arp and Childress, 1983). The hemoglobins of this species bind sulfide reversibly at a non-heme site with a very high affinity (Childress, Arp and Fisher, l984). The ratio of sulfide bound to heme is 1.3 to 1. This hemoglobin has the ability to stabilize sulfide and oxygen in concentrations of up to several mM, apparently preventing spontaneous oxidation by keeping the concentration of free sulfide and oxygen extremely low (Fisher and Childress, 1984). The affinity of this hemoglobin for sulfide is higher than that of this worm's cytochrome-c-oxidase for sulfide. As a result these hemoglobins can prevent sulfide poisoning of this enzyme and can reverse poisoning of it as well (Powell and Somero, 1983,1986). This mechanism for control of sulfide toxicity can only be effective in the long term if the hemoglobins are kept below sulfide saturation by the continuous removal of sulfide. This is achieved by the ability of the trophosome bacteria to remove sulfide from these hemoglobins. Thus the bacterial oxidation of sulfide can serve as the sink which holds the blood sulfide concentration below saturation (Fisher and Childress, l984; Belkin, Nelson and Jannasch, 1986). The sulfide binding is strongly affected by pH with maximum binding occurring at pH 7.5 and falling off rapidly at higher and lower pH values. This pH vs binding pattern is similar to the abundance of HS⁻ as a function of pH and may indicate that HS⁻ is the chemical species bound. This fall-off in binding at higher and lower pH values may function to assist in the unloading of sulfide at the trophosome.

In the living worms the concentration of sulfide in the blood is generally higher than in the coelomic fluid reflecting the differences in hemoglobin concentration between these two compartments and indicating that sulfide is readily exchanged between them (Childress, Arp and Fisher, l984). Sulfide is also found at concentrations as high as 3 mM in the trophosomes of living worms and a tissue sulfide binding factor has been demonstrated in trophosome tissue (Childress, Arp and Fisher, 1984). This sulfide binding component may be either bacterial or animal in origin and may function in a manner analogous to the functioning of myoglobin with regard to oxygen. The presence of thiotaurine in high concentrations in the troposome may indicate yet another step in the transport of sulfide to the symbionts (Alberic, l986). Once the sulfide has reached the bacterial symbionts, they may oxidize it only to elemental sulfur if sulfide is in excess or oxygen is limiting (Nelson and Castenholz, l981). Elemental sulfur is present at concentrations of up to 6% of the wet weight of the trophosome in R. pachyptila and probably serves as a non-toxic energy store which can be further oxidized as needed.

In addition to the control of sulfide toxicity by the hemoglobin sulfide binding, Riftia pachyptila also has sulfide oxidizing activity associated with its surface (Powell and Somero, 1986). These authors have suggested that this sulfide oxidizing activity of the animals plays a role in protecting the superficial animal tissues from sulfide poisoning. While it is clear that sulfide can get past these layers into the blood, the manner in which it runs this gauntlet is as yet unknown.

In summary the hemoglobins of Riftia pachyptila apparently function in the uptake and transport of both sulfide and oxygen to the bacterial symbionts. With regard to sulfide, these hemoglobins act to concentrate sulfide from the medium into the blood, transport sulfide in the blood to the symbionts, prevent the spontaneous oxidation of sulfide in the blood, control the distribution of sulfide so that it does not poison the animal cytochrome-c-oxidase or the bacterial symbionts and finally   give the sulfide up either to the bacterial symbionts directly or to a tissue sulfide binding protein. The symbiotic bacteria perform the necessary function of providing a sulfide sink to hold the hemoglobin below saturation. The animal sulfide oxidizing activity probably plays a minor role in controlling sulfide toxicity.

The astonishing properties of these hemoglobins and their roles in vestimentiferans make the absence of mouths and guts in this group seem mundane by comparison.

Calyptogena magnifica

The giant vent clam Calyptogena magnifica lives in small peipheral fissures charaterized by low rates of venting at vent sites on the Galapagos Rift and on the East Pacific Rise (Hessler and Smithey, 1983). This clam lives with its siphons extended up into the oxic ambient water while its greatly extensible foot is anchored in the fissure surrounded by sulfide-rich venting water. The gills of this clam are greatly enlarged and contain endosymbiotic chemolithoautotrophic sulfur-oxidizing bacteria (Boss and Turner, 1980; Felbeck, Childress and Somero, 1981; Cavanaugh, 1983; Fiala-Medioni, 1984) as seems to be generally true for other members of this genus and for other members of the family Vesicomyidae. About 7.7% of the gill mass is made up of bacteria (Powell and Somero, 1986). The symbionts can obtain oxygen directly from the water bathing the gills. Therefore the abundant tetrameric intracellular circulating hemoglobin probably functions only to supply the needs of the animal tissues (Terwilliger, Terwilliger and Arp, 1983). This hemoglobin has a moderate affinity for oxygen and a moderate capacity for oxygen similar to non-vent clams. It readily reacts with sulfide in vitro to form sulfhemoglobin and thereby loses its ability to reversibly bind oxygen (Arp, Childress and Fisher, 1984). Oxygen and possibly sulfide movment in the gills may be facilitated and controlled by the tissue hemoglobin which is found there (Wittenberg, 1985).

Calyptogena magnifica clearly has the ability to concentrate sulfide from the environment into its blood. Freshly captured clams often have sulfide concentrations exceeding 1 mM in their blood serum while their environmental sulfide concentrations are generally less than 0.2 mM (Arp, Childress and Fisher, 1984). The closely related C. elongata can concentrate sulfide from its environment into its blood serum reaching internal concentrations up to 20 mM sulfide and showing concentrating factors greater than2 orders of magnitude (Childress and Arp, 1984). The sulfide in the serum is reversibly bound by a large molecular weight protein which is abundant in the serum (Arp,Childress and Fisher, 1984). In both species, the high affinity of this protein for sulfide is the mechanism by which sulfide is concentrated from the environment. Since sulfhemoglobin is readily produced in vitro but is not not observed in the sulfide containing bloods taken from live clams of this genus, the sulfide binding protein must also serve to protect the hemoglobin

novum. Malacolgia: 161-194.

Cavanaugh CM (1983) Symbiotic chemoautotrophic bacteria in marine invertebrates from sulfide-rich habitats. Nature, Lond 302: 58-61.

Cavanaugh CM (1985) Symbioses of chemoautotrophic bacteria and marine invertebrates from hydrothermal vents and reducing sediments. In: Jones ML (ed) The hydrothermal vents of the eastern Pacific: An overview. Bull Biol Soc Wash 6: 373-388 .

Cavanaugh CM, Gardiner SL, Jones ML, Jannasch HW, Waterbury JB (1981) Prokaryotic cells in the hydrothermal vent tube worm Riftia pachyptila Jones: Possible chemautotrophic symbionts. Science 209: 340-342.

Childress JJ, Arp AJ (1984) Concentration of sulfide into the blood seum of the clam Calyptogena elongata. Am Zool 24: 57A.

Childress JJ, Arp AJ, Fisher CR Jr (1984) Metabolic and blood characteristics of the hydrothermal vent tube-worm Riftia pachyptila. Mar Biol 83:109-124 .

Childress JJ, Lowell W (1982) The abundance of a sulfide-oxidizing symbiosis (the clam Solemya reidi) in relation to interstitial water chemistry. EOS 63: 957.

Doeller JE (1984) A hypothesis for the metabolic behavior of Solemya velum, a gutless bivalve. Am Zool 24: 57A.

Felbeck H (1981) Chemoautotrophic potential of the hydrothermal vent tube worm, Riftis pachyptila Jones (Vestimentifera). Science 213: 336-338 .

Felbeck H (1983) Sulfide oxidation and carbon fixation by the gutless clam Solemya reidi: an animal-bacterial symbiosis. J Comp Physiol 152: 3-11.

Felbeck H, Childress JJ, Somero GN (1981) Calvin-Benson cycle and sulphide oxidation enzymes in animals from sulphide-rich habitats. Nature 293: 291-293.

Felbeck H, Somero GN and Childress JJ (1983) Biochemical interactions between molluscs and their algal and bacterial symbionts. In: Hochacka, PW (ed) The Mollusca, Vol. 2, Environmental Biochemistry and Physiology. Academic Press, New York; pp. 331-338.

Fiala-Medioni A (1984) Ultrastructural evidence of abundance of intracellular symbiotic bacteria in the gills of bivalve molluscs of deep hydrothermal vents. C R Acad Sci Paris, Ser III 298:487-492.

Fisher C R, Childress JJ (1984) Substrate oxidation by trophosome tissue from Riftia pachyptila Jones (Phylum Pogonophora). Mar Biol Lett 5: 171-183.

Hessler, R. R., and W. Smithey, 1983. The distribution and community structure of megafauna at the Galapagos rift hydrothermal vents. In: Rona PA, Bostrom K, Laubier L, Smith KL Jr (eds): Hydrothermal processes at Sea Floor Spreading Centers. Plenum Press, New York; 735-770.

Jones ML (1981) Riftia pachyptila, a new genus, new species. The vestimenteriferan worm from the Galapagos rift geothermal vents (Pogonophora). Proc Biol Soc Wash 93: 1295-1313.

Jørgensen BB (1982) Ecology of the bacteria of the sulphur cycle with special reference to anoxic-oxic interface environments. Phil Trans R Soc Lond B 298: 543-561.

Kelly DP, Kuenen JG (1984) Ecology of colourless sulphur bacteria. In: Codd GA (ed): Aspects of microbial metabolism and ecology. Academic Press, New York; pp. 211-240.

National Research Council (1979) Division of Medical Science, Subcommittee on Hydrogen Sulfide, Hydrogen Sulfide. University Park Press, Baltimore.

Nelson DC, Castenholz RW (1981) Use of reduced sulfur compounds by Beggiatoa sp. J Bacteriol 147: 140-154.

Nelson DC, Revsbech NP, Jørgensen BB (1986) The microoxic/anoxic niche of Beggiatoa spp.: A microelectrode survey of marine and freshwater strains. Appl Environ Microbiol, in press.

Patel S, Spencer CP (1963) The oxidation of sulfide by the haem compounds from the blood of Arenicola marina. J Mar Biol Ass UK 43: 167-175.

Powell EN, Crenshaw MA, Rieger RM (1979) Adaptations to sulfide in the meiofauna of the sulfide system. I. ^{35}S-sulfide accumulation and the presence of a sulfide detoxification system. J exp mar Biol Ecol 37: 57-75.

Powell EN, Crenshaw MA, Rieger RM (1980) Adaptations to sulfide in sulfide-system meiofauna. Endproducts of sulfide detoxification in three turbellarians and a gastrotrich. Mar Ecol Prog Ser 2: 169-177.

Powell MA, Somero GN (1983) Blood components prevent sulfide poisoning of respiration of the hydrothermal vent tube worm Riftia pachyptila. Science 219: 297-299.

Powell MA, Somero GN (1985) Sulfide oxidation occurs in the animal tissue of the gutless clam, Solemya reidi. Biol Bull 169: 164-181.

Powell MA, Somero GN (1986) Adaptations to sulfide by hydrothermal vent animals: Sites and mechanisms of detoxification and metabolism. Biol Bull, in press.

Reid RGB (1980) Aspects of the biology of a gutless species of Solemya (Bivalvia: Protobranchia). Can J Zool 58: 386-393.

Southward E C (1982) Bacterial symbionts in Pogonophora. J Mar Biol Ass U K 62: 889-906.

Terwilliger RC, Terwilliger NB, Arp AJ (1983) Thermal vent clam (Calyptogena magnifica) hemoglobin. Science 219: 981-983.

Terwilliger RC, Terwilliger NB, Schabtach E (1980) The structure of hemoglobin from an unusual deep-sea worm (Vestimentifera). Comp Biochem Physiol 65B: 531-535.

Terwilliger R, Terwilliger N, Bonaventura C, Bonaventura J, Schabtach E (1985) Structural and functional properties of hemoglobin from the vestimentiferan Pogonophora, Lamellibrachia. Biochim Biophys Acta 29: 27-33.

Theede H (1973) Comparative studies on the influence of oxygen deficiency and hydrogen sulfide on marine bottom invertebrates. Neth J Sea Res 7: 244-252.

Vetter RD (1985) Elemental sulfur in the gills of three species of clams containing chemoautotrophic symbiotic bacteria: A possible inorganic energy storage compound. Mar Biol 88: 33-42.

Vetter RD, Wells ME, Kurtsman AL, Somero GN (1986) Sulfide detoxification by the hydrothermal vent crab Bythograea thermydron and other decapod crustaceans. Physiol Zool, in press.

Wittenberg JB (1985) Oxygen supply to intracellular bacterial symbionts. In: Jones ML (ed) The hydrothermal vents of the eastern Pacific: An overview. Bull Biol Soc Wash 6: 301-310.

SULFIDE DETOXIFICATION AND ENERGY EXPLOITATION BY MARINE ANIMALS

Mark A. Powell, Russell D. Vetter, George N. Somero

Scripps Institution of Oceanography
University of California, San Diego
La Jolla, CA 92093, U.S.A.

The metabolism of sulfide was studied in three animals from high sulfide environments. Two of the species, the hydrothermal vent tube worm Riftia pachyptila and the gutless clam Solemya reidi, contain endosymbiotic sulfur bacteria. The symbionts of Riftia oxidize sulfide, but Solemya appears to supply its symbionts with partially oxidized form(s) of sulfur. Mitochondria isolated from Solemya coupled sulfide oxidation to oxidative phosphorylation, showing that exploitation of the energy of sulfide may be achieved by animals as well as by sulfur bacteria. Sulfide detoxification (oxidation) in the hepatopancreas of the crab Bythograea thermydron may protect the animal's aerobic respiration from poisoning by sulfide.

INTRODUCTION

Hydrogen sulfide is an energy-rich molecule that is highly toxic to aerobic organisms due to its strong inhibitory effects on the cytochrome c oxidase reaction, the terminal step in the electron transport chain. Most animals are killed by sulfide concentrations in the micro-molar range, and although some animals have been shown to tolerate sulfide concentrations of this magnitude (Powell et al., 1979; Powell et al., 1980), the adaptations conferring this tolerance have received little study.

The discovery of hot springs at deep-sea spreading centers at which geothermally-produced sulfide is present at high concentrations (up to several hundred micromolar; Edmond et al., 1982; Johnson et al., 1986) and at which dense communities of animals occur has rekindled an interest in the metabolism of sulfide. The vent fauna includes several species, notably the vestimentiferan tube worm Riftia pachyptila, the vesicomyid clam Calyptogena magni-fica, and the mussel Bathymodiolus thermophilus, that contain sulfur bacteria as symbionts in specialized tissues (Cavanaugh et al., 1981; Cavanaugh, 1983; Felbeck et al., 1981). It has been proposed that these symbionts metabolize sulfide and use some of the energy re-leased in sulfide oxidation to drive the net fixation of carbon dioxide (Calvin-Benson cycle) (Cavanaugh et al., 1981; Felbeck et al., 1981). The discovery of marine animals from other sulfide-rich habitats, e.g., areas of decomposing organic matter, that also contain bacterial

Comparative Physiology: Life in Water and on Land. P. Dejours, L. Bolis, C.R. Taylor, E.R. Weibel (eds.) Fidia Research Series, IX-Liviana Press, Padova © 1987

symbionts (Cavanaugh, 1983; Felbeck et al., 1981; Felbeck and Somero, 1982) suggested that the type of symbiosis found initially at the vents might be of widespread occurrence in the marine realm.

Our studies of sulfide metabolism in animals from sulfide-rich habitats have focused on two primary phenomena, one dealing with the tolerance to sulfide, and one with the exploitation of sulfide as an energy source. We have addressed the following questions. First, how do these animals, including species with bacterial symbionts and those lacking symbionts (e.g., the brachyuran crab of the deep-sea vents, Bythograea thermydron), avoid poisoning by sulfide? The crab and the symbiont-containing vestimentiferan and bivalves all display a capacity for aerobic metabolism (cf. Hand and Somero, 1983; Powell and Somero, 1986a), suggesting that mechanisms for detoxifying sulfide are present in all of these animals. Where is sulfide detoxified and by what biochemical mechanisms? What are the end-products of sulfide detoxification?

The second set of questions we have addressed concerns the exploitation of the energy contained in reduced sulfur compounds like sulfide. Are the sulfide oxidation reactions leading to ATP synthesis restricted solely to the bacterial symbionts, or does the animal host also possess the ability to generate ATP during sulfide oxidation? Is sulfide the only reduced sulfur compound capable of serving as a substrate for ATP synthesis, or are other sulfur compounds like sulfite and thiosulfate also used as energy sources?

Our studies have shown that animals from sulfide-rich habitats differ importantly in the manners in which sulfide is detoxified and exploited, and they further suggest that sulfide may play an important role as an energy source in both symbiont-containing and symbiont-free animals from these habitats. We illustrate these points through an examination of three animals from sulfide-rich environments whose strategies for sulfide metabolism differ strikingly.

RIFTIA PACHYPTILA, THE HYDROTHERMAL VENT TUBE WORM.

The most striking species to be discovered at the deep-sea hydrothermal vents is the vestimentiferan tube worm, Riftia pachyptila Jones (Jones, 1981, 1985). This large (up to approximately 1.5 m in length) worm lacks a mouth, gut and anus, an anatomical peculiarity that led immediately to questions about its source of nutrition. The discovery that the worm contains an internal organ, the trophosome, that houses vast numbers of bacterial symbionts led to the suggestion that these symbionts provide the worm with reduced carbon compounds that are synthesized using energy released by the oxidation of sulfide (Cavanaugh et al., 1981; Felbeck, 1981). Indeed, enzymatic studies of trophosome tissue showed the occurrence of enzymes of sulfur metabolism and of the Calvin-Benson cycle of net CO_2 fixation (Felbeck, 1981).

During the March 1985 expedition to the hydrothermal vent site near the Galapagos spreading center, we had the opportunity to test directly the hypothesis that the oxidation of sulfide and other reduced sulfur compounds (sulfite and thiosulfate) provides energy for this symbiosis. We also examined the vent clam and mussel to see if their symbiont-containing tissues, the gills, had the abilities to exploit reduced sulfur compounds, as enzymatic studies suggested was the case (Felbeck et al., 1981).

Figure 1 shows some of the results from studies in which we measured ATP production in trophosome or gill homogenates given different reduced sulfur compounds (Powell and Somero, 1986a). Our findings are consistent with those of Belkin, et al. (1986), who demonstrated stimulation of CO_2 fixation by sulfide in bacterial preparations obtained from trophosome of Riftia and by thiosulfate in bacterial preparations from gill tissue of the vent mussel. For homogenates of Riftia trophosome, both sulfite and sulfide were found to be strong stimulators of ATP synthesis; this stimulation was found only when the bacteria were lysed (by preparing the homogenates in hypo-osmotic buffer). For neither of the bivalves was sulfide stimulatory, a difference we return to below.

The ability of Riftia trophosome to metabolize sulfide is in part a consequence of the sulfide transport system the worm possesses. As detailed in the paper by Dr. Childress in this symposium, the hemoglobin of Riftia functions as an oxygen and a sulfide carrier. When sulfide is bound to the hemoglobin, the sulfide is stabilized against spontaneous oxidation, and the sulfide is rendered unable to poison aerobic respiration (Powell and Somero, 1983; 1986a). This efficient transport system, that at once protects sulfide from oxidation and protects oxidative metabolism from sulfide, enables Riftia to transport large quantities of sulfide from the ambient seawater to the trophosome bacteria without endangering the electron transport system of the mitochondria. By mechanisms that are unknown, the hemoglobin-bound sulfide is released at the trophosome where it is metabolized by the bacteria. We found that a histochemical stain for sulfide oxidizing activity, which uses the dye benzyl viologen as an electron acceptor (Powell and Somero, 1985), gave a positive test only for the bacterial component of the trophosome (Powell and Somero, 1986a).

In Riftia, then, the sulfide metabolism scheme illustrated in Figure 2A appears to be operative. The bacterial symbionts are able to exploit directly the energy in sulfide, although sulfite also is useable as an energy source for driving ATP synthesis.

SOLEMYA REIDI: ANIMAL EXPLOITATION OF SULFIDE

A different set of relationships between a marine animal from a high sulfide habitat and bacterial endosymbionts is found with Solemya reidi, a gutless protobranch clam that occurs in large numbers in marine habitats, e.g., sewage outfall zones, where high concentrations of sulfide are produced by bacteria that use sulfate as an electron sink during the decomposition of organic matter (Reid, 1980). Unlike Riftia, in which the bacteria are able to

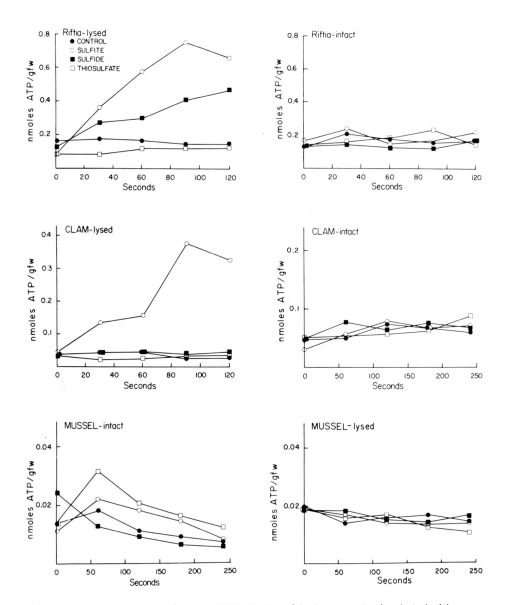

Figure 1. ATP concentrations (nmoles ATP/gFW tissue) in homogenates incubated with a reduced sulfur compound (sulfide (■), sulfite (O), or thiosulfate (□)) at 1 m$\underline{M}$ final concentration or no sulfur compound (control (●)).

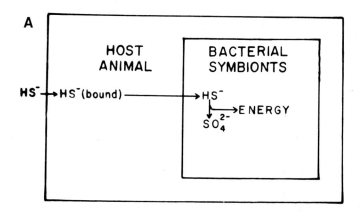

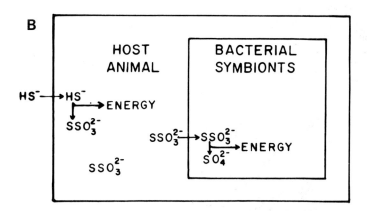

Figure 2. Two strategies for sulfide metabolism which allow exploitation of sulfide as an
energy source and avoidance of poisoning by the highly toxic sulfide. A. Proposed scheme for
sulfide metabolism in _Riftia pachyptila_. Sulfide is transported in a non-toxic protein bound
form from the site of uptake to the bacteria, where it is released for oxidation by the bacteria. B. Proposed scheme for sulfide metabolism in _Solemya reidi_. Sulfide is oxidized at
the site of uptake and the less toxic product of this reaction is transported to the bacteria
for oxidation (thiosulfate is suggested as the product of this reaction). The oxidation of sulfide is coupled to mitochondrial ATP production, thus allowing the animal to capture in a
usable form the energy released by this reaction.

oxidize sulfide, in Solemya all of the oxidation of sulfide appears to occur outside of the bacteria in the animal compartment of the symbiosis (Powell and Somero, 1985). Using enzymatic and histochemical methods for localizing sulfide oxidizing activities that are due to sulfide oxidase enzymes (Powell and Somero, 1985), we found no evidence for any sulfide oxidase activity in the bacterial endosymbionts of the gills, but we did find high activities of sulfide oxidase enzymes in bacteria-free areas of the gills and in the superficial layers of foot tissue (which lacks bacterial symbionts).

These observations suggested that in Solemya, which unlike Riftia lacks a blood transport protein for sulfide, it may be critical for the animal to detoxify rapidly any sulfide entering its tissues by oxidizing sulfide to a less toxic sulfur compound. This oxidation could also be an energy source for the animal if it were coupled to ATP production. These less toxic sulfur specie(s) could then be "fed" to the bacterial symbionts, where additional oxidation would take place to drive ATP formation and net CO_2 fixation by the reactions of the Calvin-Benson cycle that occur in the gills (Felbeck et al., 1981; Felbeck, 1983). Stimulation of CO_2 fixation by reduced sulfur compounds has been demonstrated in two species of Solemya (Cavanaugh, 1983). Thus, in Solemya, a different model for sulfide transport and metabolism was proposed (Figure 2B).

To test this hypothesis we sought to determine the precise locus of sulfide metabolism in Solemya and to establish whether this initial oxidation event was strictly for purposes of detoxification or for a dual function: detoxification and ATP synthesis. We found that the mitochondria of Solemya possess a sulfide oxidizing ability that is linked to ATP synthesis via the electron transport chain (Powell and Somero, 1986b). Addition of sulfide at concentrations up to 20 µM stimulated oxygen consumption and ATP synthesis by mitochondria from both gill and foot. Using well characterized inhibitors of the electron transport chain, we deduced that only the final site (Site 3) of the electron transport chain was involved in the sulfide driven ATP synthesis.

The oxidation of sulfide by the mitochondria of gill and foot tissue of Solemya represents a situation of fine balance. The metabolism of the clam seems to have a strongly aerobic poise, based on activities of cytochrome c oxidase in gill and foot tissues (Hand and Somero, 1983). Therefore, to exploit the energy of sulfide, an exploitation that requires a functional cytochrome c oxidase system, relatively low concentrations of sulfide must be maintained in the cell. Sulfide concentrations above approximately 20 µM strongly inhibited oxygen consumption and ATP production by mitochondria. While it is not known what the sulfide concentrations in the cells of Solemya are in situ, measurements of sulfur compounds in the blood of freshly collected Solemya revealed barely detectable concentrations of sulfide, but thiosulfate concentrations were over 500 µM (Vetter, unpublished observations).

The findings that sulfide oxidizing activity is restricted to the animal portion of the symbiosis in Solemya, and that the animal is capable of synthesizing ATP with the energy released in sulfide oxidation have important implications for the physiological and ecological characteristics of animals from high sulfide habitats. First, for symbiont-containing species, these data show that the model developed from studies of Riftia (Figure 2A) is not universally valid as a description of this newly discovered class of symbiosis. A variety of interactions between host and symbiont in terms of exchange of reduced sulfur compounds may exist. In Solemya and, possibly, in the two vent bivalves (Figure 1), sulfide may not be a substrate for the bacterial symbionts. Instead, the animal host may initially detoxify sulfide and, then, provide to its symbionts sulfite or thiosulfate, which then are metabolized by the bacteria. The type of strategy used in any given symbiosis may be determined largely by the availability of a sulfide transport system, e.g., a sulfide binding protein, that is able to transport sulfide in a safe form to the bacterial symbionts. Riftia possesses such a system; Solemya lacks such a system, and may be required to oxidize rapidly any sulfide that penetrates its cells.

The second major implication of the sulfide oxidizing abilities of mitochondria of Solemya is that animals living in sulfide-rich habitats may be able to exploit directly, i.e., without the assistance of bacterial symbionts, the energy resources offered by reduced sulfur compounds. At the present time we do not know if the mitochondria of Solemya are unique in their abilities to oxidize sulfide and use the energy to drive ATP synthesis. Our hypothesis, which we are currently testing, is that the ability first shown in our studies of Solemya is widespread, and will be found both in symbiont-containing animals and symbiont-free species from high sulfide environments. If this hypothesis is shown to be correct, then it seems likely that inorganic energy sources might contribute importantly to ATP production in animals tolerant of high concentrations of sulfide. Thus, energy flow relationships in a wide range of high-sulfide environments, including salt marshes, hypoxic basins, and geothermal springs, will require re-evaluation.

BYTHOGRAEA THERMYDRON: SULFIDE METABOLISM IN A SYMBIONT-FREE SPECIES

The metabolism of sulfide by an animal from the deep-sea vents that lacks symbiotic sulfur bacteria has been studied only for the brachyuran crab Bythograea thermydron (Vetter et al., 1986). Bythograea is an active forager observed to feed on Riftia. Thus, Bythograea encounters the same high concentrations of sulfide as the symbiont-containing tube worm and bivalves.

To determine how Bythograea, which has high activities of aerobically-poised enzymes (Hand and Somero, 1983), avoids the toxic effects of sulfide, we examined the abilities of the crab to exclude sulfide from its blood, to maintain cardiac and ventilatory functions in the presence of high sulfide concentrations, and to metabolize sulfide that enters the circulation. Our initial findings showed that sulfide diffuses into the crab's circulation very freely, and that the animal does not appear to have any barriers to the entry of sulfide from the ambient water (Vetter et al., 1986). However, even in the presence of sulfide concentrations near 1 mM, the heart rate and the frequency of scaphognathite beating of Bythograea were not less than the rates noted in the absence of sulfide (Vetter et al., 1986).

The abilities of Bythograea to sustain these physiological functions in the presence of high concentrations of ambient sulfide suggested that the crab had an internal site of sulfide detoxification. We found that the sulfide oxidizing abilities of hepatopancreas tissue were strikingly higher than those of other tissues of the crab and, in fact, were in the range noted for the symbiont-containing tissues of the vent tube worm and bivalves. The sulfide oxidizing activity of hepatopancreas appears to be due to the activities of sulfide oxidase enzymes (Vetter et al., 1986), which remain to be characterized. It is not known whether the sulfide oxidizing activities in hepatopancreas are ATP-generating, as in the case of gill and foot tissue of Solemya.

The sulfur compounds produced in hepatopancreas and released into the blood of the crab were studied using monobromobimane derivatization methods and high performance liquid chromatography (HPLC) (Vetter et al., 1986). The major product of sulfide oxidation found in the blood was thiosulfate, which is known to have a very low toxicity relative to sulfide. We propose, therefore, that animal systems for oxidizing sulfide may yield thiosulfate as a major product, and when bacterial symbionts are present, thiosulfate may be "fed" to them as an energy source, as suggested for Solemya reidi in Figure 2B. In animals like Bythograea, that lack sulfur bacteria symbionts, the fate of the thiosulfate formed during sulfide detoxification may be excretion via processes not yet characterized.

SUMMARY

The metabolism of hydrogen sulfide by marine animals is a field of study that only now is receiving a significant level of attention. The discovery of unique animals in sulfide-rich hotsprings in the deep sea led to investigations of the adaptations that provide these animals with a tolerance of sulfide and an ability to exploit the molecule's energy. Subsequent discoveries in a variety of other habitats of symbioses between marine animals and sulfur bacteria furthered interest in these questions (Felbeck et al., 1981).

Although many of the important details of sulfide metabolism remain to be uncovered, both in symbiont-containing and symbiont-free animals, work to date has provided enough information to permit the following conclusions. First, in the symbioses between marine invertebrates and sulfur bacteria, a variety of strategies for transporting and metabolizing sulfide exist. When an effective transport system is present to allow sulfide to be taken up from the ambient seawater and carried to the bacterial symbionts without dangers of spontaneous oxidation of the sulfide or poisoning of respiration by the sulfide, then the bacterial endosymbionts may perform the initial oxidation of sulfide, as well as the subsequent oxidation steps. However, when sulfide is not bound-up in a non-toxic form after entry into the body fluids, it may be critical for the animal to perform the initial oxidation step to prevent accumulation of free sulfide in the body. This accomplishes the needed detoxification of the sulfide and, if the results from studies of mitochondria from Solemya are indicative of metabolic capacities of mitochondria from other animals from sulfide-rich habitats, this oxidation may also be important in generating ATP.

The finding that animal mitochondria can exploit the energy of sulfide is unprecedented, and suggests that inorganic energy sources may be important in both symbiont-containing and symbiont-free animals from sulfide-rich habitats.

REFERENCES

Belkin S, Nelson DC, Jannasch HW (1986) Symbiotic assimilation of CO_2 in two hydro-thermal vent animals, the mussel Bathymodiolus thermophilus, and the tube worm, Riftia pachyptila. Biol Bull 170: 110-121.

Cavanaugh CM (1983) Symbiotic chemoautotrophic bacteria in sulfide-habitat marine invertebrates. Nature 302: 58-61.

Cavanaugh CM, Gardiner SL, Jones ML, Jannasch HW, Waterbury JB (1981) Prokaryotic cells in the hydrothermal vent tube worm, Riftia pachyptila Jones: possible chemoautotrophic symbionts. Science 213: 340-342.

Edmond JM, Von Damm KL, McDuff RE, Measures CI (1982) Chemistry of hot springs on the East Pacific Rise and their effluent dispersal. Nature 297: 187-191.

Felbeck H (1981) Chemoautotrophic potential of the hydrothermal vent tube worm, Riftia pachyptila Jones (Vestimentifera). Science 213: 336-338.

Felbeck H (1983) Sulfide oxidation and carbon fixation by the gutless clam Solemya reidi: an animal-bacteria symbiosis. J Comp Physiol 152: 3-11.

Felbeck H, Childress JJ, Somero GN (1981) Calvin-Benson cycle and sulfide oxidation en-zymes in animals from sulfide-rich habitats. Nature 292: 291-293.

Felbeck H, Somero GN (1982) Primary production in deep-sea hydrothermal vent organisms: role of sulfide-oxidizing bacteria. Trends Biochem Sci 7: 201-204.

Hand SC, Somero GN (1983) Energy metabolism pathways of hydrothermal vent animals: adaptations to a food-rich and sulfide-rich deep-sea environment. Biol Bull 165: 445-459.

Johnson KS, Beehler CL, Sakamoto-Arnold CM, Childress JJ (1986) In situ measurements of chemical distributions in a deep-sea hydrothermal vent field. Science 231: 1139-1141.

Jones ML (1981) Riftia pachyptila, new genus, new species, the vestimentiferan worm from the Galapagos rift geothermal vents (Pogonophora). Proc Biol Soc Wash 93: 1295-1313

Jones ML (1985) On the vestimentifera, new phylum: six new species, and other taxa, from hydrothermal vents and elsewhere. Biol Soc Wash Bull 6: 117-158.

Powell EN, Crenshaw MA, Reiger RM (1979) Adaptations to sulfide in the meiofauna of the sulfide system. I. ^{35}S-sulfide accumulation and the presence of a sulfide detoxification system. J Exp Mar Biol Ecol 37: 57-69.

Powell EN, Crenshaw MA, Reiger RM (1980) Adaptations to sulfide in the meiofauna of the sulfide system. Endproducts of sulfide detoxification in three turbellarians and a gastrotrich. Mar Ecol Prog Series 2: 169-181.

Powell MA, Somero GN (1983) Blood components prevent sulfide poisoning of respiration of the hydrothermal vent tube worm, Riftia pachyptila. Science 219: 297-299.

Powell MA, Somero GN (1985) Sulfide oxidation occurs in the animal tissue of the gutless clam Solemya reidi. Biol Bull 169: 164-181.

Powell MA, Somero GN (1986a) Adaptations to sulfide by hydrothermal vent animals: sites and mechanisms of detoxification and metabolism. Biol Bull, in press.

Powell MA, Somero GN (1986b) Hydrogen sulfide oxidation is coupled to oxidative phosphorylation in mitochondria of Solemya reidi. Science 233: 563-566.

Reid RGB (1980) Aspects of the biology of a gutless species of Solemya (Bivalvia: Protobranchia). Can J. Zool 58: 386-395.

Vetter RD, Wells ME, Kurtsman AL, Somero GN (1987) Sulfide detoxification by the hydrothermal vent crab Bythograea thermydron and other decapod crustaceans. Physiol Zool, in press.

ARSENIC TOXICITY ON SEA OR LAND

Andrew A. Benson

Scripps Institution of Oceanography, A-002
La Jolla, California 92093, U.S.A.

Arsenate absorption and detoxication by all algae and aquatic higher plants is a major marine metabolic process. They reduce and methylate arsenic by a common process ultimately releasing an arsenoribosyl glycerol. Terrestrial and aquatic plants having roots in phosphate-rich sediments reduce and methylate absorbed arsenate but like the fungi and animals they do not produce the less toxic arsenoribosides. Arsenic metabolism in plants revealed mechanisms important for understanding the probable mode of action of arsenical antiparasitic drugs.

Arsenic toxicity has frequented the thoughts and fears of man for a milennium or more. Al-Razi wrote (Elgood, 1935) in 800 A.D. of antidotes for arsenic poisoning, the magic bezoar stones (see Footnote 1) of ancient Persia. Monardes (1580) and Bauhini (1613) described these remarkable stones, their sources, their curative powers, and their market values four hundred years ago. Later the 'arsenic eaters' of Steiermark tantalized medical science (Tschudi, 1853 cf. Chevallier, 1854) by the surreptitious reports of their remarkable immunity (Maclaglan, 1864, 1875) and by the apparent tonic effects of their arsenic diet. Medical studies, that of Alexander Borodin (1858), for instance, revealed important differences and similarities between the metabolism of phosphate and arsenate in the human. Fortunately it did not affect the longevity of his Prince Igor or the Polovtsian maidens. Even at the times of Ehrlich (1909) or of Peters (1952) when arsenic biochemistry made giant strides the real masters of arsenic metabolic reactions remained unrecognized.

AQUATIC PLANTS

The earth's earliest aquatic plants, the bluegreen algae (cyanobacteria) of 3,500 million years ago knew exactly how to metabolize arsenate, how to detoxicate it, and how to excrete it so rapidly it could not accumulate in their cells. All aquatic plants, those without roots in soil or sea floor, metabolize arsenate readily, and apparently by the same general processes (Nissen and Benson, 1982). They communicate their biochemistry by the amounts and nature of their radioarsenical intermediates seen on our two-dimensional paper chromatograms and the corresponding radioautograms (Benson et al., 1950). With these as guides we have looked at the arsenate metabolism of dozens of microalgae, sea-

Comparative Physiology: Life in Water and on Land. P. Dejours, L. Bolis, C.R. Taylor, E.R. Weibel (eds.) Fidia Research Series, IX-Liviana Press, Padova © 1987

weeds, and aquatic higher plants of both fresh water and seawater. All have their major
pathway in common. All must have derived the necessary genes from their most ancient
ancestors, the bluegreen algae.

TERRESTRIAL PLANTS
Terrestrial plants, and those aquatic plants having roots in the substratum where phosphate
levels are sufficient, do not metabolize arsenate as do their aquatic relatives. Their rates
of arsenate utilization are extremely slow and the intermediates produced are different and
low in concentration. Nissen and Benson (1982) found no organoarsenical intermediates in
pine (Pinus halepensis, P. radiata) seedlings fed radioarsenate. Reduction of arsenate to
arsenite, however, was extensive in these and in the angiosperm plants which were tested.
Under stress of nitrogen and phosphorus deficiency, reduction and some methylation was
observed. Up to 51% of the reduced arsenic in tomato leaves was found in methylated
compounds, methane-arsonate, and dimethylarsinate (cacodylate). The ability to detoxicate
arsenic by conversion to the arsenoribosylglycerols seems to be lacking in "terrestrial"
plants, whether they live above or below the water surface.

A special symbiotic relationship developed in land plants which may involve arsenic metabolism. The root-symbiotic mycorrhizae serve their host by collecting phosphate through
the much more extensive surface area of their micelia. These mycorrhizal fungi collect
phosphate in their excursions through the soil and exchange it for sugars at the root-
mycorrhizal interface. The fungi are well known for their general abilities to methylate
arsenic compounds. The possibility that such arsenicals may be transferred at the root-
mycorrhizal interface should be considered.

THE COURSE OF ARSENATE DETOXICATION
Recognition of arsenocholine (Edmonds et al., 1977) of lobster, shrimp and fish and the
arsenoribosyl glycerols of marine algae (Edmonds and Francesconi, 1981) failed to suggest
the metabolic sources or pathways involved in their biosynthesis. Neither compound could
have been anticipated. Only the absence of prejudice of X-ray crystallography could have
led researchers directly to these structures.

At that time Cooney (1978, 1981) was studying the reactions of the arsenical lipid of marine
diatoms and dinoflagellate algae, the major plants of the sea, and related metabolites he
had isolated by two-dimensional paper chromatography. Base-catalyzed deacylation of the
lipid produced a compound identical to one water-soluble metabolite. Acid hydrolysis of this
produced an arsenical hydroxy compound, the same as derived from the two other arsenical
metabolites of algae. Bacterial phosphodiesterase cleaved his compound, C, in the same way
at a rate commensurate with that for glycerophosphorylcholine (Herrera-Lasso and Benson,
unpubl.). The unknown lipid was clearly a phosphatidyl ester of an arsenical hydroxy compound. With the identity of the arsenical from X-ray crystallography (Edmonds et al., 1982),

the structure of the ubiquitous arsenophospholipid of aquatic plants was readily deduced.

Phosphatidylglycerol arsenoriboside is a relatively hydrophilic phospholipid, resembling in properties, the gangliosides of animal tissues. Its hydrophilic arsenoribosyl group will be exposed to the environment when in the cell membrane (plasmalemma) and thereby subject to attack by external bacterial oxidases and consequent release to the sea.

The nature of the arsenicals of the sea has attracted skillful marine chemists. At concentrations of $2 \times 10^{-8}M$ for arsenate, and less for arsenite, the problem of analysis was not simple. Braman et al. (1977) and Andreae (1978) developed reductive and gas chromatographic methodology for analysis of the methylated arsenicals, methane arsonate and cacodylate. We find both in algae, but in rather low concentrations compared to the arsenoribosides. What, then, happens to the ribosides as an alga grows or dies?

EXPERIMENTAL OBSERVATIONS

Radioarsenic–labeled algal cultures of many species have been examined for their metabolites. The 3-O-sulfate ester 1-glyceryl dimethyl 5-C (5-deoxyribosyl-dimethylarsineoxide), compound "A", is universally predominant. Cultures of <u>Chaetoceros gracilis</u>, either bacteria-free or cultured with associated bacteria, were grown 2-14 days with arsenate-[74]As in low phosphate media. The algae (and bacteria) were centrifuged and the medium concentrated to dryness in vacuo. Residual water was removed in high vacuum at room temperature. The salt was extracted with an equal volume of warm dry methanol. The concentrated methanol extract was chromatographed two-dimensionally on paper. Radiograms and subsequent cochromatography revealed identity of the excreted arsenical as the sulfate ester of the arsenoribosylglycerol, "A."

Excretion of the anionic arsenoribosylglycerol sulfate into algal culture media--or into the sea--is by at least two orders of magnitude more important than possible release of methanearsonate or cacodylate. The stability of this compound toward bacterial metabolic attack suggests that it must be a major arsenical component of seawater.

The analytic methods which have been used for arsenicals in seawater fail to reveal the arsenoribosides. Borohydride reduction, giving methylarsines from methanearsonate and cacodylate, gives a non-volatile dimethyl-(5-deoxy-5-ribosyl) arsine. Only by oxidative degradation to arsenate or alkaline degradation to produce cacodylate can the standard analytical methods be adapted. The concentration of arsenoribosylglycerols in the sea should be an inverse indicator of phosphate levels and consequent arsenate detoxication activity.

WHY THE ARSENORIBOSIDES?

How did the arsenoribosides become Nature's choice for detoxicating arsenate in aquatic plants? Though the answer is yet equivocal we envision a course based upon extensive fact and comparative metabolic logic.

Knowles (personal communication) has delineated the initial activation of arsenate, with production of adenylyl arsenate, AMP-As. We postulate that this can be a substrate for AMP-S reductase to produce, instead of AMP-sulfite, AMP-arsenite. He has documented reduction of arsenite to arsonous acid, H-As=O, by a number of dithiol reductases, viz. glutathione reductase, lipoamide dehydrogenase, thioredoxin reductase. This insidious arsenical esterifies the first free -SH group it can find (Knowles, 1982), leaving the plant with a problem.

A laboratory rat, grown on commercial feeds, carries arsenic on the 93-Cys of 4% of its hemoglobin molecules, just because it has no effective way or need to extricate itself from the intruder. It is no wonder, perhaps, that the rat can survive in environments where other animals would be unable to resist the blood parasites.

Benson and Knowles (1981; Knowles and Benson, 1983) discovered the metabolic reduction of the herbicide, methane-arsonate, by a similar process in the chloroplast photosynthetic Hill Reaction. Methane-arsonous acid, CH_3-As=O, likewise reacted with available -SH groups—as did the dreaded war gas, Lewisite. Protein-bound Lewisite could be removed by displacement by BAL (Peters, 1952), a dithiol. Likewise, the arsenothioester formed from arsonous acid can be displaced by a dithiol to form a cyclic arsenodithio ester. When the dithiol is a critical enzyme like lipoamide dehydrogenase the result can be disastrous for the organism. This is to our advantage when the 'organism' is a trypanosome or a malarial sporozoite. But when it is the alga's own enzyme it must be removed. And it must be done rapidly—as the arsenic content of marine algae is very low. We believe that methylation and 5'-deoxyadenosylation by cobalamine and/or S-adenosylmethionine are involved in the process. Then, with sufficient ornamentation the critical disulfhydryl enzyme can be released from its arsenic bondage to produce the trialkylarsine oxide, dimethyl-5-deoxyribosylarsine oxide, derivatives found in all aquatic plants.

Excretion as the anionic sulfate ester of glyceryl dimethylarsenosoriboside, "A," is continuous through the life of the alga. The riboside appears to be stable in seawater. Its ultimate reversion to arsenate is certain but not at all understood, kinetically or metabolically.

What is the cost of survival in a sea of arsenate? The algae of most of the sea live in media where the concentrations of arsenate exceed that of essential phosphate. Though the ions are so similar in charge and size that plants cannot discriminate effectively, their chemical differences render them separable metabolically. Phosphate esters or anhydrides are not readily reduced whereas those of arsenate are reducible. Then, the metabolic paths of arsenic and phosphorus diverge. When reduced and methylated, arsenic resembles nitrogen. The ultimate derivatives, arsenocholine and arsenobetaine, so resemble their nitrogen analogs that neither the liver nor the kidney recognize the difference (Welch, 1936, Welch and Landau, 1942).

Unable to discriminate significantly, the algae absorb arsenate and must deal with it. Sargassum weed, adapted to a very low phosphate supply, differs in that it is able to capture adventitious phosphate or arsenate for later utilization. Other algae, in media containing radioarsenate, contain little free arsenate; they must detoxicate it rapidly. The metabolic cost of this is considerable. The fact that the low phosphate regions of the sea are largely equatorial and therefore solar energy-rich offers a solution to the problem. The reducing power and the energy mobilized in photophosphorylation provide the reagents and substrates needed to detoxicate the absorbed arsenate by the mechanism outlined above. If ribosyl-glycerol is lost for each arsenic absorbed the process requires fixation of at least ten molecules of CO_2 and the corresponding amounts of reduced and phosphorylated nucleotides. Arsenic detoxication is a costly process.

Phosphate and arsenate, so similar at the outset, differ greatly in their metabolism. Arsenate is readily reduced while phosphate is not. The arsonous acids are highly reactive, forming mono- and dithioesters avidly with free -SH groups. When reduced further to -1 arsenic, the arsines and arsonium compounds resemble nitrogen compounds in their proper-ties. Nature has selected these innocuous arsenical derivatives as products of arsenic metabolism in the sea.

Footnote 1. Old Persian: "to protect against poison."

REFERENCES
Andreae MO (1978) Distribution and speciation of arsenic in natural water and some marine algae. Deep-Sea Res 25: 391-402.

Bauhini C (1613) De Lapidis Bezaar. 295 p. Conr. Waldkirch, Basel.

Benson AA, Bassham JA, Calvin M, Goodale TC, Haas VA, Sepka W (1950) The path of carbon in photosynthesis V. Paper chromatography and radioautography of the products. J Am Chem Soc 72: 1710-1718.

Benson AA and Knowles FC (1981) Arsenic metabolism and photosynthetic productivity. In: Akoyunoglou (ed): Photosynthesis VI. Photosynthesis and productivity, photosynthesis and environment. Balaban International Science Services, Philadelphia; pp. 33-37.

Borodin A (1858) The analogous arsenical acids of phosphorus, their chemical and toxicologi-cal relationships. Doctoral (med.) Thesis. 33 p. Koroleva & Co. Press, St. Petersburg.

Braman RS, Johnson PL, Foreback CC, Ammons JM, Bricker JL (1977) Separation and Deter-mination of nanogram amounts of inorganic arsenic and methyl-arsenic compounds. Anal Chem 49: 621-625.

Chevallier A (1854) Arsenic eaters. Boston Med Surg J 51(10): 189-195.

Cooney RV, Mumma RO, Benson AA (1978) Arsoniumphospholipid in algae. Proc Nat Acad Sci USA 75: 4262-4265.

Cooney RV (1981) The metabolism of arsenic by marine organisms. Dissertation, University of California, San Diego. Diss Abstr Int Ser B 41: 4495.

Ehrlich P (1909) Über den jetzigen stand der chemotherapie. Ber 42: 17-47.

Edmonds JS, Francesconi KA, Cannon JR, Raston CL, Skelton BW, White AH (1977) Isolation, crystal structure and synthesis of arsenobetaine, the arsenical constituent of the Western rock lobster Panulirus longipes cygnus George. Tetrahedron Lett 18: 1543-1546.

Edmonds JS, Francesconi KA (1981) Arseno-sugars from brown kelp (Ecklonia radiata) as intermediates in cycling of arsenic in a marine ecosystem. Nature 289: 602-604.

Edmonds JS, Francesconi KA, Healy, PC, White AH (1982) Isolation and crystal structure of an arsenic-containing sugar sulphate from the kidney of the giant clam, Tridacna maxima. X-ray crystal structure of (2S)-3-[5-deoxy-5-(dimethylarsinoyl)-β-D-ribofuranosyloxy]-2-hydroxypropyl hydrogen sulphate. J Chem Soc Perkin Trans I: 2989-2993.

Elgood C (1935) A treatise on the Bezoar Stone by the late Mahmud Bin Masud, the Imad-ul-din the physician of Isphahan. Ann Med Hist 7: 73-80.

Knowles FC (1982) The enzyme inhibitory form of inorganic arsenic. Biochem Int 4: 647-653.

Knowles FC and Benson AA (1983) Mode of action of a herbicide. Johnsongrass and methane-arsonic acid. Plant Physiol 71: 235-240.

Maclagan RC (1864) On the arsenic-eaters of Styria. II. Edinburgh Med J 10: 200-207.

Maclagan RC (1875) Arsenic eaters of Styria. VI. Edinburgh Med J 21: 526-528.

Monardes N (1580) Of the Bezoar Stone. In: Joy Full Newes out of the New-found World. Fol. 121 to Fol. 132. Transl. by J. Frampton. Bonham Norton, London, 1596.

Nissen P, Benson AA (1982) Arsenic metabolism in freshwater and terrestrial plants. Physiol Plant 54: 446-450.

Peters RA (1952) Biochemistry of arsenic. J R Inst Public Health Hyg 15: 89-103; Dithiols as antidotes. ibid. 122-139.

Welch AD (1936) Utilization of the arsenic analogs of choline chloride in the biosynthesis of phospholipid. Proc Soc Exptl Biol Med 35: 107-108.

Welch AD, Landau RL (1942) The arsenic analog of choline as a component of lecithin in rats fed arsenocholine chloride. J Biol Chem 144: 581-588.

Part V

SENSORY INFORMATION AND BEHAVIOUR

ORIENTATION IN AIR AND WATER PRINCIPAL DESIGN FEATURES

Rüdiger Wehner

Department of Zoology
University of Zürich
CH-8057 Zürich, Switzerland

It is almost a truism that much of the master plan of animal design has been shaped by some very general properties of the medium
surrounding the animal. A few examples will give the flavour of
the argument: There are much larger fishes in water than there are
birds in the air; the body of a fish is more perfectly streamlined
than that of a bird; for a fish it is more difficult to extract
oxygen from its environment, but less costly to move through this
environment, than it is for a bird; many fishes possess gas bladders equipped with gas secreting and gas absorbing structures, but
birds never do. Ultimately, all these functional differences between aquatic and terrestrial animals are due to the way that
water and air differ in such general properties as density,
viscosity, oxygen content, gas diffusion rates, etc. These
fundamental physical properties of air and water are responsible
not only for how animals gain and spend their energy, but also for
how they gain the information necessary to move around, to detect
and localize objects, in short: to explore their environment.

Physical constraints, physiological adaptations ...

Just have a look at the sea. Above the water surface one can see
even the island furthest away on the horizon, but one will not receive any sounds from there. On the contrary, under water the visual scene gets blurred and obscured even a few metres ahead of the
observer, but one may readily pick up the sounds produced by the
engine of a ship too far away to be seen. The main reasons for
these striking differences between seeing and hearing in water and
air are simple and straightforward.

Let us first turn to vision. In both air and water, absorption and
scattering of light decrease the brightness and contrast of the

Comparative Physiology: Life in Water and on Land. P. Dejours, L. Bolis, C.R. Taylor, E.R. Weibel
(eds.) Fidia Research Series, IX-Liviana Press, Padova © 1987

image, respectively, but these effects are so much stronger in water than in air that under water vision is essentially a short-range affair. It is only in very clear oceans, and close to the water surface, that even large black objects can be seen up to distances of about 30 m (Lythgoe, 1986). In the turbulent sea or in turbid fresh water lakes visibility may be even down to a few centimetres.

According to the extreme low light levels prevailing in such aquatic environments many fishes exhibit retinal specializations allowing for high quantum capture rates and light sensitivities. Among these specializations are powerful spherical lenses made of optically inhomogeneous material and providing the eyes with the necessarily high focusing powers (Land, this volume), light reflectors underlying the receptor layers (tapeta lucida: Nicol et al., 1974), lack of screening pigment (Wagner and Ali, 1978), large photoreceptors (Munz and Mc Farland, 1977) often arranged in multiple layers (tiered retinae: Vilter, 1953) or in bundles forming functional multi-receptor units ("macroreceptors": Locket, 1971, 1977), and high convergence ratios of receptors and higher order interneurons. In photometric terms, the effects which most of these adaptations have on the brightness of the image seen by a fish can be worked out rather straightforwardly, but there is one big unknown: the last parameter mentioned above, that is the rate of summation of receptor outputs. This parameter might affect light sensitivity most dramatically, but at present, very little can be said about it (see e.g. Menezes et al., 1981). The situation is even worse in invertebrates where questions of receptor convergence have rarely been raised.

There is one remarkable exception to the statement just made: the case of the insect ocellus. Contrary to what is found in most eyes, in the insect ocellus retina and lens are not adjusted to keep the image focused. The image in the receptor plane is blurred. In line with this low spatial resolution is the high neural convergence found in all insect ocelli. Often several thousand receptors feed on only one or two dozens of second order interneurons. In flying insects this ocellar high-sensitivity, low-acuity system is used as a horizon detector (Stange and Howard, 1979), which discards information about fine spatial detail, but nothing is known about whether aquatic or other animals living under extreme low light conditions use such massive neural convergence to boost light sensitivity. Again, as in the vertebrate case, it is the optics rather than the properties of neural wiring that have been studied in considerable detail.

In terms of absolute sensitivity (even when based on photometric reasoning alone), the superposition compound eyes of aquatic crustaceans and nocturnal insects as well as the single lens eyes of cephalopods and nocturnal spiders by far exceed the human eye. All these types of eye are masterly described by M. Land in his 1981 review. For example, the mirror-type superposition eye of the deep-sea shrimp Oplophorus is hundred times more sensitive to light than is even the peripheral human retina. It may seem surprising that such high sensitivities are achieved by eyes not more than one or two millimetres across, for one would assume that light sensitivity is correlated with the size of the eye, especially with the size of the lens. However, the brightness of the image is proportional to the pupil aperture (and thus the size of the lens) only for point sources of light such as stars or bioluminescent objects. For the more conventional visual scenes formed by extended sources image brightness depends on the F-number of the eye (focal length divided by pupil aperture) and is proportional to F^{-2} (Kirschfeld, 1974). In this respect, the largest eye ever found, the 40 centimetre eye of an unidentified giant squid (Beer, 1897), did not exhibit higher retinal illuminances than a scaled-down version of this world record does in a more common squid. In fish and cephalopod eyes, F-numbers can be as low as 0.8 (Munk and Frederiksen, 1974; Land, 1981) and thus lower than even in the most advanced camera lenses. Nevertheless, in spite of all these optical tricks, underwater vision is restricted to the near field.

In contrast, underwater hearing extends into the far field. The propagation of sound pressure waves is almost five times faster in water than in air. This allows low frequency sounds as produced by baleen whales (about 20Hz) to attenuate very little when travelling over large distances. By calculating transmission losses at various depths, especially in those layers in which sound is effectively trapped due to particular temperature condi-

tions, Payne and Webb (1971) computed the maximum possible range
at which one fin whale might hear another. They arrived at numbers
of several hundred kilometres. One only wonders how these baleen
whales could use an acoustic communication system in modern times
when the engines of ships produce powerful sounds exactly in the
whales' frequency band.

It is also in terrestrial animals that a number of examples can be
cited in which frequency, pulse rate and other acoustic parameters
of animal sound are adapted to the particular transmission characteristics of certain kinds of environment. Tropical forest-
dwelling birds use monotonous tonal calls of frequencies around 2
kHz, whereas birds of adjacent open habitats produce trill-like
calls which change rapidly in amplitude and frequency (Morton,
1975). This is in accord with the fact that in forests rapid modu-
lations are effectively masked and blurred, while in open areas
temperature gradients and wind cause low frequency turbulences of
the air which degrade monotonous sounds more effectively than
rapidly modulated ones (Gerhardt, 1983).

In general, spatial localization can be done more precisely by
visual than by acoustic means. One is even justified in saying
that, because of the physical properties of light, vision is the
most accurate source of spatial information that an animal can
gain about the world. However, as mentioned above, vision is
strongly impaired under water. The "veil" of scattered light
between the eye and any object of the outside world reduces not
only the range but also the acuity of vision. Nevertheless
localizing a sound source under water is an even more formidable
task. Most fishes do not discriminate differences in the
directions of sound sources that are smaller than 10-20° (Hawkins
and Sand, 1977). On the one hand, the high velocities of sound
under water make the acoustic wavelengths much larger than most
fishes are. On the other, the density of a fish is nearly the same
as that of water, so that the fish itself becomes part of the
sound field. Hence, a fish cannot use interaural time and
intensity differences for localizing a source. In most fishes, the
two inner ears are so close to one another that the interaural
time difference is less than 10 µs (van Bergeijk, 1964). To make
the situation even worse, sound pressure is detected indirectly

through the motion of the walls of **one** sac of gas, the swimbladder, which contracts and expands in the pressure field. This has led some researchers to go so far as to suppose that, in fish, the ears could not subserve directional hearing. Consider, however, that sound is a pressure wave accompanied by particle motion. It is this particle motion that stimulates the otolithic apparatus of the inner ear directly and can thus be used for directional analysis (Schuijf and Buwalda, 1980). To sum up, the physical properties of air and water have led to the evolution of substantially different mechanisms for sound source localization in fish and terrestrial vertebrates.

The preceding section has shown that in turbid and murky waters, where vision is of little help, hearing cannot substitute for vision in providing the animal with detailed information about the spatial disposition of its surroundings. But what sense can? In fish two sensory systems meet these demands: the lateral-line and the electrolocation system (for the latter see Keynes, this volume). The lateral-line organs respond to relative motion between the fish and the water surrounding it (Denton and Gray, 1983) and thus are effective only in the near field. Notice, however, that it is not so much the flow field itself, but rather its spatial gradient to which this system responds. Only if the flow field varies along the fish, will the system be stimulated (Kalmijn, 1986). To ease the neural processing of such variations, the fish holds its body straight in the water so as to keep the spatial relationship of the receptors constant (Weissert and von Campenhausen, 1981). The same is true for the electrolocation system. Electrically sensitive fishes swim by gentle undulations of a long dorsal fin rather than by violent body movements (Lissmann, 1958). Again, as is the case with the lateral-line system, electrolocation is restricted to the near field and can be used only a few centimetres around the fish.

It is immediately apparent that, because of the physical properties of air and water, the two near-field detector systems present in fish and amphibians have no direct counterpart in terrestrial animals. What superficially looks like the exception to this rule is the newly discovered electric sense of a higher vertebrate, the Australian monotreme, the platypus (Scheich et al., 1986), but this exception is of a kind one would predict. The platypus is a diving mammal using its bill as an electroreceptive organ. This organ has evolved independently of the corresponding organs in the lower vertebrates, e.g. belongs to the trigeminal rather than the acoustic-lateralis system.

On the other hand, an orientation system that could be used, on purely physical grounds, in both air and water is echolocation. And it is used. Bats and dolphins are the prime examples (Keynes, this volume). Of course, because of the properties of sound transmission in air and water, some details of these sonar systems should be different in aquatic and terrestrial animals. For example, sound of constant frequency exhibits higher velocities and longer wavelengths in water than in air. With all other things being equal (which they rarely are), the same accuracy of orientation requires that the echolocative sounds are of higher frequencies in aquatic than in terrestrial animals. This is indeed what occurs. The sound frequencies of most bats lie in the range of 15-60 kHz, with only some hipposiderids exceeding 100 kHz (Neuweiler et al., 1984), while odontocete whales employ sound frequencies of 120-180 kHz. Furthermore, as deduced from some impressive series of behavioural experiments (Busnel and Fish, 1980), it is probably not too gross an oversimplification to state that bats localize objects in the air as precisely as dolphins do under water.

One information channel has been neglected so far: the chemical one. It is as if we had involuntarily overlooked the chemical senses for the very reason that they mediate the most ubiquitous and certainly oldest kind of orientation and communication in animals. Unlike light and sounds, chemical signals require some bulk movement of the medium for effective transmission. Of course, dispersion could occur by diffusion, but this is an extremely slow mechanism, even for airborne molecules. Under water the situation is much worse: the diffusion rate is down by four to five orders of magnitude. Consequently, animals broadcast their pheromones into a moving medium, be it wind or water current, and the receivers must employ some intricate behavioural mechanisms to pick up the chemical message (Atema, this volume).

... and speculations about evolution

Let me finally return to vision and focus on one important adaptation not yet mentioned: the correlation between the spectral properties of an animal's environment and the spectral absorption characteristics of the photopigments within the animal's photoreceptors (Lythgoe, 1979). Ever since Wald (1937) discovered the porphyropsins, then a new class of photopigments, this correlation has been stressed. The ambient light of freshwater lakes is relatively richer in long-wavelength radiation than that of the sea, and the porphyropsins of freshwater teleosts and larval amphibians provide their owners with increased sensitivities to just these longer wavelengths. However, the match between the spectral properties of photopigment and environment is often not as close as one would expect, and often two or more spectral types of receptor occur within the same part of the retina. In the following I shall argue that a two-receptor system need not have evolved primarily for colour coding but for less exotic and more fundamental tasks.

In terrestrial animals the background against which an object must be detected and identified can vary enormously in both colour and structure depending on the type of habitat and the line of sight. For aquatic animals the situation is quite different. As already pointed out, the heavy scattering of light under water largely destroys structural detail, so that most objects stand out against a structurally homogeneous background. Furthermore, at a given depth and for a given line of sight the spectral composition of the background spacelight is almost always the same. It does not seem too far-fetched to assume that this remarkable environmental constancy must have had important consequences for the evolution of spectrally matched photoreceptors.

But to what should the photoreceptors be matched? If they were spectrally matched to the background spacelight, as most actually are, a dark object would contrast well with the background, but a bright object would not. However, as first pointed out by J.N. Lythgoe already twenty years ago (see also Mc Farland and Munz, 1975; Levine and Mac Nichol, 1982), the situation is reversed if the photopigment did not match the spacelight (provided that the

light illuminating the object is spectrally broader than the
spacelight, as is naturally the case). As a consequence, a two-receptor system is advantageous for detecting both dark and bright
objects against a homogeneously lit background, even if the neural
interactions necessary for hue discrimination had not yet evolved.

The aquatic environment might have favoured the evolution of a
two-receptor system for other (somewhat related) reasons as well.
The differential absorption of the wavelengths of light is much
more pronounced in water than it is in air. Under water, the
spectral content of the background spacelight varies considerably
with depth, turbulence, time of day, and other factors. For any
one-receptor system the reliable detection of objects seen under
such spectrally varying spacelight conditions is a rather tricky
task. To illustrate how tricky this task is let us consider a human monochromat whose photoreceptors were all equipped with the
same common 500 nm rhodopsin. At noon, under full daylight conditions, a red cherry fruit would appear to the monochromat darker
than the surrounding green leaves, but at dusk when longer
wavelengths prevail in the ambient illumination, it would appear
lighter than its surroundings. Thus, during the course of the day,
the same object would change its perceived brightness - a very un-
comfortable state of affairs if the object is to be detected reliably. Brightness constancy is achieved, i.e. the object is made
to appear darker (or lighter) than the background irrespective of
the spectral content of the spacelight, only when at least two
types of receptor with different spectral sensitivities are used
and their outputs compared. As recently argued by von Campenhausen
(1986), brightness constancy does not necessarily imply colour
discrimination (which whould require additional neural wiring).
Hence, the need for brightness constancy might well have been a
major driving force behind the evolution of a two-receptor - but
not necessarily dichromatic - system; and under water this need is
certainly greater than in air.

References

Beer T (1897) Die Accommodation des Kephalopodenauges. Pflügers Arch ges Physiol 67: 541-586.

Bergeijk WA van (1964) Directional and nondirectional hearing in fish. In: Tavolga WN (ed): Marine Bio-Acoustics. Pergamon, Oxford; pp. 281-299.

Busnel RG, Fish JF (eds) (1980) Animal Sonar Systems. Plenum, London.

Campenhausen von C (1986) Photoreceptors, lightness constancy and colour vision. Naturwiss, in press.

Denton EJ, Gray JAB (1983) Mechanical factors in the excitation of clupeid lateral lines. Proc Roy Soc London B 218: 1-26.

Gerhardt HC (1983) Communication and the environment. In: Halliday TR, Slater PJB (eds): Animal Behaviour, Vol. 2. Blackwell, Oxford; pp. 82-113.

Hawkins AD, Sand O (1977) Directional hearing in the median vertical plane by the cod. J Comp Physiol 122: 1-8.

Kalmijn AJ (1986) Hydrodynamic and acoustic field detection in aquatic vertebrates. In: Atema J, Fay RR, Popper AN, Tavoga W (eds): Sensory Biology of Aquatic Animals. Springer, New York; in press.

Kirschfeld K (1974) The absolute sensitivity of lens and compound eyes. Z Naturforsch 29c: 592-596.

Land MF (1981) Optics and vision in invertebrates. In: Autrum H (ed): Handbook of Sensory Physiology, Vol. VII/6B. Springer, Berlin; pp. 471-592.

Levine JS, Mac Nichol EF (1982) Color vision in fishes. Scient Amer 246/2: 108-117.

Lissmann HW (1958) On the function and evolution of electric organs in fish. J exp Biol 35: 156-191.

Locket NA (1971) Retinal anatomy in some scopelarchid deep-sea fishes. Proc Roy Soc London B 178: 161-184.

Locket NA (1977) Adaptations to the deep-sea environment. In: Crescitelli F (ed): Handbook of Sensory Physiology, Vol. VII/5. Springer, Berlin; pp. 67-192.

Lythgoe JN (1979) The Ecology of Vision. Clarendon, Oxford

Lythgoe JN (1986) Light and vision in the aquatic environment. In: Atema J, Fay RR, Popper AN, Tavoga W (eds): Sensory Biology of Aquatic Animals. Springer, New York; in press.

Mc Farland WN, Munz FW (1975) The evolution of photopic visual pigments in fishes. Vision Res 15: 1071-1080.

Menezes NA, Wagner HJ, Ali MA (1981) Retinal adaptations in fishes from a floodplain environment in the central Amazon basin. Rev Can Biol 40: 111-132.

Morton ES (1975) Ecological sources of selection on avian sounds. Amer Naturalist 109: 17-34.

Munk O, Frederiksen RD (1974) On the function of aphakik apertures in teleosts. Vidensk Medd Dansk Naturhist For 137: 65-94.

Munz FW, Mc Farland WN (1977) Evolutionary adaptations of fishes to the photic environment. In: Crescitelli F (ed): Handbook of Sensory Physiology, Vol. VII/5. Springer, Berlin; pp. 193-274.

Neuweiler G, Singh S, Sripathi K (1984) Audiograms of a South Indian bat community. J Comp Physiol A 154: 133-142.

Nicol JAC, Arnott HJ, Best CG (1974) Tapeta lucida in bony fishes (Actinopterygii): a survey. Can J Zool 59: 61-81.

Payne RS, Webb D (1971) Orientation by means of long range acoustic signaling in baleen whales. Ann NY Acad Science 188: 110-141.

Scheich H, Langner G, Tidemann C, Coles RB, Guppy A (1986) Electroreception and electrolocation in platypus. Nature 319: 401-402.

Schuijf A, Buwalda RJA (1980) Underwater localization - a major problem in fish acoustics. In: Popper AN, Fay RR (eds): Comparative Studies of Hearing in Vertebrates. Springer, New York; pp. 43-77.

Stange G, Howard J (1979) An ocellar dorsal light response in a dragonfly. J exp Biol 83: 351-355.

Vilter V (1953) Existence d'une rétine à plusieurs mosaïques photoréceptrices, chez un poisson abyssal bathypélagique, Bathylagus benedicti. CR Soc Biol (Paris) 147: 1937-1939.

Wagner HJ, Ali MA (1978) Retinal organization in goldeye and mooneye (Teleostei: Hiodontidae). Rev Can Biol 37: 65-83.

Wald G (1937) Visual purple system in fresh-water fishes. Nature 139: 1017-1018.

Weissert R, von Campenhausen C (1981) Discrimination between stationary objects by the blind cave fish Anoptichthyes jordani. J Comp Physiol 143: 375-381.

PATH FINDING BY SONAR AND RADAR IN AQUATIC AND TERRESTRIAL ANIMALS

Richard D. Keynes

Physiological Laboratory
Cambridge CB2 3EG
England

Path finding systems employing echolocation by sound have been evolved independently by a number of aquatic and terrestrial organisms, and several families of fishes have acquired electric direction finding systems.

INTRODUCTION

My scientific career started with two and a half years spent at His Majesty's Anti-Submarine Experimental Establishment doing research on what the British Navy called "Asdics", later to be renamed "Sonar", which was followed by three years at the Admiralty Signals Establishment, working on gunnery radar and IFF systems. Hardly good training for a biologist, you might suppose, but I well remember how during sea trials in 1941 I became familiar with the modulated high-frequency sounds of a school of dolphins passing by, and wondered vaguely what function they served. And when in 1951 I was confronted in Carlos Chagas's laboratory in Rio with an electric eel and an oscilloscope that was not working, my wartime experience in electronics served me in good stead, both in servicing the oscilloscope and in discovering for the first time how the additive discharge of the electric organ was achieved (Keynes & Martins-Ferreira, 1953). When, therefore, it became clear in the planning of this meeting that someone should give a comparative account of direction-finding mechanisms in water and in air, I felt that I was uniquely well qualified, not as a biologist to be sure, but at least as a former naval scientist, to take on the job.

EARLY RESEARCH ON ANIMAL SONAR AND RADAR

Historically, pride of place in my account is occupied by the elegant experiments made by Lazaro Spallanzani in 1793, when he found that bats displayed a skill in avoiding obstacles in their flight path that was unaffected by blinding them, but required their ears to be unblocked and working. However, he himself could hear nothing while they flew around his experimental room, and could only attribute their path-finding ability to some kind of 'sixth sense'. At about the

Comparative Physiology: Life in Water and on Land. P. Dejours, L. Bolis, C.R. Taylor, E.R. Weibel (eds.) Fidia Research Series, IX-Liviana Press, Padova © 1987

same time (1799), Alexander von Humboldt and Aimé Bonpland visited the cave at Caripe in Venezuela, in the depths of which a large colony of the nocturnal fruit-eating oilbird *Steatornis caripensis* still has its nests. The sight of vast numbers of birds emerging from the cave at dusk impressed them as much as it does the modern visitor, but Humboldt did not, of course, speculate whether their 'shrill and piercing cries' helped them to find their way in the dark.

Then in 1920 Hartridge did some experiments on the flight of bats between two blacked-out rooms with an intervening door that he gradually closed, and suggested that they were 'diverted by a specialized sense of hearing since the sound waves of short wave-length which they are known to emit are capable of casting shadows and of forming "sound pictures"'. The high-frequency sonic transmissions of bats were first recorded by Pierce & Griffin (1938), and the term "echolocation" was introduced to biologists by Griffin in 1944, "Sonar" still being officially on the secret list. He went on to carry out the superb series of studies on the ability of various species of bat to locate and catch their prey by means of echolocation that he described in his book "Listening in the dark" (Griffin 1958).

After the end of the 1939-45 war, improvements in electronics made it much easier to detect and study not only the high-frequency sounds employed by bats and birds for path finding purposes, but also, since sound is conducted as well in water as in air, the similar transmissions of aquatic animals, and the field of animal sonar made rapid advances.

Animal radar was introduced with the exciting discovery by Lissmann in 1951 of the weak electric discharges of the African fish *Gymnarchus* and some others, which endowed them with a remarkable facility for perceiving the presence of objects in the water close by. "Radar" is strictly a misnomer for this phenomenon, because it does not involve the reflexion of energy from a distant target, but rather the detection of changes in the electrical conductivity in the immediate environment of the transmitter. It can therefore be exploited only by aquatic animals living in a medium with a relatively high electrical conductivity, and as far as I know has not evolved in any terrestrial species.

ACTIVE SONAR IN TERRESTRIAL ANIMALS

The prize performers as far as airborne sonar is concerned are, of course, the bats classified as Microchiroptera in which systems have evolved for the avoidance of obstacles and the detection, tracking and capture of prey with a high degree of precision (Schnitzler & Henson 1980). The basic echolocative sound transmissions consist of frequency modulated (FM) components alone, or various combinations of a longer lasting constant frequency (CF) component with a brief FM component. The duration of the FM components is below 5-10 msec, whereas CF components lasting from a few msec to more than 100 msec have been observed. The carrier frequencies range from about 150 kHz down to 20 kHz, so that they are generally inaudible to the human ear. The sound is generated in the larynx by the vibrations of two pairs of very thin membranes stretched over a shallow cavity, whose tension and therefore the sound frequency is adjusted by the cricothyroid muscles (Novick & Griffin 1961), innervated exclusively by the superior laryngeal nerve (Rübsamen 1980). The inferior laryngeal nerve controls the remainder of the intrinsic laryngeal musculature, and therefore commands the switching on and off of sound emission (Pye 1980). However, the whole question of the mechanism of sound production and of its motor control appears to have received relatively less attention than the sensory side of the sonar system.

There being many hundreds of species of echolocating bats whose behaviour and hunting strategies differ in a variety of ways, the characteristics of the sonar apparatus both for transmission and reception are correspondingly specialized for extraction of the information most needed in the individual species. The variability in styles of transmission and the processing of the echoes makes it difficult to generalize, but the following picture emerges from the comprehensive reviews presented by Schnitzler & Henson (1980) and Pye (1980). In free flight prior to the detection of a target, most bats emit only one pulse per wingbeat and respiratory cycle, with a repetition rate that may be as low as 4 pulses/sec. Targets are generally picked up at a range of about 2 m, and during the approach and terminal phases of the hunt, the repetition rate rises and the pulse duration falls so as to maximize the amount of information available to the bat without allowing the transmission to overlap the echo. The determination of target range depends on a measurement of the time between the emission of a pulse and the reception of the echo, with an accuracy of the order of 1 cm in distance or 60 μsec in time, though consideration of the evidence on the ability to distinguish between two closely spaced targets may suggest an even better performance.

The determination of target direction both in azimuth and in elevation depends ultimately on the detection of changes in echo intensity as a function of angular displacement, and thus involves a combination of the directionality of the receiver with that of the transmitter. As far as the latter is concerned, it is clear that as soon as a target has been acquired, the open mouth and nose-leaf of the bat are aimed directly at it, and the angular dependence of the transmitted sound intensity has been determined as -6 db at 20-50° in different species (see Pye 1980). The directional sensitivity of the individual ears is probably somewhat better. A discrimination of something like 5° appears to be achieved in both planes. It is also clear that a binaural comparison of the relative strengths of the signal at the two ears plays an important part in the perception of target direction, and that directional sensitivity is greatly reduced by putting one ear out of action. This raises the question as to what system or systems of scanning in azimuth and elevation are employed. The possibility has been considered that differences in the ongoing arrival time of the echo at the two ears might provide information about azimuth, as in the passive sonar of the barn owl (Konishi 1984), but the small size of the bat's head and the closer spacing of the ears argues against it. In the bats which emit transmissions with a long CF component, such as the horseshoe bat *Rhinolophus*, there are rapidly alternating backward and forward movements of the ears after acquisition of a target, which are more or less closely correlated with the pattern of sound emissions (Griffin *et al.* 1962; Pye *et al.* 1962; Pye & Roberts 1970), and rapid nodding of the head has also been observed. This suggests (Schnitzler 1973) that there may be scanning for elevation in these species, but the full implications of such ear movements have yet to be established.

One of the most striking specializations among the Microchiroptera involves the development in the long CF-FM bats of an ability to determine Doppler shifts in the echo frequency, resulting from a relative movement between target and bat, with an accuracy of 30-60 Hz. This is achieved through the possession by such bats of what may be termed a frequency fovea (Neuweiler 1984), in which the part of the basal turn of the cochlea, and its cortical representation, that are tuned precisely to the specific CF-frequency of the particular species, are greatly expanded, perhaps through tuning of the individual hair cells (Crawford & Fettiplace 1981). The bat then compensates for the Doppler shift by lowering its emitted pulse frequency by an amount that holds the perceived echo frequency at a constant value. This Doppler shift mechanism not only enables the bat to discriminate against stationary or slowly moving targets, but also to evaluate the wing beat frequency of a fluttering moth and thus to select appropriate targets for attack (Schnitzler & Henson 1980; Neuweiler 1984). Further discrimination against extraneous noise, clutter, and echoes from unwanted targets, is achieved by exploiting the characteristics of the FM component of the transmission.

There are various other animals which employ less sophisticated sonar systems with broad-band non-FM transmissions for pathfinding purposes (Henson & Schnitzler 1980). These are nocturnal in habit, and have a problem in avoiding relatively large obstacles in the dark, but not in locating rapidly moving prey. Among the Megachiroptera, only certain species of *Rousettus* are known to echolocate, using brief tongue clicks as their transmissions. Echolocation in cave dwelling oilbirds was first described by Griffin (1953), and their sound transmissions have been further investigated by Konishi & Knudsen (1979), clicks with a complex structure being emitted at a rate of several per second. Some cave swiftlets of the genus *Collocalia* also produce click-like sonar transmissions (Novick 1959, Fenton 1975). Both the oilbird and the cave swiftlet have been shown to be capable of avoiding obstacles around 1 cm in size, but their performance in this respect is certainly markedly inferior to that of the Microchiroptera, and it seems likely that the main function of their sonar systems is to avoid collision with the walls of the caves in which they live. How in this case they avoid serious interference with each other's signals is not obvious. It has been suggested that the European swift *Micropus apus* may use sonar in order to hunt for flying insects at night, and I have sometimes wondered when I have watched them in the daytime how, flying so fast, and being unable to use their wings and tails as scoops in the manner that a bat does (Webster & Griffin 1962), they manage to see their prey at a sufficient range to be able to change course in time to make a capture. The nightjar *Caprimulgus europaeus* also hunts insects at night, and emits short pulses of sound; but there is no evidence that either species is equipped with a high performance sonar system.

Some shrews and tenrecs have been shown to produce low-intensity clicks, and to be capable in the dark of locating objects in the 10 cm^2 size range (Henson & Schnitzler 1980). Blinded laboratory rats appear to be able to use the echoes from a variety of self-generated signals such as trains of clicks (Chase 1980), sniffs, sneezes and scratches in order to gain some appreciation of their immediate environment, and blind human subjects can learn to judge distances from quite small obstacles in a similar way. The hearing system of moths has been shown by Roeder & Treat (1957) to be capable of detecting the sounds generated by the moth's own wingbeats, so that both the transmitter and the receiver for an echolation system are available.

ACTIVE SONAR IN AQUATIC ANIMALS

Since many aquatic animals hunt their prey at considerable depths or in murky surface waters where their visual systems can be of little assistance to them, it should not be surprising to find that active sonar is sometimes employed. However, it is only in the toothed whales (odontocetes) that there is any extensive evidence for its development. An excellent review of the subject has been provided by Wood & Evans (1980).

The sounds emitted by dolphins have been known to sailors since the time of Aristotle, but it was not until after the Second World War that the suggestion was made (McBride 1956) that they might represent the transmissions of a sonar system. I do not think that the frequency-modulated squeaks with a carrier frequency of around 20 kHz, lasting as I remember for an appreciable fraction of a second, that I heard in 1941, were what are now most usually regarded as the sonar transmissions, which are clicks with a rapid onset and a duration of much less than a millisecond. Nevertheless, the general similarity in pulse structure between such squeaks and the transmissions of FM bats suggests that more than one mode of sonar may perhaps be employed by dolphins. Or since it is well known that Cetacea make considerable use of underwater sound for communication purposes of other kinds, as for example the remarkable songs of migrating humpback whales (Payne & McVay 1971), the dolphins may merely have been gossiping idly with one another as they swam.

The bulk of the studies on sonar in Odontocetes has necessarily been restricted to experiments on captive specimens of the smaller species in which their click transmissions could be recorded and correlated with their ability to detect obstacles in the water. Targets about 5 cm in diameter can be detected at ranges of 50 m or more; and dolphins are able to discriminate a 10% difference in diameter as well as changes in the shape and thickness of metallic disks (Au 1980). In the porpoises and dolphins the duration of the transmitted clicks is in the range 10-400 μsec, while in the killer whale it is 0.5-1.5 msec (Wood & Evans 1980). In the sperm whale, the clicks recorded at sea by Watkins (1980) had a duration from 2 to 30 msec or more, and occurred in small groups at a low and rather variable frequency. In dolphins studied in open water by Au (1980) the interval between clicks with no target or one at a range of 64 m was 130 msec, leaving sufficient time for echoes to be received from distances up to 100 m before the next click was transmitted. With targets at shorter ranges, Morozov et al. (1972) reported a mean pulse interval of 20 msec which was fairly constant over a distance of 40 to 4 m as the animal closed in on the target. In a single successful experiment on a blindfolded and apparently somewhat frustrated captive bottlenose dolphin presented with a live free-swimming fish, Wood & Evans (1980) recorded click frequencies as high as 1500 per second, but in the end the dolphin tossed her head, caught the fish, and returned it to the trainer several times without emitting any clicks at all, so placing the behavioural importance of her sonar in some doubt.

Although other cetaceans, both large and small, have been reported to produce clicks at sea, evidence is lacking for the echolocating function of such transmissions, and in the case of the grey seal is definitely negative (Scronce & Ridgway 1980). The same is true for a few fishes that emit clicks. For the sake of completeness, however, I should mention the possibility that the acoustico-lateralis system, whose role in coordinating the swimming of schooling fishes like the clupeids has been investigated by Blaxter, Denton & Gray (1981), may provide the sensory apparatus for detecting particle displacements and pressure waves reflected from neighbouring objects as the fish approaches them. To the extent that the sources of energy for such signals are the swimming movements of the fish itself, this would, as was pointed out by Lowenstein (1957), constitute a kind of echolocation system that might assume considerable behavioural importance under conditions where the visual system is not operative. Another example of this principle may be provided by the whirligig beetle *Gyrinus*, which according to Wilde (1941) avoids collision with the wall of a glass basin when, and only when, swimming and whirling at the surface of the water and detecting reflected surface displacements with the aid of its Johnston organ.

PASSIVE SONAR

During the First World War, the only available means of detecting the presence of a submerged submarine was to listen for its propeller noise with a somewhat crude hydrophone. Between 1918 and 1939, the main advance in techniques of anti-submarine warfare consisted in the development of echolocation systems operating in the 20 kHz waveband, which were employed primarily for finding and attacking targets at a relatively short range (Hackmann 1984). However, active sonar of this type is limited in its range and effectiveness by the slow propagation and excessive attenuation of high-frequency sound in water, and by the fact that the attacking vessel unavoidably gives its presence away in advance by its own sonar transmissions. Modern warships are therefore equipped with highly sophisticated passive sonar that enables them to detect and identify other vessels at a very much greater range.

The outstanding example of the employment of passive sonar in the animal kingdom is provided by the barn owl *Tycho alba*, which as was first shown by Payne (1971), can locate its prey in total darkness using only the sense of hearing, with an accuracy better than 1° both in azimuth and in elevation. The neural mechanisms involved have since been laid bare with great elegance by Knudsen & Konishi (1979), Knudsen (1981) and Konishi (1984). One important feature of the receptor system is the marked asymmetry of the owl's head, and in the positioning both of the facial ruffs of feathers forming the external ear and of the preaural flaps and openings of the inner ear, which results in binaural differences in the perceived loudnesses that vary with sound frequency and yield clues to the elevation of the source. Determination of azimuth, on the other hand, depends on perception of the ongoing time disparity between the signals reaching the two ears, which because of the relatively small size of the bird's head necessitates a discrimination of time differences in the microsecond range. The signals are combined in the neurons of the owl's midbrain auditory nucleus so as to form a very sharply defined receptive field. There are further subtleties in the organization of the system that I do not have time to describe, but one point in which I have a personal interest is to know whether or not the bird is in effect making use of the split-beam principle for determination of elevation. This involves a comparison of the signal strength at the cross-over point of the polar curves for two receiving aerials inclined at a slight angle to one another, in order to maximize the change in signal ratio with angle, and having been employed in the early gunnery radar sets, was first applied to sonar in experiments that I did at Fairlie in 1942. It seems possible that the combination of an inhibitory centre-surround mechanism with a tuning of space-mapped neurons to binaural time disparity could constitute such a system, but I am not sure about it.

Many aquatic and semi-aquatic animals use passive sonar by detecting the surface waves generated by their prey. An example of the high directional sensitivity achieved by the clawed toad thanks to the lateral line units in the skin surrounding its eyes has recently been given by Zittlau *et al.* (1986).

PROTECTION AGAINST SONAR

Another analogy with military echolocation systems arises in the adaptation of the hearing of certain insects, and in particular moths and lacewings, to detect the sonar transmissions of bats and enable evasive action to be taken. Thus Roeder (1962) has described the various avoidance reactions, including passive and power dives, loops, rolls and tight turns, that are observed in moths exposed to a stationary source of ultrasonic pulses; and Miller (1980) has studied the similar interactions between hunting bats and freely flying green lacewings. Attention has been paid as to what the most effective escape manoeuvre on the part of the moth might be (Altes & Anderson 1980), and conversely as to how the frequency and intensity of the sonar transmissions might be modified to the advantage of the bat (Fenton 1980; Fullard & Fenton 1980). An intriguing situation that has been studied by Roeder & Fenton (1973) arises when moths choose to overwinter at the entrances of caves that are also inhabited by hibernating bats, and are then found to show no avoidance response even though their tympanic organs maintain a normal sensitivity. It is evident that the hearing ability of insects may exert an important selection pressure on the echolocation behaviour of bats, and *vice versa*; but the evolutionary significance of this type of predator-prey interaction seems likely to prove hard to disentangle.

ELECTRIC DIRECTION FINDING IN FISHES

Several families of fishes possess weak electric organs that serve as power sources for the purpose of direction finding and electro-communication. The first weakly electric fish of this type to be described was *Gymnarchus niloticus* (Lissmann 1951), and the faculty is also found in the African mormyrids and the South American gymnotids (Szabo & Moller 1984).   Various teleosts such as *Malapterurus*, *Astroscopus*, and one of the gymnotids, *Electrophorus*, have much stronger electric organs that appear to be used for offensive or defensive purposes.   Among the marine elasmobranchs, the torpedos also have powerful electric organs, while many of the rays have weak electric organs in their tails whose precise function is still obscure.

The mode of operation of the direction finding systems of these weakly electric fishes does not depend on the reception of echoes, but rather on the ability to detect changes in the pattern of electrical conductivity of the surroundings during the pulse of current.   Thus Lissmann (1958) and Lissmann & Machin (1958) showed in training experiments that specimens of *Gymnarchus* could readily discriminate between a porous and therefore electrically transparent cup filled with aquarium water, and one containing an electrically opaque glass tube, or between glass tubes of different sizes. Experiments with moving electrostatic and magnetic fields showed that in the limit a potential gradient in the water of only 0.03 μV/cm could be detected.   Scheich & Bullock (1974) have summarized the results of similar experiments both on mormyrids and gymnotids showing that the fishes can not only discriminate between conductors and non-conductors, but will also respond to the external connexion of capacitors between two electrodes in the water.

There are basically two types of discharge pattern in the weakly electric fishes. Some fishes described as "wave species" by Scheich & Bullock (1974) are characterized by a very regular discharge of pulses at a high rate and with a nearly sinusoidal waveform. In a solitary fish, the pulse frequency is maintained within 0.1% over long periods, and varies from 50-150 Hz in *Sternopygus macrurus* up to 750-1250 Hz in *Apteronotus* (formerly *Sternarchus*) *albifrons*. If two or more fish with similar frequencies encounter one another, they are able to shift their frequencies in what is known as the jamming avoidance response, so that in studies by Bullock *et al.* (1972) of two *Eigenmannia* that when isolated were both discharging at 370 Hz, the frequency of one rose to 376 Hz when they were put together, while that of the other fell to 364 Hz. The "pulse species", on the other hand, produce brief spikes at an irregular and generally much lower rate, for example less than 1 to more than 40 pulses/sec in various mormyrids, 3-30 in *Hypopomus*, and 55-65 pulses/sec in *Gymnotus* at rest.

The anatomy and discharge mechanisms of electric organs have been thoroughly reviewed by Bennett (1971a). The majority of electric organs are modified muscle fibres whose additive discharge involves a sodium–dependent mechanism as described by Keynes & Martins-Ferreira (1953), in which the non-innervated face is sometimes inexcitable, and sometimes produces a spike whose duration differs from that of the innervated face, resulting in net potentials that may be either monophasic or biphasic in different species. In marine electric fishes such as rays and torpedos, the discharge is cholinergic rather than natrogenic, and the electroplates may be regarded as greatly overgrown motor endplates. The muscles from which the electric organs are derived embryologically are most often tail muscles, but in the torpedos are branchial muscles and in *Astroscopus* are eye muscles, while certain of the gymnotids have accessory organs with yet another origin. In *Apteronotus albifrons* and its relatives, which are the wave species with the highest discharge frequency, the myogenic electroplates have been lost, and instead the electric organ is derived from the myelinated nerve fibres that formerly innervated it. The modes of motor control of the electric organ display a corresponding variability that I do not have space to describe.

Electroreceptors are of two general types (Bennett 1971b; Kalmijn 1974; Bullock 1981), the ampullary or tonic receptors found in many types of fish, both electric and non-electric, that are tonically and fairly rhythmically active, give long-lasting responses to low frequency or DC stimuli, and have an obvious canal leading from the receptor cavity to the exterior; and tuberous or phasic receptors found exclusively in weakly electric fishes, that are sensitive to relatively high frequencies and insensitive to maintained stimuli, and lack an obvious canal communicating with the exterior. Both types are widely distributed over the body surface. The sensory input from the electroreceptors is integrated and processed in the highly developed lateral lobe and cerebellum of gymnotids and mormyrids in such a way that the sensitivity of the whole system is very much greater than could be expected for a single unit (Szabo & Fessard 1974).

That the system does have a direction finding function is not seriously questioned. Many of these fishes live in turbid water where the range of vision is restricted, and others are nocturnal in their behaviour. But there is equally little doubt that the electroreceptors in all fishes serve for the detection of small electric currents that might arise from the muscular activity of other fishes in the vicinity, and in the weakly electric species for picking up the transmissions of neighbours for purposes of social communication. It is difficult to establish under natural conditions what the precise role of the electrical pulse transmissions really is, as was once brought home to me by some experiments that I did in collaboration with Dr Joel Cohen. We constructed a maze through which an electric eel was made to swim, and recorded the small pulses of variable frequency emitted by the organ of Sachs, which were thought to serve a direction finding function substituting for vision, since the fish lacks a functional retina and is certainly blind. But alas the eel proved just as adept at negotiating the bends in our maze without touching the sides when it was electrically silent, as when it was transmitting.

THE EVOLUTION OF PATH FINDING SYSTEMS

In discussing the difficulties of his theory in *The Origin of Species*, Darwin cited the problem raised in considering the evolution of the electric organ, where he found it hard to see how the powerful electric organs of *Electrophorus* and *Torpedo* could have evolved if at the intermediate stages of their development they had no survival value. His questions have been partially answered by the discovery of the direction finding ability of the smaller gymnotids, since one can readily envisage the steps in the development of sonar and radar systems as consisting initially in the perfection of receptors for passive detection of sound and electric signals arising from external sources, and subsequently in the provision of an appropriate transmitter to go with them. In just one of the gymnotids, *Electrophorus*, the strength of the transmitter could then have been increased to the point where it could be used for long range communication, or for offensive purposes. Nevertheless, as far as *Malapterurus* and the marine electric fishes are concerned, the apparent absence of intermediate stages still has to be explained. The hearing apparatuses and electroreceptors that are necessary for the initial development of passive sonar and radar are certainly widespread in the animal kingdom, and there has even been a recent report of electroreceptors and electrolation in the platypus *Ornithorhynchus anatinus* (Scheich *et al.* 1986). Dr Lissmann informs me that he has unpublished records of weak electric discharges in another species of African catfish, probably *Clarias*, and it could be that *Malapterurus* has simply lost the radar capacity that it once possessed. However, evidence for the employment of electrolocation by marine fishes, where the range attainable would be very short because of the high electrical conductivity of sea water, is still conspicuously lacking. Perhaps in this situation the intermediate stage involves weak discharges as a component of sexually intimate breeding behaviour.

In conclusion, I have shown how electric organs have evolved independently in at least six different families of fishes, in two of which, the mormyrids and gymnotids, there has been a striking parallel development of species with similar body forms and specializations of their behaviour and direction finding systems.

I have shown how sophisticated sonar systems have arisen both in aquatic and in terrestrial mammals. And I have also shown how within one large group of echolocating bats, extreme specializations such as the accomplishment of fluttering target detection through a Doppler shift mechanism have evolved independently in species from the Old and New Worlds. I have neither the time nor the knowledge to pursue in detail all the evolutionary implications of the facts that I have presented here, but if proof were needed of the occurrence of convergent evolution in unrelated species, the story of animal sonar and radar provides it in abundance.

I am greatly indebted to Dr Hans Lissmann for setting me out on the right path in the first instance.

REFERENCES

Altes RA & Anderson GM (1980) Binaural estimation of cross-range velocity and optimum escape maneuvers by moths. In Busnel & Fish (1980), pp. 851-852.

Au WWL (1980) Echolocation signals of the Atlantic bottlenose dolphin (*Tursiops truncatus*) in open waters. In Busnel & Fish (1980), pp. 251-295.

Bennett MVL (1971a) Electric organs. Vol. V pp. 347-491 in Fish Physiology, edited by Hoar WS & Randall DJ. Academic Press, New York.

Bennett MVL (1971b) Electroreception. Vol. V pp. 493-574 in Fish Physiology, edited by Hoar WS & Randall DJ. Academic Press, New York.

Blaxter JHS, Denton EJ & Gray JAB (1981) Acousticolateralis system in clupeid fishes. In Tavolga, Popper & Fay (1981), pp. 39-56.

Bolis L, Keynes RD & Maddrell SHP (1984) (Editors) Comparative physiology of sensory systems. Cambridge University Press.

Bullock TH (1981) Comparisons of the electric and acoustic senses and their central processing. In Tavolga, Popper & Fay (1981), pp. 525-571.

Bullock TH, Hamstra RH & Scheich H (1972) The jamming avoidance response of high frequency electric fish. J. comp. physiol. Psychol. 77: 1-48.

Busnel R-G & Fish SF (1980) (Editors) Animal sonar systems. Plenum Press, New York.

Chase J (1980) Rat echolocation: correlations between object detection and click production. In Busnel & Fish (1980), pp. 875-877.

Crawford AC & Fettiplace R (1981) An electrical tuning mechanism in turtle cochlear hair cells. J. Physiol. 312: 377-412.

Fenton MB (1975) Acuity of echolocation in *Collocalia hirundinacea*, with comments on the distributions of echolocating swiftlets and molossid bats. Biotropica 7: 1-7.

Fenton MB (1980) Adaptiveness and ecology of echolocation in terrestrial
(aerial) systems. In Busnel & Fish (1980), pp. 427-446.

Fessard A (1974) (Editor) Handbook of Sensory Physiology, Vol. III/3.
Springer-Verlag, Berlin.

Fullard JH & Fenton MB (1980) Echolocation signal design as a potential
counter-countermeasure against moth audition. In Busnel & Fish (1980),
pp. 899-900.

Griffin DR (1944) Echolocation by blind men, bats and radar. Science 100:
589-590.

Griffin DR (1953) Acoustic orientation in the oil bird, *Steatornis*. Proc. Nat.
Acad. Sci. 39: 884-893.

Griffin DR (1958) Listening in the dark. Yale University Press.

Griffin DR, Dunning DC, Cahlander DA & Webster FA (1962) Correlated
orientation sounds and ear movements of horseshoe bats. Nature, Lond.
196: 1185-1186.

Hackmann W (1984) Seek and strike. Sonar anti-submarine warfare and the
Royal Navy 1914-1954. H.M.S.O., London.

Hartridge H (1920) The avoidance of objects by bats in their flight. J. Physiol.
54: 54-57.

Henson OW & Schnitzler H-U (1980) Performance of airborne biosonar systems:
II. Vertebrates other than Microchiroptera. In Busnel & Fish (1980), pp.
183-195.

Kalmijn AJ (1974) The detection of electric fields from inanimate and animate
sources other than electric organs. In Fessard (1974), pp. 147-200.

Keynes RD & Martins-Ferreira H (1953) Membrane potentials in the electroplates
of the electric eel. J. Physiol. 119: 315-351.

Knudsen EI (1981) The hearing of the barn owl. Scientific American 245: 82-91.

Knudsen EI & Konishi M (1979) Mechanisms of sound localization in the barn
owl (*Tycho alba*). J. comp. Physiol. 133: 13-21.

Konishi M (1984) Spatial receptive fields in the auditory system. In Bolis,
Keynes & Maddrell (1984), pp. 103-113.

Konishi M & Knudsen EI (1979) The oilbird: hearing and echolocation. Science
204: 425-427.

Lissmann HW (1951) Continuous electric signals from the tail of a fish,
Gymnarchus niloticus Cuv. Nature, Lond. 167: 201-202.

Lissmann HW (1958) On the function and evolution of electric organs in fish.
J. exp. Biol. 35: 156-191.

Lissmann HW & Machin (1958) The mechanism of object location in *Gymnarchus
niloticus* and similar fish. J. exp. Biol. 35: 451-486.

Lowenstein O (1957) The sense organs: the acoustico-lateralis system.

pp. 155-186 in Vol. 2 of The Physiology of Fishes, edited by Brown ME. Academic Press, New York.

McBride AF (1956) Evidence for echolocation in cetaceans. Deep Sea Research 3: 153-154.

Miller LA (1980) How the green lacewing avoids bats: behaviour and physiology. In Busnel & Fish (1980), pp. 941-943.

Morozov BP, Akapiam AE, Burdin BI, Zaitseva KA & Sokovykh YA (1972) Tracking frequency of the location signals of dolphins as a function of distance to the target. Biofisika 17: 139.

Neuweiler G (1984) Auditory basis of echolocation in bats. In Bolis, Keynes & Maddrell (1984), pp. 115-141.

Novick A (1959) Acoustic orientation in the cave swiftlet. Biol. Bull. 117: 497-503.

Novick A & Griffin DR (1961) Laryngeal mechanisms in bats for the production of orientation sounds. J. exp. Zool. 148: 125-146.

Payne RS (1971) Acoustic location of prey by barn owls (Tycho alba). J. exp. Biol. 54: 535-573.

Payne RS & McVay S (1971) Songs of humpback whales. Science 173: 585-597.

Pierce GW & Griffin DR (1938) Experimental determination of supersonic notes emitted by bats. J. Mammal. 19: 454-455.

Pye JD (1980) Echolocation signals and echoes in air. In Busnel & Fish (1980), pp. 309-353.

Pye JD, Flinn M & Pye A (1962) Correlated orientation sounds and ear movements of horseshoe bats. Nature, Lond. 196: 1187-1188.

Pye JD & Roberts LH (1970) Ear movements in a hipposiderid bat. Nature, Lond. 225: 285-286.

Roeder KD (1962) The behaviour of free flying moths in the presence of artificial ultrasonic pulses. Animal Behaviour 10: 300-304.

Roeder KD & Fenton MB (1973) Acoustic responsiveness of Scoliopteryx libatrix L., a moth that shares hibernacula with some insectivorous bats. Can. J. Zool. 51: 681-685.

Roeder KD & Treat AE (1957) Ultrasonic reception by the tympanic organ of noctuid moths. J. exp. Zool. 134: 127-157.

Rübsamen R (1980) Activity of the recurrent laryngeal nerve due to the production of ultrasonic echolocation sounds in the CF-FM bat. In Busnel & Fish (1980), pp. 969-971.

Scheich H & Bullock TH (1974) The detection of electric fields from electric organs. In Fessard (1974), pp.201-256.

Scheich H, Langner G, Tidemann C, Coles RB & Guppy A (1986) Electroreception and electrolocation in platypus. Nature, Lond. 319: 401-402.

Schnitzler H-U (1973) Die Echoortung der Fledermäuse und ihre hörphysio-
    logischen Grundlagen. Fortschr. Zool. 21: 136–189.

Schnitzler H-U & Henson O'DW (1980) Performance of animal sonar systems: I.
    Microchiroptera. In Busnel & Fish (1980), pp. 109–181.

Scronce BL & Ridgway SH (1980) Grey seal, *Halichoerus*: echolocation not
    demonstrated. In Busnel & Fish (1980), pp. 991–993.

Szabo T & Fessard A (1974) Physiology of electroreceptors. In Fessard (1974),
    pp. 59–124.

Szabo T & Moller P (1984) Neoroethological basis for electrocommunication. In
    Bolis, Keynes & Maddrell (1984), pp. 455–474.

Tavolga WN, Popper AN & Fay RR (1981) (Editors) Hearing and sound
    communication in fishes. Springer-Verlag, New York.

Watkins WA (1980) Acoustics and the behaviour of sperm whales. In Busnel &
    Fish (1980), pp. 283–297.

Webster FA & Griffin DR (1962) The role of the flight membranes in insect
    capture by bats. Animal Behaviour 10: 332–340.

Wilde J de (1941) Contribution to the physiology of the Johnston organ and its
    part in the behaviour of the Gyrinus. Arch. Neerland. Physiol. 25: 381–400.

Wood FG & Evans WE (1980) Adaptiveness and echology of echolocation in
    toothed whales. In Busnel & Fish (1980), pp. 381–425.

Zittlau KE, Claas B & Münz H (1986) Directional sensitivity of lateral line units
    in the clawed toad *Xenopus laevis* Daudlin. J. comp. Physiol. 158: 469–477.

VISION IN AIR AND WATER

Michael F. Land

School of Biological Sciences, University of Sussex,
Brighton BN1 9QG, U.K.

Terrestrial animals form images using the cornea, but in
water this is ineffective. Powerful, spherically corrected
lenses are used instead in both simple and compound eyes.
Special modifications are needed if animals are to see well
in both media, either sequentially or simultaneously.
Particularly interesting adaptations are found in animals
that hunt just below the water surface.

INTRODUCTION

As everyone knows, the unaided human eye works very badly under water because the
image it produces is so blurred as to be useless (Figure 1). However, with only
a modest alteration to its optics - in the form of a face mask - vision becomes
possible again, and it is not thereafter impaired in any important way. Thus the
transition from life in water to life in air, and the various transitions that
have gone the other way, required major changes in the optical structures of eyes
of all types, but not a great deal else.

Although far less important than the refractive index change, there are a few
minor differences in the way objects appear under water and in air which affect
visual function. Most water is somewhat turbid, and even the clearest water
absorbs light to some degree, so that both the range and clarity of underwater
vision are restricted. Waters of different origins are differently coloured -
oceanic waters tend to be blue, and freshwater lakes yellow - and this overall
cast of colour is reflected in the absorption maxima of the cone pigments of
fish (Lythgoe, 1979). Also, light below the surface tends to be less directional
than on land. The surface acts as a diffuser and at modest depth the
distribution of light around the vertical is almost symmetrical. Highlight and
shadow disappear, and this allows the use of mirror surfaces for camouflage, a
technique which will not work on land (Denton, 1970). The rest of this article
will concentrate on the optics of aquatic and terrestrial animals, as these
adaptations are by far the most important.

Comparative Physiology: Life in Water and on Land. P. Dejours, L. Bolis, C.R. Taylor, E.R. Weibel
(eds.) Fidia Research Series, IX-Liviana Press, Padova © 1987

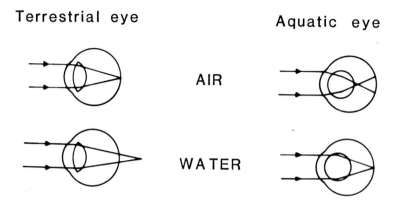

Figure 1. The performance of terrestrial and aquatic eyes in air and water.

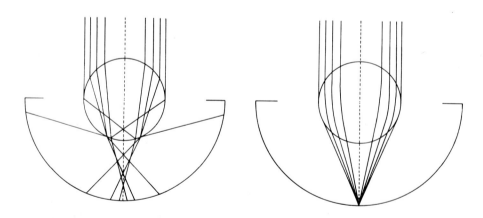

Figure 2. Ray paths through a homogeneous spherical lens (left), and a graded index "Matthiessen" lens.

IMAGING IN AIR AND WATER

Simple eyes

Vision is unusual in that the transition from water to air involved a physiological simplification rather than an increase in complexity. The presence of
air outside provided the first emerging vertebrates with a new and quite
straightforward way of forming an image. Instead of the powerful lenses
required for aquatic vision, all that was needed now was a curved transparent
interface: a cornea separating air outside from fluid within. We begin, therefore, with the more difficult problem; how to produce a good image in water,
without the help of the cornea.

The optical requirements of a simple eye (simple only in the sense of having a
single chamber) are basically the same as those of a camera; the image should be
sharp and bright and should cover a wide field. One solution to these optical
demands has emerged at least five times in evolution. It is the optically
inhomogeneous spherical lens, first described in the biological literature by
Matthiessen in the 1880's (see Pumphrey, 1961 or Land, 1981). It occurs in all
aquatic vertebrates, cephalopod and gastropod molluscs, alciopid annelid worms
and in a lone copepod crustacean (see Land 1981, 1984 a&b).

The requirements of a short focal length, a wide field and a mobile eye all
dictate that the lens must be spherical. However, a spherical lens made of
optically homogeneous material will not produce a useable image because of
spherical aberration; rays passing through the outer zones of the lens are
refracted too strongly and are focussed in front of the focus for rays closer to
the axis, and this results in a severely blurred image. The construction in
Figure 2 shows that the effect would be optically catastrophic, and there is, not
surprisingly, no known example of an animal with such a lens. The spherical
aberration of the lens can be overcome in two basic ways. One is to use non-
spherical surfaces; if the curvature of the outer zones of the lens is reduced
then rays passing through them will be bent less, and in principle all rays can
then meet at a common focus. The problem with this solution, which does occur in
a few animals (for example the copepod crustacean Pontella, see Land, 1984a),
is that the spherical symmetry of the eye is lost, and with it the wide field of
view is lost as well. This 'aspheric' correction only works for a small angular
region close to the axis, and the rest of the field of view is usually worse
than it would be without it. The second method is to use a lens of variable
refractive index, as Matthiessen surmised. Ray bending then occurs within the
lens, as rays pass through the refractive index gradients, and not at the front
and rear surfaces as in a conventional lens (Figure 2). Such lenses had actually

been suggested 25 years before Matthiessen by James Clerk Maxwell, "stimulated
apparently by the contemplation of his breakfast herring" (Pumphrey, 1961). For
an aberration-free lens with a given central refractive index, there is a unique
refractive index gradient (Fletcher et al., 1954), and although there is still
some controversy as to whether the gradient in fish lenses is exactly as
predicted (Fernald and Wright, 1983), the theoretical and observed gradients are
certainly very close.

The gradient lens offers another advantage. Its focal length is much shorter
than a homogeneous lens with the same central refractive index, about 2.5 times
the radius as opposed to about 4 times, when the central index is 1.52. This
figure of 2.5 radii is known as Matthiessen's ratio; it is diagnostic for lenses
of this type. One finds it not only in the lenses of fish eyes, but also in
all other spherical gradient lenses (Land, 1981). The real advantage of the
short focal length is that it provides a low F-number (1.25 compared with 2),
and hence an image that is 2.6 times brighter than that of the homogeneous lens.

We do not know exactly how the eye changed during the transition from water to
land, but the result was a reduction in the power of the lens as the cornea -
which of course already existed - took over. An indication of what may have
happened is provided by the metamorphosis of the eyes of anuran amphibia, which
begin their lives in water but become semi-terrestrial. In the anuran Pelobates,
the lens changes at metamorphosis from a spherical "fish" lens to a flattened
ellipsoid (Sivak and Warburg, 1983). The resulting weakening of the power of
the lens brings the eye into focus in air, but makes it severely hyperopic in
water. Thus the adult eye is thus not in fact an amphibious eye, but a truly
terrestrial one. In land vertebrates the lens is never completely lost - in
humans for example it still accounts for about 1/3 of the power of the combined
optical system - but its function changes from being responsible for the bulk of
refraction to being a device for varying the refractive state. Accommodation,
which is the action required to bring objects at different distances into focus,
is achieved either by moving the lens in the eye (all fish, amphibia and snakes)
or changing its shape (other reptiles, birds and mammals) (Walls, 1942).

Spiders are the only other terrestrial group whose main organs of sight are
simple eyes like those of vertebrates (insect larval ocelli are similar, but
are supplanted in the adult by the compound eyes). They too have a cornea as
the main refracting surface, and there is also a stationary lens which contributes
to the power of the optics (Land, 1985). In the exceptional case of the
nocturnal hunter Dinopis the lens is spherical and inhomogeneous, and reduces the
focal length so much that the eye has an F-number of only 0.6 (Blest and Land,

1977). There are no known cases of accommodation in spiders, presumably because
the focal lengths of the eyes are so short (< 1mm) that the depth of focus is
adequate for all the animals' needs (see Land, 1981 p.499). Spiders' eyes
probably evolved from pre-existing compound eyes rather than from aquatic
simple eyes as in vertebrates.

The problem of spherical aberration is not confined to lenses. The corneal
optics of terrestrial animals suffer potentially from the same problem, and the
solutions are either to make the cornea non-spherical, or to modify the structure
of the lens so that it corrects not only its own aberration (as in fish) but that
of the cornea as well. Interestingly, both methods occur in mammals. In humans
the cornea is aspheric, with a higher curvature in the centre than at the
periphery. There is no evidence that the lens compensates for the very slight
remaining spherical aberration (Millodot and Sivak, 1983). The rat, however,
has a more nearly spherical cornea and a wider aperture, so that spherical
aberration is potentially severe. Chaudhuri et al. (1983) found that the lens on
its own has pronounced <u>negative</u> spherical aberration (the opposite of a
homogeneous sphere) which compensates in whole or in part for the positive
aberration of the cornea.

Compound eyes

In the commonest kind of compound eye, the apposition type in which each group of
receptors has its own optical system, much the same considerations apply as with
simple eyes. In air the outer surface of the cornea is the principal refracting
structure, but in water this loses most of its power, so that ray bending must be
achieved in some other way. The alternatives are an inhomogeneous lens, or the
use of other curved interfaces in place of, or in addition to, the outer corneal
surface.

There are no known examples of spherical Matthiessen lenses in compound eyes.
However, there are many examples of facet lenses with an optically inhomogeneous
cylindrical structure, and these work in a very similar manner. The prototypes
of such lenses are found in the lateral eyes of the king crab <u>Limulus</u>, and their
mode of action was first worked out by Exner (1891) who invented the term "lens
cylinder" to describe them. The front surface of the cornea in <u>Limulus</u> is flat,
but the rear surface is covered in conical projections about 0.5 mm long each of
which terminates bluntly above a cluster of receptors. When the cornea is
cleaned and examined with a suitable object in front of it, a small inverted
image can be seen immediately behind the tip of each projection. Exner proposed
that each projection has a radial distribution of refractive index, highest
along the axis, and falling in a parabolic fashion towards the walls (Figure 3).

If one imagines a ray striking the outer face of such a structure at some
distance from the axis, then the ray will encounter a higher index towards the
axis, and will bend in that direction. It will continue to bend until the axis
is reached, and so will other initially parallel rays. If the gradient is the
correct one, they will all be brought to a common focus, and the structure will
behave as a lens. The application of interference microscopy to sections and
whole mounts of these optical structures has shown that the gradients Exner
postulated are indeed present (Land, 1979). A rival mechanism, in which the
images in Limulus ommatidia are supposedly formed by reflection of rays from the
surface of the projections (Levi-Setti, Park & Winston, 1975), does not seem to
be involved to any important degree. We still do not know how widespread lens-
cylinder imaging is amongst aquatic animals with compound eyes. Superposition
eyes of the refracting type (in euphausiids for example, see Land, 1981) always
use lens-cylinder optics, but Limulus is still the only well-documented case of
a lens-cylinder apposition eye.

There are alternatives. Spherical interfaces in addition to the front surface
may be involved, as in the backswimmer Notonecta, where the front surface of each
facet is almost flat. Refraction is divided between two surfaces, a strongly
curved rear one, and an intermediate one whose profile is not spherical but bell-
shaped (Figure 3). Schwind (1980) found that this profile was the right shape
to correct for the spherical aberration of the other, and hence provide an image
that was essentially diffraction limited (the measured blur circle was 2 μm
across, compared with a figure of about 10 μm which can be calculated for an
uncorrected lens). Earlier, Clarkson and Levi-Setti (1975) had postulated the
same role for the non-spherical rear surfaces of the calcite lenses of some
trilobite eyes, and had drawn attention to the similarities of these to
"Cartesian ovals", surfaces designed by Descartes and by Huygens in the
seventeenth century for producing lenses free of spherical aberration.

AMPHIBIOUS VISION

In this section we consider eyes which have to function in both air and water,
first sequentially and then simultaneously. The final section deals with the
rather special case of vision just below the surface, where the same objects may
be seen through both air and water.

Eyes that can function in either medium

There are two ways that a simple eye can be made to function in water and air. It
must either have a flat cornea that has no optical power in either medium, or, if
the cornea is convex, the lens must have unusually strong powers of accommodation

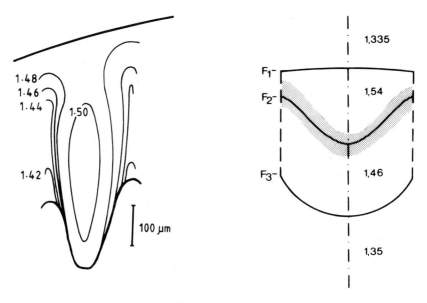

Figure 3. Lenses from aquatic compound eyes. Left: <u>Limulus</u> lens cylinder showing the refractive index gradient (data from Land, 1979). Right: <u>Notonecta</u> lens with a bell-shaped interface which reduces spherical aberration (after Schwind, 1980).

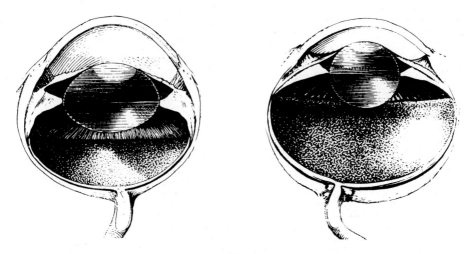

Figure 4. Terrestrial and aquatic mammalian eyes: lynx (left) and seal (right) from Soemmerring DW (1818) De oculorum hominis animalque, Vandenhoeck & Ruprecht, Göttingen. Note the spherical lens and flattened cornea of the seal.

(Sivak, 1978). If the cornea is flat, then the eye can retain a spherical
"aquatic" lens because no power is added out of water. Flat, or nearly flat
corneas do occur, in seals, penguins and in certain fish. Seals are optically
adapted to marine life, as the reversion of the lens to a spherical shape
indicates (Figure 4). The cornea is much flatter than in most mammals and
interestingly its residual curvature is different in the horizontal and
vertical meridians, the former having the greater curvature and the latter
being nearly flat. As a consequence the eye is strongly astigmatic in air,
although the astigmatism disappears in water (Piggins, 1970, Sivak, 1978). It
seems that this astigmatism is minimised by the pupil, which in air is a
vertical slit, thus admitting only bundles of rays drawn from the flat meridian
of the eye. In water, where its shape does not matter, the pupil is said to be
circular. Why should the eye not have a completely flat cornea, eliminating the
need for this elaborate compromise? Jamieson (1971) suggests that the answer is
the need for streamlining under water.

Penguins need good vision out of water where they breed, and in water where they
hunt. Sivak and Millodot (1977) studied the refractive state of penguin eyes,
and found that the cornea was nearly flat and that the eye was in focus
(emmetropic) in air. This means that the eye would be somewhat long-sighted
(hyperopic) in water, and Sivak (1978) concludes that the difference is
probably accounted for by an accommodatory mechanism in the lens.

Flying fish (Cypselurus) show an interesting variation of the flat cornea theme.
Baylor (1967) found that the cornea consists of three almost flat triangular
facets, forming a tent with its peak in the centre. Refractive measurements
through the facets showed emmetropia in air and slight hyperopia in water.
Paired flat facets have also been found in clinid fish (Figure 5) which use their
eyes in and out of water (Graham and Rosenblatt, 1970). A flat cornea has some
disadvantages: in air the periphery of the field is compressed and distorted, and
in water the field of view is reduced because the lens cannot protrude as it does
in a normal fish eye. To some extent the latter problem is overcome by the
facetted structure in flying fish, but at the cost of a disjointed image.

The alternative to a flat cornea is an eye with a normal terrestrial structure,
but with powers of accommodation in the lens that are sufficient to compensate,
in water, for the loss of corneal refraction. As we know from the inadequacies
of our own underwater vision, accommodation on this scale is beyond the
capabilities of ordinary eyes, the image to remain focussed. A young human eye
can manage about 14 dioptres of accommodation (reducing to about 1 at age 55!)
which is far short of the amount needed. (A dioptre is a measure of optical
power: the reciprocal of focal length in metres). The only animals that are

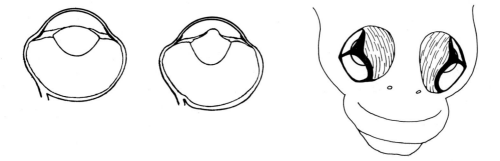

Figure 5. Two methods of making eyes amphibious. Powerful accommodation in the
merganser (left) is achieved by squeezing the lens through the iris (Sivak, 1978).
The clinid fish Mnierpes macrocephalus has flat corneal facets (drawn from
Graham & Rosenblatt, 1970).

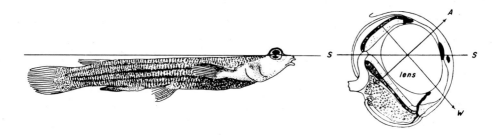

Figure 6. The "four-eyed fish" (Anableps) has one pupil looking onto air (A) and
the other into water (W) with the surface (S) between them. Combined from Walls
(1942) and Munk (1980) Hvirveldyrøjet, Berlingske Forlag, København.

known to accommodate adequately are certain diving birds. Levi and Sivak (1980)
found that in mergansers (Mergus cucculatus) the front surface of the lens was
very strongly deformed during accommodation induced by the application of
nicotine sulphate (Figure 5). The effect of the powerful ciliary muscle is to
squeeze the lens into and partially through the rigid iris, creating a locally
very high curvature. Sivak et al (1985) compared the amount of induced
accommodation in different species of duck, and found that in non-diving ducks
(wood duck and mallard) electrical stimulation of the ciliary muscle only
produced changes of about 6 and 3 dioptres. In mergansers and goldeneyes, both
diving ducks, the changes were much greater - 80 and 67 dioptres respectively -
values commensurate with the power lost from the cornea. Similar accommodative
mechanisms for achieving emmetropia in and out of water have been proposed for
some other water birds (cormorants and dippers), aquatic turtles and water
snakes, and otters, although with evidence of varying quality (Walls, 1942,
Sivak, 1978).

Simultaneous vision above and below the surface

The "four-eyed fish" (Anableps anableps) from South America, cruises with half
its eye above the surface, and half below. It has two pupils, one looking into
each medium, and a lens whose shape "combines an aquatic optical system
harmoniously with an aerial one, in a perfectly static situation" (Walls, 1942).
The compromise is achieved by the ovoid shape of the lens, with its long axis
in the direction which looks down into the water. Rays parallel to this axis
meet the strongest curvatures of the lens, and so are refracted relatively more
than rays coming from the air, which meet the weaker curvatures of the short axis
(Figure 6). The latter rays, however, are also bent by the cornea, so that the
total amount of refraction is much the same in the two cases. Sivak (1976)
measured the refractive state through the two pupils, and confirmed that the
shape of the lens did indeed compensate for the presence or absence of the
corneal contribution. It seems that this wonderful design is unique. There is
an amphibious blenny (Dialommus fuscus) with a double pupil, but in this case
the apertures are side by side and "the investigators had to give up when they
tried to interpret the eye" (Walls, 1942).

Life close to the interface

The interface between air and water has very curious optical properties,
especially if you live just below it. The oddities arise because the refraction
of rays at the water/air interface compresses the hemisphere above the surface
into 97° (this is "Snell's window"). Outside this angle underwater objects are
seen, inverted, by total internal reflection. At larger angles still, below the

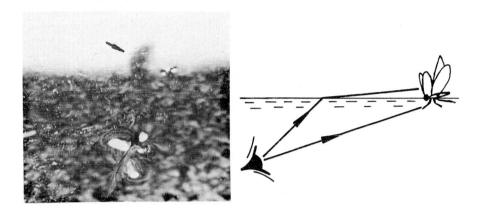

Figure 7. Photograph of a struggling insect in the water surface, seen from below. Notice that the part of the insect above the surface (arrowed) is well separated from the image of the same insect seen below the surface.

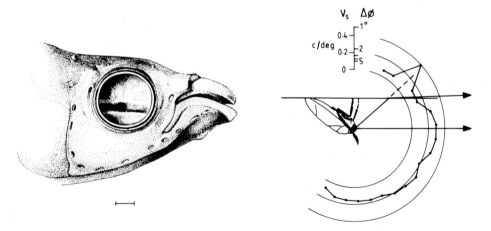

Figure 8. Animals that hunt below the surface. The fish Apocheilus lineatus has a retina with two horizontal areae (light stripes) seen here with the front of the eye removed (Munk, 1970). Scale 1 mm. The backswimmer Notonecta has two regions of high acuity (low inter-ommatidial angle ($\Delta\phi$), high sampling frequency ($\nu_s = 1/2\Delta\phi$; data from Schwind, 1980). As in the fish, these correspond to the regions just above and below the surface.

horizontal, underwater objects are seen normally. Objects in the surface film
can be seen at two separate locations: the portion above the surface will appear
inside Snell's window, but the portion below the surface will be seen in the
total reflection zone, and at a considerable angular separation from the upper
part (Figure 7). It seems that certain predatory fish and insects have special
regions of their eyes devoted to viewing these two scenes. Munk (1970) found
that in the eyes of the surface-feeding fish Aplocheilus lineatus and Epiplatys
grahami there were two horizontal band-shaped areae, with increased densities
of both receptors and ganglion cells (Figure 8). The central area is used for
lateral vision, below the water surface, but the ventral area is directed upwards
at an angle of 47° to the vertical - that is, above the surface, just inside the
margin of Snell's window. Thus, both areae look almost parallel to the water
surface, but below and above it respectively. Interestingly, the intermediate
zone, which corresponds only to very close objects in the surface, is poorly
developed and even shows evidence of degeneration.

The predatory aquatic bug Notonecta is adapted in a similar way. It ambushes
prey from beneath the water surface, hanging upside down with its feet in contact
with the surface ready to detect ripples. In this position the eyes are a few mm
beneath the surface with their ventral regions uppermost. Schwind (1980) studied
the variation of acuity within the fields of view of these eyes, and found that
they had two high acuity zones corresponding, as in the fish areae, to the region
of the aerial image just above the surface, and of the underwater image just
below the surface (Figure 8). Between these zones there was a low acuity zone,
corresponding to the close region of total internal reflection.

Notonecta is the more remarkable because it lives another life as well. To
travel to new stretches of water it must fly, and when it does so the ventral
part of the eye, which looks into air when the animal is submerged, now looks
down at the ground. Schwind (1983) found that the receptors in this part of the
eye had their microvilli arranged in such a way as to be an optimal analyser for
light polarized by water surfaces. In later papers (Schwind, 1984a,b) he
demonstrated that Notonecta is indeed sensitive to polarized light, and will
actually dive into surfaces that have the right polarization properties, whether
or not they are made of water!

REFERENCES

Baylor ER (1967) Air and water vision of the Atlantic flying fish, Cypselurus heterurus. Nature (Lond) 214: 307-309.

Blest AD, Land MF (1977) The physiological optics of Dinopis subrufus L.Koch: a fish-lens in a spider. Proc R soc Lond B 196: 197-222.

Chaudhuri A, Hallett PE, Parker JA (1983) Aspheric curvatures, refractive indices and chromatic aberration in the rat eye. Vision Res 23: 1351-1364.

Clarkson ENK, Levi-Setti R (1975) Trilobite eyes and the optics of DesCartes and Huygens. Nature (Lond) 254: 663-667.

Denton EJ (1970) On the organization of the reflecting surfaces in some marine animals. Proc R soc Lond B 258: 285-313.

Exner S (1891) Die Physiologie der facettirten Augen von Krebsen und Insecten, Deuticke, Leipzig u Wien.

Fernald R, Wright SE (1983) Maintenance of optical quality during crystalline lens growth. Nature (Lond) 301: 618-620.

Fletcher A, Murphy T, Young A (1954) Solutions of two optical problems. Proc R soc Lond A 223: 216-225.

Graham JB, Rosenblatt RH (1970) Aerial vision: unique adaptation in an intertidal fish. Science (Wash) 168: 586-588.

Jamieson GS (1971) The functional significance of corneal distortion in marine mammals. Can J Zool 49: 421-423.

Land MF (1979) The optical mechanism of the eye of Limulus. Nature (Lond) 280: 396-397.

Land MF (1981) Optics and vision in invertebrates. In: Autrum H (ed): Handbook of Sensory Physiology, v.VII/6B. Springer, Berlin & New York; pp. 472-492.

Land MF (1984a) Crustacea. In: Ali MA (ed): Photoreception and vision in invertebrates. Plenum, New York; pp. 401-438.

Land MF (1984b) Molluscs. In: Ali MA (ed): Photoreception and vision in invertebrates. Plenum, New York; pp. 699-725.

Land MF (1985) The morphology and optics of spider eyes. In: Barth FG (ed): Neurobiology of arachnids. Springer, Berlin; pp. 53-78.

Levi B, Sivak JG (1980) Mechanisms of accommodation in the bird eye. J comp Physiol 137: 267-272.

Levi-Setti R, Park DA,Winston R (1975) The corneal cones of Limulus as optimised light concentrators. Nature (Lond) 253: 115-116.

Lythgoe JN (1979) The ecology of vision, Clarendon, Oxford.

Millodot M, Sivak J (1983) Contribution of the cornea and lens to the spherical aberration of the eye. Vision Res 19: 685-687.

Munk O (1970) On the occurrence and significance of horizontal band-shaped retinal areae in teleosts. Vidensk Meddr dansk naturh Foren 133: 85-120.

Piggins DJ (1970) Refraction of the harp seal, Pagophilus groenlandicus (Erxleben 1777). Nature (Lond) 227: 78-79.

Pumphrey RJ (1961) Concerning vision. In: Ramsay JA, Wigglesworth VB (eds): The cell and the organism. Cambridge University Press; pp. 193-208.

Sivak JG (1976) Optics of the eye of the "four-eyed fish" (Anableps anableps). Vision Res 16: 531-534.

Sivak JG (1978) Vertebrate strategies for vision in air and water. In: Ali MA (ed): Sensory ecology. Plenum, New York; pp. 503-519.

Sivak JG, Millodot M (1977) Optical performance of the penguin eye in air and water. J comp Physiol 119: 241-247.

Sivak JG, Warburg MR (1983) Changes in optical performance of the eye during metamorphosis of an anuran, Pelobates syriacus. J comp Physiol 150: 329-332.

Sivak JG, Hildebrand T, Lebert C (1985) Magnitude and rate of accommodation in diving and non-diving birds. Vision Res 25: 925-933.

Schwind R (1980) Geometrical optics of the notonecta eye: Adaptations to optical environment and way of life. J comp Physiol. 140: 59-68.

Schwind R (1983) Zonation of the optical environment and zonation of the rhabdom structure within the eye of the backswimmer, Notonecta glauca. Cell Tissue Res 232: 53-63.

Schwind R (1984a) Evidence for true polarization vision based on a two-channel analyzer system in the eye of the water bug, Notonecta glauca. J comp Physiol A 154: 53-57.

Schwind R (1984b) The plunge reaction of the backswimmer Notonecta glauca. J comp Physiol A 155: 319-321.

Walls GL (1942) The vertebrate eye and its adaptive radiation, Cranbrook Institute, Michigan. Reprinted (1967) Hafner, New York.

AQUATIC AND TERRESTRIAL CHEMORECEPTOR ORGANS: MORPHOLOGICAL AND PHYSIOLOGICAL DESIGNS FOR INTERFACING WITH CHEMICAL STIMULI

Jelle Atema

Boston University Marine Program
Marine Biological Laboratory
Woods Hole, Massachusetts 02543
U.S.A.

A comparison between aquatic and terrestrial chemoreceptor organs involves a discussion of the odor transport mechanisms in water and in air. Fluid dynamics tells us that these mechanisms are surprisingly similar in the two media. Aquatic crustacea and terrestrial insects are rather well studied arthropod groups where a comparison of chemoreceptor morphology and physiology is useful. Fish and mammals give us similar material for vertebrates. In addition, vertebrate and invertebrate solutions to terrestrial and aquatic adaptation can be compared.

INTRODUCTION

A full-scale comparison of aquatic and terrestrial chemoreception is a subject on the scale of a book, not a chapter. In the small-scale version of this chapter I will focus on some of the most obvious differences as well as the surprising similarities that exist in the chemical stimulus worlds of aquatic and terres- trial animals. I will argue that this stimulus world is reflected in the morphology and physiology of the receptors designed to interact with these stimuli. I will use as examples animals from two contrasting pairs of phylogen- etic groups: among the vertebrates fish and terrestrial mammals, and among the invertebrates insects, which are mostly terrestrial arthropods, and crustacea, a mostly aquatic arthropod group. It will be seen that many of the differences are variations on a general plan. The 4-way comparison allows us to examine both the vertebrate and the arthropod plans as they have adapted to air and to water. The comparison also allows us to see the overall functional similarity of arthropod and vertebrate chemical senses in relation to the chemical stimulus world. For the purpose of this symposium volume I do not address the details of chemoreception. Interested readers are refered to recent reviews and the primary literature contained therein.

Reviews of aquatic chemoreception covering vertebrates and arthropods and appro- priately focused on the stimulus world have appeared most recently in "Sensory Biology of Aquatic Animals", edited by J. Atema, R.R. Fay, A.N. Popper, and W.N. Tavolga, published by Springer Verlag, New York, 1987; see chapters by W. Carr; J. Atema; M. Laverack; J. Caprio; T. Finger; C. Derby and J. Atema; B. Ache. Recent reviews of insect chemoreception that are particularly relevant to the present chapter can be found in Kaissling (1986) and in "Chemical Ecology of Insects", edited by W. Bell and R. Cardé, published both by Chapman and Hall Ltd., London, 1984, and by Sinauer Assoc. Inc., Sunderland, MA, 1984; in particu- lar see chapters by E. Städler; H. Mustaparta; J. Elkinton and R. Cardé; W. Bell; R. Cardé. Chemoreception in various terrestrial vertebrates has been reviewed many times, but never in a form particularly relevant to the purpose of this chapter. Specifically, there is no review that treats chemoreception from the vantage point of the natural stimulus distributions that the receptors and the animals must interact with. The present chapter will use that vantage point

Comparative Physiology: Life in Water and on Land. P. Dejours, L. Bolis, C.R. Taylor, E.R. Weibel (eds.) Fidia Research Series, IX-Liviana Press, Padova © 1987

to compare chemoreception in aquatic and terrestrial arthropods and vertebrates.

STIMULUS DISTRIBUTION PROCESSES IN AIR AND WATER

Chemical Stimuli

A chemical stimulus is a molecular compound or a mixture of compounds that causes a response in a chemoreceptor. A chemoreceptor can be defined at the level of the receptor site, i.e. the initial binding or interaction site on the receptor cell membrane such as the acetylcholine receptor of the vertebrate neuromuscular junction, or at the level of the receptor cell. The use of the term must be specified. To be effective a chemical stimulus must have the proper molecular composition (i.e. stimulus quality) to be recognized by the receptor site. The stimulus must also be presented to the receptor at an adequate rate (i.e. stimulus intensity change). For instance, receptor cells that are adapted to ambient concentrations of a stimulating mixture can only respond if a stronger (and/or a different) stimulus is applied at a rate of increase fast enough to overcome the adaptation rate of the receptor. Such dynamic response properties of the receptors have not often been studied in chemoreception but - as we will see in this chapter - they may be important temporal filters of chemical information. This will be clear when we consider the processes by which chemical stimuli are distributed in the environment. I will use the term 'odor' both in terrestrial and aquatic environments.

Chemical stimuli originate from a source of release and are carried through the environment by a number of physical processes, particularly molecular diffusion, viscous flow, and turbulent flow. The three processes interact in various ways but it will be convenient to treat them separately first. Moreover, it should be pointed out that air and water are both considered fluids. Thus, the physical rules of fluid dynamics apply equally to both media. The differences are of scale and degree, not of principle.

Scaling

At 20°C the kinematic viscosity of air ($\checkmark = 15.00 \times 10^{-6}$ m^2s^{-1}) is 15 times greater than that of fresh water ($\checkmark = 1.004 \times 10^{-6}$ m^2s^{-1}) or seawater ($\checkmark = 1.047 \times 10^{-6}$ m^2s^{-1}). Consequently, the same Reynolds number[*] will be reached at flow velocities 15 times higher in air than in water. Vogel (1981, p. 69) points out that common earthly wind speeds are often about 15X faster than common water currents, and that flying and swimming speeds have a similar relationship. Thus, despite the greater density of water ($\rho w / \rho a \approx 800$) the dynamics of motion in air and in water take place at comparable Reynolds numbers. One should keep in mind that these are not precise calculations: the Reynolds number is a scaling factor which indicates only the order of magnitude of the ratio of inertial and viscous forces involved in the process under investigation.

An intuitive sense of the importance of the length scale in fluid dynamics can be obtained by analogy with a piece of wire. A short length of it is stiff and unbending; the longer the wire the easier it becomes to bend and flex. Similarly, when floating in an ocean current we view its flow as even and homogenous over the short distance. Over several kilometers, however, the flow meanders unpredictably: on the large scale the flow is turbulent. Very small flows at the scale of bacterial movement are completely dominated by viscous forces. In addition, we know intuitively that faster flow is more likely to be turbulent:

---

[*] The Reynolds number $Re = 1U/\checkmark$, where 1 is a length scale, U velocity and $\checkmark$ kinematic viscosity; $\checkmark = \mu / \rho$, where μ is the dynamic viscosity and ρ the density of the medium. In its form $Re = \rho 1U / \mu$ one recognizes that Re represents the ratio of inertial forces (which ultimately lead to turbulence) over viscous forces (which ultimately lead to laminar flow).

increasing flow velocity has the same effect on flow stability as using a
greater length scale to view the process.

It is important to realize that large-scale flows contain within them many flows
of various smaller scales. The micro-environment of a bacterium in the Gulf
Stream is still only a few um and thus dominated by viscous forces that keep its
fluid environment coherent. Yet the entire microenvironment is carried about in
various larger scale eddies which together form a highly turbulent ocean current.
Even at the 1-10 millimeter scale of an oceanic copepod there is still no appreciable external flow to distort its surrounding fluid droplet. Unless the animal
creates currents itself, it lives in a quiet, stable shell of water. At comparable velocities air pockets of 15-150 mm diameter should be stable entities
dominated by viscous flow within, while being carried in the turbulence of larger
scale winds. However, as said above, since wind speeds tend to be (15X) larger
than water current speeds, it is likely that stable air pockets are commonly not
found until a scale of 1-10 mm. For ease of discussion I will assume that within
an order of magnitude a 1-10 mm boundary for turbulence is appropriate both for
water and air.

Odor Dispersal

Returning now to the dispersal of chemical stimuli through the medium we see that
an odor must start with a source releasing chemical substances into the environment. At size scales below 1 mm the subsequent dispersal process will be dominated by molecular diffusion. Diffusion in air is much faster than in water. At
$20^{\circ}C$ the diffusion coefficient for oxygen in air is $D = 0.196$ cm^2s^{-1}, in
water $D = 0.182$ x 10^{-4} cm^2s^{-1} (Altman 1958). However, even in water
diffusion is a remarkably efficient transport mechanism (on the order of milli-
seconds) over short (um) distances. At intermediate size scales, 1-10 mm, odor
will be transported predominantly by viscous flow. At larger size scales inertial forces begin to dominate; this leads to turbulent dispersal. A turbulent
odor plume is easily visualized by a smoke stack or by the entry of a muddy
river into another body of water. Eddies form; with time and thus distance from
the source large-scale eddies break up into smaller scale eddies, until they
reach such a small size that viscous forces once again take over. At that point
molecular diffusion finishes the dilution process, leading to homogeneous odor
distribution: the plume eventually disappears. The fluid masses are now fully
mixed. Long before this happens the odor has probably disappeared already by
means of various chemical (e.g. photo) and biological (e.g. bacterial) processes.
At the large scale of turbulent dispersal there is a significant difference in
air and water: an odor signal will bridge the distance from source to detector
(15X) faster in air than in water, since air currents tend to be faster than
water currents.

A Patch in Space Is a Pulse in Time

Seen from a distance a turbulent odor plume is typically patchy: areas of high
odor concentration are meandering through low concentration areas, or even areas
without any odor. The typical structure of a plume is such that near the source
we find patches of high concentration sharply alternating with areas of low or
zero concentration; further away, the peak concentrations are lower, the patch
boundaries have become "softened" by small-scale eddies and molecular diffusion,
and fewer areas have remained untouched by the odor. Thus, while the peaks become lower, the background rises, resulting in smaller peak to background ratios.

If we now measure odor strength in this plume at one stationary point as the
plume drifts by, our "point" detector will see a variety of odor pulses of
different length and strength alternated with periods of zero signal. Thus,
what appeared from a distance as a spatial distribution of patches at any one
instant now appears in close-up as a time series of pulses in one location. If
we choose a point close to the source we see over time a series of sharp, high

peaks against low or zero backgrounds; at points farther away we measure fewer high peaks, more soft peaks, and higher backgrounds. Considering their size (or the size of their receptor organs) relative to the odor fields of their interest, it is more likely that animals measure in the temporal domain than in the spatial domain.

Such stationary sampling within a turbulent plume has been done in air and in water. Murlis and Jones (1981) used an ion collector with a sampling area of 10 cm^2 and a response time of 1 ms to measure pulse patterns generated by the constant release of ions from a point source. They used one set of wind and terrain conditions and measured at different distances (2,5,10 and 15 m) from the source. Mean wind velocity was 3 ms^{-1}. The observed pulse patterns reveal bursts with several minor and major peaks alternating with quiet periods and periods of background "noise". The overall patterns measured correspond closely to the intuitive description given above. Their published records show that the maximum frequency of peaks within a burst is about 15 per second. One should bear in mind that we do not know yet what constitutes a significant peak for a chemoreceptor; this depends on its frequency filter and this has not been studied in detail. However, the rough estimate of 15 Hz gives an idea of an upper limit on frequency information in the stimulus plume under the standard conditions of this one experiment. Lower frequencies exist, all the way down to annual cycles in mean wind direction. The frequencies of greatest concern to chemoreceptor organs probably range from 0.01-10 Hz.

A similar experiment was carried out in water (Atema, Bryant and Murray-Brown, 1983 unpublished data; for preliminary report see Atema 1985). Here dye was released from a point source in a flume with constant flow of 3 cms^{-1} over different, turbulence-generating substrates. Dye density patterns were measured at 15, 60 and 100 cm from the source through a 1 cm^2 sampling area. Pulse patterns were similar to those measured by Murlis and Jones (1981). However, in water maximum frequencies of concentration peaks were only about 1 Hz.

While these two experiments are hardly representative of the great variety of turbulence conditions that exist in air and in water, it is interesting that both experiments were done under what were considered to be quiet and steady flow conditions. Under these conditions air flowed 100X faster and the pulse rates in air were 15X greater than those in water.

Spatial Information Contained in Different Odor Fields

In a simple diffusion field around a constantly emitting source the gradient of chemical concentration is stable and predictable. Gynogametes of marine algae provide a good example: they release a sex attractant pheromone at a constant rate. This forms a spherical diffusion field to which androgametes can orient in almost straight paths over the short distances (up to 1 mm) involved (Maier and Müller 1986). An excellent discussion of diffusion processes in relation to bacterial behavior and chemoreception has been given by Berg and Purcell (1977). In diffusion fields, one problem for animal behavior and chemoreception occurs when the source emits in a short burst. In that case the gradient decays very rapidly (Jackson, 1980). Yet, the spatial information in this diffusion field is not changed: the gradient remains centered on the source. Another problem occurs when the source moves through the surrounding fluid medium. While this complicates its geometry, the diffusion field remains stable (Lehman and Scavia, 1982).

At the intermediate scale of 1-10 mm animals can effectively control the flow and create a stable flow pattern (shear field) around themselves which remains unperturbed by the larger scaled turbulence. Copepods with body sizes of about 1 mm can set up feeding currents of some 10 mm (Strickler 1985). Such a shear field distorts the chemical diffusion fields emanating from their algal food in a predictable manner (as modeled by Andrews 1983) so that the animal has an accurate expectation of when and where the algal cell will arrive once it has sensed the algal "body odor". The Reynolds number of the feeding current indicates that inertial and viscous forces are roughly balanced (Re = 10 at the flow maximum). Thus these copepods work near the upper limit of flow predictability: faster and larger flow fields would become unstable and indeed show signs of turbulence. Then, the animal would lose control. It would no longer be informed as reliably of the spatial structure of the chemical information that enters and flows through the field. I am not familiar with similar examples in air.

Inevitably, this discussion leads to the unpredictability of the larger scale, turbulent flow fields. At size scales over 10-100 mm the animal is no longer in control of the flow field. It can no longer hide in the comfort of small predictable odor shells surrounding it. It now has to deal with environmental turbulence. While the animal is large enough to live at the size scale of turbulent eddies, it is often far too small to measure the complete odor field (i.e. plume) of which the eddies are a part. Moreover, even if they could cover the distance of the plume, animals do not often have the luxury of time in which to explore the extent of the plume. Decisions of life and death often have to be based on very brief sampling. Thus, the animal is reduced to a point detector measuring over short time periods. One might hypothesize (Atema 1987) that from the pulse patterns that can be measured in one point, the animal might reconstruct some spatial information about the plume. This information may be as little as knowing the source is near or far, or to turn left or right to reach the source. Yet this is critical information, which could be at a premium in competitive foraging. Thus one might expect to see chemosensory organ systems with filters designed to extract this temporal information.

A COMPARISON OF CHEMORECEPTOR ORGANS AND CELLS

Receptor Organs Live in Different Microenvironments

For the purpose of chemoreception we must look not only at the chemical environment in which the animal lives, but also at the "micro" environments in which its chemoreceptor organs live (Atema 1985). A clear example is, of course, our own nose and tongue. It is self-evident that receptor organs that encounter qualitatively different chemical stimuli, such as our nose and tongue, are tuned to a different "spectrum" of smell and taste substances. However, one must consider not only the qualitative aspects of the stimulus environment, but also the quantitative, i.e. spatiotemporal, aspects. The latter may well be different even for receptor organs that live in the same general environment such as the smell and taste organs of aquatic animals.

In the following comparison of receptor organs of 4 different groups of animals, marine crustacean chemoreceptors will be used as the reference: they operate in the ancestral environment, where special adaptations to osmotic and desiccation stress are minimal. Marine crustacean chemoreceptors may thus be closest to the original evolutionary stages. The receptors of the other groups will be discussed in contrast to the crustacean receptors.

Crustacean Chemoreceptors

Crustacean chemoreceptor cells are bipolar neurons, whose dendrites enter through an opening in the cuticle into a "sensillum", or receptor hair. This cuticular structure takes on many shapes in different species and even in the same species. The two best known structures are the aesthetasc sensilla of the antennules, the crustacean equivalent of the vertebrate olfactory organ, and the hedgehog or fringed sensilla of the walking legs, both studied best in large decapod crustacea: lobsters and crayfish. The antennules live in open water where they encounter mostly turbulent odor plumes. The walking legs live near the bottom where they contact patches of chemical substances in water and in the substrate; they also dig into food, which they hand over to the mouth.

The aesthetasc sensilla are perhaps the most distinct type and the most consistent in morphology among different crustacean species. In lobsters, the aesthetascs are about 1 mm long with a diameter of 30 um for most of their length. The
distal 0.8 mm portion of the sensillum has a thin (2 um) spongy cuticle and
contains the branched, modified cilia of some 400 receptor cells. Experiments
in crab aesthetascs showed that dye quickly penetrates into the lumen (Ghiradella
et al. 1968). It is thus assumed that sea water and its solutes freely diffuse
through the wall in and out of the sensillum interior. Since the fully marine
lobster is nearly isosmotic with sea water, special mechanisms may not be needed
to protect the ciliary receptor membranes from osmotic stress or from desiccation. In contrast to leg receptors, aesthetascs are also not subject to severe
mechanical abuse and do not need heavy protective cuticle. They may represent
the "purest" chemoreceptor structure known. Assuming an average of 20 ciliary
branches per receptor cell and an average branch diameter of 50 nm the total
membrane area inside one aesthetasc sensillum is 2 mm^2. With some 1,000
aesthetascs on each of the two antennules the lobster has a 4,000 mm^2 surface
area with which to interact directly with the chemistry of the surrounding sea
water. This does not include, of course, its many other chemoreceptor organs.

Critical to the focus of this chapter is the question "How does this receptor
membrane interact with the environment?" and specifically "What are the consequences of the morphological and physiological characteristics of the aesthetasc
organs for the extraction of information from chemical signals?" The first
filter is behavioral/morphological. Lobsters and most other large decapod
crustacea flick their antennules (Schmitt and Ache 1979). This functional
equivalent of vertebrate sniffing causes a brief rush of water to flow between
the densely packed aesthetasc sensilla. In between flicks odor transport is
severely impeded by the close packing of aesthetascs, which in lobsters are
separated by distances of 30-300 um. Over such short distances water is highly
viscous. The result is that the antennules "grab" a sample of surrounding water
during a flick, while the water chemistry between the sensilla remains unchanged
in between flicks. During each flick a new water sample is placed in close
contact with the aesthetasc cuticle. Diffusion through the boundary layer - the
"unstirred" layer of fluid surrounding every solid surface - and through the
spongy cuticle, a total distance of perhaps 10 um, would take in the order of 10
ms. Flicking rates in lobsters range from 0.1-4 Hz. This represents the
lobster's "digital" sampling rate. Thus, at the highest flicking rate the
receptor membranes are exposed to a new sample every 250 ms with a 10 ms latency
and rise time. Flicking is critical for efficient orientation in a plume
(Devine and Atema 1982; Reeder and Ache 1980).

The next filter is physiological. Obviously, only those chemical compounds for
which there are adequate molecular receptor mechanisms will cause responses in
receptor cells. I will not discuss this spectral filter further. The principle
is evident and the biochemical details go well beyond the scope of this chapter.
Another part of the physiological filter deals with temporal frequencies as
represented by the adaptation and disadaptation rates of the receptor cells.
This is a still poorly studied aspect of chemoreception with important consequences for the analysis of chemical gradients and odor patches. In the extreme
there are so-called phasic and tonic receptors. Phasic receptors respond only
to the rate of change, usually only to the rate of increase, of the stimulus.
They adapt and disadapt very quickly so that they can follow high pulse rates.
One might say that they have a high flicker fusion frequency, i.e. the pulse
rate at which the response to single pulses fuses. Phasic receptors would be
useful to follow high flick rates or to analyze fast pulse patterns in turbulent
odor plumes. In contrast, tonic receptors respond to the steady-state ambient
stimulus level; they do not adapt. Such receptors tend to respond more slowly
to changes. Most chemoreceptors combine phasic and tonic properties. However,
within one receptor organ one can find receptor cells with different temporal
response characteristics: some are quite phasic, others have strong tonic properties. Even within a population of receptor cells that are spectrally tuned to

only one compound, L-glutamate, some cells were mostly phasic, and others phasic-tonic (Voigt and Atema 1984). When the glutamate cells were stimulated with pulses of 1 sec duration and intervals of 5,10,20,40,80 or 160 sec, some gave 1 sec responses, followed the fastest pulse rate used (0.2 Hz), and did not show much adaptation from pulse to pulse; others gave responses lasting 5 sec or longer fusing at the 0.2 Hz rate, and showed strong interpulse adaptation up to 40 sec. Since the fastest cells might have followed higher frequencies, we can not yet say what their flicker fusion rate would be. It seems likely that they could follow the 1 Hz pulse rates measured in aquatic odor plumes described earlier.

Inside the "cutting" edge of the small claws of their first two pairs of walking legs (Derby 1982), lobsters have a single row of spined, 100-300 um long sensilla (resembling hedgehogs). Crayfish sensilla do not have spines but a jagged crest; otherwise they are similar (Altner et al. 1983). These sensilla are made of almost solid cuticle with a small 4 um wide canal inside running to a 0.5 um wide pore near the tip. In crayfish, 8 undivided receptor cilia run inside the canal; two are mechanoreceptive and the remaining 6 appear to be chemoreceptive (Altner et al. 1983). The total amount of ciliary receptor membrane in one crayfish sensillum is only about 200 um^2. This is 10,000X less than the membrane surface inside an aesthetasc. In addition, stimulus access may be at the tip pore only, or perhaps but less likely through the walls as well. This morphological design appears far less efficient in interfacing with the stimulus environment than the aesthetasc design. The hedgehog sensillum, instead, appears built to withstand mechanical abrasion and to have slow but continuous access to the chemical stimulus environment. Legs do not flick; at most they wave or rake across the substrate. In addition to hedgehog sensilla the legs have groups of 0.3 to 2 mm long, thick-walled sensilla which contain some 25 sensory cells.

In physiological experiments, leg receptors respond to a great variety of amino acids as well as other compounds (Derby and Atema 1982; Johnson et al. 1984; Derby and Atema 1987). The overall response spectrum for amino acids overlaps between antennular and leg chemoreceptors; but there are differences of degree (Johnson and Atema 1983; Johnson et al. 1984): for instance, legs have very prominent glutamate receptors and antennules have most prominent hydroxyproline and taurine receptors. Again, I will not dwell on such spectral filters; their functional significance is obvious, although their ecological connection has not been explored in detail. An interesting point is that (semi)terrestrial crustacea show spectral tuning of leg-mediated behavior to carbohydrates similar to insects, not the amino acids of marine crustacea (Trott and Robertson 1981).

The behavioral function of the legs of marine crustacea suggests that their filtering properties should be adapted to deal with the widely varying stimulus concentrations of ambient sea water, of marine substrates, and of tissue fluids of lobster food. In the case of glutamate, sea water has a mean level of 10^{-8} M and 10^{-3} M is present free in tissues. Mean ammonium levels in coastal sea water are 10^{-6} M and 10^{-2} M free in tissue fluids. Experiments with leg ammonium receptor cells show that they adapt within seconds to higher ammonium concentrations; their entire stimulus-response function shifts up with a new threshold at the new ambient background; disadaptation after strong stimuli may take minutes (Borroni and Atema 1987). Thus, at least for this cell type, emphasis may not be on speed of response, but on adaptation to widely varying background concentrations. It appears that the morphology of the leg sensilla and the behavior of the legs only allow gradual stimulus access through diffusion. This would not place a premium on adaptation and disadaptation speed for the receptor cells. However, more work is needed to show whether or not leg receptors and antennular receptors differ fundamentally in their temporal response characteristics.

In the case of crayfish there is anatomical evidence that processes of supporting cells enter the dendritic canals of the hedgehog sensilla. This suggests that the canal may be filled with receptor lymph, as found in insects. The function of receptor lymph in a freshwater animal such as the crayfish may be osmotic protection of the dendritic outer segments. Crayfish are indeed rather impervious to osmotic shock (Hanns Hatt, pers. comm.). In contrast, lobster leg chemoreceptors are not protected from severe osmotic shock: a 5 minute immersion in distilled water eliminates functional chemoreception for over one day (Derby and Atema 1982). If supporting cell processes were to be found in the hedgehog canals of lobsters, it would indicate that marine crustacea may need receptor lymph for odor transport, as is suggested in insects and vertebrates.

Insect Chemoreceptors

Insect chemoreceptor cells are similar to crustacean cells in that they are bipolar neurons, whose ciliary dendritic processes extend into the lumen of cuticular sensilla. Moreover, as in crustacea, insects have prominent concentrations of chemoreceptor sensilla on their antennae, legs, and mouthparts. The behavioral functions of these organs appear similar. And, to some extent, the environment of the different organs is similar: antennae operate in free-flowing fluid media (air or water) and legs and mouthparts operate often in contact with a substrate and/or handle food. The differences between insect and crustacean chemoreceptors seem to reflect differences in the air and water medium. Probably to avoid desiccation, insect sensilla are impermeable cuticular structures where odor access takes place through complicated pore systems. Odor access may be facilitated by active transport mechanisms. Two of the best known insect sensilla are the sensillum trichodeum of moth antennae and the tarsal hairs of blowfly feet. They are the functional and anatomical equivalent of the crustacean aesthetasc and hedgehog sensilla respectively.

There are tens of thousands of sensilla trichodea on each antenna of certain large moths. These sensilla can be up to 0.5 mm long and only 5 um thick. Usually two bipolar chemoreceptor cells and three auxiliary cells lie at the base. The receptor cells send their dendritic outer segments, unbranched modified cilia, through the lumen which is filled with receptor lymph secreted by the auxiliary cells. Odor molecules enter the lumen through a complex system of 10 nm wide pores that penetrate the sensillar shaft everywhere. Their total opening surface to the outside world is a very small fraction of the total sensillar surface. It is not known if the stimulus reaches the dendritic membrane directly via cuticular pore tubules or if it must pass through receptor lymph in addition. It is also not known if active transport mechanisms carry odor molecules to the receptor membrane or that passive diffusion is sufficient to explain the observed response latencies. However, odor binding proteins exist in this lymph; their function is still uncertain.

The sensillum trichodeum and indeed the entire male moth antenna are designed as pheromone capturing devices. It has been calculated that one pheromone molecule is sufficient to trigger a nerve impulse in the receptor cell (Kaissling and Priesner 1970). Under very high stimulus concentrations odor molecules reach the receptor membrane in 5 ms; under more natural conditions the response latency is in the order of several 100 ms. About 30% of pheromone molecules passing a cross-sectional area the size of an antenna (1-2 cm^2) can be adsorbed onto it, and of those 80% can be found on the sensilla trichodea. Within one second of stimulating with a very strong odor plume, the odor concentration in the hairs can increase 10^5 fold over the concentration in the plume (Kaissling 1986). This highly efficient capturing mechanism may be an adaptation that evolved under the extreme selection pressure experienced by male moths competing for females. However, it shows that evaporation barriers such as receptor lymph and narrow pore systems in an impermeable cuticle do not impede odor access. Unfortunately, comparable data for crustacea do not exist.

Additional evidence for the efficiency of stimulus access and removal in sensilla trichodea is given by the fact that some of their receptor cells can follow stimulus pulse rates of up to 10 Hz (Kaissling 1986; Kaissling et al. 1987). Behavioral experiments show that male silk moths orient better when exposed to odor pulses than to constant odor; this was effective up to 3 Hz pulse rates (Kramer 1986). Since insects are not flicking (but perhaps vibrating) their antennae, this result should be seen in the context of aerial odor plume pulse rates discussed earlier. Insect pheromone receptor cells may have higher flicker fusion rates than crustacean cells, although data for crustacea are still lacking. If so, this could reflect the higher odor pulse rates encountered in air. Perhaps the tolerant and slow marine environment never pressed the crustacea to develop the efficient stimulus transport mechanisms seen in insects.

The tarsal sensilla of blowfly feet are quite similar in principle to the hedgehog sensilla of crustacea. Several chemoreceptor cells and one mechanoreceptor cell send their dendrites with unbranched chemoreceptor cilia into the lumen of the rather thick-walled sensillum. There are two canals in this sensillum, both filled with receptor lymph produced by three accessory cells at the sensillum base; only the narrower of the two contains receptor cilia. A dense plug of mucopolysaccharides closes the tip pore. Small canals ($<$ 10 um diameter) seem to cross the plug. The plug ends proximally at a cuticular "sieve plate", the cupola, through which odor molecules via the receptor lymph can reach the dendrite tips which lie some 100 um under the cupola (De Kramer and Van der Molen 1980). Thus, odor access is probably by diffusion (or active transport) via receptor lymph through the plug canals and cupola, a distance of some 5 um. Tarsal sensilla are refered to as contact chemoreceptors because their normal mode of operation is to sample solutions on a substrate. In that microenvironment stimulus concentration may be relatively high. The strong diffusion gradients would allow restricted odor access and yet result in response latencies in the order of 1-5 ms.

Tuning is related to micro-environment and behavioral function. The receptors of different species are tuned to different, often food related spectra of chemical stimuli. In addition, a clear physical separation of volatile and soluble compounds exists between the stimulus environments of insect antennal and tarsal receptors, resulting in spectral separation of receptor function. In crustacea, the stimulus environment does not enforce such a separation of tuning spectra in antennular and leg receptors, although some separation exists, perhaps based on behavioral function and micro-environment. Temporal tuning has not been specifically studied for tarsal receptors, but it may be interesting to see its possible role in connection with tapping and running behavior observed in various insects (Städtler 1984). This behavior may be a functional analog to flicking of crustacean antennules.

Chemical Senses of Fish

Fish have clearly separated smell and taste organs (Atema 1980; Caprio 1987). The fish nose consists of a small pocket in front of each eye. Inside these two pockets lies a more or less folded sensory epithelium. Narrow entrances to the nasal cavity protect the nose from mechanical abuse, but also preclude efficient odor access over the millimeter distances involved just as densely packed aesthetascs impede flow and odor access over distances too large for rapid diffusion. Thus active odor transport is necessary, and this gives the animal control of the flow. The nose generates water currents either by cilia or by accessory sacs acting as pumps (Zeiske 1974; Døving et al. 1977). Ciliary flow consists of a smooth unidirectional laminar sheet of water drawn over the epithelial folds, entering through a generally small opening (naris) to the outside world and leaving through a separate, larger naris further posterior. This type of nose performs continuous sampling of one small point in space: the inflow naris. Fish have two such points, left and right. Relative to the odor plume, the position of the measuring points can, of course, be changed by swimming. In fish with accessory sacs the flow is pulsed. The sacs act like bellows and are operated often via jaw movements. Valves result in unidirectional flow over the epithelium; intake and outflow openings are generally separated. Jaw snapping in tuna excited by food odor may represent their equivalent of flicking or sniffing (pers. obs.). It is not known if there exists a relationship between odor environment and nose type. There may well be mechanical constraints such as swimming speed and drag that favor certain designs. In addition, of course, there are phylogenetic constraints.

The olfactory receptor cells are bipolar neurons similar to those of arthropods. Their dendrites send one to tens of cilia (or microvilli) into a mucus layer, which is constantly secreted by accessory cells in the epithelium. Odor molecules must diffuse through this layer to reach the receptor membranes. In principle, the mucus layer appears to have a function similar to the receptor lymph of arthropod sensilla. It will be interesting to compare the physical and chemical properties of the two secretions in detail, particularly when comparing fresh water and marine fish with crustacea from the two aquatic environments. The total receptor membrane surface area is comparable to that of arthropod olfactory organs.

Temporal filtering characteristics of receptor cells have not been measured in fish. One would expect that fish need to sample turbulent odor plumes as much as lobsters do. However, in general, fish move much faster than crustacea. Thus they can sample sequentially in the spatial domain far more easily than crustacea can. This may make sniffing less important to them, whereas flicking is important for crustacean orientation. Simultaneous spatial comparison of odor intensity is theoretically possible particularly for animals with widely spaced sampling points, e.g. spiny lobster antennules, shark nares, catfish barbels. It is not known if these animals actually make simultaneous comparisons between their two sampling points, or that they too rely more on temporal sequences. Bonnethead sharks are capable of turning toward the stronger stimulus side when two stimuli are applied simultaneously to the two nares (Johnsen and Teeter, 1985). Conditioning experiments have shown that catfish can learn to discriminate chemical intensity differences of as little as 10-30% presented simultaneously to the two lateral barbels (i.e. taste) (Johnsen and Teeter, 1980), but it is not known how much this capability is used naturally. One might expect that such animals use a combination of spatial and temporal sampling.

Fish taste organs are built with taste buds, much like our own. In addition, some species have developed spinal taste systems (Finger 1987). I will discuss only the taste bud system inside the mouth. In the harsh chewing environment of the mouth, mechanically delicate sensilla and receptor epithelia are not likely to survive. As crustacea resort to heavy cuticle and small apical access pores, so do fish use tastebuds, clusters of epithelial receptor cells embedded in the mouth lining and innervated by (cranial) nerves. This is the only chemoreceptor system that is not based on bipolar receptor neurons. Taste buds have small apical pores in which the microvilli of taste cells protrude. Mucus covers the entire lining of the mouth and pharynx. As in arthropod tarsal sensilla, the total receptor membrane surface area in the taste pores is relatively small, but some fish (carp, catfish) have large numbers of tastebuds (Finger 1987). The microenvironment of these taste buds is food (animal and plant tissues) and sometimes non-food (gravel, mud, sticks). Their behavioral task is to sort the edible from the non-edible and the poisonous. This task determines their spectral tuning. As in crustacea there is great overlap in smell and taste spectra (Caprio 1987). Chewing is the initial mechanism for stimulus access to the taste pore, where the final distance is covered by diffusion through mucus. Again, this mucus may serve roles comparable to receptor lymph in arthropod taste and smell sensilla. Temporal filtering has not been studied, but considering the microenvironment and behavioral function of mouth taste buds one would expect that flicker fusion frequencies do not need to be high, as argued for arthropod leg (taste) receptors.

In general, fish smell and taste organs are similar anatomically to those of terrestrial vertebrates, including their neuroanatomical connections through the CNS (Finger 1987). But functionally one must look at the (aquatic) crustacea for comparison.

Mammalian Chemoreceptors

The best known mammalian chemoreceptors are the olfactory organ located in the nose and the gustatory organ located in the tongue. These organs live in strikingly different microenvironments. Anatomically the mammalian olfactory organ (Moulton and Tucker 1963) resembles that of fishes: bipolar receptor neurons send ciliary extensions into a mucus layer secreted by special cells. In mammals, the mucus layer interfaces with an air stream, not a water flow. This has consequences for odor adsorption. Volatile and soluble compounds are often quite different. Partitioning from air to mucus and from water to mucus will lead to different spatial and temporal patterns across the mucosal surface (Hornung and Mozell 1981). The logical comparison for functional similarity is with insects. This extends to the possible role of odor binding (and transport?) proteins in the receptor lymph, or mucus (Kaissling 1986, Getchell 1986).

An important difference between fish, arthropod and mammalian olfactory organs is that in mammals smelling is associated with breathing, making continuous chemical monitoring not possible. Fish noses with accessory sacs function similarly, but in mammals little is yet known about nasal flow mechanisms that operate like the one-way nose valves in fish. In humans, natural sniffing results in better odor identification than obtained with artificial sniffs (Laing 1983). Sniffing rates in rodents can be high, similar to the flicking rates of crustacea. Unfortunately, in mammals nothing yet is known about the actual flow fields over the olfactory mucosa with or without sniffing. For this purpose, a large-scale model of the human nose is being developed by Mozell and coworkers (Mozell unpubl.). Mammalian olfaction resembles insect olfaction in its spectral turning to volatile compounds. Temporal filtering has not been studied. But studies on the frog olfactory epithelium (Ottoson 1956) showed cumulative adaptation properties similar to those reported in single lobster leg glutamate cells (Voigt and Atema 1987): e.g. 50% reduction of subsequent responses when inter-pulse intervals were less than 10 s. Similar results were obtained for salamander receptor cells (Getchell 1986). This is indicative of low flicker fusion rates.

Chewing in mammals is similar to chewing in fish: it breaks up food, liberates tissue juices, and squeezes the juices around the mouth, in mammals together with saliva. This brings food solutes in close contact with the tastebuds which are concentrated on the tongue. As in fish, diffusion must be the final process by which chemical stimuli get sufficiently close to the microvillar membranes of the taste cells to bind and cause receptor potentials. Rapid temporal analysis appears unimportant in this situation. However, during drinking rapid identification of poison may be critical. Humans can recognize simple taste stimuli of 50 ms duration (Kelling and Halpern 1983). Flicker fusion frequencies may lie around 5Hz (McBurney 1976). Taste appears remarkably similar in arthropods and vertebrates, both aquatic and terrestrial.

CONCLUSION

At first glance, air and water appear to be very different media for odor transport, and chemoreceptor organs of aquatic and terrestrial animals appear vastly different in morphology. Closer inspection shows that both the media and the functional morphology of receptor organs are surprisingly similar. The differences are in detail rather than in principle. Bulk flow (often turbulent) at size scales larger than 1-10 mm brings odor molecules in the vicinity of the chemoreceptor structures. Frequently, subsequent laminar flow through or over the organ is generated and controlled by the animal through stimulus acquisition behaviors such as flicking and sniffing. However, flow cannot penetrate easily into the boundary layer that surrounds epithelia and sensilla; ultimately chemical stimuli reach the receptor structures by molecular diffusion. The last stages of odor access are perhaps the most similar in aquatic and terrestrial chemoreceptor organs: chemical stimuli diffuse passively (or are carried actively by transport proteins?) through a receptor lymph secreted by accessory cells. This brings them near the receptor membrane, which is a ciliary (or microvillar) extension of the receptor cell with high densities of receptor sites. The lymph may have developed additional functions in freshwater and terrestrial animals where it serves as an osmotic buffer or a desiccation barrier respectively. The study of temporal filtering in chemoreception has only just begun. One might anticipate species and organ differences depending on differences in behavioral function and/or (micro)environment. The most dramatic difference in aquatic and terrestrial chemoreception is in spectral tuning: air and water carry different chemical compounds and this is reflected in different receptor tuning. In general, the function of temporal and spectral tuning is behavioral: extracting relevant chemical signals from a noisy background. It is the task of receptor organs and cells to interface efficiently with their stimulus environment. Since the physics of the chemical environment is not principally different in air and water, we see many convergent solutions to chemoreception in widely separated animal groups such as aquatic and terrestrial arthropods and vertebrates.

Acknowledgements

This article was written during a period of grant support from the Whitehall Foundation and the National Science Foundation (BNS 8512585).

Altman PL (1958) Handbook of respiration, W.B. Saunders Co., Philadelphia 403 pp.

Altner I, Hatt H, Altner H (1983) Structural properties of bimodal chemo- and mechanosensitive setae on the pereiopod chelae of the crayfish, Austropotamobius torrentium. Cell Tissue Res 228: 357-374.

Andrews JC (1983) Deformation of the active space in the low Reynolds number feeding current of calanoid copepods. Can J Fish Aquat Sci 40: 1293- 1302.

Atema J (1980) Smelling and tasting underwater. Oceanus 23: 4-18.

Atema J (1985) Chemoreception in the sea: adaptations of chemoreceptors and behavior to aquatic stimulus conditions. Soc Exp Biol Symp 39: 387-423.

Atema J (1987) Distribution of chemical stimuli. In: Atema J, Fay RR, Popper RN, Tavolga WN (eds): Sensory biology of aquatic animals. Springer Verlag, New York, in press.

Berg HC, Purcell EM (1977) Physics of chemoreception. Biophys J 20: 193-219.

Borroni PF, Atema J (1987) Self- and cross-adaptation of single chemoreceptor cells in the taste organs of the lobster, Homarus americanus. ISOT. In: Roper S, Atema J (eds): Olfaction and Taste IX, NY Acad Sci Press, in press.

Caprio J (1987) Peripheral filters and chemoreceptor cells in fishes. In: Atema J, Fay RR, Popper AN, Tavolga WN (eds): Sensory biology of aquatic animals. Springer-Verlag New York, in press.

De Kramer JJ, Vander Molen JN (1980) The pore mechanism of the contact chemoreceptors of the blowfly, Calliphora vicina. In: Van der Starre H (ed): Olfaction and taste VII, IRL Press Ltd, London; pp. 61-64.

Derby CD (1982) Structure and function of cuticular sensilla of the lobster, Homarus americanus. J Crust Biol 2: 1-21.

Derby CD & Atema J (1982) Chemosensitivity of walking legs of the lobster Homarus americanus: neurophysiological response spectrum and thresholds. J exp Biol 98: 303-315.

Derby CD & Atema J (1982) The function of chemo- and mechanoreceptors in lobster (Homarus americanus) feeding behavior. J exp Biol 98: 317-327.

Derby, CD & Atema J (1987) Chemoreceptor cells in aquatic invertebrates: peripheral mechanisms of chemical signal processing in decapod crustaceans. In: Atema J, Fay RR, Popper AN, Tavolga WN (eds): Sensory biology of aquatic animals, Springer-Verlag New York, in press.

Devine D & Atema J (1982) Function of chemoreceptor organs in spatial orientation of the lobster, Homarus americanus: differences and overlap. Biol Bull 163: 144-153.

Doving KB, Dubois-Dauphin M, Holley A, Jourdan F (1977) Functional anatomy in the olfactory organ of fish and the ciliary mechanism of water transport. Acta Zool Stockholm 58: 245-255.

Finger T (1987) Organization of chemosensory systems within the brains of bony fishes. In: Atema J, Fay RR, Popper AN, Tavolga WN (eds): Sensory biology of aquatic animals. Springer-Verlag New York, in press.

Getchell TJ (1986) Functional properties of vertebrate olfactory receptor neurons. Physiol Rev 66: 772-818.

Ghiradella HT, Case JF, Cranshaw J (1968) Structure of aesthetascs in selected marine and terrestrial decapods: chemoreceptor morphology and environment. Am Zool 8: 603-621.

Hornung DE, Mozell MM (1981) Accessibility of odorant molecules to the receptors. In: Cagan RH, Kare MR (eds): Biochemistry of taste and olfaction, Acad Press NY, pp. 33-45.

Jackson GA (1980) Phytoplankton growth and zooplankton grazing in oligotrophic oceans. Nature 284: 439-441.

Johnsen PB, Teeter JH (1980) Spatial gradient detection of chemical cues by catfish. J Comp Physiol 140: 95-99.

Johnsen PB, Teeter JH (1985) Behavioral responses of bonnethead sharks (Sphyrna tiburo) to controlled olfactory stimulation. Mar Behav Physiol 11: 283-291.

Johnson BR, Atema J (1983) Narrow-spectrum chemoreceptor cells in antennules of the American lobster, Homarus americanus. Neuroscience Letters 41: 145-150.

Johnson BR, Voigt R, Borroni PF, Atema J (1984) Response properties of lobster chemoreceptors: tuning of primary taste neurons in walking legs. J Comp Physiol A 155: 593-604.

Kaissling KE (1986) Chemo-electrical transduction in insect olfactory receptors. Ann Rev Neurosci 9: 121-145.

Kaissling KE, Priessner E (1970) Die Riechschwelle des Seidenspinners. Naturwissenschaften 57: 23-28.

Kaissling KE, Zack-Strausfeld C, Rumbo ER (1987) Adaptation processes in insect olfactory receptors: mechanisms and behavioral significance. In: Roper S, Atema J (eds): Olfaction and taste IX, NY Acad Sci Press, in press.

Kelling ST, Halpern BP (1983) Taste flashes: reaction times, intensity and quality. Science 219: 412-414.

Kramer E (1986) Turbulent diffusion and pheromone triggered anemotaxis. In: Payne TL, Birch MC, Kennedy CEJ (eds): Mechanisms in insect olfaction. Oxford Univ Press, pp. 59-67.

Laing DG (1983) Natural sniffing gives optimum odour perception for humans. Perception 12: 99-117.

Lehman JT, Scavia D (1982) Microscale nutrient patches produced by zooplankton. Proc Natl Acad Sci USA 79: 5001-5005.

Maier I, Müller DG (1986) Sexual pheromones in algae. Biol Bull 170: 145-175.

McBurney DH (1976) Temporal properties of the human taste system. Sensory Processes 1: 150-162.

Moulton DG, Tucker D (1964) Electrophysiology of the olfactory system. Ann New York Acad Sci 116: 357-746.

Murlis J, Jones CD (1981) Fine-scale structure of odour plumes in relation to insect orientation to distant pheromone and other attractant sources. Physiol Entomol 6: 71-86.

Ottoson D (1956) Analysis of the electrical activity of the olfactory epithelium. Acta Physiol Scand 35 Suppl 122: 1-83.

Reeder PB, Ache BW (1980) Chemotaxis in the Florida spiny lobster, Panulirus argus. Anim Behav 28: 831-839.

Schmitt BC, Ache BW (1979) Olfaction: response enhancement by flicking in a decapod crustacean. Science 205: 204-206.

Städtler E (1984) Contact chemoreception. In: Bell WJ, Cardé RT (eds): Chemical ecology of insects. Sinauer Assoc Inc, Sunderland, MA, USA, pp. 3-35.

Strickler JR (1985) Feeding currents in calanoid copepods: two new hypotheses. Soc Exp Biol Symp 39: 459-485.

Trott TJ, Robertson JR (1984) Chemical stimulants of cheliped flexion behavior by the Western Atlantic ghost crab, Ocypode quadrata (Fabricius). J Exp Mar Biol Ecol 78: 237-252.

Vogel S (1981) Life in moving fluids. Princeton Univ Press, 352 pp.

Voigt R, Atema J (1984) Chemoreceptor adaptation: responses of glutamate cells to repeated stimulation. Biol Bull 167: 534.

Zeiske E (1974) Morphologische und morphometrische Untersuchungen am Geruchsorgan viviparer Zahnkarpfen (Pisces). Z Morph Tiere 77: 19-50.

Part VI

LOCOMOTION

ENERGETICS OF LOCOMOTION IN WATER, ON LAND AND IN AIR: WHAT SETS THE COST?

C. Richard Taylor

Museum of Comparative Zoology
Harvard University
Cambridge, Massachusetts
U.S.A.

The energy cost of vertebrate locomotion varies by more than
10-fold with body size and with the environment in which the
animal moves (aquatic, terrestrial or aerial). For terrestrial locomotion, the cost of generating muscular force has
been found to vary with body size in the same manner as cost
of locomotion, whereas the cost of performing mechanical work
does not. It is proposed that the observed differences in
cost of locomotion in water, on land and in air may also result from differences in cost of generating muscular force,
and may not parallel differences in cost of performing work
against the environment.

INTRODUCTION

The metabolic cost of locomotion (expressed as the oxygen consumed to transport
one kg of body mass a distance of one km) decreases in a regular manner with body
size (Taylor, Heglund and Maloiy, 1982). For animals of the same size, swimming
has a lower cost than flying, and flying a lower cost than moving on land
(Schmidt-Nielsen, 1972). It is the energy consumed by skeletal muscles as they
are activated and generate force that sets the costs of locomotion. This paper
attempts to relate differences in cost of generating force by skeletal muscles
to observed differences in cost of locomotion.

COST OF FORCE GENERATION BY MUSCLES

Two processes account for most of the energy consumed during muscular activity:
1) force generation, which involves the breaking of the crossbridges between
actin and myosin by actomyosin ATPase; and 2) the activation-deactivation process, which involves the pumping of Ca^{2+} into the sarcoplasmic reticulum by the
Ca^{2+} transport ATPase of the sarcoplasmic reticulum. About 30% of energy utilization during activity can be attributed to activation-deactivation, and about
70% to force generation (Rall, 1985).

Comparative Physiology: Life in Water and on Land. P. Dejours, L. Bolis, C.R. Taylor, E.R. Weibel
(eds.) Fidia Research Series, IX-Liviana Press, Padova © 1987

The maximum force per cross-sectional area (stress) exerted by a mammalian muscle is independent of body size (approximately 300-400 kN/m^2 cross section of
muscle myofibril) because of the basic similarity in the structures of the contractile proteins, actin and myosin. The maximum relative shortening, or strain,
is also independent of body size (approximately 0.3). From these constants,
A. V. Hill (1950) pointed out that the maximum work (force x distance) performed
in a single contraction is also constant and independent of body size. If the
work per contraction is constant, then the work per unit time (power output) will
be proportional to the speed of shortening. Higher speeds of muscular shortening
require higher rates of cycling of the crossbridges, more frequent activation,
and therefore higher rates of energy utilization.

Many locomotory activities, like running, involve regular cyclic muscular activities where muscles stay the same length or are alternatively stretched and then
shorten, rather than simply shortening. These muscular activities occur at
higher frequencies in small animals. Higher frequencies, like higher speeds of
shortening, require higher rates of crossbridge cycling, more frequent activation
and higher rates of energy consumption.

COST OF GENERATING FORCE DURING LOCOMOTION

The metabolic cost of generating force during terrestrial locomotion has been
investigated by Taylor, Heglund, McMahon & Looney (1980). Animals were trained
to carry loads evenly distributed along their backs. The accelerations of the
center of mass of the animals and the masses of their limbs were not altered by
carrying the loads, therefore the increases in forces exerted by the extensor
muscles (responsible for > 90% of the force) were directly proportional to the
increases in the mass of the loads that were carried. Metabolic cost of generating force measured in this way exactly paralleled the changes in metabolic cost
of locomotion as a function of speed and body size (i.e., it can cost an animal
10 times as much to generate a Newton of force at a high speed as at a low speed,
and it costs a small 30 g mouse 10 times as much to generate a Newton of force as
the 100 kg pony when the animals move at the same speed).

Although the cost of locomotion changes by 10-fold with speed and with body size
at the same speed, one finds that the cost per step is a constant at speeds
where stresses (force/cross-sectional area) in the muscles and bones of the
limbs are the same. The transition speed where animals change from a trotting
gait to a gallop is a locomotory condition where muscle and bone stresses in the
limbs are independent of body size (Biewener, 1983a, 1983b). If one divides the
rate of oxygen consumption at this speed, by stride frequency, then the cost per
step (muscular event) is constant (Table 1). Thus it appears that the higher
frequencies required for generating a given level of force during locomotion

accounts for the higher demand for energy in small animals and at high speeds. Both the activation-deactivation costs and the crossbridge cycling costs should be directly proportional to frequency, thus both costs should be higher in the small animals and in the same animal at high speeds.

Table 1. The mass specific energetic cost for each stride is the same for animals of different size at equivalent speeds where muscle stress (force/cross-sectional area) is the same (Taylor, 1987).

Body mass (kg)	Equivalent Speed (m/sec)	Stride frequency at this speed (strides/sec)	Mass specific cost/stride at this speed (ml O_2/stride·kg)
0.1	0.51	8.54	0.28
1.0	1.53	4.48	0.25
100	4.61	2.35	0.28

CONCLUSIONS

Locomotion at a steady speed on land, in the water and in the air involves cyclic muscular events which consume energy. The mass specific cost per cycle is the same for terrestrial animals at speeds where muscle stresses are the same. This suggests that the lower energetic costs of locomotion for flying and swimming animals are the result of lower muscular stress and/or lower cycle frequencies required for a given level of stress for locomotion in air and in water than that on land.

REFERENCES

Biewener AA (1983a) Allometry of quadrupedal locomotion: scaling duty factor, bone curvature, and limb orientation to body size in quadrupeds. J Exp Biol 105: 147-171.

Biewener AA (1983b) Scaling relative mechanical advantage: implications for muscle function in different sized animals. Fed Proc 42: 469.

Hill AV (1950) The dimensions of animals and their muscular dynamics. Sci Prog Lond 38: 209-230.

Rall JR (1985) Energetic aspects of skeletal muscle contraction: implications of fiber types. Exercise and Sport Sciences Reviews 13: 33-74.

Schmidt-Nielsen K (1972) Locomotion: energy cost of swimming, flying and running. Science 177: 222-228.

Taylor CR (1987) Structural and functional limits to oxidative metabolism: insights from scaling. Ann Rev Physiol (in press).

Taylor CR, Heglund NC, Maloiy GMO (1982) Energetics and mechanics of terrestrial locomotion. I. Metabolic energy consumption as a function of speed and body size in birds and mammals. J Exp Biol 97: 1-21.

Taylor CR, Heglund NC, McMahon TA, Looney TR (1980) Energetic cost of generating muscular force during running. A comparison of large and small animals. J Exp Biol 86: 9-18.

PHYSICAL ASPECTS OF INSECT LOCOMOTION: RUNNING, SWIMMING, FLYING

Werner Nachtigall and Axel Dreher

Zoologisches Institut der
Universität des Saarlandes
6600 Saarbrücken, West Germany

Basic aspects of force generation, energy transmission and
momentum are described in locomotion on land and in fluids.
The importance of the wake, and why fluid locomotion is energetically demanding is demonstratet. Ring vortex generation
as a means of optimal force development is discussed using
examples of hovering and straight ahead flight.

The following sections will try to outline the basic physical principles of
locomotion i.e. running, swimming (restricted to under water rowing) and flying
(hovering and straight ahead flight) using insects as examples. Only reactive
forces due to the acceleration of masses are taken into account.

DYNAMICS OF LOCOMOTION ON LAND

Fig. 1 shows a running carabid beetle. The fore and hind legs of one side and
the middle leg of the other side form a tripod which is swung forward, contacts
the ground and pushes the body mass forward.

GENERAL ASPECTS

If, relative to a point of reference, the resting mass m_1 is set in motion,
then a force must develop between it and a second mass m_2 so that the vector $\vec{F}$
influences m_1 and $-\vec{F}$ influences m_2 (e.g. earth mass). Then m_1 (e.g. animal
mass) has, at the end of the acceleration along the path s, the kinetic energy
$$E_{kin} = \int_0^s \vec{F}\ \vec{ds} = \tfrac{1}{2} m_1\ v_1^{\ 2}$$ with $\vec{v}_1$ the speed of m_1 relative to m_2 (Fig. 2a, b).
In actual fact, such a movement will succumb to the loss of energy due to
friction (Fig. 2c), and the procedure must be repeated whereby the conditions
will have to be the same as at the beginning of the first acceleration:

1. The energy necessary for force development must be available (spring symbol
   in Fig. 2d).

Comparative Physiology: Life in Water and on Land. P. Dejours, L. Bolis, C.R. Taylor, E.R. Weibel
(eds.) Fidia Research Series, IX-Liviana Press, Padova © 1987

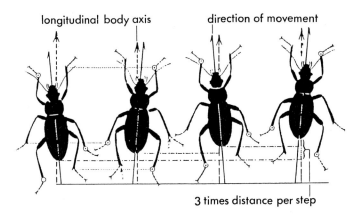

Fig. 1

Diagram of leg movements in a carabid beetle; comparable phases of 4 successive strides. Circles: legs on ground. No circles: Legs have been swung forward and just touch ground. Oscillations of the body axis have been exaggerated. (From WEBER and WEIDNER (1976))

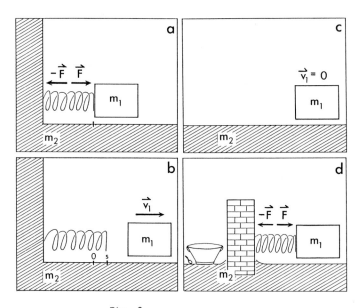

Fig. 2

Model for locomotion on land. a Effect of the force of a spring on the mass m_1 to be accelerated and its "support" m_2. b The mass m_1, accelerated over the distance s, has the velocity $\vec{v}_1$. c Due to frictien effects $\vec{v}_1$ has become zero. d A new acceleration needs a new energy source and a new support.

2. A mass to absorb the counter impulse must be available in a suitable
   position (wall symbol in Fig. 2d).

IMPULSE CONSIDERATION

Before a force developes, let the entire impulse from the mass system m_1
and m_2 be equal to zero. Thereafter it is again equal to zero, since

$$m_1 \vec{v}_1 + m_2 \vec{v}_2 = 0$$

according to the law of impulse conservation. An extraterrestrial observer,
i.e. from an inertial system, would see the relationship as follows: If a
beetle with mass $m_1 = 10^{-3}$ kg (or a model like that shown in Fig. 2a,b) runs
at a speed of $\vec{v}_1 = 2 \times 10^{-1}$m s^{-1} to the right, the earth (mass: 6 x 10^{24}kg) will
move with the speed $\vec{v}_2$ to the left. Both accelerated masses m_1 and m_2 have a
counter equal impulse; $m_1 \vec{v}_1 = - m_2 \vec{v}_2 = 2 \times 10^{-4}$ kg m s^{-1} (or N s). The
kinetical energy of the beetle is $E_{kin\ beetle} = m_1 v_1^2 = 2 \times 10^{-5}$ N m (or J).
That of the earth is $E_{kin\ earth} = m_2 v_2^2 = 3 \times 10^{-33}$ N m , i. e. almost zero.
Therefore, almost all the energy is in the beetle. In our model the energy is
transferred from the spring to m_1. For an observer on earth, v is equal to zero,
and therefore it follows that $E_{kin\ earth}$ is also zero from his point of view.
Thus one makes practically no error even then, when one leaves an inertial
system and takes on an earth bound position. Therefore, one may say, that the
energy required to accelerate a mass m_1 on earth consists practically entirely
of the kinetical energy of this mass m_1.

This simplifies matters concerning the dynamics of locomotion on land since
the energy for acceleration can be determined from the easily measured speed
of the animal's body relative to the earth. The simplification is due to the
fact that one does not have to differentiate between the relative velocity between the masses m_2 and m_1 and the relative velocity between observer and m_1.
This however, is not possible when considering the dynamics of locomotion in
fluids.

DYNAMICS OF LOCOMOTION IN FLUIDS

DIFFERENCES COMPARED TO LOCOMOTION ON LAND: SWIMMING

There are two main differences between locomotion in fluids and on land. First
of all, the propulsion organs cannot be removed from their medium (as happens
for example when the oars of a boat are lifted out of the water during rowing).
Thus forces are created in the recovery stroke which counteract the movements.
Fig. 3a shows locomotion of the great water beetle Dytiscus marginalis rowing
under water; in Fig. 3b flash photographs of its smaller relative Acilius sulca-

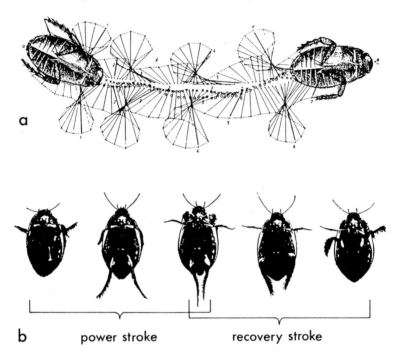

b   power stroke   recovery stroke

Fig. 3

a Movement of the rowing legs (third pair) of the water beetle Dytiscus marginalis (body length, 3,5 cm) during swimming at 10,4 cm s^{-1}. Base and tip of
a leg are connected by a straight line. (This indicates leg position during
power stroke with leg fully strechted, but not during recovery stroke as indicated in the two drawings.) Crosses: anterior body end, circles; postericer
body end. 80 frames per second. b Stroke phases of the water beetle Acilius
sulcatus. Flash photographes. (A from NACHTIGALL (1981), b from NACHTIGALL
(1964))

tus are shown to illustrate stroke positions of the rowing legs. During power
stroke the synchronously beating, fully stretched hindlegs move fast, broad
side facing the water, swimming hairs spread out. During the recovery stroke
they move slower and are held close to the body; swimming hairs are folded
together. Thus the thrust of the power stroke is higher than the counter thrust
of the recovery stroke, so that the beetle swims forward. On the other hand,
fluids do not react like a fixed wall. Another look at Fig. 3a shows that
during power stroke, the outer two thirds of the (fully stretched) hindlegs
move faster rearward than the body moves forward, which is different from land
locomotion.

Sometimes one can observe that water beetles of mass m_1 start very fast with
the velocity $\vec{v}_1$ (fig. 4a) from aquarium walls by thrusting themselves away from
these fixed walls with a normal synchronous power stroke (Fig. 4a, comp.Fig.2d).
This is equivalent to land locomotion. In free water the same power stroke
induces a vortex-like rearward directed movement ($\vec{v}_2$) of a certain water mass m_2,
called "the wake" (Fig. 4b) and the velocity $\vec{v}_1$ of the beetle is smaller.

Now, what is of interest to a biophysicist? He wishes to know the energy E
required by his experimental object with the mass m_1 to accelerate to a certain
speed $\vec{v}_1$. In land locomotion (compare Figs. 4a, 2d) he has only to determine
the speed and mass of the animal , and since the formula $E_{kin} = \frac{1}{2} m_1 v_1^2$
is valid, $\vec{v}_2$ is zero. Analysing locomotion in fluids however, he must determine
two masses and two speeds, namely that of his object (m_1 and $\vec{v}_1$) and that of its
wake (m_2 and $\vec{v}_2$). The animal must produce the following energy:
$E = E_{kin\ animal} + E_{kin\ wake} = \frac{1}{2} m_1 v_1^2 + \frac{1}{2} m_2 v_2^2$. In the latter case, with
its "movable support", E must be greater than in the case of land locomotion
with its "fixed support" similar to a railway buffer-stop. Those relationships
are demonstrated in Figs. 4c,d.

The wake is a fluid volume which cannot be clearly defined. Like the mass m_2,
it is the impulse partner to the body mass m_1 and therefore $m_1 \vec{v}_1 + m_2 \vec{v}_2 = 0$
is still valid. In land locomotion, the speed $\vec{v}_2$ of the "impulse partner" is
equal to zero, but in fluid locomotion, there is no way in which the wake may
be set at zero (Figs. 4a,c) and this is very important. Due to the particul-
arity of fluids in motion, the impulse from a developing force cannot be transmitted over "the whole earth", but only to a small part of the fluid (m_2).
Consequently a small mass m_2 demands a velocity v_2 which may not be neglected.

Presuming that the process of force development in a fluid and on land is identical, then at the end of this process just as much kinetical energy will have

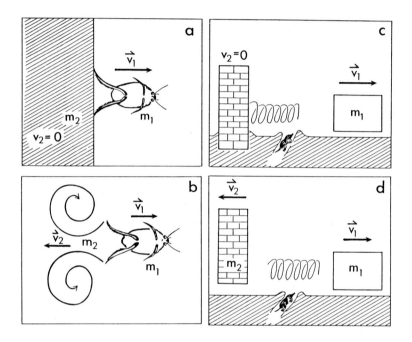

Fig. 4
a Starting push of Acilius from an aquarium wall, b Principle of wake formation of Acilius during swimming in free water, c Mechanical analogue to a,
d Mechanical analogue to b.

been created relative to the earth or total fluid as there. This time however, the kinetical energy is divided between two masses, namely m_1 and m_2: $E_{kin} = \frac{1}{2} m_1 v_1^2 + \frac{1}{2} m_2 v_2^2$. For example: If m_1 and m_2 are equal and therefore is $v_1 = v_2$ following comparison of the relative speeds of the animal's body which the same energy requirement can be calculated: The kinetical energy of the body in land locomotion is $\frac{1}{2} m_1 v_{1\ land}^2$ and in fluid locomotion it is the half: $\frac{1}{2}(\frac{1}{2} m_1 v_{1\ land}^2) = \frac{1}{2} m_1 v_{1\ fluid}^2$ which gives: $v_{1\ fluid} = \sqrt{\frac{1}{2}} v_{1\ land} = 0{,}7\ v_{1\ land}$. In the case, with the same energy requirement, the speed attained by the animal's body during locomotion in a fluid is approximately 30% less than that attained during land locomotion.

Therefore there are three principal factors of fluid locomotion which make it less suitable, as regards energetics, than land locomotion. First of all, part of the energy used is lost as kinetical energy to the impulse partner (the "wake"). Secondly, energy is lost due to pressure differences and friction on the body and the propelling organs whilst creating an impulse (power stroke). Thirdly, such losses also appear during the preparing phase for the next impulse (recovery stroke).

Solutions to the problems of propulsion in fluids should be as follows:

1. The greater energy loss due to the resistance of the fluid should be kept as small as possible.

2. The kinematics of the propelling organs should be constructed in such a way that the volume of fluid which acts as a support - the wake - should receive a strong impulse $m_2 \vec{v}_2$ - counter equal to the impulse $m_1 \vec{v}_1$ of the animal's body - but only a little kinetical energy since this has to be piped from the energy produced by the animal. When comparing the formulae for the impulse $\vec{p} = m \cdot \vec{v}$ and the kinetical energy $E_{kin} = \frac{1}{2} m v^2$, it becomes apparent that the wake will come closest to fulfilling these requirements when its own speed v_2 is as low as possible and the required impulse comes from a correspondingly large mass m_2. Technical engineering has shown us that large, slowly rotating propellers are more efficient than small, rapid ones. Less kinetic energy is required by the former for the same impulse in its wake as by the latter. This is why the thrust efficiency of the old paddle steamers with their capacity to accelerate large water masses at comparatively slow speeds was relatively good.

Thus it would seem that scientific advancement is dependent on successfully analysing the wake. This and other aspects will be demonstrated on examples taken from the biophysics of insect flight.

ANALYSIS OF WAKE; FLYING
DISCONTINUOUS RING VORTICES

Experiments on flow have shown that the wake, especially in hovering flight, consists of a column of staggered vortex rings (Fig. 5a).

When the formation of a certain structure is favoured, then this suggests that it is stable. As long as there is no obstruction along its path, a smoke ring for example, will "survive" for a long time. The stability and survival of a vortex ring on the one hand, is dependent upon it being able to move with very little inner and outer friction - the reason for this cannot be explained in a few words. On the other hand, it can be shown mathematically, that for a given impulse, the vortex ring is the structure with the least energy requirement. For this reason alone, one can be sure that biological creatures have evolved in such a way that they will form ring vortices if they move actively through a fluid. Thus a column of ring vortices will be formed when force is periodically created (Fig. 5a). In recent years, more and more proof has been obtained that such wake formations actually occur, and due to the stability of their structure can be determined even after several periods of movement (Fig. 11).

Due to its exceptionally symmetrical structure, the ideal ring vortex (Rankine-profile) can be easily analysed mathematically. There are simple formulae which use the parameters R (ring radius), K (core radius) and Γ (circulation) to determine the velocity of the ring, its impulse and its energy (Fig. 5b). From these we learn that the ring impulse is proportional to $R^2 \Gamma$ and the ring energy is proportional to $R \Gamma^2$.

One recognizes a similarity to the above mentioned theoretical impulse $\vec{p} = m\,\vec{v}$ and the kinetic energy $E_{kin} = \frac{1}{2}\,m\,v^2$ of the wake: the impulses are linearly proportional to velocity v and the kinetic energy is proportional to its square.

Thus it is easy to formulate the ideal conditions for optimal movement in fluids when considering ring vortices: for the creation of a certain impulse to the animal's body, as little energy as possible should be "put into the wake", and therefore the ring vortex should have the greatest possible radius R, but the least possible circulation Γ. The morphology and kinematics of a hovering insect should fulfill these conditions. Up to date methods for analysing the wake are unfortunately not so advanced that absolute values for E or $\vec{p}$ can be given. Thus, an assessment of the wake can only be qualitative at the present time.

HOVERING FLIGHT

The wings of large insects are curved plates. Due to the viscosity of the

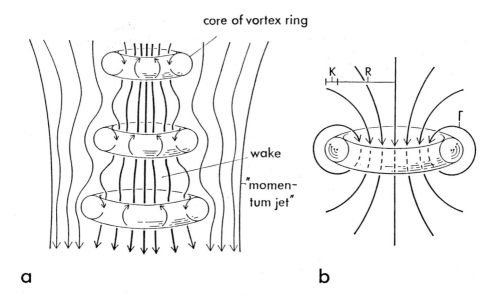

Fig. 5

Ring vortices. a Principles of a ring vortex column ("staggered ring vortices");
details hypothetical. b Characteristics of a single vortex ring (K core radius,
R torus radius, Γ circulation).

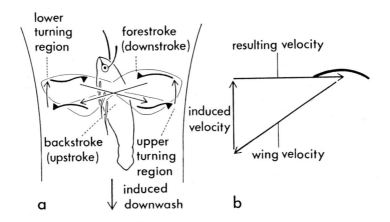

Fig. 6

Downstroke (forestroke) of the sphingid Deilephila elpenor hovering in front
of a Dianthus, body obliquely. Flash photograph, wings at upper turning point.
(Thin line: approximate position of wings at lower turning point). (From
NACHTIGALL (1969), after a photograph of H. EISENBEISS).

actual fluid, flow breaks away at the rear edge (Kutta-Joukowsky-condition), whereby a so-called starting vortex detaches itself . Its counterpart, which rotates in the opposite direction, remains as a bound vortex with the circulation Γ.

Certain hovering animals, e.g. hawkmoth "standing in the air" in front of a   flower   can move their wings in the form of a horizontal figure of eight ("normal hovering"; Weis-Fogh (1972, 1973) (Fig. 6 a). The wing path allows for the downwash by sloping downward during half of each beating phase, so that the flow to the wings is approximately horizontal (Fig. 6b). The drag from fore and upstroke neutralize one another on a time average. In this simple model the mean side force to the wings is always directed perpendicularly upward thus supplying the vertical sustaining force (in German: "Hub"). This force counteracts drag (in the downwash) and gravity. The energy carried away in the wake is equivalent to the energy lost in the downwash from the wings and body.

In the jet is equivalent to the energy loss from friction- and pressure drag, especially from the beating wings. No energy is required to create something like a "vertical sustaining force jet". It is merely a drag indicating downwash.

The momentum jet model is often used for rapid and useful estimating calculations, as for example the performance of helicopters. This does not mean however, that the principles of the fluid dynamics of vertical sustaining force creation can be explained by this method.

CIRCULATION AND WAGNER EFFECT

Fig. 7 demonstrates the special flow effects on an animal during hovering flight. Due to the pressure differences between upper and lower wing surfaces, air circulates around the wing tip so that a twisted vortex is formed behind each wing. Both these twisted vortices unite with the starting vortex from the beginning of a beating period and the bound vortex from the lower turning point of the wing movement to form a horizontal ring vortex which drifts away downwards due to its self induced movement. Since the same takes place in the following beating phases one obtains the above mentioneds ring vortex column as a phenomenon of the wake, whereby each half wingbeat period generates a ring vortex.

According to the Kutta-Joukowsky formula $L = \varrho \, b \, v_\infty \, \Gamma$, the lift L is proportional to the speed of flow v_∞ and the circulation Γ (ϱ : fluid density, b: wingspan). Circulation requires a certain time to develop . Before each new wing acceleration, at the turning point of the path in the opposite direction, v_∞ falls to zero. The "old" circulation is shed and then a "new"

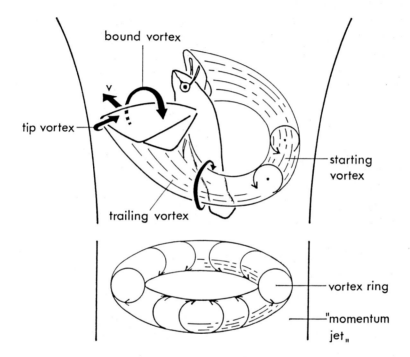

Fig. 7
Principles of ring vortex formation during a half stroke and a ring vortex shed
in the previous halfstroke in a hovering sphingid. Body from Macroglossa (inspired by RAYNER (1979b), ELLINGTON (1984)).

one is built up rotating in the opposite direction. Therefore, the bound vortex
must be thrown off by the abrupt halting of the wings; the wings are now accelerated in the opposite direction, whereby a new bound vortex (and a new starting
vortex) develop. No lift worth mentioning can be created during throw-off and
rebuilding. If one relates these facts to high frequency beating insect wings
(e.g. midges, flies, Hymenoptera etc), then approximately half of the beating
phases would be useless from an aerodynamic point of view. This is a well known
flight phenomenon known as Wagner effect. The Wagner effect vividly describes
the fact that the bound vortex on a wing, even at exceptionally high values of
accelerations of the airfoil (v "instantaneously" from zero to its end value),
can only make use of its lift-inducing circulation Γ when it is clearly separated from the starting vortex ($-\Gamma$). This is more or less the case when the air
foil has moved at least a wing-chord length from the starting point .

Thus the paths of evolutions should be such that they "outwit the laws of
physics". That is to say, the time separating the throwing off of the old bound
vortex and the creating of a "new" starting vortex must be cancelled - which
appears justifiable since both vortices have the same direction of rotation -
and that, too, means that the Wagner effect must be avoided.

The simplest mechanisms imaginable agrees with the kinematical requirement,
that the wing must be rotated around its own longitudinal axis for the following beating phase. This is demonstrated in Fig. 8 . Due to the exceptionally
rapid rotating of the wing (ca. 3o.ooo degrees s^{-1} in Phormia regina), a new
bound vortex is created at the same time as the flow breaks away, and thus the
old circulation is replaced at the beginning of the translation movement of
the new beating direction. This movement co-ordination is known as the "flex"
mechanism (Ellington 1980, 1984), described in Fig. 9 a). It must be emphasized
that, due to the "flex" mechanism the new bound vortex which also has to "peel"
itself from the starting vortex, does not begin to form gradually when the
new translation movement starts. What actually happens is, that at the turning
point of the wing movement (a), the old vortex developes, (b) the following
starting vortex is shed and (c) the new circulation is created which has almost reached its maximum value at the moment when the translation movement begins. Thus the Wagner effect is avoided.

When wing-pairs beat at such high amplitudes that they can touch each other
at the turning points, then the conditions are even better, because the starting vortex of one wing sits as a bound vortex on the other whilst the wings
are pulled apart (peeling and fling, (Fig. 9 b)) and are thus put to good use.

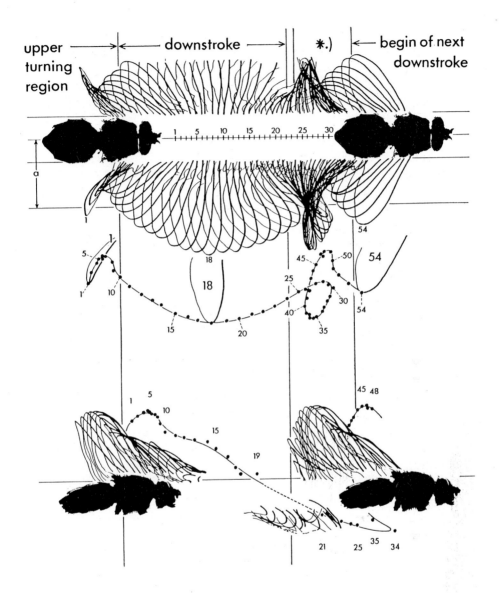

upper turning region ——→ |←—— downstroke ——→| *.) |←— begin of next downstroke

a

*·) lower turning region, upstroke and next upper turning region

Fig. 8

Fast straight-ahead flight of a calliphorid fly. Drawn from flight in front of a wind tunnel; wind velocity added. Top view and side view. 6400 frames per second. Fast wing rotation especially at lower turning point. Reconstructed from NACHTIGALL 1966.

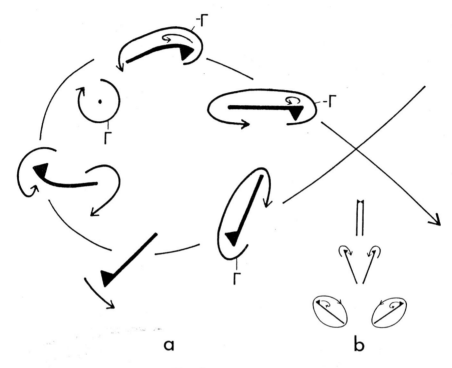

Fig. 9

Diagrams of instationary wing mechanisms. Thick lines; wing cross-section.
The triangle markes the morphological upper surface at the leading edge.
a Flex mechanism. Shedding of circulation Γ as a ring vortex and building
up a new, counter-rotating circulation -Γ at a turning region of the beating
wing, b Fling mechanism. (After ELLINGTON (1984) and NACHTIGALL (1979, 1980)).

FROM HOVERING TO STRAIGHT-AHEAD FLIGHT

Whether we are dealing with hovering or straight - ahead flight, the major part
of the force created   serves as a vertical sustaining force to compensate the
body weight. A force which acts mainly in a perpendicular and upward direction
on the propulsion organs arises, according to the Kutta-Joukowsky lift model,
 from the horizontal flow to these organs. During horizontal flight however,there
is no need to create flow by rapidly beating the wings; it is sufficient to
hold the airfoils into the airstream. The neccessity of changing the direction
of the circulation and shedding the bound vortex after each half beating period
is no longer given. In the interim period between the two beating phases, now
known as upstroke and downstroke, the starting vortices must be shed in adaption to the circulation. Their energy per unit time is equivalent to the thrust
power of the animal. Their direction of rotation is dependent upon whether the
changes in circulation around the wing are positive or negative. In order to
clarify the situation , the vortices shed under these conditions will now be
called adaption vortices. The image of the wake is just as different. A so-
called rope ladder vortex develops - its side rails are formed by the wing
tip vortices, its rungs from the adaption vortices built at the beginning of
each up and downstroke (Fig. 1o).

CONSIDERATION OF THE ENERGY COSTS

The energy in a vortex is proportional to the square of the circulation, $E \sim \Gamma^2$.
Thus the shedding of a vortex always means a loss of energy and it seems sen-
sible to avoid this process as far as possible. An aeroplane starting to fly
must invest energy to develop the bound and starting vortices. Once a constant
speed has been reached, no extra energy is required. A hovering animal however,
must as a rule spend this energy twice per beating phase. By using the principles of flex or peeling, almost half of this energy can be saved, probably because
only one vortex ring per wing beat is formed in the wake.

Changes in the circulation during straight flight are relatively small compared
to their total strenght and therefore the amount of power required is visibly
less. The bound vortex must no longer be completely shed as in hovering flight
(loss of all the energy needed for its generation). As opposed to hovering
flight, the circulation in straight-ahead flight must only be induced at the
beginning of flight. Due to drag, the propulsion organs must generate a mean
horizontal thrust which is only feasible due to their cyclic kinematics. This,
however, is problematic in flight,because the propulsion organs must generate,
simultaneously, an even greater amount of vertical sustaining force. Therefore
it is more suitable to diminish circulation at the beginning of the "recovery
stroke",i.e. upstroke. This results in less vertical sustaining force during
upstroke which must be compensated in the downstroke. Thus the horizontal compo-

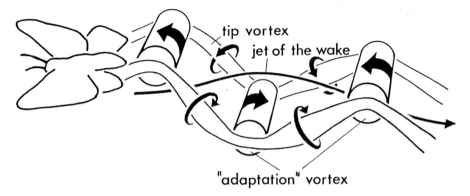

Fig. 1o

Rope ladder type vortex street behind a horizontally flying butterfly. For
the "adaptation" vortex see text. (After BRODSKIJ and IVANOV (1984)).

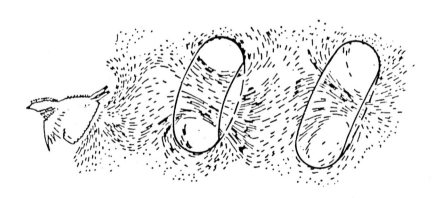

Fig. 11

Wake behind a horizontally flying Chaffinch. Flash photograph of dust particles;
ring vortices added by the authors. (From KOKSHAYSKY (1978)).

nent of lift during upstroke is also reduced. This is important because the lift vector
is directed upward-backward during every upstroke thus producing a negative horizontal thrust component. In other words, this means that, at the
turning points of the wing path in straight-ahead flight, the circulation of
the following up and downstrokes must be adapted accordingly. The difference
$\Delta\Gamma$ in the bound vortex must be taken off as nearly - $\Delta\Gamma$ in what we call "adaption
vortex". The latter build the rungs of the "rope ladder" vortex system. There-
fore one can derive the energy approximately proportional to the value $4\,\Gamma\Delta\Gamma$
required for each beating cycle in straight-ahead flight (small $\Delta\Gamma$　is pre-
requisit). This is significantly more favourable compared to hovering flight
with $E \sim \Gamma^2$.

Changing from hovering to straight-ahead flight is probably a discontinuous
process. It is possible that one or two ring vortices are shed in a horizontal direction, thus generating the thrust necessary for accelerating up to
flight speed; the animal then switches, without transition, to straight-
ahead flight kinematics.

CONCLUSION

The problem of optimizing the energetics of biological flight machines,
especially in hovering flight, can only be solved when the energy and momentum
diverted to the wake can be measured. However, the methods of data accumulation
and analysis are extremely difficult and complicated.

In order to avoid this problem, Tucker (1973), Pennycuick (1969), Rayner (1979a)
and Ellington (1984) for example, assumed that, according to the momentum jet
theory, the major part of the available energy is required to create a　verti-
cal-sustaining-force jet and that the energy lost due to drag is relatively
small. After this initial start, useful models on flight energetics and behaviour could be developed but one required relatively arbitrary correction
factors in order to compare this theory with measured data.

If biological fluid dynamics should continue to evolve, then we have no other
choice than to try once more to register and analyse the parameters, energy
and momentum , of the wake. The first quantitative measurements were made by
Spedding et al. (1984). These were analysed according to Rayner's (1979b) new
mathematical model, dispensing with the momentum jet idea. It was soon apparent that the previous values for energy and momentum, taken from the old momentum jet model, were approximately twice as high as those from Rayner's model.

Once reliable data are available, the power required by the animal to overcome the fluid drag of the downwash can be calculated from the momentum of the wake; one could call this the "unavoidable" power. The energy per unit time in the wake is equivalent to the mean total power of the flight machine. Apart from the "unavoidable" power, it may also receive the major part of that power which has to be raised, in each half beating period, in order to generate a new vortex system. The value of this power must lie between zero and the value of power calculated from the momentum jet theory using morphological parameters from the wing beating system. The former value cannot be estimated. It is the result of an evolutionary compromise between the reduction of circulation of the bound vortex in favour of a large wing span on the one hand and a reduction of the wing span in order to avoid problems of stability and inertia on the other hand. But the complex working together of the single elements of this evolutionary compromise is still veiled in obscurity.

LITERATURE

Brodskij, A.K. and V.D. Ivanov (1984): The role of vortices in insect
flight. Zool. Zh. 63, 197-208

Ellington, C.P. (1980): Vortices and hovering flight. In: Instationäre
Effekte an schwingenden Tierflügeln (ed. W. Nachtigall), pp. 64-101,
Wiesbaden: Franz Steiner

Ellington, C.P. (1984): The aerodynamics of hovering insect flight. Phil.
Trans. R. Soc. Lond. B 305, pp. 1-181

Kokshaysky, N.V. (1978): On the structure of the wake of a flying bird.
Acta XVII Congressus internationalis ornithologicus, 397-399.

Nachtigall, W. (1964): Wie schwimmen die Wasserkäfer; Umschau 64, 467-470

Nachtigall, W. (1966): Die Kinematik der Schlagflügelbewegungen von Dipteren.
Methodische und analytische Grundlagen zur Biophysik des Insektenflugs.
Z. vergl. Physiol. 52, 155-211.

Nachtigall, W. (1969): Flugmechanik der Insekten. n + m, 6 (Nr. 28), 9-21

Nachtigall, W. (1979): Rasche Richtungsänderungen und Torsionen schwingender
Fliegenflügel und Hypothesen über zugeordnete instationäre Strömungseffek-
te. J. Comp. Physiol. 133, 351-355

Nachtigall, W. (1980) Rasche Bewegungsänderungen bei der Flügelschwingung
von Fliegen und ihre mögliche Bedeutung für instationäre Luftkrafterzeu-
gung. In: Hrsg. W. Nachtigall: Instationäre Effekte an schwingenden Tierflügeln. Beiträge zu Struktur und Funktion biologischer Antriebsmechanis-
men. Akad.d.Wissensch. u.d. Lit. Mainz, Steiner-Verlag, Wiesbaden, 115-129.

Pennycuick, C.J. (1969): The mechanics of bird migration. Ibis, 111, 525-556.

Rayner, J.M.V. (1979a): A new approach to animal flight mechanics. J. Exp.
Biol. 80, 17-54.

Rayner, J.M.V. (1979b): A vortex theory of animal flight. I. The vortex
wake of a hovering animal. II. The forward flight of birds. J. Fluid
Mech. 91, 697-730; 731-763.

Spedding, G.R., Rayner, J.M.V., Pennycuick, C.J. (1984): Momentum and energy
in the wake of pigeon (Columbia livia) in slow flight. J. Exp. Biol. 111,
81-102.

Tucker, V.A. (1973): Bird metabolism during flight: evaluation of a theory.
J. Exp. Biol. 58, 689-709

Weber, H. und H. Weidner (1976): Grundriß der Insektenkunde. F. Auflage,
Fischer, Stuttgart

Weis-Fogh, T. (1972): Energetics of hovering flight in hummingbirds and in
Drosophila. J. Exp. Biol. 54, 79-104.

Weis-Fogh, T. (1973): Quick estimates of flight fitness in hovering animals.
Including novel mechanisms for lift production. J. Exp. Biol. 59, 169-230.

PHYSICAL DETERMINANTS OF LOCOMOTION

Thomas L. Daniel and Paul W. Webb[1]

Department of Zoology, NJ-15
University of Washington, Seattle, Washington 98195, U.S.A.
[1] Department of Biological Sciences and
School of Natural Resources
University of Michigan
Ann Arbor, Michigan 48108, U.S.A.

ABSTRACT.   Locomotion is a problem of momentum transfer between
animals and their environments. A momentum balance for thrust and
resistance shows that density and viscosity are the most important
physical properties affecting this transfer in air and water.
Propulsion mechanisms are categorized in terms of the forces
generated such as drag, lift, and acceleration reactions in fluids
and the reaction against the ground.   The performance of these
mechanisms is measured in terms of speed, accelerations and
transport costs.   Two major factors that affect the distribution
and performance of propulsors in all environments are (1) the
density ratio between the animal and its environment and (2) the
nature of unsteadiness in the speed of an animal.

Most studies of locomotion have focussed on movement in a particular environment, as
in pedestrian, usually terrestrial, aerial and aquatic habitats. There has been little
attempt, however, to consider if there are common elements in the various modes of
propulsion that are reflected by the physics of the environment and the physics of
propulsion. Similarly, little thought has been given to the validity of traditional
divisions of locomotion into swimming, running, or flying.

Our goal is to consider such questions. Since propulsion is a problem of momentum
transfer between animals and their environments, we seek differences and similarities in
locomotion in various environments that reflect those physical properties (Table 1) that
mediate this transfer, such as the density and viscosity of air and water, the temperature
dependence of these parameters and the elasticity of solid surfaces.

We first examine the physics of propulsion with a momentum balance for thrust and

Comparative Physiology: Life in Water and on Land. P. Dejours, L. Bolis, C.R. Taylor, E.R. Weibel
(eds.) Fidia Research Series, IX-Liviana Press, Padova © 1987

TABLE 1. Physical properties of water, air and some surfaces (McMahon and Greene 1979; Vogel 1981).

A. Fluid Properites

	Water			Air		
Temperature (C) (Salinity - %)	Density	Dynamic viscosity	Kinematic viscosity	Density	Dynamic viscosity	Kinematic viscosity
	$(kg \cdot m^{-3})$	$(kg \cdot m^{-1} \cdot s^{-1})$	$(m^2 s^{-1})$	$(kg \cdot m^{-3})$	$(kg \cdot m^{-1} \cdot s^{-1})$	$(m^2 s^{-1})$
0 (0%)	1×10^3	1.8×10^{-3}	1.79×10^{-6}	1.293	17.1×10^{-6}	13.2×10^{-6}
20 (0%)	0.998×10^3	1.0×10^3	1.00×10^{-6}	1.205	18.1×10^{-6}	15.0×10^{-6}
40 (0%)	0.992×10^3	0.65×10^3	0.66×10^{-6}	1.128	19.0×10^{-6}	16.9×10^{-6}
20 (35%)	1.02×10^3	1.09×10^{-3}	1.06×10^{-6}	--	-	-

B. Thermal Volume Expansion of Fluids

Temperature (°C)	Water	Air
5 - 10	0.53×10^{-4}	3.66×10^{-3} of
20 - 40	3.02×10^{-4}	volume at STP

C. Stiffness of Surfaces

Surface	Stiffness $(kN \cdot m^{-2})$
Concrete, asphalt	4376
Packed cinder	2916
Board track	875
Experimental wood track	100 - 195
Foam rubber track	14

resistance. This leads to a description of possible propulsion mechanisms and the
physical basis for their distribution in various environments. Therefore, rather than
separating this topic into the classic divisions of flying, running, and swimming, we
choose instead to organize our view of locomotion according to the mechanisms that
generate, and the environmental determinants of, thrust. We then attempt to identify
performance criteria by which we may judge, not only the physical feasibility of various
propulsion mechanisms, but also their relative utility to animals that move steadily,
periodically, or impuslively. We ask how effective each mechanism is in each environment
in bringing about motions that ensure energy economy or successful outcomes in predator
prey encounters. Finally, we examine the effects of temperature on our measures of
performance.

MOMENTUM BALANCE

Translation through the environment is the result of work done on the environment.
In the strictest sense, translation arises from the transfer of momentum from some part of
an organism to its environment. At the same time, however, momentum is transferred from
the environment to the organism. This reaction in its environment, together with the
inertia of the animal, resists translation of the entire organism.   Therefore, any study
of locomotion is a study of the mechanisms and devices by which momentum transfer is
mediated and the balances struck between these propulsive and resistive forces. This is
formalized by a general momentum balance that equates these forces.

All organisms move in fluid environments where the forces resisting their motion can
be divided into three components:

$$\text{Resistance} = \text{Drag} + \text{Acceleration Reaction} + \text{Body Inertia} \qquad [1]$$

$$= \frac{1}{2} \rho_e S_b C_{db} u_b^2 + \quad a_b \rho_e V_b (du_b/dt) + \quad \rho_a V_b (du_b/dt) \qquad [2]$$

(all variables are defined in the appendix). Drag depends only upon the ablsolute speed
of the animal while the latter two terms depend only upon the change in an animal's
velocity and thus resist acceleration or deceleration. The second term on the right hand
side of equation 2 is called the acceleration reaction and arises from the force required
to increase the kinetic energy of the fluid about a body whose speed is increasing.
Resistive forces are balanced by propulsive forces that may arise from any one or a
combination of the following:

$$\text{Thrust} = \text{drag} + \text{lift} + \text{acceleration reaction} + \text{ground reaction} \qquad [3]$$

$$= \frac{1}{2} \rho_e S_p C_{dp} u_p^2 \quad + \quad \frac{1}{2} \rho_e S_p C_{lp} u_p^2 \quad + \quad a_p \rho_e V_p (du_p/dt) \quad + \quad F_x(t) \qquad [4]$$

Thrust arising from the reaction in the fluid to limb or body motion is usually neglected

for animals that move in contact with the ground. Similarly ground reaction forces are
neglected where no such contact occurs, and fluid dynamic forces dominate. Contact with
the ground, therefore, poses a discontinuity in our simple momentum balance for locomotion
(Schmidt-Nielsen, 1984), actually simplifying equation 4 for each of the dominant
situations.

The only explicit environmental factor in equations 2 and 4 is density. Indeed, most
of the terms in this equation are linearly dependent on density. The effect of viscosity
is expressed implicitly in the behavior of the coefficients C_d and C_l. Each of these
depends upon the Reynolds number (ul/ν), a major determinant of the relative importance of
inertial (density dependent) to viscous forces in the environment (see Vogel, 1981;
Alexander, 1983). Both coefficients genenerally decrease with increasing Reynolds
numbers. Thus the equation explicitly incorporates one environmental parameter (ρ_e) as
well as size (in S and V) and implicitly incorporates viscosity, size, and shape in the
coefficients C_d, C_l, and a.

A major theme developed below, following equations 2 and 4, is the importance of the
density ratio (ρ_a/ρ_e) which affects propulsion possibilities and performance criteria in
different environments. This is seen by equating equations 2 and 4 and dividing
throughout by ρ_e. Thus density remains only as the ratio, and only associated with the
term for the inertia of the animal. As an example of its importance, large values for this
ratio suggest that body inertia can potentially overwhelm thrust producing reaction forces
in the fluid. In air, where this ratio is approximately 1000, we may expect constraints
on the suite of propulsive mechanisms that can be used, because reaction forces in the
fluid of similar magnitude to the inertia of an animal's body are difficult to obtain.

PROPULSION MECHANISMS

The mechanism by which momentum transfer occurs depends upon the motions of the
propulsors and the reaction in the environment to such motions. For pedestrian animals,
appendages exert stresses on the ground and the reaction to such stresses depends upon the
elastic response of the ground to the fall of an appendage and the coefficient of friction
between the appendage and the ground.

Similarly, in fluid environments, stresses from reactions in the environment arise as
viscous or inertial forces due to the motion of propulsors. The relative importance of
these forces is related to the viscosity and density of the medium and the kinematics of
the propulsor by the Reynolds number. Reynolds numbers are low and viscous forces
dominate for slowly moving, tiny animals. Equations 2 and 4 then reduce to drag terms
only and, indeed, at very low Reynolds numbers, drag can be expressed in terms of
viscosity and speed. The neglect of density in this so-called "Stokesian" realm has
little consequence for the magnitude of thrust forces (Weihs, 1980; Vogel, 1981;
Childress, 1981). Thus drag-based propulsors dominate at low Reynolds numbers. At higher
Reynolds numbers (>500), lift forces begin to be effective in both air and water.

Lift- and drag-based mechanisms assume reasonably steady flows, when at the same time the acceleration reaction would be small. In reality, wheels are virtually absent in living organisms (Gray, 1968; LaBarbera, 1982) so that propulsors typically oscillate or undulate so that their motions are time-dependent. Sometimes the forces and energy requirements of such time-dependent motions can be described with a quasi-steady analysis that integrates the time varying steady-state fluid reactions. Many time-dependent flows, however, cannot be approximated with steady state analogs. The extent to which unsteadiness contributes to the balance of forces acting on oscillated or undulated propulsive devices is indicated by a dimensionless parameter, the reduced frequency parameter, $\sigma = \omega l/u$. It can be visualized as a measure of the flow in the direction of body movement relative to the flow perpendicular to body motion. Alternatively, σ is the ratio of two Reynolds numbers: one for forward motion Ul/ν and the other for oscillatory motion $\omega l^2/\nu$. Here, ωl is the angular, rather than forward, velocity of the propulsor. The ratio of the oscillatory Reynolds number to the forward Reynolds number is $\omega l/u$.

Unsteadiness in the motion of propulsors has two important consequences for the generation of thrust forces depending on whether these forces act normal or perpendicular to the axis of motion. The former are propulsors that generate lift. In this situation the reduced frequency parameter reflects a lag between the circulation developed on a wing and the motion of the wing. This lag results from the fact that circulation takes time to develop. For rapid flapping motions, there is insufficient time to develop the full amount of the steady state circulation. The effects of σ on lift-based propulsors is given formally by Theodorsen's function (Figure 1). As σ exceeds a value of about 0.1, unsteady effects are no longer negligible. At this value there is a reduction of about 20% in the lift developed by a propulsor and a phase shift of about 30 degrees between the peak lift and the peak speed of the wing. Increasing σ still further leads to greater reductions in lift and greater phase shifts between the lift and motion of the wing. It is important to note that biology has chanced upon ways to overcome the problem of creating sufficient circulation in time-dependent flows. The clap-and-fling mechanism in insects (Weis-Fogh, 1975), the rapid wing pronations of hovering animals (Ellington, 1984), and the flip of pteropod parapodia (Satterlie et al, 1985) all provide novel mechanisms that promote the rapid development of circulation.

A second type of unsteadiness is associated with forces parallel to the direction of propulsor movements. This form of unsteadiness, the acceleration reaction, results from changes in the kinetic energy of the fluid about an accelerating animal or appendage. Thus, rapid accelerations generate forces in addition to steady state forces such as lift and drag. At very high reduced frequency values, accelerational forces may overwhelm any state forces so that the latter can be neglected. Indeed Lighthill's (1975) slender body theory and Wu's (1971a,b,c) two-dimensional flexing wing theory, that provide the basis for modern studies of axial fish propulsion and aquatic flapping wing locomotion, are based on unsteady forces arising from such accelerational flow.

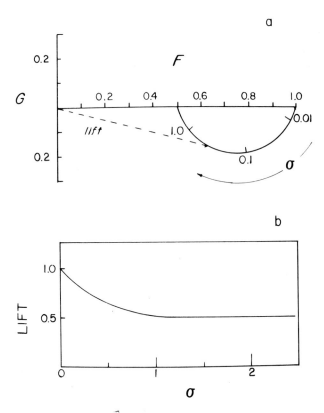

Figure 1. Reduced frequency parameters in relation to Reynolds number for some swimmers and flyers using lift-based propulsors. The cetaceans Tursiops, Solalia (Videler and Kamermans, 1985) and Lagenorhyncus (Lang and Daybell, 1963; Lang, 1966) and the skipjack tuna Katsuwonis (Magnuson, 1978) all swim in the thunniform mode, using a high aspect ratio caudal fin powered by body undulations. The fishes Cymatogaster (Webb, 1973) and Coris (Geerlink, 1983) swim in the labriform mode using the pectoral fins. Data for the flyers are taken from combined interspecific data sets in Greenewalt(1975) and Rayner (1979). The line for the passeriform birds 'and ducks are for a mass range of 1 g to 10 kg. Dipterans range from 0.1 to 5 g.

There is a major difference between the consequences of unsteadiness of propulsor motion in the two situations. Unsteady lift-based propulsors experience time lags and lift reductions mediated by the action of viscosity in the development of circulation. In contrast, the acceleration reaction is always present, even in inviscid fluids, and acts to increase propulsive forces. As a result the very large differences between the dynamic viscosity of air and water might suggest large differences in the reduced frequencies of aerial and aquatic lift-based swimmers and fliers. In practice, the limited data available show little or no difference (Figure 2). Note, however, that Figure 2 shows that σ normally exceeds 0.1 so that forces on propulsors in fluids have large, often dominant, unsteady components.

PROPULSION MECHANISMS AND ENVIRONMENT

The forgoing discussion suggests that the feasibility and diversity of various propulsors used in aerial, aquatic, and terrestrial environments depends on a small set of common physical mechanisms that reflect the physical properties of the environment and the kinematics of the propulsor. Table 2 provides a summary of these physical relationships, including the equations for thrust generation and the constraints relating to the properties of the medium and the size of the animal.

Distribution of propulsors. An obvious feature of Table 2 is that some mechanisms for producing thrust are conspicuously present in some environments and absent in others, as specified by the density ratio. Thus, drag-based propulsors are common in aquatic environments but are relatively rare in aerial environments. Similarly, pedestrian locomotion is common in terrestrial environments but relatively rare in aquatic benthic environments. These differences follow from the density ratio which now appears to be the single most important physical property influencing the methods used by animals in transferring momentum to the environment. The density ratio is always close to unity in water so that dynamically generated buoyant forces (lift) are small or unecessary. As a result, greater diversity in propulsion mechanisms is possible, with a wide range of propulsors either moving or generating forces largely parallel to the direction of motion. For example, cilia and drag-based paddles such as beetle hind-limbs and decapod pleopods beat in the plane of mean body motion, generating thrust forces in the same plane (Lochhead, 1977; Nachtigall, 1980; Wu, 1977). Flagellae and axial motions of small anguilliform fish are based on waves that are propagated in the axis of mean motion, again generating thrust forces parallel to the direction of mean body and wave motion. Thus, the diversity of all these drag-based mechanisms arise in the absence of the requirement for supporting the mass of the animal in its environment.

Lighthill's (1975) slender body theory and Wu's (1971a,b,c) waving plate theories used to describe axial fish propulsion show that thrust arises from the reaction in the fluid to periodic accelerations of body segments perpendicular to the direction of mean motion. This reaction force, in turn, is proportional to the added-mass of accelerated

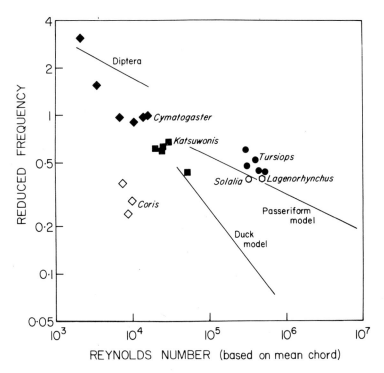

Figure 2. (a) The lift on an oscillating wing is proportional to Theodorsen's function
Θ = F + iG where F and G are, respectively the components of lift in phase and out of phase
with the motion of the propulsor. F and G both depend upon the reduced frequency
parameter (σ). Notice that very subtle values for the reduced frequency parameter can
lead to large decrements in the lift produced by unsteady wing oscillations. (b) The
magnitude of the lift vector is plotted against the reduced frequency value.

TABLE 2. Equations, ranges and limits for propulsion are arranged according to mechanism rather than traditional divisions of locomotion. ρ_R is the density ratio between the animal and its environment (ρ_o/ρ_e). All symbols are defined in the appendix.

MECHANISM	RANGE (size or Re)	EXAMPLES	EQUATION FOR THRUST	MORPHOLOGICAL OR KINEMATIC LIMITS	DENSITY RATIO ρ_R	AUTHORITY
Drag						
distributed viscous drag (envelope model)	Re << 1	Ciliates	$T = \frac{8}{3}\pi\mu k \ r_o a_o b_o$ (ciliate sphere)	swimming speed maximized with appropriate choice of metachronal wave parameters a_o and b_o.	near 1	Childress, 1981; Brennan, 1975
undulations	Re << 1	Flagellae	$T = (v - u)[(K_T - K_N)\beta l + K_N l]\cdot \frac{K_T}{K_N}\ l$ (planar waves)	efficiency and speed maximized for appropriate values of head and tail drag coefficients.	near 1	Childress, 1981
rowing	no clear restrictions	Pectoral fins, beetle hind limbs	$T = \frac{1}{2}\rho S \ C_d u^2$		near 1	Blake, 1983; Nachtigall, 1980.
Lift						
quasi-steady or steady	Re >> 1	Lunate tails, bird wings.	$T = \frac{1}{2}\rho S \ C_l u^2$		unrestricted	Alexander, 1982; Pennycuick, 1972.
unsteady	Re >> 1	Bird and insect wings, ray fins	$\bar{T} = \rho s \ c \ u^2 A^2 \cdot \pi c \ \alpha A$ $\pi \omega^3 c^2 \rho \ \alpha^2$ $A = u[(\omega\alpha(p - c/2) + \|(u\alpha - \omega h)] \times [(F + iG) - \omega\alpha c]$	thrust and efficiency maximized for appropriate values of pitching axis (p) and angle (α) and frequency (ω)	unrestricted	Wu, 1971a,b,c; Lighthill, 1971
Inertia						
unsteady jets	Re >> 1	Medusae, squid	$\bar{T} = \overline{u_{JET} \ (dm/dt)}$	thrust and efficiency maximized with appropriate jet cycle durations	near 1	Johnson et al., 1973; Daniel, 1983.
undulatory	Re >> 1	salmon, trout and other large fish	$\bar{T} = (1+\alpha)m[(\partial h/\partial t)^2 - u\ (\partial h/\partial x)^2]_{TAIL}$ (slender-body theory)	optimal body waveforms yield maximum thrust or maximum efficiency for given forward speed	near 1	Lighthill, 1975; Weihs and Webb, 1983
Ground Reaction						
pedestrian movement	no clear restrictions	ants to elephants	$T = F_{x,G}$	Optimal gates exist that minimize energy costs and maximize stability. Froude number must be less than one for ballistic walking.	>> 1	Alexander, 1982; McMahon, 1984
adhesive crawling	no clear restrictions	slugs	$T = A \ T_{FOOT}$	the weight of the animal divided by the area of the foot must be greater shear strength of mucus for vertical crawling	unrestricted	Denny, 1981

segments and thus it is also proportional to the density of the environment. Again, the density ratio plays a crucial role in determining the utility of this propulsive mechanism in various media: acceleration reaction forces of this sort are generally smaller in air than in water.

Jet propulsion and other inertia based mechanisms of locomotion also are possible only when the density ratio is small. Here, force is the product of the mass and acceleration of displaced fluid. Therefore thrust could be generated by high accelerations of low density fluid or by low acceleration of high density fluid. Alexander (1982), however, has pointed out that the latter is extremely inefficient. An even greater limitation to jet propulsion mechanisms is their periodic nature. Since jets generate forces in the direction parallel to motion, and since they are produced discontinuously, there are periodic acclerations of the body in the direction of motion. The balance of forces on the animal suggest that thrust forces should at least equal the accleration forces resisting changes in the animal's velocity. The latter are proportional to the density of the animal and the medium, the former are proportional only to the density of the medium. Thus, in water where the density ratio is near unity, thrust and acceleration reaction forces will be of the same order. In air, where there is a 1000 fold decrement in the density of the medium, thrust forces may be much smaller than the force required to overcome the inertia of the body.

The density ratio also determines the utility of pedestrian locomotion in air or water. Unlike unsteady swimmers and flyers, however, a high density ratio makes the use of the substrate more effective. This result follows from two factors. First, the propulsive force generated by appendage contact with the ground is proportional to the coefficient of static friction between the appendage and ground and the horizontal component of the force exerted by the appendage. The latter is, in turn, proportional to the effective weight of the animal. In air, this weight is simply the product of mass and gravitational acceleration. In water, where the density ratio approaches unity, the effective weight clearly tends towards zero and use of the ground becomes increasingly less effective. Therefore, mobile benthic invertebrates should have higher densities than their pelagic counterparts. Such differences are apparent at least among the arthropods and molluscs where pelagic representatives of these phyla show special adaptations for density reductions such as inclusion of lipids in copepods and reduced shell mass in pteropods.

Secondly the high density ratio of pedestrian animals imposes a limit on forward speed. Using a simple model of locomotion called ballistic walking, Alexander (1980,1982) and McMahon (1984) show that the ratio of the forward speed to the square root of the product of limb length and gravity must always be less than unity (this ratio is called a Froude number). This argument follows from the fact that forward motion results from periodically "falling forward" about a leg pivoting against the ground. Such a model would suggest that, for a given limb length, speed is not limited by the environment. The gravitational term, however, represents the legward fall of the animal, or the radial acceleration of the limb toward the pivot on the ground. In aquatic environments, this inward acceleration is greatly diminished by the low density ratio where the maximum speed, in turn, will be diminished to keep the Froude number less than unity. Mobile benthic invertebrates are, generally, much slower than their terrestrial counterparts with the possible exception of crawling molluscs for whom locomotor forces arise only from the stresses established in pedal mucus between the foot and substrate (Denny, 1981). These stresses are presumably independent of the density ratio.

Animals using limbs for locomotion, such as arthropods, appear to use subtly different mechanisms in air and water. In water, there may be a reduction in the number of limbs used to propel the body (Pond, 1975; Clarac, 1981) and, due to the decrease in effective weight, there should be a change in the components of force from predominantly vertical to predominantly horizontal, the latter being forces that essentially pull the animal forward. The shift to predominantly horizontal components of force may also follow from the greater drag resisting forward motion in water. The smallest terrestrial animals may also be affected by drag, but in this case by environmental flows rather than by flows induced by locomotion. Thus Alexander (1982) argues that the larger surface to volume ratio of small terrestrial animals results in proportionately greater drag forces imposed by environmental flows. These relatively larger forces, may impose a requirement for a change in stance with limbs spread laterally to support the drag on the animal.

Size limitations. Thus far, the density ratio is seen as the parameter of overwhelming importance in determining the feasibility of various propulsion mechanisms in various environments. Viscosity, while not explicitly important in the forgoing discussion, plays a crucial role that is brought out by Table 2, seen as the size range (Reynolds number) to which each of the propulsive mechanisms apply. For example, drag-based mechanisms appear not to be highly size restricted while lift and inertia based mechanisms are found only among larger organisms. Such differences in size range follow from the role played by viscosity in each propulsive mechanism.

The tiniest of swimming organisms live in the realm of low Reynolds numbers. For such organims, viscous drag on cilia, flagellae, setulated arthropodan appendages, and other such devices generates forward thrust. Larger organisms, however, are not excluded from using drag per se as a propulsive mechanism, but the nature of the stress is different. As size, and thus Reynolds number, increases the forces acting on propulsive devices shift from predominantly viscous to predominantly pressure stresses. Fore-aft asymmetries in such pressure stresses, however, result from the effects of viscosity as seen in flow separation and wake formation. Thus at both low and high Reynolds numbers drag is a feasible mechanism for thrust production. Note, however, that while viscosity plays a crucial role in the development of all drag forces, viscous drag mechanisms are limited only to the slow and tiny while pressure drag mechanisms are limited to the large and fast (Vogel, 1981).

The apparent lack of tiny lift-based propulsors follows from a similar argument. Lift is the result of a pressure asymmetry perpendicular to the direction of motion. Just as with presssure drag, lift may arise only at higher Reynolds numbers where viscous losses are unimportant.   At low Reynolds numbers, viscosity prevents the development of circulation (ultimately the pressure asymmetry) about a wing or wing-like appendage. Thus, small organisms face a two-fold problem in using lift for propulsion: (1) low Reynolds numbers prevent the formation of a pressure differential across the surface of the wing and (2) the time required to develop this pressure difference increases with decreasing Reynolds numbers. There is no viscous analog to lift and thus the very tiny low Reynolds number animals are barred from using lift-based mechanisms.

A lack of tiny lift based propulsors, however, does not mean that tiny organisms cannot propel themselves in air. Thrips, for example, may be using their wings more as paddles than as lifting surfaces, although Ellington (1984) suggests that their wing kinematics ironically imply a lift-based mechanism. Equivalently, the rapid downward fling of lepidopteran wings during take-off maneuvers may represent a drag based mechanism of propulsion (Ellington, 1984). Those small animals that do use lift based propulsion in air have a variety of morphological and behavioral adaptations for augmenting the development of circulation about their wings (Ellington, 1984) such as pronations, suppinations and flexing of wings in hovering animals and the "clap and fling" method of insect flight and other examples noted above.

Propulsion by purely inertial mechanisms, such as acceleration reaction forces in slender body locomotion or jet reaction mechanisms, are also subject to size constraints. These mechanisms occur for nearly ideal flows that can be treated as inviscid. Thus, small size and hence small Reynolds numbers preclude the development of large reaction forces of this nature. In addition, reaction mechanisms are also rare among the largest flyers and swimmers such as the albatros and whales. The reduced frequency values for these animals are so low that they approach steady-state propulsion. Reaction mechanisms thus appear contrained to animals of moderate size and are replaced by steady-state, or quasi-steady forces at extremes of small or large size. In the former, the size constraint is set by the overwhelming importance of viscosity in determining the balance of forces on propulsors. The reasons for an upper size constraint on the use of the acceleration reaction follows from the balance of forces in equation 4. Thrust forces may be normalized by some measure of area, or l^2 , where l is some characteristic length. Thrust is itself determined by the muscles, and muscle stress (force per unit area, proportional to l^2) is roughly a constant (McMahon, 1984). Then:

$$\text{Thrust}/l^2 \propto \quad \rho_e C_d u^2 \quad + (1+a)\rho_e l \ (du/dt) \ = \text{const} \qquad\qquad [5]$$

The second term on the right hand side of equation 5 combines both the inertia of the body and its acceleration reaction. Very large animals move at a Reynolds number where C_d

is essentially constant and independent of size. Therefore, the drag term, normalized by 1^2 is also a constant. As a result, body acceleration is inversely proportional to 1, essentially because body volume and inertial forces scale with 1^3 while the driving muscle stress scales with 1^2 . Jet propelled animals may also be subject to size constraints, although these relate more to the energetics of swimming with a periodic jet (Daniel, 1983). Nevertheless, the arguments suggested above should apply equally well to explain the rarity of tiny jet propelled animals, and the apparent lack of large ones.

PERFORMANCE CRITERIA

Thus far we have examined the physical mechanisms underlying various modes of locomotion, identifying, where possible, the limits to the use of each mechanism for animals of different sizes in different environments. But, while a certain mechanism may be physically feasible, its level of performance may be so low as to make it impractical. For example, ballistic walking may be feasible in aquatic environments, but its rarity probably results from the energy required to overcome drag, or the limit to speed imposed by the Froude number for walking. Similarly, drag-based propulsors in aquatic environments may be feasible for animals of all body size, but the maximum speeds attainable by such a mechanism may prevent its use in animals requiring rapid forward speed. Therefore, understanding the distribution of various propulsors requires additional criteria that reflect the performance capabilities of the propulsive mechanisms in various environments.

Animal motions can be described in terms of speed, acceleration, and turning radius. These three variables are the most suitable kinematic measures of performance in different environments. In addition, animals must accumulate some surplus energy for growth and reproduction, and cost and efficiency of locomotion are criteria that reflect the consequences of movement to the overall energy budget of an animal. These three kinematic and two energetic variables are performance measures for comparing propulsors in different environments.

Kinematic performance criteria.   Maximum speed, and acceleration, and minimum turning radius are likely to be important only during escape or attack maneuvers (Howland, 1974; Webb, 1976, 1986). They may be used only rarely, but since they determine, to a large extent, the success or failure of an animal in a predator-prey interaction, these measures of performance are of great selective importance. Indeed, in many organisms such as fish (Webb, 1976) or shrimp (Daniel, unpublished), the amount of muscle devoted to these rarely used events is far greater than that associated with the slower routine movements that characterize the majority of an animal's activities.

Turning radius is subject to rather strong environmental and morphological constraints. A sharp turn requires a rapid change in the direction, and thus a high centripetal acceleration, of the center of mass of an animal. Such motions can only be accomplished when there is sufficient reaction to side-slip in the environment and when

TABLE 3. Linear and rotational accelerations for animals in various environments.
Acceleration values are reported as fractions of the earth's gravitational accleration.
Wing span is used as the length measure for flying animals.

A. Linear acceleration.

	mass (g)	length (cm)	acceleration (fraction)	authority
LAND.				
Stellio stellio	10		0.83	Huey and Hertz, 1984
(lizard)	20		· 1.04	(initial acceleration
	50		1.42	on level ground)
	100		1.78	
Jumping				
rabbit flea	3.2×10^4	0.10	204	⎤ Bennett-Clarke, 1977;
human flea	4.9×10^4	0.10	245	⎦ Schmidt-Nielsen,, 1984
springtail		0.15	988	Christian, 1978
click beetle	0.04		382	⎤ Bennett-Clarke, 1977;
locust	3	1.7	15	Schmidt-Nielsen, 1984
bushbaby	200	16	14	
man	7×10^4	120	1.5	
leopard		147	1.6	⎦
AIR.				
Chysopa carnea		1.0	1.3 – 2.6	Miller and Olsen, 1979
(lacewing)				(response to bat attack)
WATER.				
Diaptomus fransiscanus	0.06		1.2	Lehman, 1977
Lapidocerca trispinosa	0.08 – 0.19		1.0	Vlymen, 1970
Daphnia pulex	0.12		0.4	Lehman, 1977
Cyclops scutifer	0.15		0.56	Strickler, 1975
Euphausia superba	0.45		0.56	Kils, 1979
Loligo vulgaris	3		1.01	⎤ Packard, 1969
(squid)	100		1.43	Johnson et al, 1972
	240		0.90	⎦
Sepia (squid)	250		0.65	Trueman and Packard, 1968

TABLE 3 cont.	mass (g)	length (cm)	acceleration (fraction)	Authority
Pandalus danae (shrimp)		9.10	3.39	Daniel (unpublished)
		9.62	2.18	(mean of first 50 ms)
		10.32	3.84	
Oronectes virilis (crayfish)	8.3	18.3	5.2 (0.67)	Webb, 1979 maximum value with mean in parentheses.
Slamo gairdneri (trout)	7.2	9.6	3.4 (1.6)	Webb, 1976
	79	20	3.3 (1.5)	maximum value with mean
	562	39	4.1 (1.8)	in parentheses.
bottlenose dolphin		235 – 265	0.01 – 0.15	Kayan et al., 1978

B. Centripetal acceleration (powered turns only).

	mass (g)	length (cm)	speed of turn (cm/s)	radius of turn (cm)	centripetal acceleration (fraction)	authority
LAND.						
cockroach	–	3.5	60.6	2.6	1.44	Camhi and Tom, 1978
	70×10^4	200	464	375	0.59	McMahon, 1984
			575	609	0.55	
			671	1081	0.42	
			741	1875	0.30	
AIR.						
long-eared bat	9.5	25.5	1	8– 10	0.5	Rayner and Aldridge, 1985
WATER.						
smallmouth bass	23.6	158	250	2.6	24.5	Webb, 1983
Salmo gairdneri	25.7	169	250	4.6	13.9	
(trout)	7.2	9.6	108	1.7	7.0	Webb, 1976
	79	20.7	118	3.1	4.6	
	562	38.7	201	7	5.9	
Sphenicus humboldti (Humboldt penguin)		58	240	14	1.4	Hui, 1985

the propulsor can generate sufficient moments about the center of mass. At very low Reynolds numbers, the reactions in the environment are predominantly viscous and body inertia (and hence side-slip) is, for all practical purposes, negligible. Thus the only limitation to rotational movements imposed on such animals will be determined by the ratio of moments generated by propulsive structures to the moments imposed by viscous resistance over the body.

For larger animals, the body inertia may become large relative to environmental forces. Turning radius is then determined not only by the balance of resistive and reactive moments acting on the body relative to those produced by the propulsor, but also by the inertia of the body. In fluid environments, therefore, sharp turns may arise when (1) the reaction in the fluid to sideslip of the body is high, (2) the reaction in the environment to rotation of the body is low, and (3) the magnitude of propulsive moments is sufficiently large to overcome the rotational inertia of the body. Satisfying these conditions depends again on the density ratio. For aerial organisms, with large density ratios, the rotational inertia of the body can be large relative to inertial propulsive forces. In addition, lift-based propulsive forces that scale with the surface area of an animal may be small relative to rotational forces that scale with the volume of the animal. Thus sharp turns are difficult in aire. In contrast, both propulsor forces and side-slip reactions of aquatic animals with low density ratios can be high relative to the rotational inertia of the body. The reaction in water to rotational motion, however, might also be high. But, anterio-posterior compression of an animal can minimize rotational reactions without jeopordizing reactions to side-slip (Alexander, 1967; Daniel, 1984). Slide-slip reactions may be augmented by lateral compression and enhancement of depth through appendages extending dorso-ventrally. Both of these morphological changes increase the added-mass of the animal associated with lateral motions. Such changes correlate well with the differences in overall morphology between highly maneuverable coral reef fish and less maneuverable cruising predatory fish (Webb, 1984). In general, therefore, we may expect aquatic animals to accomplish sharper turns than aerial animals, as implied by the limited data available in Table 3.

For terrestrial animals, a slightly different picture emerges. The rotational inertia of the body must be overcome by the reactions of propulsors with the ground rather than with the environmental fluid, although a high density ratio is again required to minimize side-slip. Because the ground provides a reaction that is equal and opposite to that imposed by propulsors, turning radius at a given velocity will be limited by the mass of an animal (Howland, 1974), and the ability of the limb to withstand large bending moments normal to the direction of motion.

Environmental and related morphological limits to acceleration performance are similar for animals using lift-, drag-, or inertia-based mechanisms -- the density ratio determines their ability to accomplish rapid linear accelerations with give appendage motions. Again, the very large density difference between aerial animals and their environment precludes the development of large body accelerations. By the same token, aquatic animals should be able to develop larger accelerations (Table 3).

Similarly, the magnitude of acceleration parallel to the ground for terrestrial animals requires a sufficiently large friction force, as is required for rotational acceleration in turning. Terrestrial animals can, however, generate very large forces normal to the ground during walking, hopping, and jumping (Alexander, 1980,1982,1983; McMahon, 1984). Jumping, in particular, requires vertical forces that are sufficiently large to accelerate the mass of an animal against gravity. It is interesting to note that the large vertical accelerations of flying animals during take-off maneuvers are powered, not by wings, but by legs (e.g. Wyman et al., 1985), once airborne, accelerations are likely to be weak. We know of no observations of underwater jumping, but the large acceleration reaction forces on aquatic pedestrian animals should preclude the development of large, vertical, body accelerations. In general, we suspect that aquatic animals should show the largest body accelerations parallel to the direction of motion, terrestrial animals should show largest vertical accelerations, and aerial animals the lowest body accelerations. Limited data support this guess (Table 3), but observations are rare.

In addition to this density-dependent environmental limitation, there is also an intriguing size constraint to acceleration performance that follows from similar reasons to those developed above for size limits to propulsion mechanisms using the acceleration reaction (equation 5) where the relative magnitude of viscous and inertial terms affects acceleration performance for a given muscle stress. Returning to the stress balance of equation 5, (1) at high Reynolds numbers where C_d is independent of both size and velocity, larger body size implies lower acceleration performance; (2) at low Reynolds numbers (Re < 1) where inertial terms are small compared to viscous terms and C_d is inversely proportional to both body size and speed, larger body size implies greater acceleration; and (3) at moderate Reynolds numbers where neither inertial nor viscous terms can be neglected, and there is an inverse relationship between C_d and the Reynolds number, there exists a unique body size that maximizes acceleration. Thus, if acceleration per se is the criterion by which we choose to evaluate performance during escape or attack movements, then there is a clear size constraint to such behaviors. For any one group of organisms, this constraint can be identified from the scaling relationships of the musculature associated with escape, the Reynolds number dependence of the drag coefficient for the body, and the allometry bewteen surface area and volume. These relationships have been worked out for one genus of shrimp, Pandalus danae, for which all of the above have been measured along with escape performance (see Fig. 3).

Environmental limits to maximum sprint or sustainable speed for animals moving in fluids are not clear at present, while the data base is large, at least for some groups (e.g. Beamish, 1978). In steady-state locomotion, body inertia plays a negligible role in the balance of forces on an animal. Thus, for a drag-based or lift-based propulsor in air

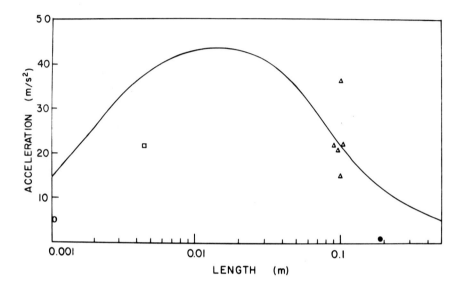

Figure 3. The stress balance (equation 5) is solved analytically to predict mean
acceleration in the first 50 ms of escape for crustaceans. A maximum muscle stress of
2500 N m (lobster fast remoter (Mendelson, 1969)), and added mass coefficient of 1.0
(Daniel, 1984) and a drag coefficient that varies as $1 + 10$ Re^{-1} (Daniel, unpublished) were
assumed in the computations. Triangles represent the mean acceleration in the first 50 ms
of escape by the coonstripe shrimp Padalus danae (Daniel, unpublished). The square and
circle are, repectively, reported acceleration values for Euphausia superba (Kils, 1979)
and Cylops scutifer (Strickler, 1975) The solid circle is the mean acceleration reported
for the crayfish Oronectes virilis (Webb, 1979) .

or water, resistive and propulsive forces depend equally on the density and viscosity of the medium. Hence, while fish may experience greater drag forces due to the greater density of the fluid, so too may they produce greater propulsive forces. Limits to speed may therefore be set by maximum power output of muscle and comparisons of speeds of animals should be made with physiologically equivalent performance levels, as argued by Heglund et al. (1982) for gaits of terrestrial animals.

In contrast environmental limits to the speed of pedestrian animals are set by the Froude number, assuming ballistic walking is used. The greater drag imposed on aquatic pedestrian animals may also set severe limits to speed. Experimental data for some animals (crayfish, rock lobsters: Pond, 1975; Clarac, 1981), however, suggest that such a generalization may not always apply as some animals move faster in water than on land using the same propulsors. This speed difference is presumably due to the additional requirements that such animals must veritcally support and periodically accelerate their body mass in air; requirements not normally met in their aquatic environment.

Performance levels and energy. Perhaps the two most commonly used measures for performance are cost of locomotion and efficiency. The former, viewed either as the total power required to move a unit weight of animal at a unit speed or as the total energy required to move a unit mass of animal through a unit distance, represent what Gabrielli and von Karman (1950) refer to as the "economy of transport". A similar measure of performance is the overall efficiency of locomotion, defined as the ratio of useful power output to total power input. Low costs and high efficiencies are presumably favorable characteristics for migratory animals· and cruising predators (Weihs and Webb, 1983).

Two generalizations emerge on costs of transport, primarily from the work of Taylor et al. (1970), Tucker (1975), and Schmidt-Nielsen (1977): (1) across many genera, and even many phyla, cost decreases with increasing body size and (2) locomotion on land is more costly than in air which, in turn, may be more costly than locomotion in water. The factors underlying the first observation are not, at present, well understood for all modes of locomotion.

The differences, however, in cost for the three modes of locomotion relate to the energy requirements imposed by the different environments. On land, locomotion is accompanied by periodic vertical accelerations and decelerations of the body. Thus energy is expended (wasted) for motion in a direction normal to that of the mean body motion. Similarly, for flying animals, energy is expended to maintain the vertical position of the body in air and recent observations of Bilo et al.(1982) indicate that large vertical forces also occur and will contribute to energy losses in flight. For both modes of locomotion, however, the density difference between the animal and its environment imposes extra energy requirements.

The situation is less clear for aquatic animals. On the basis of measured costs of transport for salmonid fish (Brett, 1964; Schmidt-Nielsen, 1984) we might expect all swimming animals would experience lower costs of transport than flyers or runners of

equivalent body mass. Recent theoretical (Weihs and Webb, 1983; Daniel, 1983) and
experimental (Daniel, 1985; Hargreaves, 1981) studies of aquatic locomotion, however,
indicate that the cost of transport of many swimming animals may be either much lower or
much greater than initially thought. This apparent paradox arises from the almost
inevitable unsteady nature of animal locomotion. Within this broad category, we identify
at least two types of unsteadiness: periodic propulsion and accelerational locomotion. In
the former, periods of swimming alternate with an unpowered glide or coast phase, and in
the latter, the instantaneus velocity varies greatly during a propulsion cycle. Examples
of such motion are the accelerational pulsatile swimming movements of medusae (Daniel,
1983,1985) and the periodic sink and swim modes of larval fish or the burst and coast
modes of larger fish (Weihs, 1973,1974).

The two swimming modes have important consequences for the cost of locomotion.
Significant energy savings may occur for sink and swim or burst and coast locomotion due,
basically, to the reduced drag on a body in the glide phase of locomotion (Weihs,
1973,1974). This result follows from the argument that propulsion by body undulations
leads to greater drag than that experienced by rigid bodies. Rayner (1985) has also shown
a similar result for bounding flight of birds in which alternate flapping and tucked wing
positions lead to large energy savings. Daniel (1985), however, has shown that, in
pulsaltile locomotion, when the acceleration forces are large compared to drag forces and
when deceleration is the result of an active process, the energy requirements may far
exceed those of steady-state or burst and coast swimmers. Thus the nature of
unsteadiness, reflected by the time course of oscillations in velocity is an important
determinant of the cost of locomotion, determining when the total energy requirements are
greater or less than the steady-state equivalent.

The effect of unsteadiness in increasing locomotion costs in water has similarities
with terrestrial locomotion where the high cost is associated with fluctuations in body
velocity in the vertical plane. Therefore, we conclude that any time locomotion is
accompanied by variations in speed about a mean value, energy costs will be greater than
the steady-state equivalemt as long as energy is expended in both accelerating and
decelerating the body mass, irrespective of the medium or contact with the ground. The
cost of locomotion may be smaller than the steady-state equivalent if deceleration is
strictly passive, and if the net resistance to forward motion remains unchanged (or
decreases).

Transport costs (or work) may also be relevant criteria for assessing performance
during escape or attack when the total work done in moving the body, or achieving a
maximum distance in minimum time could be limiting. The choice of a specific parameter,
however, does not affect the qualitative conclusions reached by the above analyses in the
discussion of kinematic performance : there still should exist a constraint on the
environment in which accleration can be usefully accomplished and there still should exist
a size constraint. The quantitative results (e.g the body size at which a particular

parameter is maximized), however, will depend upon which parameter is chosen.

Thermal characteristics.   All the motions we have discussed so far assume a
stationary frame of reference in the medium at rest relative to the animal. The
appropriate measures of performance for swimmers and flyers, however, is motion relative
to the ground which is influenced by winds, thermals, and other currents moving relative
to the ground. These flows arise, in part, from temperature-induced density changes in
air and water. Density changes are small in air since its volume coefficient for thermal
expansion is 1/273 or 0.004, approximately one order of magnitude greater than that of
water (Table 1). Nevertheless, density gradients in air are sufficient to drive major
world winds and, indirectly, ocean currents. The greatest advantages accrue to ground
speed for animals moving in the directions of winds and currents. Moving against such
currents, however, presents problems that are the same as moving upstream and against
tides (Weihs, 1978). In this case, viscous stresses against the ground set up velocity
gradients that can provide a partial flow refuge, as for example, used by salmonid fry and
adults (Brett, 1983). In addition, these velocity gradients can be exploited by
dynamically soaring birds (Cone, 1962; Pennycuick, 1972).

Special opportunities for using temperature-induced density gradients and flows are
possible in aerial, rather than aquatic, environments because the thermal capacity of air
is small compared to water so that large density changes can occur over short periods of
time, especially near the ground. The ground also constrains the direction of resulting
flow to create thermals (von Mises, 1945) that are exploited by birds to remain airborne
throughout the day and for rapid cross-country movements at low energetic cost
(Pennycuick, 1972).

Temperature-induced changes in the viscosity of water would have their greatest
effect at low Reynolds numbers where resistive and propulsive forces are proportional to
the viscosity, rather than density, of the medium. Such changes appear to affect the
speed and acceleration of recently hatched larval fish (Fuiman, 1986). Reduced viscosity
at higher temperatures has been implicated as a drag reducing mechanism in tuna who
dissipate heat around their corselet and as an explanation for cyclomorphosis (seasonal
changes in body shape) in Daphnia, but the evidence supporting such claims is equivocal
(Vogel, 1981).

CONCLUSIONS

We have chosen to examine locomotion according to mechanisms that generate thrust,
rather than the traditional divisions of swimming, flying, and pedestrian (terrestrial)
categories, to seek the underlying physical bases for differences and similarities in

locomotor methods in various environments. Our momentum balance for thrust and resistance shows that density and viscosity are the most important physical properties for movement in fluids permitting propulsion by lift, drag, and acceleration reaction forces, while ground reactions are most important to pedestrians. This difference, based on the dominant forces in the presence or absence of ground contact during locomotion, is a discontinuity in the momentum balance.

Irrespective of the dominant forces generated, propulsors in all environments inherently move periodically. Reduced frequencies of propulsors in fluids and periodic vertical accelerations of pedestrians, are such that unsteady, acceleration-related forces dominate the force balance and energy requirements in all environments over most of the size range of animals, usually increasing transport costs, except for specialized swimming and flying behaviors. For very small and very large animals, propulsor motions approach steady-state conditions where drag and lift dominate the force and energy requirements.

The density ratio is the single most important factor structuring the feasibility of using various propulsion mechanisms in different environments, and thus determines the distribution of propulsors among various environments. This ratio also underlies many differences in kinematic and energetic performance levels, while the balance of steady and unsteady forces determines scale effects.

Acknowledgements.   This paper grew out of research funded by the National Science Foundation, Regulatory Biology Grant No DCB-8408132 to TLD and PCM-8401650 to PWW. TLD thanks E. Meyhofer for help in preparing the manuscript. Each author attributes all errors of fact or interpretation to the other author.

BIBLIOGRAPHY

Alexander RMcN (1967) Functional design in fishes, Hutchinson Univ. Library, London.

Alexander RMcN (1980) The mechanics of walking. In: Elder HY, Trueman ER (eds): Aspects
    of animal movement. Cambridge Univ. Press, Cambridge; 221-234.

Alexander RMcN (1982) Locomotion of animals, Blackie, London.

Alexander RMcN (1983) Animal mechanics, Blackwell Scientific, London.

Beamish FWH (1978) Swimming capacity. In: Hoar WS, Randall DJ (eds): Fish physiology.
    Academic Press, New York; pp. 101-107.

Benett-Clarke HC (1977) Scale effects in jumping animals. In: Pedley TJ (ed): Scale effects
    in animal locomotion. Academic Press, New York; pp. 185-201.

Bilo D, Lauk A, Wedekind F, Rothe W, Nachtigall W (1982) Linear accelerations of a pigeon
    flying in a wind tunnel. Naturwissenschaften 69:345-346.

Blake RW (1983) Fish locomotion, Cambridge Univ. Press, London.

Brennan C (1975) Hydromechanics of propulsion for ciliated micro-organisms. In: Wu TY-
    T, Brokaw CJ, Brennan C (eds): Swimming and flying in nature.   Plenum Press, New
    York; pp. 235-251.

Brett JR (1964) The respiratory metabolism and swimming performance of young sockeye
    salmon. J Fish Res Board Can 21:1183-1226.

Brett JR (1983) Life energetics of sockeye slamon, Oncorhyncus nerka. In: Aspey WS
    and Lustick SI (eds): Behavioral energetics. Ohio Univ. Press, Ohio; pp. 29-63.

Camhi JM, Tom W (1978) The escape behavior of the cockroach Periplaneta americana I.
    Turning response to wind puffs. J Comp Physiol 128A:193-201.

Childress S (1981) Mechanics of swimming and flying, Cambridge Univ. Press, Cambridge.

Christian E (1978) The jump of the springtails. Naturwissenschaften 65:495-496.

Clarac F (1981) Decapod crustacean leg coordination during walking. In: Herried CF II,
Fourtner CR (eds): Locomotion and energetics in arthropods. Plenum Press, New York;
    pp. 31-71.

Cone CD (1962) The soaring flight of birds. Scientific American 206:130-140.

Daniel TL (1983) Mechanics and energetics of medusan jet propulsion. Can J Zool
    61:1406-1420.

Daniel TL (1984) Unsteady aspects of aquatic locomotion. Am Zool 24:121-134.

Daniel TL (1985) Cost of locmotion: unsteady medusan swimming. J Exp Biol 119:149-164.

Denny M (1981) A quantitative model for adhesive locomotion in a terrestrial slug,
    Ariolimax columbianus. J Exp Biol 91:195-217.

Ellington C, (1984) The aerodynamics of hovering insect flight (I - VI). Phil Trans Roy
    Soc B 305:1-181.

Fuiman LA (1986) Burst swimming performance of larval zebra danios and the effects of
    diel temperature fluctuations. Trans Amer Fish Soc 115:143-148.

Gabrielli G, vonKarman T (1950) What price speed ? Specific power required for propulsion vehicles. Mech Engr 72:775-781.

Geerlink PJ (1983) Pectoral fin kinematics of Coris formosa. Netherlands J Zool
    33:515-531.

Gray J (1968) Animal locomotion, Weidenfeld and Nicholson, London.

Greenewalt CH (1975) The flight of birds. Trans Amer Phil Soc 65:3-67.

Hargreaves BR (1981) Energetics of crustacean swimming. In: Herried CF II, Fourtner CR
    (eds): Locomotion and energetics in arthropods. Plenum Press, New York; pp. 453-490.

Heglund NC, Fedak MA, Taylor CR, Cavagna GA (1982) Energetics and mechanics of
    terrestrial locomotion. IV. Total mechanical energy changes as a function of speed
    and body size of birds and mammals. J Exp Biol 97:57-66.

Howland HC (1974) Optimal strategies for predator avoidance: the relative importance of
    speed and manoeuverability.. J Theor Biol 47:333-350.

Huey RB, Hertz PE (1984) Effects of body size and slope on acceleration of a lizard
    (Stellio stellio). J Exp Biol 110:113-123.

Hui CA (1985) Maneuverability of the Humboldt penguin (Spheniscus humboldti) during
    swimming. Can J Zool 63:2165-2167.

Johnson WP, Soden PD, Trueman ER (1973) A study in jet propulsion: An analysis of the
    motion of the squid, Loligo vulgaris. J Exp Biol 56:155-165.

Kayan VP, Kozlov LF, Pyatetskil VE (1978) Kinematic characteristics of the swimming of
    certain aquatic animals. Fluid Dynamics 13:641-646.

Kils U (1979) Swimming speed and escape capacity of Antarctic krill, Euphausia superba.
    Meeresforsch 27:264-266.

LaBarbera M (1982) Why the wheels won't go. Am Nat 121:395-408.

Lang TG (1966) Hydrodynamic analysis of cetacean performance. In: Norris KS (ed): Whales,
    dolphins and porpoises. Univ. California Press, Berkeley; pp. 410-432.

Lang TG, Daybell DA (1963) Porpoise performance tests in a seawater tank. US Naval
    Ordenance Test Station Tech Rep 3063:1-50.

Lehman JT (1977) On calculating drag characteristics for decelerating zooplankton.
    Limnol Oceanogr 22:170-172.

Lighthill MJ (1975) Mathematical biofluiddynamics, Soc Ind Appl Math, Philadelphia.

Lochhead JH (1977) Some unsolved problems of interest in the locomotion of crustacea.
    In: Pedley TJ (ed): Scale effects in animal locomotion. Academic Press, New York;
    pp. 257-268.

Magnuson JJ (1978) Locomotion by scombrid fishes: hydromechancis, morphology, and
    behavior. In: Hoar WS, Randall DJ (eds): Fish Physiology. Academic Press, New York;
    pp. 239-313.

McMahon TA, Greene PR (1979) The influence of track compliance on running. J Biomech
    12:893-904.

McMahon TA (1984) Muscles, reflexes, and locomotion, Princeton Univ. Press, Princeton.

Mendelson MJ (1969) Properties of very fast lobster muscle. J Cell Biol 42:548-563.

Miller LA, Olsen J (1979) Avoidance behavior in green lacewings I. Behavior of free
    flying green lacewings to hunting bats and ultrasound. J Comp Physiol 131:113-120.

Nachtigall W (1980) Mechanics of swimming in water beetles. In: Elder HY, Trueman ER
    (eds): Aspects of animal movement. Cambridge Univ. Press, London; pp. 107-124.

Packard A (1969) Jet propulsion and the giant fibre response of Loligo. Nature
    221:875-877.

Pennycuick CJ (1972) Animal flight, Arnold, London.

Pond CM (1975) The role of walking legs in aquatic and terrestrial locomotion in the
    crayfish Austropotamobius pallipes (Lereboullet). J Exp Biol 58:725-744.

Rayner JMV (1979) A new approach to animal flight mechanics. J Exp Biol 80:17-54.

Rayner JMV (1985) Bounding and undulating flight in birds. J Theor Biol 117:47-77.

Rayner JMV, Aldridge DJN (1985) Three-dimensional reconstruction of animal flight paths
    and the turning flight of micropteran bats. J Exp Biol 118:247-265.

Satterlie RA, LaBarbera M, Spencer AN (1985) Swimming in the pteropod mollusc, Clione
    limacina. J Exp Biol 116:189-204.

Schmidt-Nielsen K (1977) Locomotion: energy cost of swimming, flying, and running.
    Science 177:222-228.

Schmidt-Nielsen K (1984) Scaling, Cambridge Univ. Press, Cambridge.

Strickler JR (1975) Swimming of planktonic Cyclops species (Copepoda, Crustacea):
    Pattern, movements and their control. In: Wu TY-T, Brokaw CJ, Brennan C (eds):
    Swimming and flying in nature. Plenum Press, New York; pp. 599-613.

Taylor CR, Schmidt-Nielsen K, Raab JL (1970) Scaling of energetic cost of running to body
    size in mammals. Amer J Physiol 319:1104-1107.

Trueman ER, Packard A (1968) Motor performance of some cephalopods. J Exp Biol
    49:495-507.

Tucker VA (1975) The energetic cost of moving about. Amer Scientist 63:413-419.

Videler J, Kamermans P (1985) Dolphin swimming performance: differences between the
    upstroke and downstroke. Aquat Mammals 11:46-62.

Vlymen WJ (1970) Energy expenditure of swimming copepods. Limnol Ocenaog 15:348-356.

Vogel S (1981) Life in moving fluids, Princeton Univ. Press, Princeton.

von Mises R (1954) Theory of flight, Dover, New York.

Webb PW (1973) Kinematics of pectoral fin propulsion in Cymatogaster aggregata.
    J Exp Biol 59:697-710.

Webb PW (1976) The effect of size on the fast-start performance of rainbow trout (Salmo
    gairdneri Richardson) and a consideration of piscivorous predator-prey interactions.
    J Exp Biol 65:157-177.

Webb PW (1979) Mechanics of escape responses in crayfish (Oronectes virilis, Hagen)
    J Exp Biol 79:245-263.

Webb PW (1983) Speed, accelerationation and manoeuvrability of two teleost fishes.
    J Exp Biol 102:115-122.

Webb PW (1984) Form and function in fish swimming. Scient Amer 251:72-82.

Webb PW (1986) Effect of body form and response threshold on the vulnerability of four species of teleost prey attacked by largemouth bass (Micropterus salmoides). Can J Fish Aquat Sci 43:763-771.

Weihs D (1973) Mechanically efficient swimming technique for fish with negative buoyancy. J Mar Biol 31:194-209.

Weihs D (1974) Energetic advantages to burst swimming of fish. J Theor Biol 48:215-229.

Weihs D (1978) Tidal stream transport as an efficient method for migration J Cons Int Explor Mer 38:92-99.

Weihs D (1980) Energetic significance of changes in swimming modes during growth of anchovy larvae, Engraulis mordax. Fish Bull US 77:597-604.

Weihs D, Webb PW (1983) Optimization of locomotion. In: Webb PW, Weihs D (eds): Fish biomechanics. Praeger Scientific, New York; pp. 339-371.

Weis-Fogh T (1975) Flapping flight and power in birds and insects, conventional and novel mechanisms. In: Wu TY-T, Brokaw CJ, Brennan C (eds): Swimming and flying in nature. Plenum Press, New York; pp. 729-762.

Wu TY-T (1971a) Hydromechanics of swimming propulsion. Part 1. Swimming of a two-dimensional flexible plate at variable forward speeds in an inviscid fluid. J Fluid Mech 46:337-355.

Wu TY-T (1971b) Hydromechanics of swimming propulsion. Part 2. Some optimum shape problems. J Fluid Mech 46:521-544.

Wu TY-T (1971c) Hydromechanics of swimming propulsion. Part 3. Swimming and optimum movements of slender fish with side fins. J Fluid Mech 46:545-568.

Wu TY-T (1977) Introduction to the scaling of aquatic animal locomotion. In: Pedley TJ (ed): Scale effects in animal locomotion. Academic Press, New York; pp. 203-232.

Wyman RJ, Thomas JB, Salkoff L, King DG (1985) The Drosophila giant fiber system. In: Eaton RC (ed): Neural mechanisms of startle behavior. Plenum Press, New York; pp. 133-161.

APPENDIX

Symbol	Definition	Units
A	area of a slug foot	m^2
a_o	cilium beat amplitude parallel to body motion	m
b_o	cilium beat amplitude perpendicular to body motion	m
C_d	drag coefficient (subscript b for body, p for propulsor)	
C_l	lift coefficient (subscript b for body, p for propulsor)	
c	half chord length of a wing	m
F	in-phase component of Theodorsen's function	
F	force on ground in direction of thrust	N
G	out-of-phase component of Theodorsen's function	
h	amplitude of wing oscillation	m
K_N	drag coefficient of a cylinder normal to flow	
K_T	drag coefficient of a cylinder tangential to flow	
l	characteristic length of animal	
m	mass	kg
p	pitching axis of a wing	
Re	Reynolds number	
S	surface area (subscript b for body, p for propulsor)	m^2
s	span of a wing	s
T	thrust	N
u	velocity (subscript b for body, p for propulsor)	$m \cdot s^{-1}$
u_{JET}	velocity of ejected fluid	$m \cdot s^{-1}$
V	volume (subscript b for body, p for propulsor)	m^3
v	wave speed of propulsor	$m \cdot s^{-1}$
α	added-mass coefficient (subscript b for body, p for propulsor)	
β	mean square length of a flagellum	m
Θ	Theodorsen's function	
μ	dynamic viscosity	$kg \cdot m^{-1} \cdot s^{-1}$
ν	kinematic viscosity	$m^2 \cdot s^{-1}$
ρ	density (subscript a for animal, e for environment)	$kg \cdot m^{-3}$
σ	reduced frequency parameter	
τ	shearing stress	$N \cdot m^{-2}$
ω	circular frequency	s^{-1}

COST OF TRANSPORT AND PERFORMANCE NUMBER, ON EARTH AND OTHER PLANETS

Colin J. Pennycuick

Department of Biology, University of Miami, Florida 33124, U.S.A.

Schmidt-Nielsen's (1972) "cost of transport" is closely related to the reciprocal of the lift-to-drag ratio traditionally used to characterise flight performance. With some precautions, it can be converted to an absolute, dimensionless "performance number", which applies to any form of locomotion. This is independent of gravity in flight and walking, but proportional to gravity in swimming. Range is inversely proportional to gravity in flight and walking, but independent of gravity in swimming. The maximum mass for flying animals depends on gravity and air density. Fossil evidence of giant flying and walking animals supports independent evidence that gravity may have varied widely over geological time.

THE "COST OF TRANSPORT" CONCEPT

The notion of "cost of transport" was introduced by Schmidt-Nielsen 1972 in order to compare metabolic measurements made on different animals, using different methods of locomotion, and travelling at different speeds. The idea was to isolate the amount of energy expended by an animal in moving a unit amount of its body weight horizontally through a unit distance. In practical experiments, energy consumption and distance both appeared in the form of their time rates, that is, metabolic rate and speed. The "cost of transport", C_m, took the form

$$C_m = P_m/WV, \tag{1}$$

where P_m is the metabolic rate as measured from rate of oxygen consumption, V is the speed, and W is the body weight. At the time, it was still common for energy in metabolic measurements to be expressed in calories, consequently, C_m was expressed in cal km^{-1} kg^{-1}. The use of this curiously heterogeneous combination of units obscures the physical nature of "cost of transport". In the first place the use of the gram-force as a unit of weight (although legitimate in the engineering metric system) has led to confusion with the gram of the SI system. The latter is a unit of mass, not weight. Substituting mass for weight implies, in effect, that the process being studied is considered to be unaffected by a change in the strength of gravity, which is usually not a safe assumption in studies of locomotion. To minimise the possibility of confusion, an animal's weight is best represented as the product of its mass (m) and the acceleration due to gravity (g), so that equation 1 becomes

$$C_m = P_m/mgV. \tag{2}$$

P_m is a power, having the dimensions of energy divided by time. The denominator of equation 2, being the product of a force and a speed, is also a power, having the same dimensions. C_m is therefore the ratio of two powers, and is dimensionless. Substituting mass for weight is equivalent to omitting g from equation 2, thus assigning to C_m the inappropriate dimensions of acceleration. The dimensionless character of C_m is further obscured by expressing the metabolic rate, P_m, in units which are not compatible with those used for the variables in the denominator. If m, g and V are expressed in kg, m s^{-2} and m s^{-1} respectively, then the product of all three is a power expressed in watts.

Comparative Physiology: Life in Water and on Land. P. Dejours, L. Bolis, C.R. Taylor, E.R. Weibel (eds.) Fidia Research Series, IX-Liviana Press, Padova © 1987

To calculate C_m correctly, P_m must also be expressed in watts, not in unrelated units based on calories.

C_m so calculated, though dimensionless, is not the ratio of two like quantities, since P_m, the power determined from metabolic measurements, includes some elements of power expenditure that have no counterpart in the denominator. The latter, being the product of weight and speed, is a purely mechanical quantity, whereas the numerator is the rate of chemical energy consumption needed to support the mechanical power output of the muscles. The subscripts indicate that C_m is a "metabolic" cost of transport, derived from the "metablic power", P_m. We can multiply P_m by a "conversion efficiency" (η), where $\eta < 1$, to obtain that portion of the chemical power consumption that is converted into mechanical power. η is usually thought to be between 0.20 and 0.25 in steady locomotion, on the basis of classical measurements of heat and work production by isolated muscles (reviewed by McMahon 1984). A value of 0.23 is adopted here, from the experimental determinations by Bernstein et al (1973) and Tucker (1972). We can estimate the mechanical power (unsubscripted P) generated by the muscles as

$$P = \eta P_m \qquad\qquad (3)$$

and we can then write an unsubscripted "cost of transport" (C) as

$$C = \eta C_m = \eta P_m/mgV = P/mgV. \qquad\qquad (4)$$

"Performance number" related to "Cost of Transport"

The dimensionless number C has a simple and direct meaning in flight, most easily understood in the special case of a fixed wing aircraft (figure 1a), flying horizontally, at a constant speed (V), so that the forces on it are balanced. The weight (mg) is balanced by the lift on the wings (L). An inescapable aerodynamic drag force (D) would cause the aircraft to decelerate, were it not balanced by a thrust (T), provided by the engine. The power required from the engine is

$$P = TV. \qquad\qquad (5)$$

The cost of transport, C from equation 4, is therefore

$$C = TV/mgV = T/mg. \qquad\qquad (6)$$

Since the weight (mg) is the same as the lift, and the thrust is the same as the drag, this can be written

$$C = D/L. \qquad\qquad (7)$$

In this simple arrangement, the "cost of transport" ratio is the same as the ratio of drag to lift. In the case of flapping wings, no steady lift or drag forces can be identified, but the lift and drag can be seen as time averages of the vertical and horizontal components of force, over a complete wingbeat cycle. This simple and direct representation of cost-of-transport in terms of lift and drag forces is not so helpful in the case of idealised walking (in which there is no drag) or of idealised swimming (in which there is no lift), but the representation in terms of weight, speed and power, as in equation 4, can be used for any type of locomotion.

The concept of "lift to drag ratio" dates from the nineteenth century beginnings of aeronautics, and is a much older concept than "cost of transport". To express it in a general form, which applies to any type of locomotion, it may be renamed the "performance number" (N), defined as

$$N = mgV/P, \qquad\qquad (8)$$

where P is the total power, mechanical or equivalent, expended by an animal or machine of mass m travelling horizontally at a speed V. N is closely related to the inverse of Schmidt-Nielsen's "cost of transport", and, in the case of flight, is the same as the effective lift-to-drag ratio. N explicitly involves the acceleration due to gravity (g). For a given value of the performance number, more

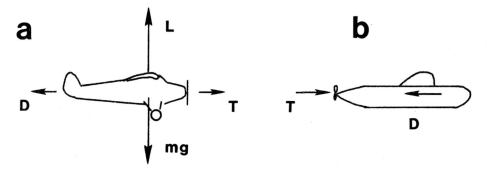

Figure 1. Balance of forces in (a) flight and (b) submerged swimming.

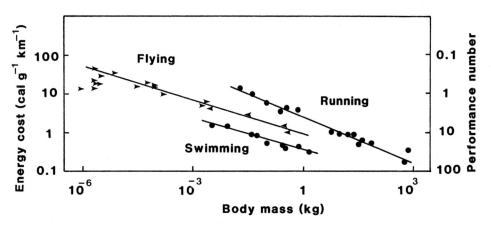

Figure 2. "Cost of transport" (left) re-expressed as performance number (right). After Schmidt-Nielsen (1972).

power is needed to propel an animal of mass m in high gravity than in low.

The results given by Schmidt-Nielsen (1972) can easily be converted into the absolute form represented in equation 4. These numbers are shown on the scale at the right of figure 2, with Schmidt-Nielsen's original scale at the left. It can be seen that performance numbers for medium-sized birds are around 10, which is approximately what one would expect from mechanical measurements (discussed by Pennycuick 1975a). Values of 30 or more are reported for the larger swimming and running animals. Schmidt-Nielsen's results show an upward trend of performance number with body mass for all types of locomotion.

Relation of range to performance number

An animal's "range" is defined as the distance it can travel, without feeding, before all its stored fuel is used up. Various authors have approached this by separately estimating speed and "metabolic rate" (i.e. total chemical power consumption), but a more general formulation can be made in terms of the performance number (N). A treatment developed for flying animals by Pennycuick (1969, 1975a) gives the range (R) as

$$R = (kN/g) \ln(1/(1-F)), \hspace{2cm} (9)$$

F is the "fuel fraction", that is, the fraction of the body mass consisting of consumable fuel when the animal sets out on its travels. k is the amount of mechanical work obtained when unit mass of the animal is consumed, that is, it is a form of fuel energy density, which takes account of conversion efficiency, and also of incidental mass loss due to loss of water etc. Equation 9 was developed in relation to the range of migrating birds, but actually it applies to any form of sustained locomotion, since the only assumption it contains is that the rate of diminution of the animal's body mass is proportional to the total power consumption (k being the constant of proportionality). It says that, other things being equal, range is proportional to N/g. The effect of increasing gravity, if N remains unchanged, is to reduce the range in proportion.

Basal metabolism - mechanical equivalent

In studies of flying and swimming, the mechanical power (P) can often be estimated directly, without having to take the circuitous route through metabolic measurements. For calculations of range, it is necessary to include all components of power consumption in P, including the basal metabolism, which can be thought of as a constant "overhead", continuing independently of any other energy expenditure that the animal may be incurring. An estimate of basal metabolism can only be obtained from metabolic measurements, and therefore has to be multiplied by η to convert it into a "mechanical equivalent", before adding it to the mechanical power.

VARIATION OF BODY MASS, GRAVITY AND FLUID DENSITY

Schmidt-Nielsen (1972) considered the scaling of his "cost of transport" in relation to body mass, as revealed empirically by metabolic experiments. He later extended this analysis (Schmidt-Nielsen 1984) including additional data, and taking a more analytical approach. Such an approach can be extended further to cover the effects of varying other variables besides body mass. It is used below to consider the effects on performance number and range of varying body mass, gravity and fluid density. Attention is restricted to three narrowly defined and artificially restricted modes of locomotion. The conclusions, though not general, indicate some major contrasts between flying, swimming and walking, at least in large animals.

1. Flight at high Reynolds Numbers

This is defined as locomotion in an inviscid fluid whose density is much less than the animal's, so that nearly all of the animal's weight has to be supported by hydrodynamic forces. In all forms of locomotion the ratio of speed to power is itself a function of speed, and therefore so also is the performance number. In the case of flight in the earth's atmosphere, there is a well-defined speed (the "maximum range speed") at which the performance number reaches a maximum. This results from the "U" shape of the curve of power required versus speed (figure 3a). We may begin by noting that the power required to fly at the maximum range speed is not a fixed multiple of the basal metabolic rate, as some physiologists appear to believe. By specifying a large bird, we may ensure that the basal metabolic rate is small compared to the mechanical power, and safely neglect it. Then, the equations given by Pennycuick (1975a) may be slightly adapted to represent the speed and power at the maximum range speed as

$$V = a_1\ m^{1/2}\ g^{1/2}\ \rho^{-1/2}\ A^{-1/4}\ b^{-1/2}, \tag{10}$$
$$P = a_2\ m^{3/2}\ g^{3/2}\ A^{1/4}\ \rho^{-1/2}\ b^{-3/2}. \tag{11}$$

a_1 and a_2 are constants whose values are not affected by changes in the density of the fluid or the strength of gravity, or by scaling the animal. m is the body mass, g is the acceleration due to gravity, ρ is the air density, b is the wing span, and A is the "effective cross sectional area" of the body, that is, the actual frontal area multiplied by the body's drag coefficient. "Scaling the animal" means changing its size without altering its geometrical form or its density. In this case the effective cross-sectional area of the body will vary with the two-thirds power of the mass, and the wing span with the one-third power, which may be written

$$A \propto m^{2/3},\ b \propto m^{1/3}. \tag{12}$$

Substituting these proportionalities in equations 10 and 11, we get

$$V \propto g^{1/2}\ \rho^{-1/2}\ m^{1/6}, \tag{13}$$
$$P \propto g^{3/2}\ \rho^{-1/2}\ m^{7/6}. \tag{14}$$

We can now find out how the performance number (N) depends on the same three variables,

$$N = mgV/P \propto g^0\ \rho^0\ m^0. \tag{15}$$

The range (R) is proportional to the performance number divided by g, so

$$R \propto N/g \propto g^{-1}\ \rho^0\ m^0. \tag{16}$$

Performance number and body mass

To comment first on the effect of scaling, the conclusion that N is scale-independent is at variance with Schmidt-Nielsen's results (figure 2), which show a progressive increase of N with increasing body mass. Because of the U-shaped power curve, flying animals, unlike those that run or swim (below), have a well-defined maximum-range speed, and they are obliged to fly at or near it if range is to be maximised. If metabolic measurements are used to estimate the power, they must be made at or near the maximum-range speed, a precaution which physiologists have not invariably observed. The slope of Schmidt-Nielsen's line is probably caused partly by an upward bias in "cost" for the smaller species, due to making power measurements at inappropriate speeds. The "flight metabolism" of insects and small birds is most easily measured in hovering or at low speeds, where power requirements are high. The use of such measurements in equation 1, possibly combined with a cruising speed that is too low, results in an

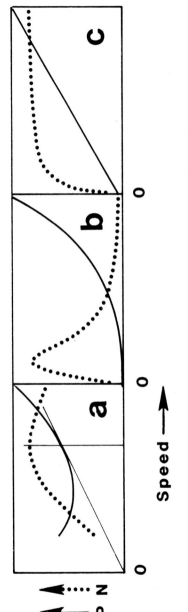

Figure 3. Power required (solid line), and performance number (dotted) as a function of speed for (a) flight, (b) swimming and (c) walking and running.

overestimate of "cost", and underestimate of the performance number. There are also two other effects that produce an upward trend in performance number with body mass. First, the larger animals fly at higher Reynolds numbers. This results in lower coefficients for body drag and wing profile drag, with a corresponding increase in performance. Second, wing span increases allometrically with body mass over a very wide range, in the sense that larger species have relatively longer wings (Greenewalt, 1962). This also leads to better performance for larger species.

The mass-independence indicated by proportionality 15 is not exactly correct, but is probably nearer the truth than the slope of Schmidt-Nielsen's line in figure 2 would suggest, at least in the larger animals (vertebrates). As Schmidt-Nielsen (1984) remarks, long, non-stop migrations are characteristic of small as well as large birds, whereas among walking and swimming animals, only large species migrate long distances. This tallies with the notion that performance number (and hence range) is more nearly mass-independent in flying animals than it is in walking or swimming animals.

Effect of air density and gravity

If the air density decreases, which occurs when a migrating bird climbs, both the maximum range speed, and the power required to fly at that speed, increase by the same factor, so that the performance number remains unchanged (proportionality 15). This is only true provided that the bird has sufficient spare muscle power available to meet the increased power requirement at the higher speed. Likewise, an increase in the strength of gravity would require both the speed and the power to be increased by the same factor. The performance number would not be affected, provided that the muscles were capable of supplying the extra power. The range, however, is proportional to the performance number divided by g, and therefore decreases as g is increased, even though the performance number remains the same.

2. Swimming at high Reynolds numbers

This is defined as locomotion in an inviscid fluid whose density is approximately the same as the animal's, so that nearly all of the animal's weight is supported hydrostatically. Unlike a hovering bird, a fish in hydrostatic equilibrium can hang motionless in the water without exerting any power beyond basal metabolism. The curve of power versus speed is therefore not U-shaped. The total power output at zero speed is equal to the basal metabolic rate, and increases ever more steeply with increasing speed. Figure 3b shows a curve for the case of a streamlined body, whose drag is balanced by the thrust from a propeller (figure 1b), which is a realistic representation of a submarine, and not too unrealistic for fast-swimming pelagic fishes such as tunas (Scombridae). In this case the drag (D) is

$$D = (\rho A V^2)/2, \qquad\qquad (17)$$

where ρ is the water density and A is the effective cross-sectional area of the body, varying, as before, with the two-thirds power of the body mass. The mechanical power is

$$P = DV = (\rho A V^3)/2. \qquad\qquad (18)$$

The only comprehensive data on the scaling of cruising speed in fishes are those of Brett (1965), who found that the maximum aerobically sustainable speed in salmon of different sizes varied with the one-sixth power of the body mass. This is an intra-specific comparison between individuals of the same species at different stages of growth. We may assume provisionally that the same rule applies to interspecific comparisons between animals adapted to a similar mode of swimming, although this is an uncertain extrapolation. In this case

$$V \propto m^{1/6},\tag{19}$$

so, from equation 18,

$$P = (\rho A V^3)/2 \propto m^{2/3} \times m^{1/2} = m^{7/6}.\tag{20}$$

On this assumption, the speed and power required for cruising scale similarly with body mass in both flight and swimming (proportionalities 13 and 14 above). However the relationship of performance number and range to gravity is different.

$$N = mgV/P \propto g \; m^0,\tag{21}$$

$$R \propto N/g \quad \propto g^0 \; m^0.\tag{22}$$

A flying animal has to support its weight, and therefore has to work harder if gravity is increased, but the weight of a swimming one is supported by the fluid. The result, which applies whatever assumption is made about the scaling of cruising speed, is that performance number is directly proportional to gravity. This may seem strange, as the strength of gravity is not easily detected by an animal in hydrostatic equilibrium. It means that the animal does not have to work harder to swim, even if gravity is increased. The power required to swim at a given speed, and also the range, are unaffected by a change in the strength of gravity.

Schmidt-Nielsen's (1972) metabolic results show a progressive increase of performance number with body mass. As with flight, this is probably at least partly due to a progressive increase in drag coefficient, due to the larger species swimming at higher Reynolds numbers. The drag coefficient should level off at Reynolds numbers above a few millions, in which case proportionalities 21 and 22 would represent a fair approximation for animals larger than those represented in Schmidt-Nielsen's data.

Performance number is low at speeds near zero because of basal metabolism, then increases sharply to a maximum at some low speed (figure 3b). In large swimming animals, the speed at which the maximum occurs will be too low to define a "maximum range speed" corresponding to that in flight. Practical cruising speeds will be well above the maximum, where performance number eventually becomes approximately proportional to the inverse square of the speed. The faster the fish elects to swim, the lower is N. It is not realistic to assume that cruising speed is selected to maximise N (as in flying animals), and this makes it difficult to predict how N should scale in swimming animals. For gill-breathers, there must be an upper limit to the cruising speed, set by the maximum rate of oxygen supply from the gills. The animal may indeed cruise at this speed, if it has urgent reasons to complete its journey as quickly as possible. On the other hand, if fuel economy is more important than speed, the animal has the option to increase its performance number and range, by selecting a lower cruising speed, nearer to the maximum in the curve. Where maximum speed is not required, there is probably a "comfortable" cruising speed related to the optimum strain rate, and contraction frequency of the aerobic locomotor muscles. Considerations of this kind are discussed by Pennycuick and Rezende (1984), but it is not clear what the implications would be for the scaling of swimming speed with body mass. The discrepancy between proportionality 21 and Schmidt-Nielsen's observed increase of performance number with body mass could be due to the speeds at which the metabolic measurements were made scaling in some other way than as indicated by proportionality 19.

3. Walking and running

This is defined as locomotion using reciprocating legs on a hard, horizontal surface, on which the feet do not sink in or slip, in a medium which does not offer appreciable hydrodynamic drag. A walking animal, like a swimming one, can stand still without exerting any mechanical power at all, and therefore also has a power curve that is not U-shaped, but increases progressively with speed. There is a great volume of empirical measurements of oxygen consumption of animals on a treadmill (Taylor et al, 1970). The mechanical power consumption of walking and running animals, unlike that of swimming animals, is usually directly proportional to speed over quite a wide range (figure 3c). The performance number is low at very low speeds, because of basal metabolism, but quickly levels off to an almost constant value. Once again, no clearly defined optimum speed can be identified in terms of maximising performance number.

The theory of Alexander and Jayes (1983) supplies a basis for estimating the way in which the speed an animal selects for prolonged cruising depends on its body mass and the strength of gravity. They propose that animals of different size exhibit geometrically similar motion when running at the same Froude number (F), where

$$F = v^2/gl, \tag{23}$$

l is the leg length, measured from the hip or shoulder joint. For example, quadrupeds which have different gaits such as walk, trot or canter, shift from one gait to another at a fixed value of the Froude number. From equation 23, this means that the speed at which different-sized animals shift gait scales as the square root of the leg length, or the one-sixth power of the mass. If the same rule governs cruising speed in a given gait (walk, say), then the relation of cruising speed to body mass and gravity would be

$$V \propto g^{1/2} m^{1/6}. \tag{24}$$

Alexander and Jayes (1983) also deduced that the power required by different animals to cruise at a given Froude number is

$$P = KmgV, \tag{25}$$

where K is a dimensionless constant. Combining this with proportionality 24 indicates that

$$P \propto g^{3/2} m^{7/6}. \tag{26}$$

Proportionalities 24 and 26 lead to the same rules for performance number and range as apply to flight, namely,

$$N = mgV/P \propto g^0 m^0, \tag{27}$$

$$R \propto N/g \propto g^{-1} m^0. \tag{28}$$

There are difficulties in reconciling the notion that performance number and range are scale-independent in walking and running with Schmidt-Nielsen's (1972) metabolic results (figure 2). Schmidt-Nielsen (1984) gives a simple argument leading to the conclusion that the power required to walk at a constant Froude number must scale with a power of the mass that is less than one. If the motion is geometrically similar in different animals, the stress exerted by their locomotor muscles should be constant, and so should the active strain, which is equivalent to saying that the work done in each contraction is directly proportional to the mass of muscle (and also to the body mass in geometrically similar animals). The mechanics underlying this argument are derived from Hill (1950). The mechanical power output is equal to this work multiplied by the contraction frequency. There is field evidence from Pennycuick (1975b) that the latter varies with the minus one-sixth power of the mass in wild quadrupeds in cruising locomotion. Thus

$$P \propto m^{5/6}. \tag{29}$$

This argument does not address the variation of power with gravity, although an increase of gravity would presumably increase the stress required in the muscles, and hence also the power. If we assume that the power remains proportional to gravity, as in Alexander and Jayes' theory, then it follows that

$$N = mgV/P \propto g^0 \, m^{1/3}, \tag{30}$$

$$R \propto N/g \quad \propto g^{-1} \, m^{1/3}. \tag{31}$$

The relationship of N and R to gravity are the same as under Alexander and Jayes' theory, but the relationship to body mass tallies better with the results of metabolic experiments. Taylor et al (1970) found that the latter indicate that N scales with the 0.40 power of the body mass. Some such relationship is suggested by the general observation that long, sustained, walking migrations are more characteristic of large animals than of small ones.

The above two theories postulate different relationships between power output in cruising locomotion and body mass, and are difficult to reconcile with each other, although both assume that cruising speed varies with the one-sixth power of the mass, and that power is proportional to speed. Actually no assumption about the scaling of cruising speed is necessary if power is proportional to speed, because an individual animal's performance number is then independent of speed, and a function of its mass only. The scaling of performance number with speed, as determined experimentally by Taylor et al (1970), is independent of any assumption about the scaling of cruising speed.

Table 1. Relationship of performance number and range to gravity, body mass and fluid density.

	Flying			Swimming		Walking	
Performance number	g^0	m^0	ρ^0	g^0	m^0	g^0	$m^{1/3}$
Range	g^{-1}	m^0	ρ^0	g^0	m^0	g^{-1}	$m^{1/3}$

Effects of changing gravity

The conclusions of the previous three sections are summarised in Table 1. This shows that performance number is independent of gravity in flying and walking, but proportional to gravity in swimming. Thus the fauna of some other planet, with much higher or lower surface gravity than that of Earth, should show a corresponding difference in the performance numbers of swimming animals, while those of walking and flying animals would be much the same as on Earth. It seems to be a coincidence that the line for swimming animals in Schmidt-Nielsen's (1972) graph (figure 2), falls anywhere near those for walking and flying animals. Only on a planet whose surface gravity is near that of Earth would this be true. On a planet with much higher or lower surface gravity, the line for swimming animals would move up or down in proportion, whereas those for walking and flying animals would stay where they are. The effect of gravity on range would be the reverse. The range of swimming animals would be unaffected by higher or lower gravity, whereas that of walking or flying animals would be decreased in higher gravity, and increased in lower gravity.

UPPER LIMITS OF BODY MASS

Scaling arguments show that there has to be an upper limit to the body mass of animals that are able to fly by muscle power, although such arguments cannot predict the absolute value of the limit (Pennycuick, 1969, 1975a). This type of argument addresses the trend of variation in some performance variable, when body mass is varied, without involving the absolute values of variables. The method can easily be extended to cover other variables besides body mass, such as the strength of gravity, and fluid density.

Upper limit for flying animals

The equations for power required to fly, given by Pennycuick (1975a), involve gravity and air density as well as body mass. If attention is confined to these three variables, one can write for the power required (P_r):

$$P_r \propto m^{7/6} g^{3/2} \rho^{-1/2}. \qquad (32)$$

This has to be matched by the power available from the muscles, which may be represented as the product of the work done in one contraction by the flight muscles (Q), and the contraction frequency (f):

$$P_a = Qf. \qquad (33)$$

The present argument is restricted to large animals, exerting maximum power in short bursts of anaerobic activity. In this case Q can be regarded as proportional to the mass of muscle, and hence to the body mass, although this ceases to be a good assumption in the case of muscles which are operating aerobically or at very high strain rates (Pennycuick and Rezende 1984, Ellington 1985). Q is not affected by the strength of gravity or the air density.

The contraction frequency (f) declines with body mass, but the way in which it does so depends on the conditions. Hill's (1950) assumption, that the maximum frequency is limited by the breaking stress of the tendons, leads to the relationship

$$f \propto m^{-1/3}, \qquad (34)$$

and is appropriate to maximal anaerobic exertion. In the case of cruising flight, Pennycuick (1975a) argues that in geometrically similar animals cruising at a fixed value of the lift coefficient, the flapping frequency would vary as

$$f \propto m^{-1/6}. \qquad (35)$$

These two assumptions can be taken as extreme cases for calculating the relationship of power available to body mass.

For maximal exertion: $\qquad\qquad P_a \propto m \times m^{-1/3} = m^{2/3}, \qquad (36)$

whereas for cruising flight: $\qquad P_a \propto m \times m^{-1/6} = m^{5/6}. \qquad (37)$

The effects of gravity and air density on the scaling of muscle power available are somewhat more conjectural. Of course, an individual animal probably would respond with changes of wingbeat frequency if subjected to changes in g or ρ. It does not necessarily follow that a species which evolved to be adapted to a different gravity would have a different wingbeat frequency, and if it did, the nature of the relationship is not easy to predict. We assume provisionally that power available is unaffected by changes of g or ρ. The subsequent argument can easily be modified to accommodate different assumptions. For the moment we may expand proportionality 36 thus:

$$P_a \propto m^{2/3} g^0 \rho^0. \qquad (38)$$

If the "power margin" (M) is defined as the ratio of power available to power required, we can write:

$$M = P_a/P_r \propto m^{-1/2} g^{-3/2} \rho^{1/2}, \qquad (39)$$

or, in a more convenient logarithmic form:

$$\log M = -1/2 \log(m/m_0) - 3/2 \log(g/g_0) + 1/2 \log(\rho/\rho_0) + K, \qquad (40)$$

where K is a constant. K can be eliminated if suitable values are chosen for m_0, g_0 and ρ_0. For example log M can be defined to be zero for a bird which just has sufficient power to sustain horizontal flight at the minimum power speed in maximal anaerobic exertion. Condors and other large birds are capable of doing this in Earth's gravity and sea-level air density. Living birds have been reported with masses up to 16 kg, although it is not entirely clear whether they can still fly horizontally when so loaded. We may guess that 15 kg is the limit, though as usual this is provisional, as the argument is being used to establish trends rather than absolute values. Suppose that: m_0 = 15 kg, g_0 = 9.81 m s^{-2}, ρ_0 = 1.23 kg m^{-3}. In that case, if m = m_0, g = g_0 and ρ = ρ_0, we know empirically that log M = 0 and K = 0. We can now reorganise equation 40 to show how g varies as a function of m and ρ, while still meeting the condition that the bird is just capable of flying horizontally.

$$\log(g/g_0) = 1/3 \log(\rho/\rho_0) - 1/3 \log(m/m_0). \qquad (41)$$

This form represents maximal exertion. If the argument is repeated, using proportionality 37 instead of proportionality 36 as the starting point, the corresponding result for cruising flight is

$$\log(g/g_0) = 1/3 \log(\rho/\rho_0) - 2/9 \log(m/m_0). \qquad (42)$$

Equations 41 and 42 are plotted in figure 4. The line for equation 41 passes through the origin, reflecting the starting assumption that the power margin is zero for an animal of mass m_0 (15 kg), in Earth gravity and sea-level air density. The line slopes downwards to the right of the origin, at a gradient of minus one-third, meaning that animals of greater mass can still achieve zero power margin in the same air density, provided that gravity is reduced in proportion to the cube root of the body mass.

To plot the corresponding line for cruise (equation 42, and dashed lines in figure 4), we start by guessing that a 1.5 kg animal ($m_0/10$) is the largest that has enough muscle power to cruise at its maximum range speed. The line therefore crosses the y-axis (representing g = g_0), at log m/m_0 = -1, and then slopes downward to the right at a gradient of minus two-ninths. On these assumptions, the two lines converge at a body mass about 100 m_0 (1500 kg). The upper line indicates that a creature of this impressive size would still just be able to fly at its minimum power speed, provided that gravity were reduced by a factor of $100^{-1/3}$, or 0.22. This would require a planet somewhat larger than the moon, but with an atmosphere of the same density as on Earth at sea level. However the convergence of the two lines means that this animal would also be flying at zero power margin under cruising conditions. Any further increase of body mass would mean that flight at the minimum power speed would no longer be possible, not because of structural considerations, but because of inability to beat the wings fast enough to generate enough aerodynamic force for lift and propulsion. Flight therefore becomes impractical for still heavier animals, however low the acceleration due to gravity.

In smaller animals (to the left on figure 4), the contraction frequency progressively increases, and so would the power margin if our starting assumptions were true. However empirical data on the power output of insect flight muscles, reviewed by Ellington (1985), suggest that the work done in each contraction (Q, above) is no longer proportional to the mass of muscle in insects with high wingbeat frequencies. One reason is that the strain, through which the muscle can shorten, is less in high-frequency muscles than in those of larger animals, and this in turn may be due to inherent limitations on the maximum strain rate. The difference in stress between shortening and lengthening may also decline at high frequencies, and this too reduces the work generated by unit mass of muscle. It is not very clear whether these effects lead to a simple rule for scaling power available with body mass. It seems possible from Ellington's results that the power available scales in about the same way as power required, over quite a wide

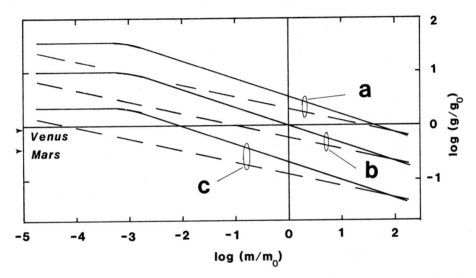

Figure 4. Gravity versus body mass for a flying animal with zero power margin, in atmospheric densities similar to (a) Venus, (b) Earth and (c) Mars.

range of body mass, i.e. that the power margin becomes scale-independent in small
animals. This is represented in figure 4 by showing the upper line curving round
at about 10^{-3} m_0, and becoming horizontal for lower masses. The lower
("cruising") line is far below the maximal-exertion line at this point, but
converges on it to the left, in the region of $10^{-5.5}$ m_0. The left-hand end of
the diagram is more conjectural than the rest, as it represents flight at Reynolds
numbers so low that the notion of "cruising at constant lift coefficient" is no
longer meaningful.

Maximum value for gravity, and effect of air density

   It can be seen that if gravity is increased, the condition that power margin be
zero can still be met if the body mass is reduced, so long as the upper line
continues to slope upwards to the left. In gravity 10 times Earth's, and the same
air density, flight would still just be possible for animals with mass about
10^{-3} m_0, i.e. about 15 g on present assumptions. Flight would still be
possible in insects, and in small birds and bats, but not in larger vertebrates. A
further increase in gravity, such that log g/g_0 > 1, would be above the
horizontal part of the curve, meaning that flight would be impossible at any body
mass, unless the whole curve can be raised. Equations 41 and 42 show that
increasing the air density has this effect, raising both the "maximum exertion"
and the cruise curves by a fixed amount. The upper pair of lines (figure 4a) have
been raised to correspond to the surface density on Venus, which is about 35 times
that on Earth. The value of Venus' gravity, which is slightly less than that of
Earth, is also marked (arrow at left). If we neglect the inhospitable temperature
and chemical composition of the Venusian atmosphere, flight with maximal exertion
would be possible for a Venusian super-condor up to a value for log m/m_0 about
1.5, i.e. up to a body mass of around 470 kg. The lower pair of lines (figure 4c)
are similarly plotted for the thin atmosphere of Mars, whose surface density is
about one hundredth that of Earth, with surface gravity about 0.38 times that of
Earth (arrow at left). On this basis flight at maximal exertion on Mars would be
possible up to a value of log m/m_0 around -0.8, that is, for a body mass up to
about 2.4 kg, whereas if Earth gravity were combined with the air density of Mars,
the largest animal able to fly would be about $m_0/100$, or 150 g.

   The hypothetical Martian animal flying in an air density one hundredth that of
Earth would have to fly ten times as fast as an animal of the same mass on Earth,
if it had the same wing geometry. Behind the assumption of proportionality 38
(that flapping frequency is not affected by air density), lies the requirement
that the wing geometry of a Martian animal would have to differ from that of
terrestrial animals in such a way that its wingbeat frequency would be same as
that of an animal of the same mass (but different geometry), adapted to fly on
Earth. This may not be a practicable requirement. If it were modified, the effect
would be a change in the vertical spacing of the three sets of lines in figure 4,
but the underlying argument would still apply. One may also note that changing the
air density does not alter the body mass at which the "maximum exertion" and
"cruise" lines intersect at the right of the diagram, unless further assumptions
are introduced. This means that a body mass of about 100 m_0 should be attainable
at any air density, provided gravity is suitably adjusted, but that no combination
of gravity and density will permit an animal heavier than this to fly.

LOCOMOTION IN EXOTIC FLUIDS

   The sharp distinction made here between flying and swimming results from the
fact that air and water differ in density by a factor of around 800 on Earth at

sea level. However, fishes such as fast-swimming pelagic sharks are appreciably denser than sea water, and support their weight by a combination of hydrostatic and hydrodynamic forces. Other planets may exist with fluids of intermediate density, such that an organism's weight is only partly supported by hydrostatic forces. For example the hydrogen atmosphere of Jupiter apparently merges imperceptibly into a liquid hydrogen "ocean", without a recognisable surface (Ingersoll, 1981). In such cases the narrow definitions of "flight" and "swimming" used here represent special cases in which the fluid density is either negligible in comparison with that of the animal, or approximately equal to it. A more general analysis would be needed to assess the effects of gravity on performance number and range for intermediate forms of locomotion. One can also envisage locomotion in a fluid much denser than the animal, as an extension of the locomotion of slightly buoyant animals such as penguins.

PHYSICAL CHANGES IN THE EARTH

Field observations of very large birds such as condors make it clear that they have little if any power margin in the present conditions of air density and gravity (Pennycuick and Scholey 1984). However, much larger flying animals have existed during at least two past epochs. The body masses of the Cretaceous giant pterosaurs are difficult to estimate because there are no living representatives for comparison, but certainly exceeded the mass of any living bird (Bramwell and Whitfield 1974). More recently, the Miocene bird _Argentavis magnificens_ was quite similar to a condor, but with about twice the linear dimensions, and hence (presumably) eight times the mass (Campbell and Tonni 1983). Equation 41 indicates that such a bird would have the same power margin as a condor if the air density were increased by a factor of 8, for instance by additional carbon dioxide or nitrogen. The surface density of the Earth's atmosphere certainly has varied in the past, but the increase required is unduly large. Alternatively, the same result would be attained if gravity were reduced by a factor of 2. It may not be coincidental that very large terrestrial animals occurred at much the same times as giant flying animals. The dinosaurs reached their largest size in Cretaceous times, while the largest known mammal, _Baluchitherium_, was a Miocene perissodactyl whose size was comparable with that of the larger dinosaurs (Romer 1945). Although there is no simple argument setting an upper limit to the size of walking animals, it seems likely that larger animals would be possible in reduced gravity, if only through simple scale effects related to the compressive stress in the leg bones. Episodes of reduced gravity in the Cretaceous and Miocene would account for the simultaneous appearance of giant walking and flying forms.

Before rejecting the notion of past changes in the strength of gravity as absurd, it may be noted that Casey (1976) has assembled a large volume of evidence that the Earth's radius has increased by about 20% since the breakup of the continents in the Jurassic. Casey lists no less than five separate mechanisms which might account for this, some of which imply major changes in surface gravity while others do not. While Casey's hypothesis of progressive expansion would not account for temporary episodes of reduced gravity, it does indicate that the possibility of drastic changes in the Earth's physical characteristics over geological time should not be lightly dismissed.

I am most grateful to Fred Schaffner, John Chardine and Laurel Duquette for their helpful comments on a preliminary draft of the manuscript.

REFERENCES

Alexander RMcN, Jayes AS (1983) A dynamic similarity hypothesis for the gaits of quadrupedal mammals. J Zool (Lond) 201:135-152.

Bernstein MH, Thomas SP, Schmidt-Nielsen K (1973) Power input during flight of the Fish Crow Corvus ossifragus. J exp Biol 58:401-410.

Bramwell CD, Whitfield GR (1974)Biomechanics of Pteranodon. Phil Trans Roy Soc Lond B 267:503-592.

Brett JR (1965) The relation of size to rate of oxygen consumption and sustained swimming speed of sockeye salmon (Oncorhynchus nerka). J Fish Res Bd Canada 22:1491-1497.

Campbell KE, Tonni EP (1983) Size and locomotion in teratorns (Aves: Teratornithidae). Auk 100:390-403.

Casey SW (1976) The expanding earth. Elsevier, Amsterdam.

Ellington CP (1985) Power and efficiency of insect flight muscle. J exp Biol 115:293-304.

Greenewalt CH (1962) Dimensional relationships for flying animals. Smithsonian Misc Collns 144, No 2, pp.1-46.

Hill AV (1950) The dimensions of animals and their muscular dynamics. Sci Progr (London) 38:209-230.

Ingersoll A (1981) Jupiter and Saturn. In: Beatty, JK, O'Leary, B, Chaikin, A (eds): The New Solar System. Cambridge University Press, Cambridge. pp. 117-128.

McMahon TA (1984) Muscles, reflexes and locomotion. Princeton Univ Press, Princeton.

Pennycuick CJ (1969) The mechanics of bird migration. Ibis 111:525-556.

Pennycuick CJ (1975a) Mechanics of flight. In: Farner DS, King JR (eds): Avian Biology. Academic Press, New York. Vol 5, pp. 1-75.

Pennycuick CJ (1975b) On the running of the gnu (Connochaetes taurinus) and other animals. J exp Biol 63:775-799.

Pennycuick CJ, Scholey KD (1984) Flight behaviour of Andean condors Vultur gryphus and turkey vultures Cathartes aura around the Paracas Peninsula, Peru. Ibis 126:253-256.

Pennycuick CJ, Rezende MA (1984) The specific power output of aerobic muscle, related to the power density of mitochondria. J exp Biol 108:377-392.

Romer AS (1945) Vertebrate Paleontology 2nd Ed. University of Chicago Press, Chicago.

Schmidt-Nielsen K (1972) Locomotion: energy cost of flying, swimming and running. Science 177: 222-228.

Schmidt-Nielsen K (1984) Scaling: why is animal size so important? Cambridge Univ Press, Cambridge.

Taylor CR, Schmidt-Nielsen K, Raab JL (1970) Scaling of energetic cost of running to body size in mammals. Amer J Physiol 219:1104-1107.

Tucker VA (1972) Metabolism during flight in the laughing gull, Larus atricilla. Amer J Physiol 222:237-245.

Part VII

ENERGY METABOLISM, TEMPERATURE AND EVOLUTION

LIVING IN TWO WORLDS: THE MARINE IGUANA, *AMBLYRHYNCHUS CRISTATUS*

George A. Bartholomew

Department of Biology, University of California
Los Angeles, California 90024
U.S.A.

Although its undersea feeding habits make the marine iguana
unique among lizards, it is strikingly similar to other herbivorous iguanids in morphology, aerobic and anaerobic energy
metabolism, preferred body temperatures, acid-base balance,
cost of terrestrial locomotion, kidney function, and capacity
to control rates of change in body temperature. Its aquatic
feeding despite very modest swimming ability, its persistant
behavioral maintenance of high body temperatures while out the
water, and the remarkable osmoregulatory capacity of its nasal
salt glands appear to be the key elements in its unique
amphibious mode of life.

INTRODUCTION

The marine iguanas of the Galapagos Islands are unique among lizards in
their behavioral adjustments to the marine littoral habitat. They feed
exclusively in subtidal and intertidal areas, primarily on marine algae, spend
the night and most of the day on shore and lay their eggs in the sand above
the high tide line. They presumably evolved their uniquely amphibious mode of
life after rafting or floating from South America to the Galapagos Islands,
the age of which is variously estimated at 3 to 9 million years (Durham and
McBirney, 1975; Bailey, 1976). No other members of the Family Iguanidae feed
underwater, although many iguanids are effective swimmers and dive to escape
terrestrial predators. The closest relative of the marine iguana, the
Galapagos land iguana (Conolophus subcristatus) which is also endemic to the
Galapagos, is completely terrestrial.

Amblyrhynchus is an attractive subject not only for the study of physiological adjustments to amphibious life, but for an evaluation of the relative
roles of physiology, morphology, and behavior in the evolution of an amphibious mode of life. It is important to note that, unlike other vertebrates
--sea turtles, sea snakes, crocodilians, seals, sea otters, penguins, and
volant sea birds-- which have secondarily invaded the sea, marine iguanas have

Comparative Physiology: Life in Water and on Land. P. Dejours, L. Bolis, C.R. Taylor, E.R. Weibel
(eds.) Fidia Research Series, IX-Liviana Press, Padova © 1987

become amphibious in the geologically recent past. Indeed, their adjustments
to amphibious life may be evolving right before our eyes.

AQUATIC VERSUS TERRESTRIAL BEHAVIOR

The Galapagos Archipelago is on the equator. However, its islands are
surrounded by cool, nutrient-rich, up-welling water and exposed for more than
half the year to the boisterous winds and heavy seas characteristic of the
southeast tradewinds. This combination of physical circumstances, in concert
with the amphibious habits of marine iguanas, results in a series of contrasting environmental conditions that are probably the most extreme encountered
by any large reptile on a daily basis.

These contrasts are best exemplified on the higher islands where the lee-
ward coasts are adiabatic deserts, usually consisting of barren lava flows.
The iguana colonies scattered along these rocky shores often contain aggregations of hundreds of individuals that are exposed hour after hour to the full
intensity of the equatorial sun while sprawled out on rocks with surface temperatures in excess of 50 C. If they are to avoid lethally high body temperatures they must employ postural adjustments that minimize both solar input and
contact with the substrate, while maximizing convective cooling (Bartholomew,
1966; White, 1973). If one prevents them from making these postural adjustments they die of overheating within a few minutes.

Marine iguanas are strictly diurnal. They usually remain within 5 to 20
meters of the water's edge. They feed exclusively in intertidal and subtidal
areas which they usually enter for an hour or so every day or on alternate
days. The rocky reefs that support the most extensive algal growth are often
exposed to heavy surf, particularly in the tradewind season. When the waves
are high iguanas may remain ashore for several consecutive days.

The diving of Amblyrhynchus differs from that of other iguanids in being
an integral component of feeding behavior rather than a mechanism for escape
from terrestrial predators. Marine iguanas dive with inflated lungs and while
swimming near the surface must work to overcome their buoyancy. However, when
feeding, even in shallow water, their buoyancy presents only a minor problem
because they cling to the rocky substratum with their long and recurved claws.
In water more than a few meters deep the air in their lungs is compressed and
they become neutrally buoyant.

Large iguanas of both sexes (body mass > 1.8 kg) feed underwater
(Trillmich and Trillmich, 1986). However, the smaller members of the population are much less aquatic. Hatchlings feed exclusively in the intertidal and
never enter the water; juveniles confine their activities mostly to tidal
flats and shallow pools; medium-size animals (females and young males) often

feed underwater, but remain near shore and in shallow water. This feeding pattern appears to be a direct consequence of body size. Individuals of all sizes feed willingly on the Ulva beds that are exposed at low tide, but these algae are heavily grazed in open areas near the iguana colonies and are concentrated in cracks and crevices that only the hatchlings are small enough to enter. Iguanas of intermediate size feed on algae in tide pools and in the wave-washed flats of the lower intertidal. The subtidal algal beds are the least heavily grazed, in part because of the extensive areas involved, and in part because they are used only by large animals which constitute a small segment of the population. These large individuals sometimes swim as much as 400 m, both offshore and along shore, to feed.

Diving Depths.--Only anecdotal information on the depths to which marine iguanas dive has been published. Skin divers have commonly seen them at depths down to 5 meters (Carpenter, 1966). Scuba divers have observed them on the bottom in substantially deeper water (Hobson, 1965). In avoiding pursuit they will dive to depths of at least 20 m (Hobson, 1969). In July, 1978 David Vleck and I captured 36 adult male iguanas after dark in a colony about 1 km southeast of Punta Espinosa on Isla Fernandina,. We marked each one with a spot of paint, weighed it, attached a disposable depth indicator to its nape with surgical sutures, and released it at the point of capture. The depth indicators consisted of glass capillary tubes dusted on the inside with powdered talc. One end of each tube was flame sealed. When submerged, water entered through the open end of the tube to a distance determined by the water pressure. The tubes were calibrated by lowering them from shipboard to a known depth, and then measuring the length of tubing from which the talc had been washed. Forty-eight hours after attaching the depth indicators we returned, recaptured the experimental animals and retrieved 20 intact depth indicators. The mean depth of the deepest dives made by the males (mass 2.71 kg +0.51 S.D.) while carrying depth gauges was 5.1 m +2.6 S.D. The deepest dive recorded was 11.7 m.

Dive Duration.--The durations of dives of 12 large males (mass, 2.1 to 3.4 kg) marked with distinctive patches of yellow road-marking paint were determined visually and timed with a stop watch by David Vleck and me. Eight of the marked animals were also fitted with intraperitoneal telemeters, and these individuals were tracked both visually and by radio as they swam and dived within 150-200 m of the colony mentioned above. We were able to obtain satisfactory data on sequences of dives by ten individuals (total dives, 52). Some of the dives appeared to be exploratory and lasted only 10 to 20 seconds and 9 dives lasted less than one minute. Excluding dives of less than 1 minute, the frequency distribution of dive duration is normal and both the mean and the median are 5.2 minutes. The mean length of all dives was 4.48 minutes + 2.4 S.D. The longest dive was 10.4 minutes. While feeding, the iguanas spent almost twice as much time under water as on the surface. The time in minutes (T) spent at the surface of the water between dives increased linearly with the duration (D) of the previous dive; $T = -1.1 + 4.5D$ ($r^2 = 0.72$, P<.001, N = 32).

Swimming.--Except while feeding or attempting to escape, swimming iguanas float at the surface with head out of water. They are graceful swimmers, but so slow that a human swimmer can capture them by hand. Their swimming is anguilliform (eel-like). They hold their limbs against their sides and develop thrust by lateral undulations of the entire body. Their morphological modifications for swimming are minor--the terminal third of the tail is somewhat flattened, and there is some webbing at the base of the toes, but these features are only slightly more developed than in other iguanids.

The tail makes a major contribution to the generation of thrust. One might predict, therefore, that the relative length of the tail would be greatest in the large individuals because they are the most dependent on swimming. However, the change in relative body proportions during growth is negligible. The linear regressions of the log-transformed values for snout-vent length, tail length, and total length on body mass all scale to the 1/3 power of body mass from 20 grams to over 4000 grams (Bartholomew, Bennett, and Dawson, 1976).

Marine iguanas swim in two modes, burst and cruise, with burst being approximately twice as fast as cruise. Like other reptiles their stamina is limited. They normally cruise at a leisurely pace and employ burst swimming only when attempting to escape. Burst swimming can be sustained for only a couple of minutes and entails substantial increases in blood lactate (see below). During burst swimming iguanas remain submerged. After tiring they either cling to the bottom or return to the surface and cruise swim. Cruise swimming is aerobically supported and can be maintained indefinitely (Bartholomew et al., 1976). While they are feeding on the bottom they cling to the rocks and, like grazing mammals, shift position by walking.

Marine iguanas swim much more slowly than do fish of the same size and their capacity to escape from aquatic predators by speed alone is extremely limited. The mean cruising speed of undisturbed adult males (body length ca. 1.0 m, body temperature ca. 21 C) is 0.42 m/s. Their mean frequency of undulation is 0.97/s, making their stride length about 40% of body length. During both burst and cruise swimming, velocity and absolute distance traveled per undulation increase directly with size, whereas body lengths/s and frequency of undulation are inversely related to body size. The cruising velocity of a marine iguana is less than 1/5 that of a sockeye salmon

(Oncorhynchus nerka) of the same size. Its burst speed is less than 1/16 that
of a wahoo (Acanthocybium solanderi) of the same length. Even when compared
with fish of similar length that swim in the anguilliform mode the burst speed
of a marine iguana is extremely modest--about 40% of that of an eel (Anguilla
sp.).

   Body temperature while swimming.--Marine iguanas do most of their feeding
at low tide when depth of water over the bottom is minimal, but a few animals
can often be seen offshore during the daytime at any stage of the tide
(Hobson, 1965; Bartholomew, 1966; Trillmich and Trillmich, 1986). If iguanas
enter the sea near sunrise their body temperatures approximate that of the
water. However, during most of the day while on shore they maintain body temperatures between 35 and 38 C (Bartholomew, 1966; White, 1973) which is at
least 10 C and sometimes 20 C higher than that of the cool, upwelling waters
in which they swim while foraging. When an iguana enters the water to feed it
shows no hesitation about moving from the air into the relative cold sea. The
temperatures of animals leaving the water after feeding are about the same as,
or slightly above, that of the water. It is a matter of both physiological
and ecological interest to determine how rapidly they cool after entering the
water because this determines the body temperatures at which they swim and
feed. Many lizards have the capacity to modulate rates of change in body temperature. Typically they heat up more rapidly than they cool which maximizes
the amount of time that they can remain near their preferred body temperature
(see Bartholomew, 1982 for a review). In both air and water the cooling rates
of marine iguanas are about half the heating rates (Bartholomew, 1966), a more
pronounced difference than usually found in lizards. This temperature hysteresis is associated with differences during heating and cooling in heart rate
(Bartholomew and Lasiewski, 1965), cutaneous vascular supply (Morgareidge and
White, 1969), and to a lesser extent rates of oxygen consumption (Bartholomew
and Vleck, 1979). The metabolic rates of marine iguanas appear to be as sensitive to temperature as those of strictly terrestrial members of the family
Iguanidae (Bennett, Dawson, and Bartholomew, 1975).

   The cooling rate of only one telemetered individual (mass 2.9 kg) under
free-ranging conditions is available. While swimming in water with a temperature of 22 C its deep body temperature fell from 36.2 to 24.9 C in 36 minutes
(Bartholomew and Vleck, unpubl). This observation is consistent with the continuously measured rates of decline in body temperature of captive iguanas of
similar size in water (Bartholomew and Lasiewski, 1965; Trillmich and
Trillmich, 1986). Within half an hour of entering the water the deep body
temperature of even the largest iguanas will have fallen to within a few
degrees of water temperature. Because of Q_{10} effects and the fact that

aerobic scope for activity of the marine iguana peaks near 35 C (Bennett, Dawson, and Bartholomew, 1975), the low body temperatures of swimming iguanas can have substantial energetic consequences (see below).

ENERGETICS OF AMPHIBIOUS ACTIVITY

A priori one could reasonably assume that the swimming and submarine feeding of Amblyrhynchus would involve special physiological adjustments including the ability to sustain prolonged periods of anaerobiosis. However, the energy metabolism of this species is very similar to that of other iguanids of similar size (Bennett et al., 1975). When attempting to avoid danger free ranging iguanas may remain submerged for as long as 30 minutes (Hobson, 1965), and under captive conditions individuals will voluntarily remain under water for 50 minutes (Bartholomew and Lasiewski, 1965), However, the longest feeding dives under natural conditions last about ten minutes. When they are forced to remain under water for 30 to 60 minutes, or forced to sustain either burst swimming or intense terrestrial activity, they develop high and immobilizing levels of blood lactate (Bartholomew, Bennett, and Dawson, 1976). However, during the normal course of daily activity, including walking, running, swimming, diving and feeding, lactate production does not appear to contribute significantly to energy expenditure of free-living marine iguanas. Iguanas returning to shore following a period of subtidal feeding have blood lactate concentrations no higher than those of resting ones (Gleeson, 1980a).

Under natural conditions the walking speed of adult marine iguanas is about 1 km/h, although they are capable of bursts of about 9 km/h. As in other vertebrates their oxygen consumption increases linearly with running velocity. At a body temperature of 35 C they can aerobically sustain walking speeds of 1 km/h for at least 20 minutes on a treadmill. A marine iguana with a mass of 2.9 kg walking 1 km/h consumes 1862 ml O_2/h (Gleeson, 1979).

The cruise swimming speed of adult marine iguanas with body temperatures near 23 C is about 1.5 km/h. This performance is aerobically supported and can be maintained for at least an hour. At this swimming speed an iguana one meter long with a mass of 2.8 kg consumes 840 ml O_2/h (Vleck, Gleeson, and Bartholomew, 1981). Thus, while cruise swimming, adult iguanas travel half again as fast as they do when walking on land and consume slightly less than half as much oxygen per unit time.

The minimum cost of transport (oxygen consumed per unit mass times distance traveled) of marine iguanas on land at a body temperature of 35 C does not differ from that of other lizards (liters O_2/kg km = $0.67M^{-0.25}$ where M is kg). Therefore, one can reasonably infer that the adaptations of marine

iguanas to underwater feeding have not diminished the efficiency of their terrestrial locomotion. The minimum cost of transport of marine iguanas with a
body temperature of 23 C while swimming (liters O_2/kg km = $0.31M^{-0.56}$
where M is kg) is substantially greater than that of fish (Vleck et al.,
1981). This is not surprising in view of their modest swimming performance,
but it is much less than their cost of walking. It requires 2.2 times as much
energy for a 2.5 kg marine iguana to walk a given distance as to swim it.

Foraging Economics.--Expressed as ml O_2/kg, the energy spent on swimming (the major cost of foraging for a marine iguana) equals the distance swum
(d) x ($0.31M^{0.44}$) where d is meters traveled and M is mass in kg. If the
quantity of algae, and hence energy, that can be harvested on one trip scales
proportionally to body mass (a reasonable assumption in view of the fact that
capacity variables such as stomach volume generally do so), the ratio of
energy gained during foraging to energy expended on swimming is directly proportional to the 0.56 power of body mass and inversely proportional to the
distance to the feeding grounds (see Vleck et al., 1981, for details). Consequently, if a 2.5 kg adult male and a 60 g hatchling were to swim the same
distance to feed, the adult, although it is is 60 times larger than the hatchling, would have to eat only 5.2 times as much algae as the hatchling to meet
the cost of swimming. Clearly, aquatic feeding is much more economical for
large iguanas than for small ones.

Based on time budgets and energy expenditures while resting, walking, and
swimming Gleeson (1979) estimated that adult marine iguanas spend only 10% of
their total daily energy budget on foraging. This remarkably low figure has
been substantiated by Nagy and Shoemaker (1984) who determined rates of food
intake and metabolism in the field using doubly labeled water. They found
that the foraging costs of adult marine iguanas were only about 8% of the
daily energy budget and that foraging efficiency (metabolizable energy gained
/ foraging energy expended) was 14.7. This value is 6 to 10 times that
reported for insectivorous lizards, but similar to that of other herbivorous
lizards, including the chuckwalla (Sauromalus obesus), a large terrestrial
iguanid (Nagy and Shoemaker, 1975).

The low total cost and the high efficiency of foraging is impressive in
view of the marine iguana's unique submarine feeding habits. The low cost of
their aquatic locomotion, which contributes to this situation, results not
only from the inherent economy of slow undulatory swimming, but the fact that
swimming marine iguanas have low body temperatures and hence relatively low
rates of energy metabolism. Whatever the contributory factors, the underwater
foraging behavior of marine iguanas is no more costly energetically than the
feeding behavior of terrestrial herbivorous lizards. From the standpoint of

energetics, the selective pressures for physiological and anatomical changes
in feeding biology should be no more intense on marine iguanas than on terrestrial herbivorous lizards.

Cost of Basking.--Basking, which envolves a series of stereotyped thermoregulatory postures (Bartholomew, 1966; White, 1973), is the most time-
consuming activity of marine iguanas. Paradoxically, the energy cost of maintaining these postures, together with the elevation of energy metabolism
associated with the high body temperatures during basking, make it much more
expensive energetically than foraging. A one kg iguana spends 56 kJ per day
while basking. Of this 30 kJ is the metabolic cost of resting at 35 C. Thus
the actual cost attributable to basking per se is about 26 kJ or 37% of the
daily energy budget of a marine iguana. This is four-and-a-half times the
daily cost of its foraging. Moreover, at night marine iguanas sometimes form
dense clusters, often several animals deep. This could signicantly reduce the
rate of nocturnal decline in body temperature of the participating individuals
(White, 1973; Boersma, 1983).

The functional significance to marine iguanas of basking cannot be
assigned with total confidence. It must be related to high body temperature
and the elevated metabolic rates associated with it. Over a range of body
masses from 70 to 4000 grams oxygen consumption of motionless iguanas during
the daytime at 35 C is 2.5 to 3.3 times that at 25 C (Bartholomew and Vleck,
1979). It was suggested (Bartholomew, Bennett, and Dawson, 1976) that basking
could favor the rapid elimination of a lactate burden if that burden were
associated with the oxygen debt commonly observed in lizards after intense
activity. Gleeson (1980b) showed that in marine iguanas recovering from
exhaustive activity, oxygen debt remains long after lactate disappears from
the blood, and suggested that total metabolic recovery from intense activity
rather than lactate removal is the function that should be favored by the high
temperatures of basking. In any event, removal of lactate burden as a routine
benefit of basking can be ruled out because of the infrequency with which
iguanas depend on anaerobiosis. However, the high temperatures of basking
should enhance digestion and perhaps also could facilitate rapid removal of
electrolytes by the nasal salt glands.

Nasal salt glands.--The electrolyte content of the diet of marine iguanas
is high, particularly in Na and Cl, and inevitably they ingest some sea water
while feeding. However, their kidney function is similar to that of other
iguanids; they have little renal capacity for concentrating electrolytes and
they excrete nitrogen predominantly as uric acid and to a lesser extent as
ammonia. Moreover, marine iguanas do not seem to differ from other lizards in

their control of acid-base balance (Ackerman and White, 1980). Their osmoregulation and electrolyte balance depend primarily on their large and active
nasal salt glands (Schmidt-Nielsen and Fange, 1958) which have the greatest
concentrating ability known in lizards (Dunson, 1969). The functional utility
of the extrarenal excretory capacity of marine iguanas has been quantitatively
assayed by feeding trials on captive animals combined with isotopic determinations of water flux in free-ranging animals (Shoemaker and Nagy, 1984).

Marine iguanas do not drink sea water, but the water influx of adults in
the field is substantial, averaging 48.2 ml/kg day. Of this, 38% is sea water
swallowed while feeding, 58% comes from the succulent algae on which they
feed, and 4% is oxidatively produced. The electrolyte output of the salt
glands is impressive. The values in m mol/kg day for K^+, Na^+, and Cl^-
were 3.6, 14.5, and 15.7 respectively. In captive marine iguanas fed marine
algae, salt gland excretion eliminated 25% of the water input, 95% of the
Na^+ and Cl^- input, and 80% of the K^+; the mmol concentrations of electrolytes were respectively, NA^+, 1170; K^+, 280; Cl^-, 1330. The quantities of Na^+ and Cl^- excreted by the salt glands of marine iguanas are
largest reported for any lizard. For example, the CL^- output of the salt
gland of marine iguanas is 1.48 mEq/100 g day which is 7 times that of the
chuckwalla, a large herbivorous desert iguanid studied by Nagy and Shoemaker
(1975). The salt glands of marine iguanas discriminate between Na^+ and K^+
—the quantity of K^+ which they excrete is typical of that of other
herbivorous lizards. Clearly, the nasal salt glands of marine iguanas constitute an effective osmoregulatory mechanism and supply one of the key functional capacities which supports their amphbious mode of life.

EVOLUTIONARY CONSIDERATIONS

Ideally science is self-correcting. When one's own work needs correcting, it is particularly satisfying to be able to do the job one's self. Ten
years ago William Dawson, Albert Bennett and I published an evolutionary
appraisal of the aquatic adaptations of the marine iguana (Dawson et al.,
1977). Our central conclusion was that the marine iguana did not occupy its
amphibious niche because of unique physiological capacities. Rather, it did
so by virtue of being a member of a widely adapted and extremely successful
reptilian lineage the functional capacities of which preadapted it for
exploiting the special combination of ecological circumstances that exist in
the Galapagos Islands--a rich and productive algal flora, a tropical terrestrial climate, and a dearth of terrestrial competitors and predators. Its
success was attributable not to the precision of its physiological adaptations, but to its effectiveness in assembling from functional capacities

shared by other members of the family Iguanidae a new, coherent, and adequate
pattern of performance.

A large amount of new information about the physiological functions of
marine iguanas (and also about the diving physiology of other vertebrates) has
accumulated during the past decade. How does our earlier interpretation look
today? Our central conclusion concerning the importance of preadaptation in
the marine iguana's performance still seems reasonable, but we accepted the
then current dogma concerning diving physiology and over estimated the importance of anaerobiosis in their under water feeding. We did not appreciate the
energy efficiency of their submarine foraging, or the magnitude of their
energy expenditure on basking, and we did not sufficiently emphasize the
importance of their impressive capacity for extrarenal osmoregulation.

The salt glands of herbivorous lizards normally excrete more K^+ than
Na^+. Presumably this a function of the relatively larger quantities of K^+
in their diet. In marine iguanas the relative intakes of these two ions is
reversed. The salt glands excrete most of the dietary K^+, but in addition
excrete the large quantities of Na^+ contained in the algae and in the sea
water swallowed while feeding underwater. The ability of the salt glands of
the marine iguana to accommodate the special osmoregulatory problems associated with its submarine feeding, osmoregulatory problems associated with its
submarine feeding, especially their ability to discriminate between Na^+ and
K^+ and to match output to input, is clearly a key physiological adaptation
of this species.

The processing of the large quantities of algae that marine iguanas
ingest is a problem of similar magnitude to obtaining it in the first place.
It involves not only digestion of a large volume of material, but also a very
substantial amount of osmoregulation. It seems probable that the prolonged
daily basking with its associated high levels of temperature and energy metabolism finds its primary role in facilitating digestion and osmoregulation.

The most important adjustment of marine iguanas to their amphibious mode
of life is their willingness, despite being relatively weak swimmers, to enter
the open sea, sometimes through heavy surf, and feed underwater at low body
temperatures in a situation where they are exposed to the hazards of pre-
dation. This performance represents an impressive extension of the aquatic
behavior found in other iguanids. Underwater feeding is fully developed only
in large adults. The possibility exists that it is largely a learned
response, but the relative contributions of learning and genetics in this
behavioral pattern, which is the key adaptation of the species to amphibious
life, is unknown.

REFERENCES

Ackerman RA, White, FN (1980) The effects of temperature on acid-base balance and ventilation of the marine iguana. Resp Physiol 39:133-147.

Bailey K (1976) Potassium-argon ages for the Galapagos Islands. Science 192: 465-467.

Bartholomew GA (1966) A field study of temperature relations in the Galapagos marine iguana. Copeia 1966: 241-250.

Bartholomew GA (1982) Physiological control of body temperature. In: Gans C, Pough FC (eds): Biology of the Reptilia Vol 12. Academic Presss, London and New York; pp. 168-211.

Bartholomew GA, Bennett AH, Dawson WR. (1976) Swimming, diving and lactate production of the marine iguana, Amblyrhynchus cristatus. Copeia 1976: 709-720.

Bartholomew GA, Lasiewski RC (1965) Heating and cooling rates, heart rate and simulated diving in the Galapagos marine iguana. Comp. Biochem. Physiol 16: 573-582.

Bartholomew GA, Vleck D (1979) The relation of oxygen consumption to body size and to heating and cooling in the Galapagos marine iguana, Amblyrhynchus cristatus. J Comp Physiol 132: 285-288.

Bennett AF, Dawson WR, Bartholomew GA (1975) Effects of activity and temperature on aerobic and anaerobic metabolism in the Galapagos marine iguana. J Comp Physiol 100: 17-329.

Boersma PD (1983) The benefit of sleeping aggregations in marine iguanas, Amblyrhynchus cristatus. Burghardt GM, Rand RS (eds): Iguanas of the world. Noyes Publ., New Jersey, pp. 292-300.

Carpenter CC (1966) The marine iguana of the Galapagos Islands, its behavior and ecology. Proc Calif Acad Sci 34: 329-376.

Dawson WR, Bartholomew GA, Bennett AF (1977) A reappraisal of the aquatic specializations of the Galapagos marine iguana (Amblyrhynchus cristatus). Evolution 31: 891-897.

Dunson WA (1969) Electrolyte excretion by the salt gland of the Galapagos marine iguana. Amer J Physiol 216: 995-1002.

Durham JW, McBirney AR (1975) Galapagos Islands. In: Fairbridge RW (ed): The encyclopedia of world regional geology. Pt 1. Dowden Hutchinson and Ross, Stroundsberg, Pa.; pp. 285-290.

Gleeson TT (1979) Foraging and transport costs in the Galapagos marine iguana, Amblyrhynchus cristatus. Physiol Zool 52: 549-557.

Gleeson TT (1980a) Metabolic recovery from exhaustive activity by a large lizard. J Appl Physiol: Resp Environ Exercise Physiol 48: 689-694.

Gleeson TT (1980b) Lactic acid production during field activity in the Galapagos marine iguana, Amblyrhynchus cristatus. Physiol Zool 53: 157–162.

Hobson ES (1965) Observations on diving in the Galapagos marine iguana, Amblyrhynchus cristatus (Bell). Copeia 1965: 249–250.

Hobson ES (1969) Remarks on aquatic habits of the Galapagos marine iguana, including submergence times, cleaning symbiosis, and the shark threat. Copeia 1969: 401–402.

Morgareidge KR and White FN (1969) Cutaneous vascular changes during heating and cooling in the Galapagos marine iguana. Nature 223: 587–591.

Nagy KA, Shoemaker VH (1975) Energy and nitrogen budgets of the free living desert lizard, Sauromalus obesus. Physiol Zool 48: 252–262.

Nagy KA, Shoemaker VH (1984) Field energetics and food consumption of the Galapagos marine iguana, Amblyrhynchus cristatus. Physiol Zool 57: 281–290.

Schmidt-Nielsen K, Fange R (1958) Salt glands in marine reptiles. Nature 182: 783–785.

Shoemaker VH, Nagy KA (1984) Osmoregulation in the Galapagos marine iguana, Amblyrhynchus cristatus. Physiol Zool 57: 291–300.

Trillmich GK, Trillmich F (1986) Foraging strategies of the marine iguana, Amblyrhynchus cristatus. Behav Ecol Sociobiol 18: 259–266.

Vleck D, Gleeson TT, Bartholomew GA (1981) Oxygen consumption during swimming in Galapagos marine iguanas and its ecological correlates. J Comp Physiol 141: 531–536.

White FN (1973) Temperature and the marine iguana--insights into reptilian thermoregulation. Comp Biochem Physiol 45A: 503–513.

STRATEGIES FOR REGULATING BRAIN AND EYE TEMPERATURES: A THERMOGENIC TISSUE IN FISH

Barbara A. Block

Duke University Zoology Department
Durham, North Carolina 27706
U.S.A.

The ability to regulate the temperature of the blood perfusing the brain and eyes is of vital importance to many species of terrestrial and aquatic vertebrates. In marine fish several strategies for brain and eye warming are apparent. The warm fish such as the tunas and mackerel sharks have elevations from 4-15°C above water temperature in the brain and eyes. To keep the brain warm these fish conserve metabolic heat with countercurrent heat exchangers in the circulation to the brain and eyes. Cold-bodied fish such as marlins and sailfish have a specialized thermogenic tissue that warms both the brain and eyes. The thermogenic tissue is a modified eye muscle that is structurally and functionally specialized for generating heat.

INTRODUCTION

Many species of mammals, birds, reptiles, and fish are able to keep the temperature of the brain and eye at a different temperature from the body core (Kilgore et al. 1973; Baker 1982; Carey 1982; Block and Carey 1985; Block 1986). For example, temperatures in the cranial cavity of the swordfish, Xiphias gladius, can remain relatively constant while the body temperature changes up to 14°C during deep descents into cold waters (Carey 1982). Many terrestrial mammals and birds have specialized vascular arrangements in the cerebral and orbital circulation that cool the blood flowing to the brain when body core temperatures rise (Baker 1979; Kilgore et al. 1979). The central nervous system in most vertebrates is more sensitive to temperature variations than other tissues, and chilling or heating the brain often has adverse effects that can be lethal (Burger et al. 1964; Prosser and Nelson 1981). Selective regulation of brain and eye temperature appears to be a common strategy in vertebrates to buffer the CNS from temperature variations. Brain and eye warming systems have evolved independently in many species of large pelagic fish. This paper examines how and why fish keep their brain and eyes

Comparative Physiology: Life in Water and on Land. P. Dejours, L. Bolis, C.R. Taylor, E.R. Weibel (eds.) Fidia Research Series, IX-Liviana Press, Padova © 1987

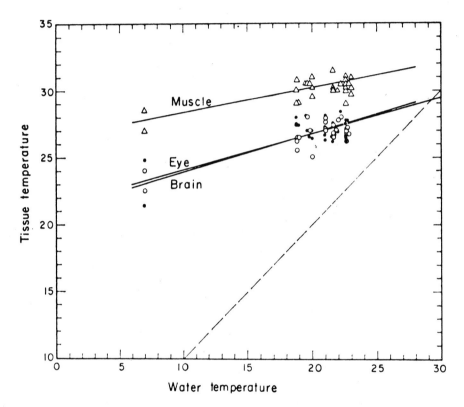

Figure 1 Relationship between tissue temperatures and water temperatures in the northern bluefin tuna, <u>Thunnus thynnus</u>. Most fish have tissue temperatures equivalent to the ambient water temperature (dashed line). Maximum muscle temperature (△), brain temperature (●), and eye temperature (o) from Linthicum and Carey (1972).

warm.

The ability to warm the brain and eyes is associated both with warm fish such as the tunas and mackerel sharks and with other species that are conventionally thought of as ectothermic, such as the istiophorid billfish (marlins, sailfish, and spearfish) and the swordfish. The tissues of most fish remain the same temperature as the water in which they swim. This is due primarily to the heat loss associated with their mode of respiration. Fish gills have a large surface area that is physically arranged to promote gas exchange by diffusion and convection. Given the high heat capacity and low oxygen content of water, the gills also act as heat exchangers. To raise tissue temperatures significantly above the ambient water temperature a fish must regulate the convective transfer of heat to the gills.

WARM BRAIN AND EYES IN TUNAS AND SHARKS

Tunas and mackerel sharks are well known for their ability to maintain warm muscle temperatures (Carey and Teal 1966; Carey 1983). In the tunas and sharks, local elevations in tissue temperatures are maintained by having vascular countercurrent heat exchangers in the circulation between the tissues and the gills. The heat exchangers are formed by retia mirabilia, specialized vascular beds arranged to provide a large area of contact between the venous and arterial blood flows. The heat exchangers are usually located around tissues such as the red swimming muscles of tunas, that have a high aerobic potential.

Warm brain temperatures in excess of 5°C above ambient water temperatures were first reported in the skipjack tuna Katsuwonus pelamis (Stevens and Fry 1971). Other species of tunas also have elevated brain and eye temperatures and the warmest measurements (up to 18°C above water temperature) have been reported by Linthicum and Carey (1972) from the northern bluefin tuna, Thunnus thynnus (Fig. 1). Temperatures in the brain and eyes of mako and porbeagle sharks (Lamnidae) are also warmer than the water they swim in (Block and Carey 1985). Maximum brain and eye temperature elevations of 6.5°C above water temperature have been recorded in the porbeagle Lamna nasus, while most pelagic sharks have brain and eye temperatures within 0.1°C of water temperature (Fig. 2a,b).

Arterial blood supplying the brain and eyes of tunas and mackerel sharks comes from the first gill arch and is thus separated from the warm muscle and visceral tissues. Given that blood passing to the brain should be cool after passage through the gills, how are these fish able to maintain such large gradients between the brain and eye temperatures and the water temperature? In bluefin tuna, Linthicum and Carey (1972) described a carotid rete supplying

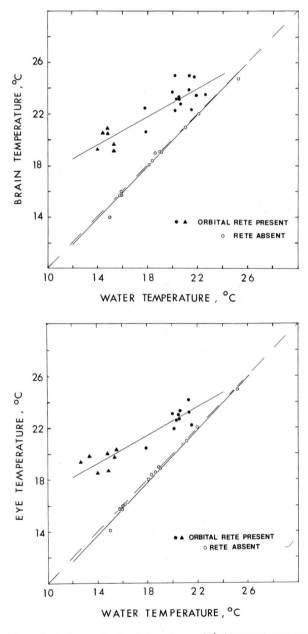

Figure 2 Relationship between brain (a) and eye (b) temperatures and surface water temperatures in mackerel sharks with an orbital rete (mako ●, por- beagle ▲) and sharks without an orbital rete (o, included here are six species of pelagic sharks). Dashed line denotes brain temperature equivalent to water temperature. Data is from Block and Carey (1985).

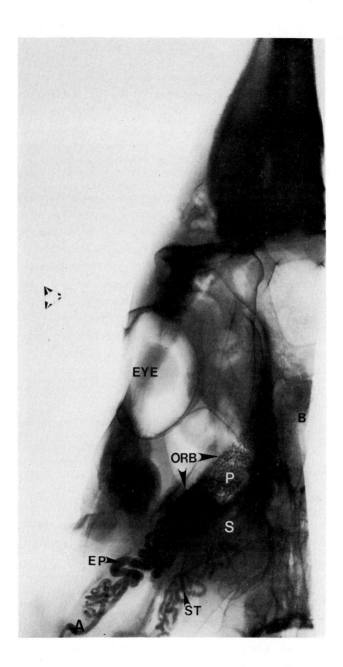

Figure 3 Radiograph of the left half of a porbeagle (<u>Lamna nasus</u>) shark in dorsal view. The arteries of the orbital rete were injected with a radio-opaque silicon and are situated in the orbital sinus. B brain; EP efferent pseudobranch; ORB orbital rete; P pseudobranchial plexus; S stapedial plexus; ST stapedial artery.

blood to the brain and suggested that this unusual vascular supply played a
key role in heat conservation in the tuna brain. The carotid rete is found in
several tuna species but is not found in other pelagic fish. Circulation to
the brain and eyes in the mackerel sharks is also different from most other
sharks (Block and Carey 1985). A large arterial plexus, the orbital rete,
passes through the orbital sinus and supplies blood to both the brain and eyes
(Fig. 3). The structure of the shark orbital rete, an arterial plexus within
a venous sinus, is remarkably similar to the arrangement of the carotid rete
in mammals. Warm brain and eye temperatures have been recorded only in the
sharks that have the orbital rete.

The elevated brain temperatures along with the elaborate vascular anatomy
in the heads of tunas and mackerel sharks indicate that heat is being
conserved. One question that has only been partially answered is where does
the heat come from? The brain and eyes are active metabolic tissues and it is
possible that the heat exchangers are conserving heat being produced in these
tissues.   In the mackerel sharks metabolic heat generated in the red swimming
muscles may also be contributing to the warm temperatures in the head region.
An unusual vascular arrangement is present and a prominent vein courses from
the red muscle into the orbital sinuses with the arterial rete (Wolf and Carey
1984). A similar "red muscle vein," allowing for the convective transfer of
warm blood from the muscle to the head, has not been found in the tunas. The
carotid rete of tunas is embedded in an unusual looking pigmented tissue that
does not appear to have thermogenic properties (Block et al. 1982). The head
region of tunas is probably warmed by conduction of heat from the large volume
of warm red muscle that comprises the body musculature.

A THERMOGENIC TISSUE IN FISH

A thermogenic tissue has recently been found in the heads of large oceanic
fish commonly known as billfish (3 genera, 9 species) and swordfish (Carey
1982; Block 1983; Block 1986). The billfish and swordfish are not warm-bodied
fish like the tunas and mackerel sharks. Only the brain and eyes are warm,
while the rest of the body remains at the ambient water temperature.   The
warm tissue temperatures in the heads of these fish are associated with the
presence of an eye muscle that is modified for heat production. Temperature
measurements in the cranial cavity of free-swimming swordfish indicate that
the "heater" tissue is capable of warming the brain up to $14^{\circ}C$ above water
temperature (Carey 1982). Temperatures in the cranial regions of billfish
captured on hook and line range from 2 to $8^{\circ}C$ above water temperature (Block
1986).

The eyes of most billfish species are quite large and the muscles which

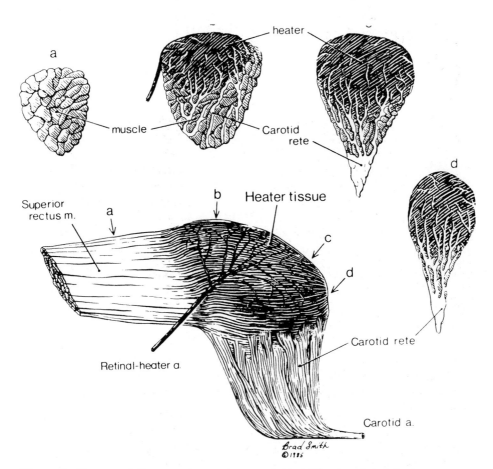

Figure 4 Diagram of the superior rectus muscle and the associated heat-
producing tissue and carotid rete from a black marlin, <u>Makaira indica</u>. Cross-
sections a-d indicate the change in morphology that occurs as the eye muscle
courses into the orbit toward the base of the braincase. This is a lateral
view and the arrows along with the letters indicate the position from which
the sections were taken. The carotid rete is a countercurrent heat exchanger
and prevents the convective dissipation of the heat being produced in the
heater tissue. The retinal-heater artery is the major outlet for blood
leaving the heater tissue. This artery supplies warm blood from the heater
tissue to the eye (from Block 1986).

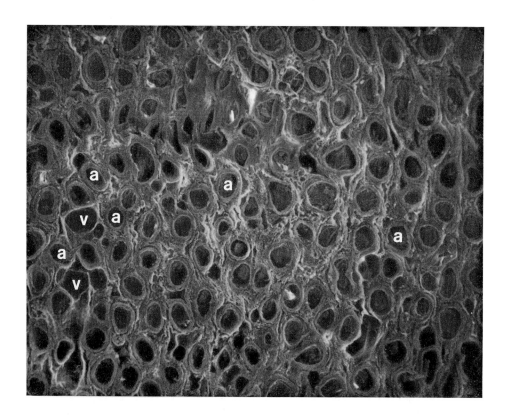

Figure 5 Scanning electron micrograph of a transverse section through the carotid rete of a blue marlin, M. nigricans. The arteries (a) have been injected with silicon prior to being fixed for electron microscopy. Most of the veins (v) have collapsed but a few can be distinguished by their thinner walls. The arterial diameters in this fish are 110 μm ± 0.18μm.

move the eye are robust and have a deep red coloration. The eye muscle associated with thermogenesis (the superior rectus) has a normal insertion on the eye; but as it courses into the orbit toward the brain, there is a structural and functional change in the muscle tissue (Fig. 4). The thermogenic portion of the eye muscle has the appearance and texture of a glandular tissue. The muscle in this region is composed of modified striated muscle cells that have few, if any, contractile proteins. The two tissues, muscle and heater, are clearly discernable but difficult to separate at their junction. A vascular countercurrent heat exchanger supplies the thermogenic portion of the muscle with blood and reduces heat loss from the brain and eye region (Fig. 5).

The change in morphology evident in the superior rectus eye muscle can be easily observed in histological preparations. The proximal portion of the eye muscle contains normal cross-striated eye muscle fibers (Fig. 6a-c) while the muscle tissue closest to the brain is packed with the heat-generating cell type. The most prominent feature of the heater cells is the dense packing of mitochondria found throughout the cell. The mitochondria can occupy up to 70% of the cell volume (Fig. 7a). The cells also have abundant smooth endoplasmic reticulum that is often stacked in multiple layers between the mitochondria (Fig. 7b,c). The membrane system is an interconnecting network of tubules and cisternae that has many structural similarities to the sarcoplasmic reticulum of normal muscle cells. Contractile filaments are rarely found in heater cells, and when they are present they are usually in disarray.

HEAT PRODUCTION IN THE FISH THERMOGENIC TISSUE

Tissues specialized solely for heat production are not common in the animal kingdom. The fish heater tissue and mammalian brown fat are the only animal tissues known to have thermogenesis as their primary function. Both tissues have large mitochondrial volumes and capillary supplies. While the two tissues share similar functions, the biochemical pathways for generating heat appear to be quite different. This is most likely related to the evolutionary origin of both tissues; brown fat is an adipose tissue and the billfish heater tissue is a modified striated muscle. Brown fat mitochondrial metabolism is distinct because oxidative energy from substrates is not stored as ATP but is entirely released as heat (Nicholls and Locke 1984). In contrast to brown fat mitochondria, the billfish heater tissue mitochondria do conserve oxidative energy as ATP (Block 1984; 1986). How then does the fish heater tissue generate heat?

As a muscle tissue homologue, heat production in the fish heater tissue may be associated with the accelerated turnover of ATP. ATP hydrolysis,

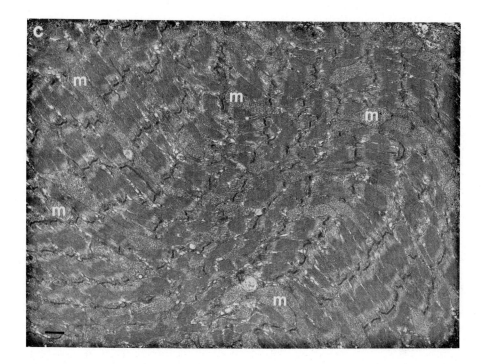

Figure 6 Electron micrographs a,b from the superior rectus muscle of a blue marlin, M. nigricans, and c from the same muscle in a sailfish, I. platypterus, prior to the transition into the heat-generating tissue. This portion of the eye muscle contains normal striated muscle fibers. As in other vertebrate eye muscles there are many different fiber types present. The sarcoplasmic reticulum, SR (arrows), is more prevalent in the fast fibers (a,b) while mitochondria, m, are the most abundant feature of the tonic oxidative fibers (c). Scale bars in a-c are 1 μm.

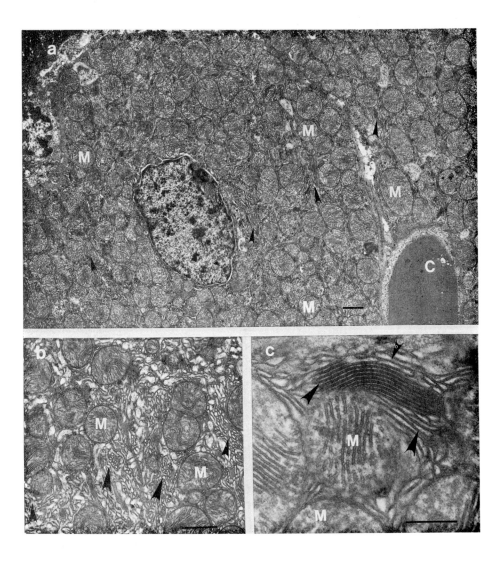

Figure 7 Electron micrographs of the heater tissue from sailfish, I.
platypterus (a,c) and a blue marlin, M. nigricans (b). The heater cells have
both mitochondria, m, and an extensive smooth endoplasmic reticulum (arrows)
but generally lack contractile proteins. The membrane system in heater cells
appears ultrastructurally and biochemically to be similar to muscle
sarcoplasmic reticulum. Occasionally the membranes are found in stacks (c).
Scale bars in a,b are 1 μm and 0.5 μm in c.

resulting in increased substrate oxidation, is a common means of generating heat in striated muscle (Hochachka 1974). Shivering thermogenesis in birds, mammals, insects, and reptiles primarily utilizes the actomyosin ATPase for heat production. In mammals, thyroid thermogenesis also generates heat in muscle because of increases in ATP turnover by the sodium, potassium ATPase (Jansky 1973; Edelman 1974). The calcium ATPase of the sarcoplasmic reticulum has also been implicated in muscle thermogenesis in several heat-generating muscular disorders (Britt 1978).

Heat production in the billfish heater tissue cannot be due to shivering thermogenesis. The loss of most of the contractile filaments in the modified muscle cells indicates that the actomyosin ATPase is contributing little to heat production. Emphasis, however, may be shifted to the heat-generating potential of the calcium pump protein (calcium ATPase) located in the sarcoplasmic reticulum (SR) of normal muscle cells. Several studies on the heater tissue provide evidence for a specific heat generating pathway involving both the mitochondria and the extensive membrane system found in heater cells.

Biochemical assays indicate that the extensive smooth membrane system (SER) of heater cells is similar to the SR of normal muscle cells (Block 1985, Block 1986). The SR of striated muscle cells regulates the contraction-relaxation cycle of muscle by rapidly releasing and reaccumulating calcium (Tada et al. 1978). Calcium uptake is mediated by the calcium ATPase that accounts for up to 90% of the total membrane protein of the SR (Meissner 1983). When SR is isolated from skeletal muscle homogenates by differential centrifugation, it forms vesicles that retain the ability to rapidly accumulate calcium against a concentration gradient when ATP, calcium, and magnesium are present (Meissner and Fleischer 1971). Isolated membrane vesicles prepared from the SER of heater cells are able to sequester calcium in a comparable fashion to isolated muscle SR prepared from the billfish superior rectus (Fig. 8). This is indicative of the high calcium-stimulated ATPase activity measured in the heater cell membrane fraction (Block 1986). Gel electrophoresis of the membrane fraction prepared from heater tissue also indicates that the membrane system has a prominent protein that runs similarly to the calcium ATPase protein of normal muscle SR (Fig. 9).

The smooth membrane system of the heater cells appears to be homologous to the sarcoplasmic reticulum of normal muscle cells. The heater smooth membranes have a large amount of the calcium pump protein. The presence in heater cells of a prominent ATPase situated next to mitochondria that can generate ATP indicates that ATP hydrolysis is the probable source of heat in the fish thermogenic tissue. To generate heat all that is necessary is to cycle calcium across the heater cell membrane system at the expense of ATP.

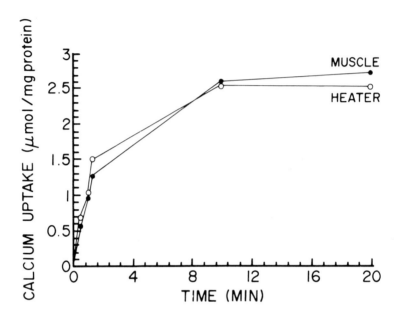

Figure 8   Calcium uptake in membrane vesicles prepared from the sarcoplasmic
reticulum of muscle (o) and the smooth endoplasmic reticulum of heater cells
(o). Calcium uptake was measured in the presence of oxalate and inhibitors of
mitochondrial respiration. Heater and muscle data are means of nine billfish.

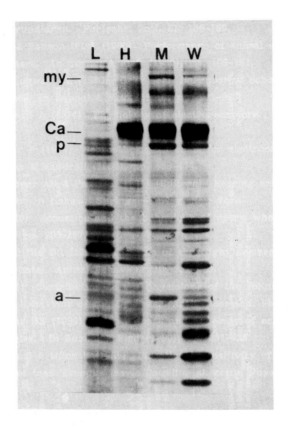

Figure 9 One-dimensional, sodium dodecyl sulfate, polyacrylamide gel electrophoresis of protein profiles prepared from the smooth endoplasmic reticulum fraction of heater and liver tissues (H,L) and the sarcoplasmic reticulum fraction of white epaxial muscle (W) and the superior rectus eye muscle (M). All samples contain the same amount of protein and are from a blue marlin, M. nigricans. Heater tissue contains a prominent band of 100 kD that runs similarly to the calcium ATPase protein band found in muscle (Ca). The prominent protein band in the heater tissue corresponding to the calcium ATPase correlates with the high calcium uptake and calcium-stimulated ATPase activity found in this tissue. Other non-contractile (mitochondria-rich) tissues such as liver do not have a similarly running protein band. Heater tissue and liver fractions do not have myosin and actin contamination (my and a). Phosphorylase (p), is prominent in muscle tissue but virtually absent in the heater tissue.

Heat production would be enhanced in this tissue, if in response to a
stimulus, the SER of the heater cells became more permeable (or leaky) to
calcium.

WHY HAVE WARM BRAIN AND EYES?

Fishermen are well aware that fish are acutely sensitive to environmental
factors such as temperature. Many fishermen use water temperature gauges to
locate water masses with the species of fish they want to catch. Changes in
temperature often influence biochemical and physiological processes in aquatic
and terrestrial animals. In fish, temperature changes and exposure to thermal
extremes have a pronounced effect on the higher functions of the fish central
nervous system (Prosser and Farhi 1965; Prosser and Nelson 1981). Rapid
cooling disrupts synaptic transmission and many fish die when exposed to acute
temperature changes (Cossins et al. 1977; Friedlander et al. 1976).   Whereas
water temperature limits the distribution of many fish, pelagic species such
as billfish, swordfish, tunas, and mako sharks do not appear to be as
restricted in their movements by temperature.   Telemetry records from
swordfish, marlins, tunas, and mako sharks indicate that these fish are able
to go through rapid temperature changes without being adversely affected
(Carey and Robison 1981; Carey 1982; Laurs et al. 1984; Holts and Dizon
personal communication). Brain and eye warming systems in fish appear to
increase their range of movement in the ocean. This is of great importance in
the open ocean where food is often scarce. Warming of the brain and eyes
enables these pelagic fish to remain active and responsive predators
throughout the wide range of temperatures they encounter in their vertical and
migratory movements (Block and Carey 1985).

The warming of the central nervous system may be the main role of the
heat generating tissue and heat conservation mechanisms in the heads of marine
fish. Billfish and tunas are active visual predators, and the warming of the
eye may confer some visual advantage. Many studies have shown that fish
vision is sensitive to temperature changes (Ali 1975; Block, Brill and
Bushnell unpublished data). Warming of the retina may decrease the visual
threshold, or in some other manner improve the ability of these fish to detect
a visual stimulus.

CONCLUSIONS

Brain and eye warming systems found in marine fish fall into two
categories. In the warm-bodied sharks and tunas, vascular heat exchangers are
found in the brain and eye regions without a specialized heat-generating

tissue. The vascular anatomy permits the conservation of heat in the cerebral and ocular tissues. A high energetic cost is associated with the elevated body temperatures in the warm fish. The absence of a thermogenic tissue in these fish is probably due to the fact that they are generating sufficient metabolic heat and need only to conserve this heat to maintain warm tissue temperatures. The billfish and swordfish do not maintain warm body temperatures. In these fish only the brain and eyes remain warm while the rest of the body is at water temperature. A highly modified eye muscle functions as a brain and eye heater tissue in these "ectothermic" fish.

Brain and eye warming systems may not be limited to tunas, billfish, and mackerel sharks. Thresher sharks (Alopiidae) and opah (Lamprididae) have peculiar vascular arrays located behind the large eyes that appear to have a function similar to the orbital and carotid retia mentioned above. The butterfly mackerel (Scombridae) has heater tissue associated with the lateral rectus eye muscle (Carey 1982). Cetaceans and some species of marine birds also have retia located within the circulation to the brain and eyes (Frost et al. 1975; McFarland et al. 1979). When a penguin makes an excursion into cold polar seas, or when a whale dives into cool deep waters in search of prey; the head becomes a major avenue of heat loss. The presence of prominent vascular mechanisms in the heads of endothermic marine birds and mammals suggests that they may also serve the valuable function of conserving heat in the brain and eyes.

This work has been supported by NIH grant HL02228 to Knut Schmidt-Nielsen. Financial support also was provided by the Lerner-Gray Fund for Marine Research, The Pacific Gamefish Research Foundation, Sigma-Xi, The Explorer's Club, and Duke University.

REFERENCES

Ali MA (1975) Temperature and vision. Rev Can Biol 340:131-136.

Baker MA (1979) A brain-cooling system in mammals. Sci Am 240:130-139.

Baker MA (1982) Brain cooling in endotherms in heat and excercise. Ann
Rev Physiol 44: 85-96.

Block BA (1983) Brain and eye heaters in the billfish. Am Zool 23:936A.

Block BA (1984) A thermogenic tissue in billfish: Do fish have brown fat?
Am Zool 24:98A.

Block BA (1985) Ca^{2+} ATPase in the billfish heater tissue. The
physiologist 28:271.

Block BA (1986) Brain and Eye Warming in Billfishes (Istiophoridae): The
Modification of Muscle into a Thermogenic Tissue. Doctoral
Dissertation, Duke University, North Carolina.

Block BA (1986) Structure of the brain and eye heater tissue in marlin
sailfish and spearfishes. J Morph, in press.

Block BA, Copeland E, Carey FG (1982) Fine structure of tissue warming the
brain and eye in tuna. Biol Bull 163:356.

Block BA, Carey FG (1985) Warm brain and eye temperatures in sharks. J
Comp Physiol 156:229-236.

Britt BA (1978) Malignant hyperthermia: A review. In: Milton AS (ed),
Handbook of Experimental Pharmacology. Springer Verlag:Germany, pp. 547-
515.

Burger FJ, Fuhrman FA (1964) Evidence of injury by heat in mammalian
tissues. Am J Physiol 206:1057-1061.

Carey FG (1983) Warm fish. In: Taylor CR, Johansen K, Bolis L (eds): A
Companion to Animal Physiology, Cambridge University Press, Cambridge,
pp. 216-233.

Carey FG (1982) A brain heater in the swordfish. Science 216:1327-1329.

Carey FG, Teal JM (1966) Heat conservation in tuna fish muscle. Proc Natl Acad Sci USA 56:1464-1469.

Carey FG, Robison BH (1981) Daily patterns in the activities of swordfish, Xiphias gladius, observed by acoustic telemetry. Fish Bull US   79:277-292.

Cossins AR, Friedlander MJ, Prosser CL (1977) Correlations between behavioral temperature adaptations of goldfish and the viscosity and fatty acid composition of their synaptic membranes. J Comp Physiol 120:109-121.

Edelman IS (1974) Thyroid thermogenesis. N Engl J Med 290:1303-1308.

Friedlander MJ, Kotchabhakdi N, Prosser CL (1976) Effects of cold and heat on behavior and cerebellar function in goldfish. J Comp Physiol 112:19-45.

Frost PH, Siegfried WR, Greenwood PJ (1975) Arterio-venous heat exchange systems in the Jackass penguin, Spheniscus demersus. J Zool London 175: 231-241.

Hochachka PW (1974) Regulation of heat production at the cellular level. Fed Proc 33:2162-2169.

Jansky L (1973) Non-shivering thermogenesis and its thermoregulatory significance. Biol Rev 48:85-132.

Kilgore DL, Bernstein MH, Schmidt-Nielsen K (1973) Brain temperature in a large bird, the rhea. Am J Physiol 225:739-742.

Kilgore DL, Boggs DF, Birchard GF (1979) Role of the rete mirabile ophthalmicum in maintaining the body-to-brain temperature difference in pigeons. J Comp Physiol 129:119-122.

Laurs RM, Fiedler PC, Montgomery DR (1984) Albacore tuna catch distributions relative to environmental features observed from satellites.

Linthicum DS, Carey FG (1972) Regulation of brain and eye temperatures by the bluefin tuna. Comp Biochem Physiol 43A:425-433.

McFarland WL, Jacobs MS, Morgan (1979) Blood supply to the brain of the dolphin, Tursiops truncatus, with comparative observations on special aspects of the cerebrovascular supply of other vertebrates. Neurosci Biobehavior Rev Suppl 3:1-93.

Meissner G (1983) Monovalent ion and calcium ion fluxes in sarcoplasmic reticulum. Molec Cell Biochem 55:65-82.

Meissner G, S Fleischer (1971) Characterization of sarcoplasmic reticulum from skeletal muscle. Biochim Biophys Acta 241:356-378.

Nicholls DG, Locke RM (1984) Thermogenic mechanisms in brown fat. Physiol Rev 64:1-65.

Prosser CL, Farhi E (1965) Effects of temperature on conditioned reflexes and on nerve conduction in fish. Z Vergl Physiol 50:91-101.

Prosser CL, Nelson CO (1981) The role of nervous systems in temperature adaptation of poikilotherms. Annu Rev Physiol 43:281-300.

Stevens ED, Fry FJ (1971) Brain and muscle temperatures in ocean caught and captive skipjack tuna. Comp Biochem Physiol 38A:203-211.

Tada M, Yamamoto T, Tonomura Y (1978) Molecular mechanism of active calcium transport by sarcoplasmic reticulum. Phys Rev 58:1-77.

Wolf NG, Carey FG (1984) Warm muscles to warm brains in lamnid sharks. Americ Zool 24: 47A.

EVOLUTION OF THE CONTROL OF BODY TEMPERATURE: IS WARMER BETTER?

Albert F. Bennett

School of Biological Sciences
University of California
Irvine, California
U.S.A.

The evolution of increased locomotor and activity
capacities has been linked with the evolution of
endothermy. Do relatively high and stable body
temperatures improve locomotor performance?
Studies on acclimation in salamanders and
interspecific adaptation to low temperatures in
lizards suggest that relatively little compensation
for the depressing effects of low temperature has
been developed. In regard to locomotion, warmer
appears to be better.

INTRODUCTION

The advantages and costs of endothermy are topics of considerable
biological interest. Metabolic regulation of body temperature
permits both a high and stable thermal regime for physiological
function; however, it does so at considerable energetic cost. The
balance of cost and benefit and the presumptive selective agents
and pathways during the evolution of endothermy have been the
subject of much speculation. Several authors (e.g., Bakker, 1971;
Crompton et al., 1978; Bennett and Ruben, 1979) have pointed out
the increased locomotor capacity associated with the development
of endothermy in several vertebrate lineages and suggested that
these constituted a significant factor in its evolution. Other
authors (e.g., Heinrich, 1977) have pointed out the greater
catalytic capacity of biochemical and physiological systems
operating at high temperatures, a capacity which in itself may
have been an important factor in setting thermostats in endotherms
at relatively high levels. In this paper, I wish to discuss

Comparative Physiology: Life in Water and on Land. P. Dejours, L. Bolis, C.R. Taylor, E.R. Weibel
(eds.) Fidia Research Series, IX-Liviana Press, Padova © 1987

briefly the intersection of these hypotheses, specifically, to examine the influence of body temperature on locomotor capacity. Is locomotor performance improved by higher body temperature or can performance be maximized at relatively low body temperatures? Do animals necessarily have increased speed and stamina with increasing body temperatures? In short, in regard to locomotor behavior, is warmer better? If so, then the locomotor advantages and the advantages associated with regulation of high temperature may not be completely separate but may have acted as synergistic selective pressures during the evolution of endothermy. I will examine the general influence of temperature on locomotor performance in ectothermic vertebrates and then examine two specific case studies of adjustment to low body temperatures and its locomotor consequences.

THE EFFECT OF BODY TEMPERATURE ON LOCOMOTOR CAPACITY

Locomotor capacity, as discussed here, refers to the highest performance (e.g., greatest burst or sustained speed) of which an animal is capable at a particular body temperature. Maximal locomotor capacity is the greatest level of performance observed at any body temperature. These locomotor capacities set an envelope of limits within which all behavior must occur and consequently form ultimate constraints on the entire behavioral repertoire of an animal. The question examined in this paper is whether maximal locomotor capacity can be obtained at low body temperatures. Endurance and burst speed capacities are considered separately.

Endurance (stamina), performance that can be sustained for more than a few minutes, is limited by the maximal oxygen consumption of an animal. If the demand for energy exceeds that which can be supplied by maximally mobilized aerobic sources, supplemental anaerobic metabolism is activated and endurance decreases rapidly. Maximal oxygen consumption in ectotherms is highly temperature dependent, with temperature coefficients (Q_{10}) generally between 2 and 3. Consequently, maximal aerobic speed, the greatest speed that can be sustained for more than a few minutes, is likewise temperature sensitive and is greatly restricted at low body temperatures. The lizard Dipsosaurus dorsalis, for example, at normally preferred and field active body temperatures of 40°C, can maintain speeds of over 13 m/min. At body temperatures of 25°C, however, it can sustain only 5 m/min (Q_{10} = 1.9) (John-Alder and Bennett, 1981). A similar thermal dependence of endurance is evident in other reptiles (Bennett, 1982) and fish (Beamish, 1978), although in the latter group, endurance may plateau or decline at the highest temperatures measured. Equivalent locomotor performance at any sustainable speed requires more energy at higher body temperatures, but this increased cost must be weighed against expanded endurance capacities.

Burst speed capacity sets the limits on escape or pursuit behavior. The thermal dependence of burst speed in the lizard Dipsosaurus dorsalis is reported in Fig. 1. At low body temperatures (25°C and below), burst speed is apparently limited by muscle twitch kinetics (Marsh and Bennett, 1985) and has a Q_{10} of >2.0. At higher body temperatures (30 to 40°C), the thermal dependence of burst speed is lower (Q_{10} = 1.3 to 1.4) but is still significant. A similar temperature effect has been found in other lizards (e.g., Bennett, 1980; Hertz et al., 1982). Burst speed in swimming animals, especially fish, may be less thermally dependent (see Beamish, 1978), but it has not been intensively investigated.

Ectotherms active at high body temperatures, such as Dipsosaurus, may thus be able to achieve nearly maximal locomotor performance. At preferred or field active body temperatures, maximal burst and aerobic speeds are attainable. But what about species active at low body temperatures? For instance, the lizard Gerrhonotus multicarinatus is normally active at 20 to 25°C, but maximal running speed and distance running capacity are attained only above 35°C (Bennett, 1980). Can animals active at low

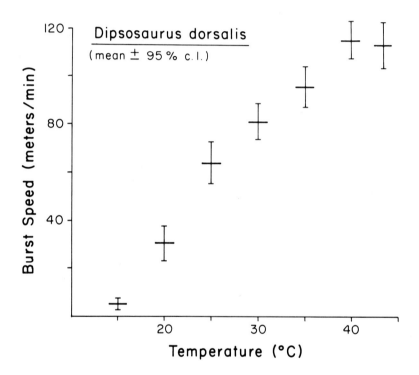

Fig. 1.   Burst speed as a function of body temperature in the
lizard <u>Dipsosaurus dorsalis</u> (data from Bennett, 1980).

temperatures achieve maximal performance at those temperatures?
The following case studies examine this issue in two different
taxa of ectotherms naturally active at low temperatures.

ACCLIMATION TO LOW BODY TEMPERATURE BY A SALAMANDER

The salamander Ambystoma tigrinum is the most geographically
widespread amphibian in North America. Adults overwinter
underground and migrate to ponds in the very early spring,
sometimes through snow. In these ponds, which still may be ice-
covered and have water temperatures of 0 to 10°C, they court,
mate, and lay eggs. Courtship behavior involves intense physical
activity, with males attempting to displace one another from a
female (Arnold, 1976). Thus locomotor performance at low
temperatures is crucial to successful reproduction in this
species.

Paul Else and I investigated whether these salamanders are able to
achieve maximal locomotor capacity at low body temperatures and
whether they undergo any behavioral or physiological compensation
(acclimation) when maintained at low temperatures. We maintained
groups of animals at 10 and 20°C for approximately one month and
measured burst speed and endurance capacities on land and in water
at 10 and 20°C for each group. Burst speeds were analyzed from

Table 1. Thermal dependence of locomotion in Ambystoma tigrinum
(unpublished data, Else and Bennett).

Performance	Q_{10} (10 to 20°C)	Probability of Acclimation
Running burst speed	1.35	0.65
Swimming burst speed	1.00	0.96
Running endurance	1.57	0.56
Swimming endurance	1.66	0.59

videotapes. Endurance was measured with a treadmill or water
tunnel; speed was increased step-wise until animals became
exhausted.

Locomotor performance capacity is generally greater at 20°C than
at 10°C (Table 1). Terrestrial burst speed and endurance in water
and on land improve significantly with increasing temperature.
Only burst swimming speed shows no significant thermal dependence
over this temperature range. Further, no significant acclimation
effect is apparent for any of the variables examined: animals
maintained at cold temperatures do not compensate with altered
locomotor abilities (Table 1). In this salamander, there is no
maximization of locomotor performance at low temperature: cold
exposure results in a significant depression of locomotor
capacities that is not compensated by long-term physiological
adjustment. These results concur generally with most other
studies on fish and amphibians (e.g., Beamish, 1978; Putnam and
Bennett, 1981; Renaud and Stevens, 1983).

EVOLUTION OF LOWER THERMAL PREFERENDA BY LIZARDS

Locomotor capacity appears to acclimate poorly if at all in
individual animals. Perhaps, however, over substantially longer
time periods, it is possible for different species to adapt to low
body temperatures and attain maximal performance at those
temperatures. In investigating such a possibility, it is
important to examine an assemblage of closely related species.
Comparative studies have been correctly criticized (e.g., Gould
and Lewontin, 1979) when they draw inferences from observations on
very distantly related animals, as historical patterns of adaptive
or developmental constraints may be significant in influencing the
results obtained.

An ideal group in which to examine interspecific thermal
adaptation is the scincid lizards of Australia, consisting of over

200 species of skinks all belonging to the same subfamily. Thermal preferenda of these species vary broadly, from 24 to 36°C (Bennett and John-Alder, 1986). Ancestral thermal preference of the group was determined by a minimum-evolution analysis based on parsimony to be approximately 32°C (Huey and Bennett, unpublished data). Consequently, several species of skinks have evolved lower preferenda, and we can ask whether these species have successfully adapted locomotor capacity to attain maximal speeds at these low temperatures.

Ray Huey and I (unpublished data) examined burst speed in 12 species of Australian skinks in 6 different genera with a computer-controlled, photocell-timed raceway. These species had thermal preferenda ranging from 24 to 36°C and were run at body temperatures ranging from 15 to 40°C. Some results of this study are reported in Table 2. The temperature at which maximal burst speed is attained has indeed decreased during the evolution of lower thermal preferenda. However, it has not changed equivalently ($T_{max\ burst\ speed} = 26.4 + 0.25\ T_p$; p slope = 1.0 is <0.001): for every 4°C decrement in preferred temperature, temperature of maximal burst speed declines only 1°C. Consequently, cryophilic species do not attain maximal burst performance at naturally experienced temperatures. Genera with

Table 2. Thermal preference (T_p) and performance in Australian skinks (unpublished data from Huey and Bennett).

Genus	# spp.	T_p (°C)	T of Maximal Burst Speed	% Maximal Speed at T_p
Ctenotus	3	35.4	35.0	94
Egernia	1	33.7	35.3	93
Leiolopisma	2	33.2	34.0	93
Sphenomorphus	3	29.8	32.1	93
Eremiascincus	1	24.4	34.0	50
Hemiergis	2	24.2	31.7	62

preferenda of >29°C attain over 90% of their maximal speeds at
their thermal preferenda; genera with preferenda <25°C achieve
only 50 to 62% of their capacity at these temperatures. Evolution
and adaptation in this group have not resulted in maximizing
locomotor performance at low temperatures. Species with low
thermal preferenda must pay a behavioral price of low locomotor
capacity at normally experienced body temperatures.

SUMMARY: WARMER IS BETTER

Relatively little information on the topic of thermal adaptation
of locomotor performance is available and consequently any
conclusions must be very tentative. At a minimum, it is safe to
conclude that complete compensation of locomotor performance at
low body temperature is not universal. It may be very rare or
absent. It was not found in individual salamanders acclimated to
different temperature. Only partial compensation was found among
different lizard genera, resulting in submaximal performance at
body temperatures normally chosen by cryophilic animals.

These observations and others in the literature suggest that, in
regard to locomotor performance, warmer may indeed be better. In
other words, animals generally do not or can not maximize
locomotor speeds at low body temperatures, even if exposure to
these temperatures is prolonged. Consequently, the evolution of
high and stable body temperatures had important consequences for
both endurance and burst speed capacities. The high levels of
performance that can be attained only at these temperatures may
have represented very significant selective advantages during the
evolution of endothermy.

ACKNOWLEDGMENTS

Financial support for this work was provided by grants from the National Science Foundation (DCB85-02218 and BSR86-00066).

REFERENCES

Arnold SJ (1976) Sexual behavior, sexual interference and sexual defense in the salamanders Ambystoma maculatum, Ambystoma tigrinum and Plethodon jordani. Z Tierpsychol 42: 247-300.

Bakker RT (1971) Dinosaur physiology and the origin of mammals. Evolution 25: 636-658.

Beamish FWH (1978) Swimming capacity. In: Hoar WS, Randall DJ (eds): Fish physiology, Vol. 7. Academic Press, New York; pp. 101-187.

Bennett AF (1980) The thermal dependence of lizard behaviour. Anim Behav 28: 752-762.

Bennett AF (1982) Energetics of activity in reptiles. In: Gans C, Pough, FH (eds): Biology of the Reptilia, Vol. 13. Academic Press, New York; pp. 155-199.

Bennett AF, John-Alder HB (1986) Thermal relations of some Australian skinks (Sauria: Scincidae). Copeia 1986: 57-64.

Bennett AF, Ruben JA (1979) Endothermy and activity in vertebrates. Science 206: 649-654.

Crompton AW, Taylor CR, Jagger JA (1978) Evolution of homeothermy in mammals. Nature 272: 333-336.

Gould SJ, Lewontin RC (1979) The spandrels of San Marco and the panglossian paradigm: A critique of the adaptationist programme. Proc Roy Soc London B 205: 581-598.

Heinrich B (1977) Why have some animals evolved to regulate a high body temperature? Am Nat 111: 623-640.

Hertz PE, Huey RB, Nevo E (1982) Fight versus flight: Body temperature influences defensive responses of lizards. Anim Behav 30: 676-679.

John-Alder HB, Bennett AF (1981) Thermal dependence of endurance, oxygen consumption, and locomotory energetics in the lizard Dipsosaurus dorsalis. Am J Physiol 241: R342-R349.

Marsh RL, Bennett AF (1985) Thermal dependence of isotonic contractile properties of skeletal muscle and sprint performance of the lizard Dipsosaurus dorsalis. J Comp Physiol 155: 541-551.

Putnam RW, Bennett AF (1981) Thermal dependence of behavioural
    performance of anuran amphibians. Anim Behav 29: 502-509.
Renaud JM, Stevens ED (1983) The extent of long-term temperature
    compensation for jumping distance in the frog, <u>Rana</u> <u>pipiens</u>,
    and the toad, <u>Bufo</u> <u>americanus</u>. Can Zool 61: 1284-1287.

BRAIN GANGLIOSIDES: NEURO-MODULATORS FOR ENVIRONMENTAL ADAPTATION

Hinrich Rahmann

Zoological Institute
University of Stuttgart-Hohenheim
7000 Stuttgart 70 (Hohenheim)
FRG

Comparative investigations of vertebrate brain gangliosides reveal distinct correlations between concentration and composition on the one side, and the phylogenetic and ontogenetic level of nervous organization on the other. In addition to this, the brain ganglioside composition is correlated with the state of environmental adaptation, among which temperature is the most effective factor. - Physico-chemical data concerning specific interactions between gangliosides and Ca^{2+} give evidence to support the hypothesis that Ca^{2+}-ganglioside complexes act as modulatory compounds for regulating short-term synaptic transmission of information and long-term environmental adaptation within the nervous system.

INTRODUCTION

The ability of the vertebrates to adapt to fluctuations in their environment is mainly based upon adaptive alterations within the nervous system, in which the synaptic terminals were found to be the most sensitive structures. Following normal use, disuse or neuronal potency the synapses show characteristic morphological adaptations (= synaptic plasticity): On ultrastructural level the nerve endings in a fishbrain, for example, following thermal adaptation, show significant alterations in the number of synaptic vesicles and in the formation of the synaptic membranes (length of contact zone, width of post-synaptic densities; Sester et al. 1984).

It is well known, that almost every stage of neuronal activity, especially electroresponsiveness, depends upon the presence of extracellular calcium. By means of newly developed electron-microscopical techniques (Probst 1986, Probst and Rahmann 1986) a parti-

cular accumulation of Ca^{2+} within the extracellular space of the
very local zone of synaptic contact was found. In addition to this
significant differences in the presence of Ca^{2+}-ATPases within the
synaptic terminals of summer-active versus winter-inactive fishbrain were detected (Sester et al. 1986), giving evidence for a
differentiated activation of Ca^{2+} depending upon neuronal activity.

On the basis of these and electrophysiological findings, the interest of neurochemical research during the last years became focussed
more and more on those compounds, which due to their physico-chemical properties, especially Ca^{2+}-interactions, and occurrence in the
CNS, might be specifically involved in Ca^{2+}-mediated neuronal processes, particularly in the short-term event of synaptic transmission of information and in long-lasting adaptive neuronal processes
(e.g. transition from aquatic to terrestrial life, thermal adaptation).

With regard to this gangliosides (glycosphingolipids containing
different numbers of negatively charged sialic- = neuraminic acids,
NeuAc; Fig. 1), are considered to be essentially involved in synaptic transmission and adaptive neuronal processes (Rahmann 1983;
Rahmann et al. 1982). This is due to both the high accumulation and
specific composition of gangliosides in synaptic membranes (Fig. 2)
and to characteristic physico-chemical properties of the amphiphilic lipids in combination with Ca^{2+} (Probst et al. 1984).

In order to obtain further evidences to support the theory of an
essential involvement of gangliosides as modulatory substances in
synaptic transmission and long-term neuronal adaptation, several
experimental approaches were undertaken in order to investigate the
occurrence, concentration and composition of brain gangliosides in
various vertebrates having developed different strategies during
phylogeny for a survival under different environmental conditions
like e.g. aquatic vs. terrestrial life or extreme temperatures.
Additionally, various in-vitro-experiments were performed in order
to study the influence especially of temperature changes on the
interaction of individual ganglioside fractions or ganglioside mixtures derived from the brain of warm or cold adapted animals with
Ca^{2+}.

MATERIALS AND METHODS

Gangliosides from whole brains (or single brain regions) were ex-

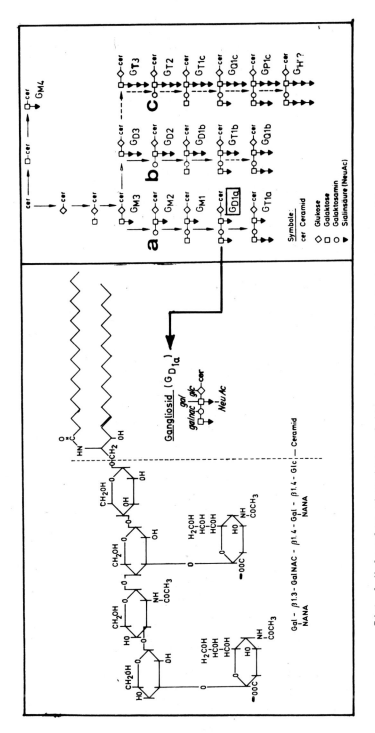

Fig. 1 Molecular structure of the major vertebrate brain ganglioside GD1a, and pathway of biosynthesis of different polar ganglioside fractions.

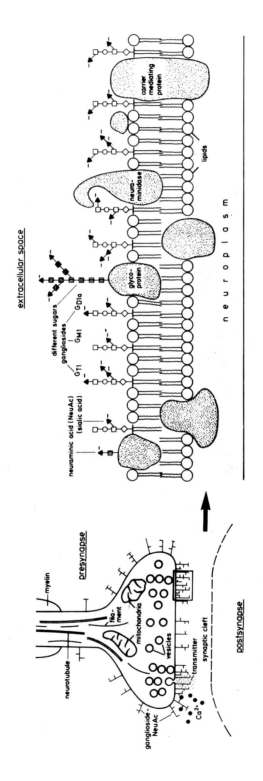

Fig. 2 Model of a synapse, demonstrating the localization of different polar gangliosides and sialo-glycoproteins in the outer lipid layer of the membrane. Gangliosides, by means of their negatively charged sialic acid residues (NeuAc), are able to form molecular clusters by cross linkages with Ca^{2+}-ions.

tracted, according to common procedures, from more than 100 vertebrate species belonging to the classes of lampreys, cartilaginous
fishes, bony fishes, amphibians, reptiles, birds and mammals. From
fish cold and warm acclimated groups and from mammals, both normo-
thermic species and hibernators were investigated. Neonatal, still
heterothermic developmental stages of homeothermic birds and
mammals were also tested. Quantitative estimations of ganglioside-
and glycoprotein-bound neuraminic (sialic) acid (NeuAc) were carried
out. Next, individual ganglioside fractions were separated by HP-
TLC (Rahmann and Hilbig 1983). For comparative calculations, the
different ganglioside fractions were arranged into three groups according to their content of 3, less than 3 or more than 3 NeuAc-
residues, which roughly corresponds with polarity. The degradations
of membrane-bound gangliosides were tested following treatment with
neuraminidase (0.15 Units) from Clostridium perfringens.

The effects of temperature changes in dwarf hamsters (Phodopus sun-
gorus) and fishes were measured more specifically analyzing physico-
chemical properties (Ca^{2+}-binding, surface pressure isotherms) from
individual ganglioside fractions and from brain ganglioside mixtures by means of monolayer studies (Probst et al. 1984).

RESULTS

Brain ganglioside composition in vertebrates which have developed
strategies for survival under different environmental conditions

During phylogeny of vertebrates, the concentration of brain gangliosides increases with progress of nervous organization and function.
In lower cold-blooded fishes, in amphibians and in reptiles, the
content varies between 110 and 800 μg ganglioside bound NeuAc/g
fresh wt. of brains. In warm-blooded birds and mammals, however, it
ranges between 400 and 1500 μg. Also, a change in the preference of
one of the three pathways of ganglioside biosynthesis (a-, b-, or c-
route) occurred and the complexity of the ganglioside composition
was significantly reduced: Among lower vertebrates, a larger number
of complex and highly polar fractions containing 4 to 5 and even
more sialic residues prevails, whereas in birds and mammals only
few, less polar fractions constitute the membrane pattern (Rahmann
and Hilbig 1983; Fig. 3).

During ontogenetical development in all vertebrate species investigated so far, striking changes in amount and composition of brain

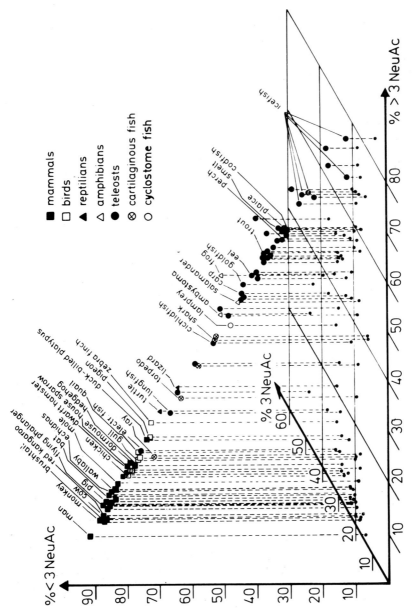

Fig. 3 Three-dimensional arrangement of brain ganglioside composition of 81 vertebrate species in relation to the polarity (3NeuAc
residues, more or less than 3). For reasons of clarity only, some
representatives have been assigned.

gangliosides were found (ref. Rahmann 1983). In amphibians minor
alterations are correlated with metamorphosis and by this with the
transition from aquatic to terrestrial life. In the axolotl (Ambystoma mexicanum), however, the changes in the brain ganglioside
composition can be artificially induced by hormone stimulated metamorphosis following long-term applications of thyroxine into the
water (Fig. 4). This data shows that the ontogenetic changes in the
brain ganglioside pattern more likely are due to endogenously induced alterations than to the transition from aquatic to terrestrial
life.

Besides these phylogenetic and ontogenetic trends, however, clear
correlations became evident when comparing the ganglioside composition to the state of thermal adaptation: "The lower the environmental ($\hat{=}$ body) temperature is, the higher the polarity ($\hat{=}$ degree
of sialylation, N-instead of N-O-acetylation) of brain gangliosides
is". This general rule had been proved correct for species which
were ecologically adapted as opposed to species in habitats with
extreme temperatures (antarctic versus tropic fish), during seasonal
acclimatization (carp, rainbow trout), hibernation (fat dormouse,
European hamster), periods of daily torpor (dsungarian hamster),
and during neonatal, still heterothermic development of birds and
mammals (Fig. 5).

In hibernators, for instance, the ratio between the two major brain
ganglioside fractions GD1a and GT1b with values of about 0.7 generally is much lower than that of homeothermic species (1.2). This,
too, indicates a higher polarity of the ganglioside mixture in
hibernators. In dwarf hamsters, this ratio shows seasonal fluctuations especially in basal brain and cortex. but not in cerebellum.
Additionally, the brain gangliosides during winter are more sensitive against neuraminidase.

From this data, it is concluded that variations in the membranous
composition of gangliosides may induce alterations in the physico-
chemical properties of the membrane. Thus the processes of synaptic transmission may become modulated in adaptation to changes in
the environment, of which temperature plays the most prominent role.

Physico-chemical aspects of Ca^{2+}-ganglioside-complexes as modulatory
systems

When discussing any modulatory properties of neuronal membrane-bound

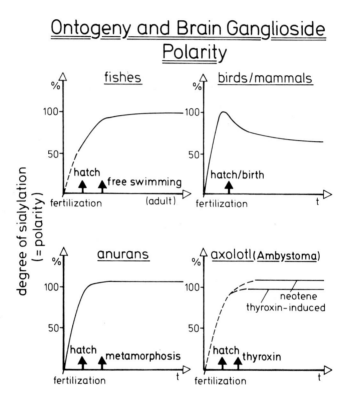

Fig. 4 Time course of changes in brain ganglioside polarity ($\hat{=}$ degree of sialylation during ontogeny of vertebrates from fertilization to adult stage. In the neotene Ambystoma mexicanum metamorphosis had been artificially induced following long-term application of 40 µg tetrajod-thyroxine-sodium salt per l of tank water.

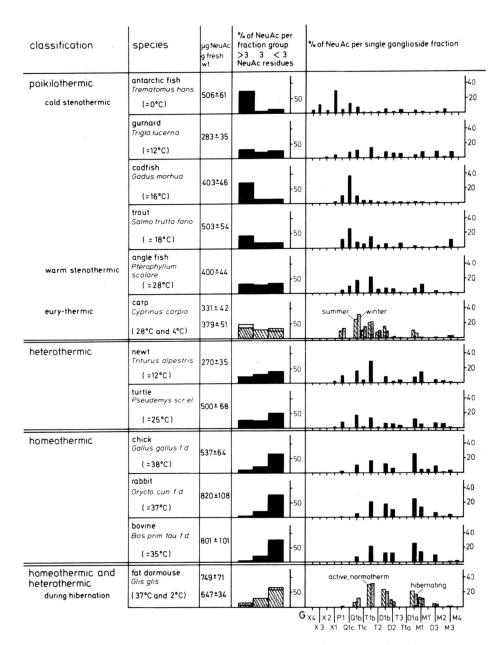

Fig. 5 Concentration, relative composition of major polarity groups
of gangliosides (comp. Fig. 3) and individual ganglioside fractions
(GM4-GX4) from several vertebrates having developed different
strategies for survival in the cold.

compounds, which may be involved in the process of synaptic transmission, it has to be emphasized that gangliosides reveal high and
specific abilities to complex with Ca^{2+}-ions, after which their
amphiphilic character is changed (Probst et al. 1984).

In addition to this, in Ca^{2+}-binding studies with individual ganglioside fractions and differently composed ganglioside mixtures
extracted from the brain of various temperature adapted fishes or
freshly-hatched, still heterothermic versus adult homeothermic
chickens, great differences in the thermo-sensitivity of Ca^{2+}-gan-
glioside-complexes were found (Probst and Rahmann 1980). In mono-
layer-studies using various membrane lipids, individual ganglioside
fractions and mixtures from summer-versus winter-adapted hamsters,
it was shown that gangliosides, in contrast to the other membrane
lipids become specifically modulated by temperature and/or Ca^{2+}-
changes (Fig. 6; Probst et al. 1984).

DISCUSSION

The phenomenal data on adaptive changes in the ganglioside composition particularly of differently temperature-adapted vertebrates
complements one another in terms of the physico-chemical results on
the thermosensitivity of Ca^{2+}-ganglioside-complexes. The results are
also in full agreement with ultrastructural adaptations in synaptic
terminals with regard to the number of transmitter vesicles, length
of synaptic contact zone, width of postsynaptic densities (Sester
et al. 1984) and with the results on differences in the occurrence
of Ca^{2+}-ATPase in pre-synapses of warm-versus cold-adapted fish
brains (Sester et al., 1986). Furthermore, this data corresponds
with the fact that, long-term adaptive changes in fish were observed in post-synaptic amplitudes of the EEG (Reckhaus and Rahmann
1983) and in the regain of lost learning ability following temperature changes (Rahmann et al. 1980).

Therefore, it is concluded that neuronal gangliosides under natural,
in vivo-conditions act as modulatory compounds at the very local
zone of synaptic contact by changing the polarity of the synaptic
membrane over longer periods of time by altering their conformational status in connection with Ca^{2+} during the short-term process
of synaptic transmission (Fig. 7). These conformational changes most
probably help to guarantee an undisturbed transmission process although the environmental conditions have changed. In summary, the
high adaptability of gangliosides in combination with Ca^{2+} as op-

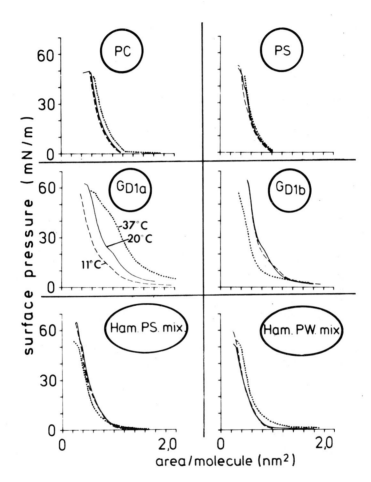

Fig. 6 Surface pressure/area isotherms of monolayers from phospha-
tidyl-choline (PC), phosphatidyl-serine (PS), GD1a- and GD1b-gan-
gliosides and ganglioside mixtures from pons of summer- and winter-
adapted dwarf hamsters at different temperatures and Ca^{2+} in the
subphase.

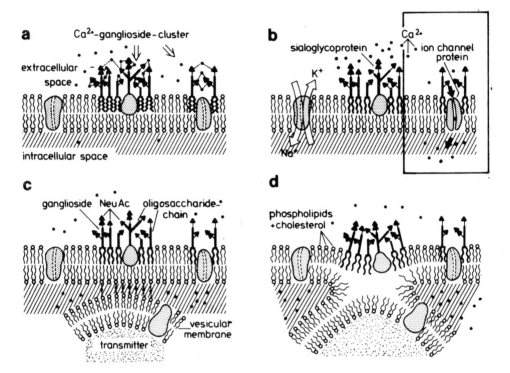

Fig. 7 Functional model of the involvement of gangliosides in synaptic transmission at the very local zone of the presynaptic membrane (comp. Fig. 2). a. Ion channels in the membrane are tightened by means of Ca^{2+}-ganglioside-complexes; b. release of Ca^{2+} by electrogenic (ionic) changes: Ca^{2+}-ions enter presynapse; c-e. fusion of transmitter vesicles with presynaptic membrane; f. activation of

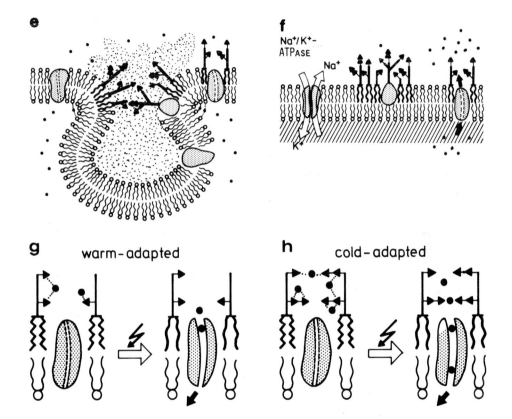

ATPases for transport; g and h. ion-channel enlarged: in warm-adapted animals 2 to 3 Ca^{2+}-ions are needed for quantal transmitter release; in cold-adapted animals 4 to 5 Ca^{2+} are necessary to guarantee this function (arrow indicates action potential) (From: Rahmann 1983).

posed to ambient changes, obviously reflects a very efficient phylogenetic strategy of the vertebrates on the molecular level within
the nervous system to adapt to changes in their environment.

REFERENCES

Probst W (1986) Ultrastructural localization of calcium in the CNS
of vertebrates. Histochemistry 85: in press.

Probst W, Rahmann H (1980) Influence of temperature changes on the
ability of gangliosides to complex with Ca^{2+}. J therm Biol 5:
243-247.

Probst W, Rahmann H (1986) Endogenes Calcium im ZNS von Wirbeltie-
ren: Ultrastrukturelle Lokalisierung. Verh Dtsch Zool Ges 79:
in press.

Probst W, Möbius D, Rahmann H (1984) Modulatory effects of different
temperatures and Ca^{2+}-concentrations on gangliosides and phospholipids in monolayers at air/water interfaces and their possible
functional role. Cell Mol Neurobiol 4:157-176.

Rahmann H (1983) Functional implication of gangliosides in synaptic
transmission (critique). Neurochem Int 5:539-547.

Rahmann H, Hilbig R (1983) Phylogenetical aspects of brain gangliosides in vertebrates. J Comp Physiol 151:215-224.

Rahmann H, Schmidt W, Schmidt B (1980) Influence of long-term acclimation on the conditionability of fish. J therm Biol 5:11-16.

Rahmann H, Probst W, Mühleisen M (1982) Gangliosides and synaptic
transmission (review). Japan J Exper Med 52:275-286.

Reckhaus W, Rahmann H (1983) Long-term thermal adaptation of evoked
potential in fish brain. J therm Biol 8:456-457.

Sester U, Probst W, Rahmann H (1984) Einfluß unterschiedlicher
Akklimationstemperaturen auf die Ultrastruktur neuronaler Synapsen
von Buntbarschen (Tilapia mariae; Cichlidae, Teleostei). J Hirn-
forsch 25:701-711.

Sester U, Probst W, Rahmann H (1986) Ultrastruktureller, enzymhisto-
chemischer Nachweis von Ca^{2+}-ATPasen im ZNS von Vertebraten. Verh
Dtsch Zool Ges 79: in press.

PHYSIOLOGICAL AND ECOLOGICAL CORRELATES
OF TEMPORALLY LIMITED RESOURCES
IN THE POLAR REGIONS

William M. Hamner

Department of Biology
University of California
Los Angeles, CA 90024
U.S.A.

The short pulse of primary production in polar regions is
followed by a disproportionately long period when energy
is unavailable. Phyletically diverse animals that depend
on seasonally available resources have converged on a
limited number of physiological, morphological, and
behavioral adaptations to this cycle. Topographic and
oceanographic features of each polar region have strongly
influenced specific adaptations to polar climates.

INTRODUCTION: Most physiological considerations of temperature and energy have a
diurnal time base because in the tropical or temperate habitats where most of us
work the dominant temperature signal is diurnal. In the polar regions the
temperature cycle of importance has an annual periodicity and for polar animals
diurnal physiological adjustments to changes in temperature are far less
important than are annual patterns of acclimatization or adaptations of
morphology and behavior (Clarke 1983). Annual time budget analysis requires
balancing an annual expense account of energy gained each summer against the cost
of survival each winter (Walsberg 1983), but time budget analysis has seldom been
applied to polar species over an entire year. I have not computed annual energy
budgets, but rather I indicate for which groups of organisms I think those
calculations would prove most interesting. I compare the physical and biological
characteristics of the north and south polar regions and then address the
question of how polar animals adjust to the brief but intense pulse of energy
each short summer and how they then survive the very long winters.

COMPARISON OF THE POLAR REGIONS: Seasonal changes in solar energy produce a
dramatic temperature cycle at both poles. The shape of this cycle is skewed by
the unequal latent heats of fusion and evaporation of ice and its amplitude is
damped by the high heat capacity of sea water. Persistent ice shades the
surface, resulting in a brief growing season. The north and south polar regions

Comparative Physiology: Life in Water and on Land. P. Dejours, L. Bolis, C.R. Taylor, E.R. Weibel
(eds.) Fidia Research Series, IX-Liviana Press, Padova © 1987

are thus climatically similar but topographically quite different (Table 1).

	ARCTIC		ANTARCTIC
1.	Ocean enclosed by land	1.	Ocean surrounds a continent
2.	Mostly isolated from oceans to south	2.	Convergent with oceans to north
3.	Surface flow predominantly eastward	3.	Surface flow circumpolar
4.	Pack ice thick ($\pm$3.5 m), multiyear	4.	Pack ice thin ($\pm$1.5 m), 1 year old
5.	Pack ice cover varies 20% seasonally	5.	Seasonal pack ice varies up to 10X
6.	Broad continental shelf	6.	Narrow continental shelf
7.	Rivers dilute surface waters	7.	No rivers
8.	Strong stratification; water stable	8.	Limited stratification; much mixing
9.	Nutrient levels high but patchy	9.	Nutrient levels high, not limiting
10.	Primary productivity high, patchy	10.	P.P. locally high; follows ice
11.	Zooplankton assemblages varied	11.	Euphausia superba dominant
12.	Epipelagic fish diverse, abundant	12.	Epipelagic fish not important
13.	Many pelagic-benthic interactions	13.	Limited pelagic-benthic interaction
14.	Many terrestrial animals	14.	Almost no terrestrial animal life
15.	Homeotherms abundant, important	15.	Homeotherms abundant, important

Table 1. Comparison of Arctic and Antarctic regions (modified
from Hempel 1985, Knox and Lowry 1977, Nemoto and Harrison 1981)

Antarctica is a continent covered by glaciers and surrounded by sea. The
Antarctic Convergence delimits the northern extent of the Antarctic region (von
Arx 1962). During the winter sea ice extends hundreds of miles north across the
Southern Ocean, occasionally blocking half the Drake Passage, but in summer the
retreating ice exposes vast expanses of open ocean. In contrast the Arctic is
predominantly an ocean almost permanently covered by sea ice. Arctic waters are
contained by continental boundaries, although there is limited exchange with the
Bering Sea and the North Atlantic Ocean. In winter the sea ice in the western
Arctic extends southward into the northern Bering Sea and in the eastern Arctic
southward to Spitzbergen, Labrador, and the Gulf of Saint Lawrence, not reaching
as far south in the eastern Atlantic because of the Gulf Stream. In summer much
ice melts but the bulk of the Arctic Ocean remains ice covered all year.

Not only is the areal extent of the sea ice different but the quality of the ice
also differs. Most Arctic pack ice does not melt annually but remains thick and
multilayered. Light does not penetrate this ice as readily as Antarctic ice and
epontic ice algal assemblages underneath Arctic ice are less well developed. In
spring the surface waters of the Arctic Ocean, diluted by nutrient-rich rivers,
stratify strongly, and primary productivity is shallow, patchy, and brief. In
the Arctic summer the extent of open water is modest, the fetch reduced, and the

Arctic Ocean is relatively calm. In the Antarctic the West Wind Drift has an
infinite fetch around the continent and the Southern Ocean is usually well mixed,
but in spring fresh melt water temporarily stabilizes the water column,
permitting a burst of primary production that follows the rapidly retreating ice
southward (Smith and Nelson 1986). Only those animals that can migrate by flying
or swimming can track and feed continuously on this travelling bloom (Frazer and
Ainley 1986; Kanda et al. 1982).

The deep parts of the Arctic Ocean are beneath the permanent polar ice cap. When
sea ice retreats in summer it is primarily the water column over the shallow
continental shelves that is exposed to sunlight. The organisms that inhabit
these shallow seas must be adapted not only to pelagic conditions but also to the
nearby benthic environment. In contrast the narrow continental shelf around
Antarctica is mostly covered with permanent ice and pelagic-benthic interactions
are not as significant because of the great depth of the Antarctic water column.

One of the more striking faunistic differences between the polar regions is the
virtual absence in the Antarctic of terrestrial animals. The absence of
terrestrial predators provides a refuge above water for many species of aquatic
vertebrates. Polar bears and Arctic foxes would devastate the Antarctic
ecosystem. A second important faunistic difference relates to the large number
of species and families of bony fishes which occur in the Arctic but which are
absent from the Antarctic ....cods, sculpins, herring, flounders, salmon, and
eelpouts. The rich Arctic ichthyofauna exhibits much of the character of the
North Pacific and North Atlantic and endemism at the Family level is low. In
contrast Antarctic waters contain many endemic, quite primitive, bony fish.
Almost all of these are benthic (Targett 1981). Only two species are epipelagic
and neither commands much ecological importance.

The most important faunistic difference between the polar regions, however,
relates to fundamental differences in their zooplankton communities (Kawamura
1980; Longhurst et al. 1984). Phytoplankton growth is limited to the surface in
the Arctic Ocean, mixing is low, and herbivorous zooplankton must come to the
surface to feed. Most Arctic copepods, euphausiids, and larval herring and
capelin stay near the surface during the short polar summer in schools or dense
swarms, forming vast patches of plankton that ripple the surface, producing
Scoresby's "meadows of brit" (1820); other species stratify at depth (Longhurst
et al. 1984). Gelatinous predators (ctenophores and medusae) are important in
the Arctic (Zenkevitch 1963; Hamner pers. obs.), partly because continental
shelves are ideal habitats for animals like coelenterates that alternate between
benthic polyps and pelagic medusae. In Antarctic waters the dominant species in

the water column is <u>Euphausia superba</u>, the Antarctic krill, which aggregates in dense patchy schools near the surface (Hamner 1984; Hamner et al. 1983). This domination of the Antarctic's water column by E. <u>superba</u> is the single most dramatic difference between the northern and southern polar seas. Other euphausiids and copepods also occur in the Southern Ocean, yet none of these so markedly affects the food chain.

ADAPTATIONS TO THE POLAR ENVIRONMENT: The most singular faunistic similarity between the polar regions is the abundance of homeothermic vertebrates... sea birds, seals, and cetaceans (Laws 1977; Remmert 1980). Many of these are carnivores that exploit foods that are available in the polar regions throughout the year, such as squid, but other animals are absolutely dependent on seasonally available foods from the base of the food chain. The annual time budgets of these seasonal consumers are of particular interest because they must be able to ingest enough food in a period often as short as two months in order to survive without food for the rest of the year. This is not always easy. Animals can increase their ingestion rates by increasing the amount that they swallow, by increasing body size, and by processing foods at the lowest possible level of the food chain. However, food can be ingested only as fast as it can be processed, and ultimately assimilation rates and growth rates limit the amount consumed.

Irrespective of how clever they may be at acquiring food during the summer, polar animals that depend on seasonally available foods inevitably must survive a long time before food is available again. These animals can survive the winter in only four ways: migrating, starving, shifting diets, or changing state (torpor, diapause, alternation of generations). Topography places constraints on which of these adaptations is possible and the evolutionary history of specific animal groups has further limited the types of possible adaptations. For example, only birds, insects, and aquatic animals routinely engage in long-distance migrations. These groups inhabit 3-dimensional media that permit directional movement relatively unimpeded by topographic barriers, and it is not surprising that birds and marine mammals are disproportionately represented in the polar regions. Topographic barriers create so many detours that it is usually not worth the energy to migrate by walking. The 3-dimensional nature of the sea and the air and the 2-dimensional nature of land is a contrast so obvious that it seldom receives attention, yet the spatial dimensionality of the environment has been critical to the evolution of physiological and ecological polar adaptations. Evolutionarily only birds, bats, and insects (two phyla) can fly, but at sea animals from every phylum migrate freely, both vertically and horizontally, physically constrained only by size, cross-sectional frontal area, and Reynold's number. The interrelations of migration and periods of fasting are

shown in Figure 1. At the top are depicted the extent and phases of the Arctic
ice cover, the radiation that reaches the water, and seasonal cycles of salts
(nutrients), phytoplankton, and zooplankton. Below, six months out of phase, is
the Antarctic ice cover and in between are the vertical migrations of the Arctic
copepod, Calanus glacialis, and the horizontal migrations of the Antarctic fin
whale, Balaenoptera physalis.

Polar copepods differentiate sexually during the winter at depth and in spring
migrate to the surface to spawn (Tandel et al. 1985). The copepodites grow
rapidly but there is not enough time the first summer both to complete the life
cycle and to accumulate enough energy reserves for overwintering. In the fall
the copepodites return to deep water and spend the winter metabolically quiescent
as Stage IV copepodites. During the second summer Stage IV copepodites do not
mature but instead accumulate a very large wax reserve to survive the next winter
and to provide energy for the development of eggs and sperm. Thus two years are
required to complete one developmental cycle... the first devoted mostly to
protein metabolism and the second almost entirely to wax metabolism. Annual time
budget analyses of polar copepods have not received much attention (but see
Conover 1962; McLaren 1963).

The annual energy budget of baleen whales is qualitatively similar to that of
polar copepods (Mackintosh 1965). Southern fins copulate in the subtropics in
the austral fall. Pregnant females migrate to the Antarctic feeding grounds
early and leave late, accumulating fat as well as the proteins necessary for
fetal development. On returning north females give birth and remain in warm
water for several months longer than in the prior year, suckling calves before
returning south with the young. In Antarctic waters calves are weaned and then
feed themselves; lean females recover body fats in preparation for the next
breeding cycle. Most species of baleen whales live off their fat for much of the
year. Brody (1975) and Lockyer (1976) have published annual time budgets for
baleen whales, but their arguments are necessarily tentative because of the
difficulty in collecting data for such enormous animals. A doubly-labelled water
study at current prices could cost $11,400,000 per injection (Nagy pers. comm.).

Many polar animals do not store fats. Coelenterates and ctenophores respond to
starvation by active degrowth of the entire body. During degrowth the animal is
not physically distorted but simply becomes smaller and physically
indistinguishable from much younger individuals actively engaged in normal
constructive anabolic processes. Catabolism usually refers to "destructive"
processes but degrowth is neither destructive nor undirected. Degrowth differs
from the processes associated with starvation-induced fat metabolism because in

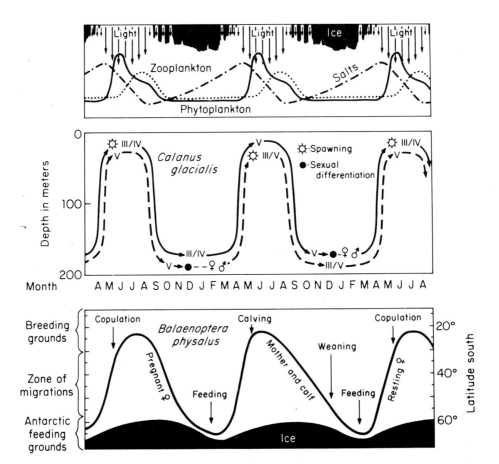

Figure 1. Top: Cartoon of 2.5-year cycle of ice, light, and production in the Arctic (redrawn from Zenkevitch 1963). Middle: Vertical migrations and life history of an Arctic copepod (redrawn from Tandel et al. 1985). Bottom: Seasonal latitudinal migrations of the southern fin whale and annual extent of Antarctic pack ice (redrawn from Mackintosh 1965).

degrowth the organism remains fully functional although all of the tissues and organ systems in the body are reduced proportionately in both number and type. Degrowth of the medusa Aurelia, a cosmopolitan genus, illustrates degrowth (Hamner and Jenssen 1974; Figure 2). Degrowth in Aurelia was first investigated by deBeer and Huxley (1924). They observed that "... the medusa Aurelia, when left without food, underwent a remarkable decrease in size...," and that "... Our case shows very clearly the extraordinary power possessed by many lower organisms of living on their own tissues - using their capital as income - when starved." As we will see, degrowth is not limited to "lower" organisms, but is also important to "higher" invertebrates. Medusae can also engage in a second pattern of winter survival by alternation of generations, rapidly growing as jellyfish during summer and overwintering as hydroids or scyphistomae, tiny stages with modest metabolic requirements. Alternation of generations between a pelagic and a benthic phase works most effectively if the benthos is not too far from the surface, as is true in the Arctic with its wide continental shelves and correspondingly rich pelagic coelenterate fauna. In contrast, Antarctic waters are deep and do not contain many medusae. A bioenergetic approach to the subject of alternation of generations would be fascinating.

Many species use a combination of adaptations to survive the winter. For example, the Antarctic krill, Euphausia superba, utilizes three different adaptations to survive the winter ... degrowth, change of diet, and migration. Krill do not store fats during the winter but, like jellyfish, can degrow when starved, getting progressively smaller with each molt yet retaining all the attributes of a normal animal (Ikeda and Dixon 1982; Figure 3). Ikeda (1985) believes degrowth to be important, but he argues also that krill must supplement their diet during the winter. If krill starved every winter, the upper size asymptote for the theoretical population growth curve, the level where growth in summer would just match degrowth in winter, would be only 4 cm length. But at sea E. superba regularly reach adult lengths longer than 6 cm and consequently these individuals must eat alternative foods during winter when the water column is devoid of phytoplankton.

The dominance of Euphausia superba in the Southern Ocean is clearly related to the flexibility of its feeding behavior. In summer krill capture algae by engulfment with a filter basket (Barkley 1940) formed by the thoracic legs covered with setae and setules, with a spacing between the setules of about 7 , permitting krill to retain even quite small diatoms (McClatchie and Boyd 1983). Antarctic krill are also omnivorous, eating copepods and even each other. In the Antarctic winter, however, when planktonic diatoms are not present, juvenile and adult krill forage on algae growing on the undersurface of sea ice both in

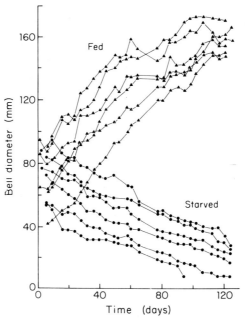

Figure 2. Growth when fed and degrowth during starvation of the medusa <u>Aurelia</u> <u>aurita</u> in the laboratory (after Hamner and Jenssen 1974).

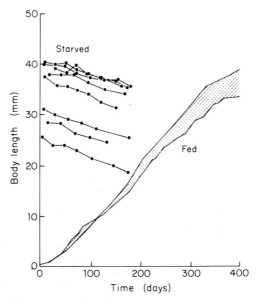

Figure 3. Degrowth and growth of <u>Euphausia</u> <u>superba</u> in the laboratory (after Ikeda 1985, Ikeda and Dixon 1982).

aquaria (Hamner et al. 1983) and in the sea (pers. obs.). In the laboratory
adult krill can harvest enough epontic ice algae in ten seconds to satisfy their
metabolic requirements for an entire day. Clearly ice algae are an important
alternative diet during the Antarctic winter. The dominance of Antarctic krill
is also related to their seasonal migrations. When the sea ice retreats, the
seasonal pulse of phytoplankton moves poleward and so must herbivores if they are
to continue to feed. Krill schools can follow the bloom because they are
constantly moving horizontally, travelling steadily at 10–20 cm sec^{-1} (Hamner
et al. 1983) and moving 8 to 17 km day^{-1} (Kanda et al. 1982), great distances
for animals only 6 cm long.

In terms of its biomass, perhaps greater than the entire world's fishery for all
other marine species combined, Euphausia superba may be the dominant organism on
earth. Our comparison of physical attributes of the Arctic and Antarctic regions
(Table 1) indicates why this particular crustacean so dominates the Southern
Ocean. The continuous circulation of the Southern Ocean, the thinness of the
Antarctic pack ice, the presence of epontic ice algae, and the almost complete
seasonal melt of the pack ice each year produce a set of environmental conditions
that can be best exploited by a large, schooling herbivore which can track the
summer bloom long distances via horizontal migration and which can utilize
alternative foods in winter. An annual energy budget for krill certainly is
needed, but this will be most difficult to compute. Degrowth makes age
determination by size class analysis impossible and schooling makes it difficult
to assess individual metabolic needs.

For krill and copepods sufficiently detailed data to compute an annual
time-energy budget may prove impossible to obtain, partly because we can't follow
such small individual animals at sea for long periods and partly because time
budget calculations for these organisms will probably always be statistical
statements derived from measurements of populations. For polar animals of any
size, however, realistic annual time budget analysis demands detailed information
in space and time of the physiological requirements and behavior of individuals
(Walsberg 1983). In order to achieve this end we need much more information
about the physiology of polar species (Clarke 1983).

REFERENCES

Barkley E (1940) Nahrung und filterapparat des walkrebschens E. superba Dana. J
    Fisch Beiheft 1: 65–156.
Brodie PF (1975) Cetacean energetics, an overview of intraspecific size
    variation. Ecology 56: 152–161.

Clarke A (1983) Life in cold water: The physiological ecology of polar marine
  ectotherms. Oceanogr Mar Biol Ann Rev 21: 341–453.

Conover RJ (1962) Metabolism and growth in Calanus hyperboreus in relation to its
  life cycle. J Cons Int Expl Mer 153: 190–197.

deBeer GR, Huxley JS (1924) Studies in dedifferentiation. V. Dedifferentiation
  and reduction in Aurelia. Quart J Microscop Sci 68: 471–479.

Fraser WR, Ainley DG (1986) Ice edges and seabird occurrence in Antarctica.
  BioScience 36: 258–263.

Hamner WM (1984) Aspects of schooling in Euphausia superba. J Crust Biol 4 (spec
  no 1): 67–74.

Hamner WM, Hamner PP, Strand SW, Gilmer RW (1983) Behavior of Antarctic krill,
  Euphausia superba: Chemoreception, feeding, schooling, and molting. Science
  220: 433–435.

Hamner WM, Jenssen RM (1974) Growth, degrowth, and irreversible cell
  differentiation in Aurelia aurita. Am Zool 14: 833–849.

Hempel, G (1985) On the biology of polar seas, particularly the Southern Ocean.
  In: Gray JS, Christiansen ME (eds): Marine biology of polar regions and effects
  of stress on marine organisms. Wiley & Sons, New York; pp. 3–33.

Ikeda T (1985) Life history of Antarctic krill, Euphausia superba: A new look
  from an experimental approach. Bull Mar Sci 37: 599–608.

Ikeda T, Dixon P (1982) Body shrinkage as a possible over-wintering mechanism of
  the Antarctic krill, Euphausia superba Dana. J Exp Mar Biol Ecol 62: 143–151.

Kanda K, Takagi K, Seki Y (1982) Movement of the larger swarms of Antarctic
  krill, Euphausia superba population off Enderby Land during 1976–77 season. J
  Tokyo Univ Fish 68: 25–42.

Kawamura A (1980) A review of food of balaenopterid whales. Sci Rep Whale Res
  Inst 32: 155–197.

Knox GA and Lowry JK (1977) A comparison between the benthos of the Southern
  Ocean and the North Polar Ocean with special reference to the Amphipoda and the
  Polychaeta. In: Dunbar MJ (ed): Polar oceans. Arctic Inst North Am, Calgary;
  pp. 423–462.

Laws RM (1977) The significance of vertebrates in the Antarctic marine ecosystem.
  In: Llano GA (ed): Adaptations within Antarctic ecosystems: Proc 3rd SCAR symp
  Antarctic biol. Smithsonian Inst, Washington; pp. 411–438.

Lockyer C (1976) Estimates of growth and energy budgets of large baleen whales
  from the southern hemisphere. UN FAO, Scientific Consultation on Marine
  Mammals, Bergen, Norway, 31 Aug–9 Sept 1976, doc ACMRR/MM/SC/41.

Longhurst A, Sameoto D, Herman A (1984) Vertical distribution of Arctic
  zooplankton in summer: Eastern Canadian archipelago. J Plankton Res 6: 137–168.

Mackintosh NA (1965) The stocks of whales, Fishing News (Books), London.

McClatchie S and Boyd CM (1983) Morphological study of sieve efficiencies and
    mandibular surfaces in the Antarctic krill, Euphausia superba. Can J Fish Aquat
    Sci 40: 955–967.
McLaren IA (1963) Effects of temperature on growth of zooplankton, and the
    adaptive value of vertical migration. J Fish Res Board Can 20: 685–727.
Nemoto, T, Harrison G (1981) High latitude ecosystems. In: Longhurst AR (ed):
    Analysis of marine ecosystems. Academic Press, New York; pp. 95–126.
Remmert H (1980) Arctic animal ecology, Springer-Verlag, Berlin.
Scoresby W (1820) An account of the Arctic regions with a history and
    description of the northern whale fishery, vols. 1, 2. 1969 edition, David &
    Charles Reprints, London.
Smith WO Jr, Nelson DM (1986) Importance of ice edge phytoplankton production in
    the Southern Ocean. BioScience 36: 251–257.
Tandel KS, Hassel A, Slagstad D (1985) Gonad maturation and possible life cycle
    strategies. In: Gray JS, Christiansen ME (eds): Marine biology of polar regions
    and effects of stress on marine organisms. Wiley & Sons, New York; pp. 141–155.
Targett TE (1981) Trophic ecology and structure of coastal Antarctic fish
    communities. Mar Ecol Prog Ser 4: 243–263.
von Arx WS (1962) An introduction to physical oceanography, Addison-Wesley,
    London.
Walsberg GE (1983) Avian ecological energetics. In: Farner DS, King JR, Parkes KC
    (eds): Avian biology, vol 7. Academic Press, New York; pp. 135–158.
Zenkevitch, L (1963) Biology of the seas of the USSR, George Allen & Unwin,
    Winchester, Mass.

A REAPPRAISAL OF DIVING PHYSIOLOGY: SEALS AND PENGUINS

Gerald L. Kooyman

Physiological Research Laboratory
Scripps Institution of Oceanography
University of California, San Diego
La Jolla, California
U.S.A.

Three major topics are discussed briefly: effects of compression, hypometabolism, and heart rate. The main topic is blood flow distribution, with special emphasis on muscle circulation. Evidence for the pattern of muscle blood flow consists of variations in blood lactate concentration [LA], and on the decay in specific activity of injected isotopes. High blood [LA] declines at different rates, which appear to be dependent on the dive behavior of the seal. Specific activity decay shows a pattern during dives that is more similar to forced submersions than to resting animals. A model is proposed in which blood flow distribution is controlled differently for extended and short duration dives.

INTRODUCTION

For the past 20 years, studies of reptiles, and more recently birds and mammals, have steadily eroded the traditional view of the dive response. It has been argued that the dive response is not a broad, on/off cardiovascular and metabolic "master switch," but rather a complex graded response dependent on the intentions of the diver or conditions to which it is exposed (Kooyman G.L., 1981; Elsner R., Gooden B., 1983). Some of the skepticism and disagreements among investigators are due probably to the animals studied, the divergence of responses to breathholding that may exist, and the biased interpretation from studies of particular animals and experimental procedures. Consequently, there is the risk of conflicts among muskrat to manatee investigators, or the dabbling duck to penguin point of view. Nevertheless, the diversity of evidence and opinions has added new vigor and excitement to this subject of comparative physiology.

Some degree of the differences in opinion is due to semantics. Therefore, three

Comparative Physiology: Life in Water and on Land. P. Dejours, L. Bolis, C.R. Taylor, E.R. Weibel (eds.) Fidia Research Series, IX-Liviana Press, Padova © 1987

terms will be defined, and I will subsequently adhere to them. 1) <u>Forced</u>
<u>submersion</u> is an investigator-controlled breathhold to which the animal is
subjected, often without previous experience or training. A mild variation of
this procedure is breathholding on command, which is not perhaps as emotionally
violent to the animal, but rather a "friendly persuasion." Nevertheless, it is a
psychological restraint in which the animal may be wondering, "How long do I have
to do this before I get a fish?" 2) <u>Resting submersion</u> is a sleeping or resting
apneusis in which the animal periodically rises to the surface for a breath.
Most reports on this subject appear to be from studies of reptiles. 3) <u>Diving</u> is
an active effort to forage or pursue prey. Shallow pool diving probably fits
here, but as a rather limited expression of this broad category. Also, reference
has been made to the conflict between diving and exercise (Castellini M.A.,
1985), but it is no more a conflict than running is to track and field sports.
Diving <u>is</u> exercise; the question seems to be more a matter of pace. Further,
pacing may be an important aspect of the two basic types of dives: a) the
short-term, aerobic dive, and b) the long-term, exploratory dive. I address
these types of dives in more detail below, but first I would like to comment
briefly on four major areas of inquiry into diving. All except one of these deal
directly or indirectly with management of metabolic resources or end products
such as oxygen, blood glucose, tissue glycogen, carbon dioxide, or lactic acid.

MAJOR TOPICS OF DIVING

The <u>effects</u> of <u>compression</u> have been dealt with in both seals and penguins, as
well as a few other diving vertebrates. It is an important issue but not central
to the theme I want to pursue, nor is there much disagreement in how seals, at
least, deal with this constant stress of diving. It has been shown, both during
forced submersions in the laboratory and while voluntarily diving under natural
conditions, that blood and tissue N_2 remain low and within tolerable mammalian
levels, no matter what the depth of dive (Falke K.J., et al., 1985; Kooyman G.L.,
et al., 1972).

<u>Hypometabolism</u> is the strategy of stretching the breathhold time by lowering the
"furnace" output, so to speak. There is some uncertainty about the amount of
metabolic reduction during forced submersions, and little information on this
topic in voluntary diving. Much of the evidence comes indirectly from the
mismatch in calculated O_2 deficit of the forced submersion and the
post-submersion O_2 debt repayment (Scholander P.F., 1940), and from direct
calorimetry (Pickwell G.V., 1968).

<u>Heart rate</u> is an area of disagreement, in which some of the problems are due

probably to experimental design. It is closely related, but subsidiary, to the main topic I will pursue. Very briefly, the only study on voluntary diving in penguins was limited to a small, shallow pool. Under this circumstance, the birds displayed little difference between resting and diving heart rates (Butler P.J., Woakes A.J., 1984). In contrast, Weddell seals hunting under sea ice responded to diving with a moderate lowering of heart rate, compared to resting animals. During extended exploration dives, a much more apparent bradycardia occurred (Hill RD, et al., 1984; Kooyman G.L., Campbell W.B., 1973), which suggests some significant alterations in blood flow.

During forced submersions, there is an almost universal and considerable reduction in heart rate and cardiac output and major alterations in <u>blood flow distribution</u>. Whether a consistent response of this kind occurs during voluntary dives, other than the exceptional extended dive, is less certain, and it is the major topic of this paper.

Some evidence of blood flow distribution comes from studies of renal clearance of inulin in Weddell seals. Over the course of several hunting dives, the average renal clearance of inulin, which is directly correlated with blood flow, continued at a rate equal to that of the resting seal (Davis R.W., et al., 1983). However, the method employed was not discrete enough to discern whether blood flow and thus clearance was pulsatile or steady. If it were pulsatile, and flow was reduced during the dive, then the recovery rate during the short period on the surface would have to be high enough to compensate for the reduced flow and clearance during the dive, to yield an average resting value. This possibility seems unlikely based on recent behavioral information.

In both northern elephant seals (LeBoeuf B., et al., 1986) and leatherback sea turtles (Eckert S.A., et al., 1986), diving is continuous for at least 10 to 20 days. During this period only about 10% of the time is spent at the surface. It is likely that during ventilation, cardiac output increases about 3 times above the dive rate. This assumes a constant stroke volume and a post-dive tachycardia similar to that of Weddell seals in which it is about 3 times that of the aerobic dive rate, and 2.4 times the resting rate (Kooyman G.L., Campbell W.B., 1973). Simple arithmetic indicates that visceral blood flow might be able to decline by only 15% during the dive regime of seals and turtles and still meet the needs of a resting animal. The estimates would be even less for an animal that has consumed a large meal and requires a greater visceral blood flow. Furthermore, a majority of the flow between dives probably does not go to the viscera, but rather to muscle.

A key to how the diving response and the inter-dive ventilation promotes efficient management of internal resources is the nature of <u>skeletal muscle blood flow</u>. It is this sub-topic of blood flow distribution to which I will devote the rest of this report. This is for the important reasons that skeletal muscle represents 30 to 40% of the total body mass, and during exercise it is most responsible for the many times increase in metabolic rate.

MUSCLE CIRCULATION

The evidence to follow is drawn from blood lactic acid concentrations [LA] obtained mainly in the post-dive recovery. Blood [LA] is a valuable clue to blood flow distribution, particularly of muscle, because it is the major source of LA. Presumably, post-dive blood [LA] rises and falls due to the previous perfusion condition, and level of dependence on anaerobic glycolysis. Thus, it provides a historical record of metabolism of the dive. It is a qualitative record because the post-dive recovery is so dynamic that blood [LA] may not equilibrate with muscle concentration, and muscle may turn over a substantial amount of the LA (Stanley W.C., et al., 1986).

One of the first bits of evidence that most dives of Weddell seals (Kooyman G.L., et al., 1983) were aerobic was the analysis of blood [LA] after the dive (Guppy M., et al., 1986; Kooyman G.L., et al., 1980; Kooyman G.L., et al., 1983). When these values are plotted against the previous dive duration, there is little difference in [LA] from resting until the dive durations exceed a time that is rather consistent among animals of similar size. Such post-dive data do not give specific evidence for the characteristics of blood flow distribution during the dive. Blood flow to muscles in both the shorter and longer dives may both be greatly restricted, and the O_2 store of the muscle is simply enough to allow 15 to 20 min dives before the store is exhausted. For example, if it is assumed that the resting muscle O_2 consumption is 2.2 ml O_2 min-1 kg-1 in a 450 kg seal (Kleiber M., 1961), and there is 60 ml O_2 kg-1 of muscle, then the resting O_2 muscle supply could last for a remarkable 27 min. On the other hand, because of the expectation to surface before blood O_2 depletion, there may be little restriction in flow during aerobic dives as the seal rapidly utilizes its blood O_2 store, and the muscle remains fully oxygenated.

Some further clues are hidden in the post-dive recovery blood [LA]. If a normal, undisturbed post-dive recovery (Figure 1A) is compared to some variations in recovery, provocative differences are apparent. Soon after surfacing from a 53 min dive, an adult seal dived for 10 min. Blood [LA] continued to fall during the dive, but rose again after the dive to nearly the pre-dive 10 min

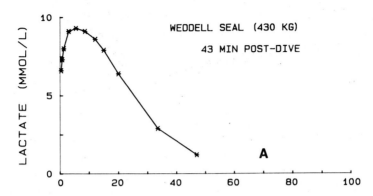

Figure 1A. Post-dive variation in arterial blood lactate concentration after a
43 min dive. The seal remained at the surface for the entire recovery (from
Kooyman et al., 1980).

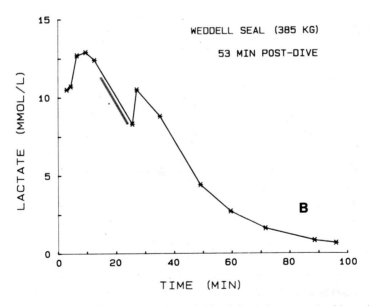

Figure 1B. Post-dive variation in arterial blood lactate concentration after a
53 min dive. The seal made one 10 min dive, indicated by the hatched bar (from
Kooyman et al., 1980).

concentration (Figure 1B). In contrast to this decline and rise in [LA], there was a steady decline in [LA] after a 33 min dive of a subadult, which continued to dive so regularly that the pattern suggested foraging (Figure 2). The maximum depth of the first dive was 220 m, and the maximum depth of the other 9 dives was 190 m. This same type of recovery occurred after a dive of 25 min. When these two curves, and one in which the animal made only one 11 min dive soon after surfacing from a 28 min dive, are compared, it can be seen that all had about the same rate of decline in [LA] (Figure 2). Thus a seal can and will continue to dive with a [LA] burden. If the subsequent dives remain within the seal's aerobic dive limit the [LA] will decline at a rate similar to a non-diving animal. This is information difficult to obtain because divers usually rest for a long period after an extended dive. In this particular subadult seal, as soon as the [LA] after the 25 min dive returned to a resting concentration, the 28 min dive followed.

These are the secure parts of the discussion, and recalling the admonition of Pascal, "It is a wise ignorance that knows itself," I will now take these data an uncertain step further. These bits of evidence suggest to me that quite different circulatory patterns existed in the dive that interrupted the recovery from the 53 min dive, and in the dives following the 33 and 25 min breathholds. Broad restriction in muscle blood flow persisted throughout the 10 min dive that interrupted the recovery from the 53 min dive. Since Weddell seals seldom make a dive so soon after such a long dive, it may have been a startled reaction to another seal, or the investigators drawing blood samples, which resulted in a moderate to strong dive response. Thus the large pool of high LA concentration in the muscle did not decline, or declined at a much slower rate than the circulating pool of steadily declining [LA] (Figure 1A & B). In contrast, the steady decline in [LA] during continuous diving seems more complex and may have been due to: 1) adequate flow in the subsequent dives to maintain aerobic conditions and steady disappearance of LA, 2) muscle Mb was loaded with enough O_2 that even without perfusion aerobic conditions prevailed throughout the subsequent dives and muscle LA turnover continued, or 3) discrete sampling smooths the curve, which actually alternates between slow and fast disappearance rates of LA. Evidence from other methods suggests that flow restriction may be the most plausible general condition.

In measurements of specific activity (SA) of injected isotopes of glucose and FFA, Castellini et al. (1985) first noted the marked difference in decay rates between swimming and steadily breathing seals, and those forcibly submerged. The differences in pattern indicated marked alteration in blood flow distribution in which the isotopes were taken up much more slowly during forced submersions.

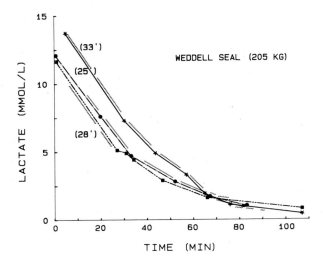

Figure 2.   Decline in blood [LA] after extended dives; durations are in parentheses. Hatched marks indicate dives (from unpublished observation of Castellini, Davis and Kooyman).

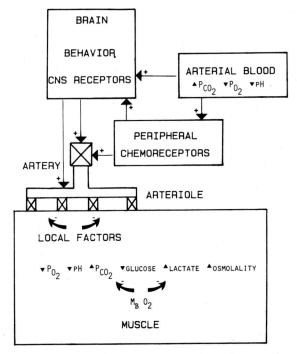

Figure 3. Model of blood flow control to muscle during dives. Triangles next to the metabolites indicate rising or falling tension or concentration. The sign next to the arrow heads indicate positive or negative input to reduction in flow.

Noting this result, the procedure was later followed in experiments on freely
diving Weddell seals (Guppy M., et al., 1986). The SA of LA, glucose and
palmitate were found to qualitatively appear to be more similar between short and
prolonged dives than either were to resting periods; that is, the dives showed a
slow decline in SA's of bolus injections of isotope, as if circulation were
restricted. It is from these various pieces of evidence that I would like to
conclude by proposing a model of muscle blood flow during the two basic types of
dives.

CONCLUSION

The evidence presented above add support to the hypothesis that the dive response
is a complex suite of neural and metabolic activities which are modified by the
circumstances and the objectives of the diver. Some of what follows has been
perhaps alluded to in other reports, but I do not know of any specific
declarations. There are no new physiological mechanisms from the general
mammalian pattern proposed, but rather the divers just do it better--so to
speak. The serial, short-duration dives suggest the greatest possible
differences from forced submersions, or extended dives, and are at the greatest
level of complexity because of the degree of variability. The model relies
primarily on local muscle O_2 and fuel stores, but without excluding access to
external O_2 and metabolites (Figure 3). Myoglobin (Mb) concentrations are
about 10 times higher in most aquatic mammals than in terrestrial forms, and a
key reason for the model is to emphasize the importance of Mb as an O_2 store.
For Mb to participate in this role, flow must be pulsatile so that O_2 is
stripped from the Mb molecule during the no flow condition. If it were otherwise
and perfusion were continuous, Mb would remain saturated because of its much
greater affinity for O_2 than is Hb. This type of O_2 utilization is different
from terrestrial mammals in magnitude of the no flow condition, which lasts only
during the muscle contraction phase in an exercising terrestrial mammal, but
possibly for several minutes in the diving vertebrate.

The circulatory pattern proposed varies somewhat from the general mammalian theme
in the different levels of blood flow restriction. As the diver submerges in a
hunting dive, muscle flow ceases due to vasoconstriction of arterioles. Duration
of flow restriction is dependent on the work level of the muscle, and the
concentration and saturation of Mb. During most of this time, metabolism is
oxidative and the fuel is triglycerides and fatty acids. As muscle O_2
declines, glycogen consumption, anaerobic glycolysis, and [LA] concentration
increase. Such local factors as well as falling pH and increasing P_{CO_2},
LA, K^+ and osmolality act on arteriole myogenic tone and cause the resumption

of blood flow and exchange of metabolites and gases. This pattern of flow would account for a means of utilizing the O_2 store within the muscle, but what happens if by intent, or by accident, a seal makes an extended dive?

At the onset of a planned extended dive, a general vasoconstriction at the arterial level occurs similar to that of a forced submersion. This restriction in blood flow persists throughout the dive, and circulating blood O_2 and other resources are conserved mainly for obligate aerobic tissue. This restriction is by no means perfect, as indicated by the rise in blood [LA] in forced submersion experiments. However, based on the rate of rise, the restriction appears to be better in seals than penguins (Scholander, 1940). Furthermore, during extended dives some factor or factors may augment the input to arterial vasoconstriction. Such thresholds have been noted during forced submersions in both seals (Daly M., et al., 1977) and ducks (Jones D.R., et al., 1982).

Most information that has formed the basis of this report comes from one species, the Weddell seal. So little reference to penguins in this report, although it was part of the assignment, is due to the limited information on freely diving birds. However, most of the results presented I suspect are general phenomena, but comparative studies are needed for substantiation. Hopefully there is more correctness than error to the model, but most importantly it provides details for some possible tests. Until recently, the specific experiments now required for a greater understanding of internal resource management and blood flow control during breathholding and hypoxia seemed intractable. However, the rapid evolution of new techniques for diving experiments has raised the prospects in this sphere of research.

Acknowledgements. This work was supported by NSF grant DPP 79-23623 and USPHS HL 17731. I am also grateful for the many discussions with M. Castellini, R. Davis, P. Ponganis and F. White on this subject.

REFERENCES

Butler PJ, Woakes AJ (1984) Heart rate and aerobic metabolism in Humboldt penguins, Spehniscus humboldti, during voluntary dives. J Exp Biol 108: 419-428.

Castellini MA (1985) Closed systems: resolving potentially conflicting demands of diving and exercise in marine mammals. In: Gilles R (ed): Circulation, respiration, and metabolism: current comparative approaches. Springer-Verlag, Berlin; pp. 219-226.

Castellini MA, Murphy BJ, Fedak M, Ronald K, Gofton N, Hochachka PW (1985)
Potentially conflicting metabolic demands of apnea and exercise in seals. J Appl
Physiol 58: 392-399.

Daly M, Elsner R, Angell-James JE (1977) Cardiorespiratory control by carotid
chemoreceptors during experimental dives in the seal. Am J Physiol 232:
H508-516.

Davis RW, Castellini MA, Kooyman GL, Maue R (1983) Renal GFR and hepatic blood
flow during voluntary diving in Weddell seals. Amer J Physiol 245: 743-748.

Eckert SA, Nellis DW, Eckert KL, Kooyman GL (1986) Diving patterns in two
leatherback sea turtles (Dermochelys coriacea) during internesting intervals at
Sandy Point, St. Croix, U.S. Virgin Islands. Herpetologica, in press.

Elsner R, Gooden B (1983) Diving and asphyxia: a comparative study of animals and
man. In: Physiological Society Monograph 40. Cambridge University Press,
Cambridge, England.

Falke KJ, Hill RD, Qvist J, Schneider RC, Guppy M, Liggins GC, Hochachka PW,
Elliott RE, Zapol WM (1985) Seal lungs collapse during free diving: evidence from
arterial nitrogen tensions. Science 229: 556-558.

Guppy M, Hill RD, Schneider RC, Qvist J, Liggins GC, Zapol WM, Hochachka PW
(1986) Microcomputer-assisted metabolic studies of voluntary diving of Weddell
seals. Am J Physiol 250: R175-187.

Hill RD, Schneider RC, Zapol WM, Liggins GC, Qvist J, Falke K, Guppy M, Elliott R
(1984) Microprocessor controlled monitoring of bradycardia in free diving Weddell
seals. Antarct J 19: 150-151.

Jones DR, Milsom WK, Gabbott GRJ (1982) Role of central and peripheral
chemoreceptors in diving responses of ducks. Am J Physiol 243: R537-545.

Kleiber M (1961) The fire of life. An introduction to animal energetics, Wiley
and Sons, Inc., New York 454 pp.

Kooyman GL (1981) Weddell seal: consummate diver, Cambridge University Press,
Cambridge, England 135 pp.

Kooyman GL, Campbell WB (1973) Heart rate in freely diving Weddell seals (Leptonychotes weddeli). Comp Biochem Physiol 43: 31-36.

Kooyman GL, Castellini MA, Davis RW, Maue RA (1983) Aerobic dive limits in immature Weddell seals. J Comp Physiol 151: 171-174.

Kooyman GL, Kerem DH, Campbell WB, Wright JJ (1973) Pulmonary gas exchange in freely diving Weddell seals (Leptonychotes weddelli). Respir Physiol 17: 283-290.

Kooyman GL, Schroeder JP, Denison DM, Hammond DD, Wright JJ, Bergman WP (1972) Blood nitrogen tensions of seals during simulated deep dives. Am J Physiol 223: 1016-1020.

Kooyman GL, Wahrenbrock EA, Castellini MA, Davis RA, Sinnett EE (1980) Aerobic and anaerobic metabolism during diving in Weddell seals; evidence of preferred pathways from blood chemistry and behavior. J Comp Physiol 138: 335-346.

Le Boeuf, Costa DP, Huntley AC, Kooyman GL, Davis RW (1986) Pattern and depths of dives in northern elephant seals. J Zool 208: 1-7.

Pickwell GV (1968) Energy metabolism in ducks during submergence asphyxia: assessment by a direct method. Comp Biochem Physiol 27: 455-485.

Scholander PF (1940) Experimental investigations on the respiratory function in diving mammals and birds. Hvalradets Skrifter Norske Videnskaps:Akad Oslo 22: 1-131.

Stanley WC, Gertz EW, Wisneski JA, Neese RA, Morris DL, Brooks GA (1986) Lactate extraction during net lactate release in legs of humans during exercise. J Appl Physiol 60: 1116-1120.

PHYSIOLOGICAL CORRELATES
OF REINVASION OF WATER BY REPTILES

Roger S. Seymour

Department of Zoology, University of Adelaide
Adelaide, S.A. 5001
Australia

As reptiles radiated into aquatic habitats, they further
modified what was a preadapted physiology. In response to
hydrostatic pressure, aquatic snakes evolved low blood
pressure, poor barostatic reflexes, central hearts and long
vascular lungs. They also developed non-pulmonary gas exchange
organs, supplied with unsaturated blood derived mainly from
powerful right-to-left shunts. Deep diving snakes could take
advantage of O_2 from both the lung and sea water, even at
depths where pulmonary gas tensions are very high, while
avoiding N_2 bubble formation during decompression.

INTRODUCTION

In the course of vertebrate evolution, the reptiles arose from the amphibians
and assumed lifestyles that were not tied to water. Most biologists believe that
a water-tight integument, uricotelism, and the amniote egg opened up terrestrial
habitats into which the first reptiles quickly radiated. Early fossil reptiles
were morphologically well adapted to land, so it appears that most major groups
of aquatic reptiles represent reinvasions of water. There are about 500 species
of extant reptiles that are amphibious or totally aquatic. The most successful
recent groups are turtles, crocodiles and certain snakes, but there are a few
primarily aquatic lizards such as the marine iguana. By comparing living
species, we can obtain an idea of the physiological changes that occurred when
reptiles reinvaded water. Their characteristics provide clues to the environmental factors that had important selective influence on the evolution of modern
groups.

In some respects, returning to aquatic lifestyles was easy for reptiles because
they already possessed a suite of physiological characteristics that preadapted
them for diving. For example, the basic reptilian design includes ectothermy,
bradymetabolism, high capacity for anaerobic metabolism, tolerance of acid-base

Comparative Physiology: Life in Water and on Land. P. Dejours, L. Bolis, C.R. Taylor, E.R. Weibel
(eds.) Fidia Research Series, IX-Liviana Press, Padova © 1987

disturbance and a pattern of ventilation involving breathholding. It is not surprising, therefore, that most aspects of the physiology of terrestrial and aquatic reptiles are quite similar. There are almost no recognised correlations between diving behavior and blood O_2 capacity, blood volume, hemoglobin-O_2 affinity, muscle myoglobin concentration, muscle enzyme activity, anaerobic and aerobic metabolic scope, blood buffering or tolerance to anoxia (Seymour, 1982).

Despite their preadaptations, however, some aquatic reptiles show distinctive physiological adaptations to environmental factors involving gravity, hydro-static pressure and absence of air. In this brief paper, I limit my discussion to the regulation of blood pressure and cutaneous gas exchange. More extensive treatments of the physiology of terrestrial and aquatic reptiles are available (White, 1976; Seymour, 1982; Butler and Jones, 1982; Glass and Wood, 1983).

EFFECT OF GRAVITY ON CIRCULATION

The blood in the vascular systems of terrestrial animals is affected by gravita-tional force which tends to make blood flow to lower parts of the body. A change in posture from horizontal to vertical results in a well-studied suite of baro-static adjustments that help maintain cardiac output and cerebral blood flow in the face of increased hydrostatic pressure of the blood column above the heart and pooling of venous blood below it (Gauer and Thron, 1965). One important conclusion of these studies, however, is that immersion in water eliminates the cause of the responses, namely the changes in transmural pressures throughout the cardiovascular system. In water, the internal hydrostatic pressure of the blood column is almost equalled by the external hydrostatic pressure of the water and practically no changes in transmural pressure occur during changes in orientation.

Because hydrostatic pressure depends on the absolute vertical length of the vascular system, it is of little consequence in most reptiles that are either small, compact or horizontal in body attitude. However, snakes are inordinately subject to the effects of gravity because they are long and many species climb. Aquatic snakes evolved independently from several families of typically terrestrial species. Despite a phylogenetically diverse origin, aquatic snakes show a quite consistent set of cardiovascular adaptions. The details may be found in recent reviews by Lillywhite (1987) and Seymour (1987).

Regulation of arterial blood pressure in response to tilting in air is extremely poor in aquatic snakes, better in semi-aquatic species, and best in terrestrial and arboreal species. Cephalic blood pressure drops to zero and the vasculature presumably collapses during head-up tilting in some aquatic species. However, if the snake is tilted in water, there is no measurable change in blood pressure. The poor performance of aquatic snakes results mainly from three factors. First, they have inherently low arterial blood pressure, even when horizontal. The heart does less work circulating the blood at low pressure. The higher pressure of terrestrial snakes helps stabilize blood flow during postural changes, but it is more energetically costly. Second, blood pools and edema develops in the dependent parts of tilted aquatic snakes, causing a drop in cardiac output. In terrestrial species, pooling is minimized and venous return is facilitated by lateral undulations of the body. Third, the responses of the autonomic barostatic system to tilting, hemorrhage, and injected catecholamines are very weakly developed in aquatic species. The limited regulatory ability that remains, however, reminds us that gravity is not the only selective influence on barostatic reflexes. Some regulation of blood pressure is necessary in a reptile because it is instrumental in controlling circulatory pathways, including central vascular shunts.

The heart is located farther back in the body in aquatic snakes (25-45% of the body length) than it is in terrestrial and arboreal species (15-25%). The necessity of having the heart close to the head to stabilize cephalic blood flow is absent in water, and a centrally located heart does less work perfusing the whole body.

The vascular portion of the lung also tends to be very long in aquatic species, but short in terrestrial ones. When tilted in air, the lungs of sea snakes show extensive edema and capillary rupture, but tilted in water, the dependent part of the lung collapses completely and prevents dangerous transmural pressure gradients.

NON-PULMONARY GAS EXCHANGE AND DIVING

Reliance on non-pulmonary gas exchange. Evolutionary emancipation from water by the reptiles was apparently accompanied by the appearance of a thick, cornified skin that reduced evaporative water loss. Because the major gas exchange organs of animals are usually moist and permeable to water, it is often thought that effective gas exchange and limitation of water loss are mutually exclusive, although the functional connection has never been clear. The water tightness of reptilian skin does not depend on the presence of scales (Bennett and Licht, 1975), but is linked to lipids in the skin (Roberts and Lillywhite, 1980).

Because respiratory gases are lipid soluble, the layers responsible for the
barrier to water may have a minimal effect on gas exchange. This accounts for
the occurrence of cutaneous gas exchange in many reptiles (Seymour 1982; Feder
and Burggren, 1985). Non-pulmonary gas exchange in snakes and lizards is
entirely cutaneous, but the hard shells of aquatic turtles may have led to
evolutionary development of pharyngeal and cloacal gas exchange.

Not surprisingly, non-pulmonary gas exchange tends to be higher in aquatic
species than in terrestrial species of the same mass (Seymour, 1982). High
capacity occurs in sea snakes that have a high concentration of superficial
capillaries in the skin (Rosenberg and Voris, 1980). In reptiles living in
shallow water, especially if it is hypoxic, the close availability of air may
have placed less selective advantage in non-pulmonary gas exchange. Thus several
semi-aquatic turtles and snakes (also crocodilians; Wright, 1986) have low
cutaneous gas exchange capacity.

Although shallow diving, certain turtles (e.g. Trionyx) can satisfy all of their
resting requirements through non-pulmonary means if the water is well oxygenated
(Dunson, 1960). Whether or not sea snakes can do this is not known for certain,
but rates of cutaneous O_2 uptake come close to the minimum metabolic rates of
reptiles in general and early methods may have underestimated the exchange
capacity of the sea snake skin. Under certain natural conditions in which the
water is well oxygenated and moved across the snakes by currents, it is likely
that cutaneous gas exchange is sufficient for resting metabolism. This accounts
for the conspicuous absence of certain species at the surface during the part of
their diurnal cycle when they can be found resting on the sea bed.

When sea snakes become active, however, they return periodically to the surface
to breathe. Even deep diving species remain aerobic during most foraging dives
(Seymour, 1979) and submergence times in the field average about 15 min
(Heatwole, 1975). Because a snake feeding at 100 m depth would require about 10
min to swim to the surface and back (Heatwole et al., 1978), there is an obvious
advantage in cutaneous gas exchange in lengthening the time spent at depth. In
fact, the deeper diving hydrophiine sea snakes tend to have higher rates of
cutaneous gas exchange than do the shallow diving laticaudines (Heatwole and
Seymour, 1975, 1978).

Surface breathing not only replenishes pulmonary O_2 stores, but helps maintain
neutral buoyancy at depth and thereby reduces the cost of swimming. Unlike the
swimbladders of fish, the snake lung changes volume as O_2 and N_2 are removed.
Pelamis platurus adjusts its lung volume according to the depth of the sub-

sequent dive (Graham et al., 1987). It maintains neutral buoyancy during the dive by gradually ascending to keep lung volume constant, but eventually it must return to the surface to refill.

Effect of body size. Reliance on non-pulmonary gas exchange decreases in larger reptiles. Cutaneous O_2 uptake in reptiles appears to be largely limited by diffusion (Seymour, 1982). Therefore the rate is strongly affected by Fick's law and depends on the ratio of the surface area and effective thickness of the skin, although other factors, such as capillary recruitment, may affect this relationship (Feder and Burggren, 1985). Because surface area of animals is proportional to the body mass raised to the 0.67 power ($M^{0.67}$), and the skin thickness is approximately proportional to $M^{0.4}$ in some reptiles (Smith, 1979), the rate of cutaneous gas exchange should scale to about $M^{0.67}/M^{0.4} = M^{0.27}$. This exponent is in fact similar to those describing non-pulmonary O_2 uptake in turtles and terrestrial squamates ($M^{0.23}$, $M^{0.32}$; Dunson, 1960; Seymour, 1982), but it is far below the exponent for metabolic rate in reptiles generally ($M^{0.80}$; Bennett and Dawson, 1976). Therefore large aquatic reptiles such as marine turtles would not be expected to rely extensively on cutaneous gas exchange. Loss of effective non-pulmonary exchange should not disadvantage large diving reptiles, however. The loss is compensated for by relatively larger O_2 storage in the lung and blood. Scholander (1940) proposed that O_2 stores of divers scale to approximately $M^{1.0}$, considerably above the exponent for metabolic rate, so larger animals should be able to dive for longer periods with the O_2 that they carry with them.

Role of shunting. Cutaneous circulation in reptiles is derived exclusively from the intersegmental arteries carrying systemic arterial blood (Feder and Burggren 1985). This arrangement does not facilitate gas exchange because the blood supplied to the skin may be already highly saturated with O_2. In sea snakes this problem is lessened by taking advantage of the pattern of right-to-left shunting typical of non-crocodilian reptiles. In most squamates, there is some degree of venous bypass of the pulmonary gas exchange surface, either through an intra-ventricular shunt (White, 1976) or functional venous admixture in the lung itself (Seymour, 1978). This effect dilutes O_2-rich blood equilibrating in the lung with O_2-poor blood from the systemic veins.

Although pulmonary bypass in most reptiles depends somewhat on the phase of the ventilatory cycle, the average right-to-left shunt is generally not great, and the level of arterial O_2 saturation is fairly high. Reasonable estimates indicate a venous component of the systemic arterial blood ranging up to about 30% (see Seymour, 1982). The shallow diving sea snake, Laticauda colubrina, also

has good separation of venous and arterial circuits (Seymour, 1978). However, in
several species of deep diving hydrophines, venous blood represents about 50 to
70% of the systemic cardiac output, even when they are breathing voluntarily at
the surface. The enormous pulmonary bypass of these sea snakes results in low
levels of $Hb-O_2$ saturation and relatively low P_{O_2} in the systemic arteries. In
Hydrophis, for example, the arterial blood is only 60 – 70% saturated at a P_{O_2}
of about 5.3 kPa (40 torr) (Fig. 1). The importance of maintaining a low saturation in the cutaneous blood supply cannot be overemphasized. A favorable P_{O_2}
gradient between water and blood depends on maintaining low P_{O_2} along the length
of the capillary. If the blood enters the skin practically saturated, then its
P_{O_2} quickly rises with only a small uptake of O_2, regardless of its initial P_{O_2}.
In this regard, it is significant that species with high O_2 saturation of
arterial blood typically have relatively low cutaneous gas exchange capacity.

If some aquatic reptiles can exchange all of their requirements through the skin
at rest, they essentially become unimodal water breathers, somewhat like fish. A
basic principle in comparative respiratory physiology states that the P_{CO_2} in
water breathers is very low (Rahn, 1966). This occurs because Krogh's diffusion
coefficient for CO_2 is about 20–25 times higher in water than in air. To obtain
enough O_2 by diffusion through the skin, the animal will certainly lose considerable CO_2 from the blood perfusing the gas exchanger. However, arterial and
venous P_{CO_2} in sea snakes is usually around 2.4–4.5 kPa (18–34 torr) which is
much higher than expected if these animals were fish (Seymour and Webster, 1975;
Seymour, 1978). The parallel arrangement of the cutaneous gas exchanger and the
systemic circulation (Fig. 2), tends to retain CO_2 in the animal. This may be
important in acid–base regulation in aquatic reptiles that have evolved from
air–breathing progenitors that had high internal P_{CO_2}.

Effect of deep diving. Maintenance of an effective P_{O_2} gradient into the animal
is made more difficult, the deeper it dives. Certain hydrophiine sea snakes feed
at depths at least to 100 m and thus are the only deep diving vertebrates that
certainly rely on both pulmonary and cutaneous gas exchange at depth. This
capability depends on the ability to take advantage of O_2 stored in the lung at
high pressure while maintaining a favorable gradient for cutaneous O_2 uptake
from the sea.

The elongate lung of sea snakes is the most important site of O_2 storage.
Because the blood volume and O_2 capacities are normal and the level of O_2
saturation is low, the lung holds about three times the O_2 as the blood at the
beginning of a dive (Seymour, 1982). Sea snakes usually take a single breath
upon surfacing and the tidal volume is about 80% of total lung volume (Seymour

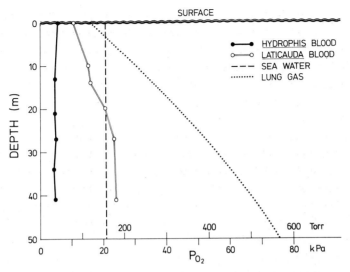

Fig. 1. Oxygen tensions in the lung and arterial blood of sea snakes during a dive in water saturated with air. Data are mean values from <u>Hydrophis</u> [●] and <u>Laticauda</u> [o]. Lung P_{O_2} is calculated according to hydrostatic compression and the O_2 uptake from the lung due to the metabolic cost of swimming to depth (Seymour, 1978, 1982).

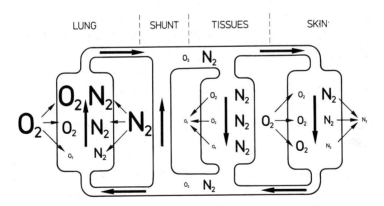

Fig. 2. Schematic diagram of the circulation in a sea snake that has dived to depth with air in its lung. The blood obtains O_2 and N_2 in the lung, but a powerful shunt dilutes the pulmonary blood with O_2- and N_2-poor from the systemic veins. O_2 is lost to the tissues but gained from the sea water, while N_2 in blood passing through the tissues is diluted by N_2-poor blood returning from the skin. The sizes of the symbols for O_2 and N_2 indicate differences in partial pressure, but they are not quantitatively accurate.

and Webster, 1975; Seymour et al. 1981). Thus a snake begins a dive with a pulmonary P_{O_2} of about 16 kPa (120 torr) which increases to about 30 kPa (225 torr) at a depth of only 10 m (Fig. 1). If the P_{O_2} of well mixed sea water is about 21 kPa (157 torr) at all depths, diving quickly causes an outward gradient from the lung to the sea. Venous admixture tends to keep arterial saturation low, however, and its level depends on the degree of shunting, the metabolic rate and the cardiac output.

Direct measurements from _Laticauda colubrina_ show that, despite a moderate venous shunt averaging 28% of the cardiac output at the surface, the P_{O_2} of arterial blood begins to exceed that of sea water at a hydrostatic pressure equivalent to 20 m depth (Fig. 1). If the snake were to dive deeper, it could lose O_2 to the sea. However, the natural distribution of this species seems to be confined to shallower water.

On the other hand, _Hydrophis_ species are deep divers. Arterial P_{O_2} and saturation appear practically independent of applied pressure (Fig. 1). Even at a depth of 41 m, when the pulmonary P_{O_2} reaches 66 kPa (500 torr), the arterial P_{O_2} remains at about 5.3 kPa (40 torr) and the saturation at about 60%. Therefore these snakes maintain a favorable gradient for continued O_2 uptake at all diving depths. The independence of arterial saturation apparently stems from an increase in the level of shunting, presumably due to hydrostatic collapse of the lung. Effective shunting increases if either the collapsed portions continue to be perfused, or if an increasing resistance in the pulmonary circuit increases right-to-left intraventricular shunting.

Deep diving in air-breathing animals creates other problems of high gas tensions in the body caused by hydrostatic compression of the lungs. Oxygen toxicity, nitrogen narcosis and decompression sickness may affect man, but apparently do not occur among natural divers, including reptiles. Most fresh water reptiles are shallow divers and are not in danger but the marine turtles and snakes that dive to considerable deaths may be.

Earlier I have stressed the importance of central shunting and non-pulmonary gas exchange in protecting deep diving reptiles from decompression sickness (Seymour, 1978). Enough N_2 is present in one lung volume to saturate a sea snake's body above the level for bubble formation. If the animal dives repeatedly, the N_2 could build up higher. However, these snakes appear to be protected by basically the same mechanism that keeps arterial O_2 low. The pulmonary bypass dilutes N_2-rich blood equilibrating in the lung with N_2-poor blood from the venous side.

At first it seems paradoxical that the snake at depth can eliminate N_2 while taking up O_2 through the skin. However, there is an important difference between these gases (Fig. 2). Unlike O_2, N_2 is not consumed by the body so it tends to increase in both sides of the circulation during a dive. Exactly how much it increases depends on the level of shunting and the rate it is lost through the skin. Although shunting tends to keep arterial P_{N_2} low, it eventually builds up until the P_{N_2} gradient across the skin causes the rate of cutaneous loss to equal the rate of uptake from the lung. At this point the arterial P_{N_2} can increase no further and its level remains well below the threshold for bubble formation after dives of at least 100 m in two <u>Hydrophis</u> species (Seymour, 1978). The shallow diving <u>Laticauda colubrina</u>, on the other hand, is less protected and might form bubbles in its blood after dives below about 30–70 m, but it normally does not dive that deep. <u>Hydrophis</u> species generally have higher shunts and cutaneous gas exchange capacity than does <u>Laticauda</u>. Protection from bubble formation depends on both factors; bubbles can be produced in the blood if either cutaneous N_2 loss or central shunting is experimentally altered.

The problems of hydrostatic pressure have not been well studied in the marine turtles that are known to make extremely deep dives. Green turtles, <u>Chelonia mydas</u>, dive as deep as 290 m (Landis, 1965), and leatherback turtles, <u>Dermochelys coriacea</u>, dive at least to 475 m (Eckert et al., 1986). Recent time–depth recordings from leatherbacks indicate that the dives are made aerobically (average dive times about 12–15 min) and in quick succession (average surface time 5–10 min) (Eckert et al. 1986). Repeated dives to only 25 m can cause decompression sickness in man (Paulev, 1965), and experimental compressions equivalent to a depth of 180 m can kill green turtles (Berkson, 1967). It is not clear how large marine turtles avoid the problem in nature, but die in hyperbaric chambers. Do they exhale enough gas before a natural dive to limit the N_2 saturation of the body? Or do they reduce N_2 recruitment from the lung during a dive by squeezing the pulmonary gas into non-exchange areas, as in marine mammals? Or, like sea snakes, could they increase pulmonary bypass and eliminate N_2 through non-pulmonary gas exchangers? Marine turtles clearly have evolved an effective mechanism, but only further research will discover what it is.

REFERENCES

Bennett AF, Dawson WR (1976) Metabolism. In: Gans C, Dawson WR (eds): Biology of the reptilia, Vol. 5. Academic Press, New York; pp. 127–223.

Bennett AF, Licht P (1975) Evaporative water loss in scaleless snakes. Comp. Biochem. Physiol. 52A: 213–215.

Berkson H (1967) Physiological adjustments to deep diving in the Pacific green turtle (Chelonia mydas agassizii). Comp. Biochm. Physiol. 21: 507–524.

Butler PJ, Jones DR (1982) The comparative physiology of diving in vertebrates. Adv. Comp. Physiol. Biochem. 8: 179–364.

Dunson WA (1960) Aquatic respiration in Trionyx spinifer asper. Herpetologica 16: 277–283.

Eckert SA, Nellis DW, Eckert KL, Kooyman GL (1986) Diving patterns in two leatherback turtles (Dermochelys coriacea) during internesting intervals at Sandy Point, St. Croix, U.S. Virgin Islands. Herpetolgica (in press).

Feder ME, Burggren WW (1985) Cutaneous gas exchange in vertebrates: design, patterns, control and implications. Biol. Rev. 60: 1–45.

Gauer OH, Thron HL (1965) Postural changes in the circulation. In: Hamilton WF, Dow P (eds): Handbook of Physiology, Sect. 2, Circulation, Vol. 3. American Physiological Society, Washington; pp. 2409–2439.

Glass ML, Wood SC (1983) Gas exchange and control of breathing in reptiles. Physiol. Rev. 63: 232–260.

Graham JB, Gee JH, Motta J, Rubinoff I (1987) Subsurface buoyancy control by the sea snake Pelamis platurus. (MS).

Heatwole H (1975) Voluntary submergence times of marine snakes. Marine Biol. 32: 205–213.

Heatwole H, Seymour RS (1975) Pulmonary and cutaneous oxygen uptake in sea snakes and a file snake. Comp. Biochem. Physiol. 51A: 399–405.

Heatwole H, Seymour RS (1978) Cutaneous oxygen uptake in three groups of aquatic snakes. Aust. J. Zool. 26: 481–486.

Heatwole H, Minton SA Jr, Taylor R, Taylor V (1978) Underwater observations on sea snake behaviour. Rec. Aust. Mus. 31: 737–761.

Landis AT Jr (1965) Research: new high pressure research animal? Undersea Tech. 6: 21.

Lillywhite HB (1987) Circulatory adaptations of snakes to gravity. Amer. Zool. (in press).

Rahn H (1966) Aquatic gas exchange: theory. Respir. Physiol. 1: 1–12.

Roberts JB, Lillywhite HB (1980) Lipid barrier to water exchange in reptile epidermis. Science 207: 1077–1079.

Rosenberg H, Voris HK (1980) Cutaneous capillaries of sea snakes and their possible role in gas exchange. Amer. Zool. 20: 758.

Scholander PF (1940) Experimental investigations on the respiratory function in diving mammals and birds. Hvalradets Skr. 22: 1-131.

Seymour RS (1978) Gas tensions and blood distribution in sea snakes at surface pressure and at simulated depth. Physiol. Zool. 51: 388-407.

Seymour RS (1979) Blood lactate in free-diving sea snakes. Copeia 1979: 494-497.

Seymour RS (1982) Physiological adaptations to aquatic life. In: Gans C, Pough FH (eds): Biology of the reptilia, Vol. 13. Academic Press, New York; pp. 1-51.

Seymour RS (1987) Scaling of cardiovascular physiology in snakes. Amer. Zool. (in press).

Seymour RS, Spragg RG, Hartman MT (1981) Distribution of ventilation and perfusion in the sea snake, Pelamis platurus. J. Comp. Physiol. 145: 109-115.

Seymour RS, Webster MED (1975) Gas transport and blood acid-base balance in diving sea snakes. J. Exp. Zool. 191: 169-182.

Smith EN (1979) Behavioral and physiological thermoregulation of crocodiles. Amer. Zool. 19: 239-247.

White FN (1976) Circulation. In: Gans C, Dawson WR (eds): Biology of the reptilia, Vol. 5. Academic Press, New York; pp. 275-334.

Wright JC (1986) Low to negligible cutaneous oxygen uptake in juvenile Crocodylus porosus. Comp. Biochem. Physiol. (in press).

Part VIII

WATER, IONIC EXCHANGES, OSMOREGULATION MECHANISMS

VOLUME CONTROL AND ADAPTATION TO CHANGES IN IONS CONCENTRATIONS IN CELLS OF TERRESTRIAL AND AQUATIC SPECIES: CLUES TO CELL SURVIVAL IN ANISOSMOTIC MEDIA

Raymond Gilles

Laboratory of Animal Physiology, University of Liège,
22, quai Van Beneden, B-4020 Liège, Belgique

This brief overview compares the physiological responses of cells from homeosmotic and euryhaline poecilosmotic species to osmotic challenges. The data presented led to the idea that adaptation and survival of cells to changes in the osmolality of their environmental medium cannot be related solely to their possibility of volume regulation. Their possibility to cope with the disrupting effects changes in intracellular concentrations of inorganic ions can have on the structure and activity of different macromolecules must also be taken into consideration. In this respect, the amino-compounds usually found in rather large amounts in cells of marine euryhaline poecilosmotic species could have an important part to play. They indeed seem to have a "stabilizing" effect on protein structure that can oppose the "disrupting" effects of inorganic ions. Further, their use as major osmolytes in the cells of euryhaline species allows a control of the intracellular level of inorganic monovalent ions otherwise used as important osmotic effectors. The cells can thus avoid the important disrupting effects of changes in these ions concentration.

INTRODUCTION

This conference being about adaptations to terrestrial and aquatic life, I would like to consider that homeosmoticity (and the very effective mechanisms of blood osmolality control it implicates) has been evolved by terrestrial species, aquatic forms being essentially poecilosmotic. From then on I would like to speculate on the incidence this type of evolution has had on the possibility for cells to withstand osmotic challenges. In homeosmotic species indeed, cells are not experiencing osmotic challenges anymore and they

Comparative Physiology: Life in Water and on Land. P. Dejours, L. Bolis, C.R. Taylor, E.R. Weibel (eds.) Fidia Research Series, IX-Liviana Press, Padova © 1987

apparently have lost the possibility to survive to them. In this
respect, it is interesting to recall that changes of only 20 to 40
mOsm/l. in blood osmolality induced <u>in vivo</u> in a mammalian species
either as part of an experimental procedure or during the development of a pathology or accidentaly (too rapid rehydration following
dehydration for instance) will cause large functional damages. These
damages have usually been described under the general term of "water
intoxication". A clear understanding of "water intoxication" is actually lacking; no doubt however that this phenomenon is associated
with cell disfunction in anisosmotic media. Usually, cells of poecilosmotic species can withstand without any problem osmotic challenges of much larger amplitude. This possibility is particularly well
developped in the so-called euryhaline marine invertebrates. The
cells of such species can indeed easily cope with changes in blood
osmolality of 500 mOsm/l or more in the course of acclimation of the
animals to media of different salinities. In the euryhaline chinese
crab *Eriocheir sinensis* for instance, adaptation from sea water to
fresh water results in a decrease in blood osmolality from some
1100 mOsm/l to about 550 mOsm/l (Gilles, 1974a). Such rather large
changes in blood osmolality have been currently described in a variety of euryhaline invertebrates acclimating to media of different
salinities (for reviews see for instance Potts and Parry, 1964;
Gilles, 1975, 1986; Spaargaren, 1979). In one of the most extreme
euryhaline blood osmoconformer known, the serpulid worm *Mercierella*
enigmatica which can tolerate natural waters of salinities between
1 and 55‰ S, blood osmolality will vary between 84 and 2304 mOsm/l
upon acclimation to these extreme media (Skaer , 1974). In these
conditions, the cells are adapting without any problem to a change
of more than 2000 mOsm/l !

This raises the questions of knowing 1) what are the causes of the
cell disfunction responsible for "water intoxication" in mammalian
cells and 2) what are the mechanisms present in euryhaline invertebrate cells unabling them to cope with these problems.

Cells withstanding changes in blood osmolality are in fact faced
with 3 types of problems. 1) Development of swelling pressures and
thus of membrane tensions that will affect, if not controlled, all
membranes activities, 2) changes in spatial organization of intra-

cellular macromolecular structures that will be induced by the changes in cell volume; 3) adverse effects that changes in the concentration of inorganic ions will induce at the level of the plasma membrane and of the intracellular macromolecular architectures.

The two first problems are directly related to changes in cell volume. Up to now the data about volume control in hyperosmotic conditions have remained far too scanty for general statements to be made. We will thus concentrate on the reactions of cells to hypoosmotic challenges.

COPING WITH VOLUME CHANGES - VOLUME REGULATION

Most cell types react to a hypo-osmotic challenge by a volume regulation process (figure 1). This phenomenum, so-called "volume regulatory decrease" (VRD), has been extensively studied during the past 15 years on cells from both homeosmotic and poecilosmotic species. The data have been reviewed many times in the recent years (euryhaline invertebrate cells : Gilles, 1974a, 1975, 1979, 1986 - other cells : Cala, 1983, 1985; Ellory et al., 1985; Grinstein et al., 1985; Hoffmann, 1977, 1985 a,b, 1986; Kregenow, 1981; Lauf, 1985) and we will thus only briefly summarize the main facts at hand.

1) VRD, as studied in in vitro experiments on isolated cells and tissues, can last from a few minutes to several hours depending on the tissue (or cell type) and on the species considered. There is no significant systematic difference in the volume regulation efficiency depending on the origin of the cells : poecilosmotic or homeosmotic species.

2) In cells from homeosmotic species as well as in cells of poecilosmotic ones, VRD appears to be related to a decrease in amount of different intracellular osmotic effectors. Their change in concentration leads to an adjustment of the osmolality of the intracellular fluid to the one of the external medium and therefore to a decrease in swelling pressure allowing volume regulation. Other possible mechanisms, such as cell mechanical resistance and elasticity or changes in water structure have only been barely considered up to now and will not be discussed in this brief overview.

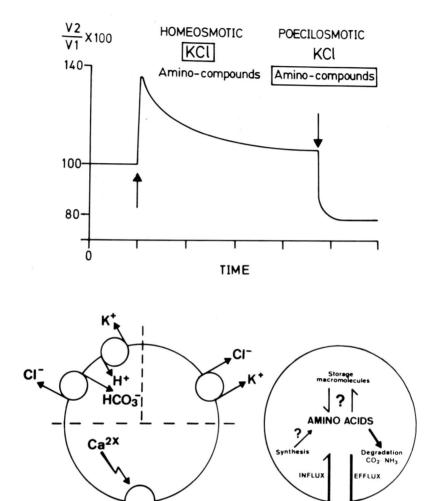

<u>Figure 1</u>

Upper part : the volume regulatory decrease process occurring in
most cell types after osmotic swelling in hypo-osmotic media. The
time scale is dependent on the cell types and species considered.
At the first arrow, cells are placed in the hypo-osmotic medium, at
the second one, they are placed back in the control saline.
Lower part : the most important mechanisms considered to be implicated in the control of the amount of the major intracellular osmotic effectors (KCl and non-essential amino acids) during volume
regulatory decrease.

3) In both cell types, the intracellular osmotic effectors at work
are essentially the same. In most cases they are the inorganic ions
K^+ and Cl^- and some amino-compounds such as the non-essential amino
acids alanine, aspartate, glutamate, glycine, proline and serine,
the amino taurine and the quaternary ammonium derivatives glycine-
betaine and trimethylamine oxide.

4) In both cell types, the mechanisms at work in the control of the
amount of these osmotic effectors are essentially the same (figure 1).

Cells from homeosmotic species can thus regulate their volume as
effectively as cells of poecilosmotic ones by using essentially
the same mechanisms. This makes it obvious that euryhalinity, and
the successful survival of cells to anisosmotic media it depends on,
cannot only be related to the capabilities of cells for volume control. In the same line of thought, we have already drawn attention
in previous reviews on the fact that some marine osmoconforming
species, easily surviving for long periods important dilution of
their blood and external medium, have only extremely limited power
of cell volume regulation (Gilles and Péqueux, 1983; Gilles, 1986).
This is for instance the case of the stone crabs *Cancer pagurus*
(Wanson et al., 1983) and *C.irroratus* (Moran and Pierce, 1984) as
well as of some other crustaceans (Freel, 1978).

To what kind of mechanisms are thus basically related the "water
intoxication" phenomenon seen in cells of homeosmotic species and
the successful survival of cells of poecilosmotic ones to osmotic
challenges ? In fact cells withstanding anisosmotic media have not
only to cope with changes in water amounts but also with changes in
levels of inorganic ions. I would like to concentrate in this brief
overview only on the effects that changes in the intracellular concentration of the monovalent inorganic ions found as major osmotic
effectors in both blood and cells (Na^+, K^+, Cl^-) can have on the
structure and activity of different intracellular macromolecules.

COPING WITH CHANGES IN IONS LEVELS - COMPENSATORY SOLUTES

The effects of monovalent inorganic ions on the structure of macromolecules have been studied since more than twenty years especially
by the Von Hippel group (Von Hippel and Wong, 1963, 1965; Von Hippel

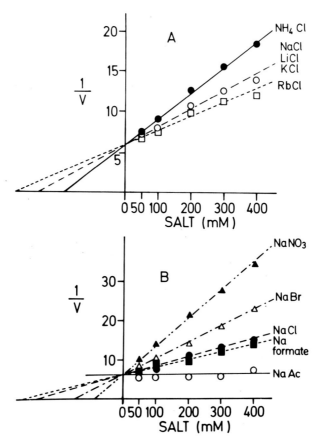

Figure 2

Effect of different salts of monovalent ions on the activity of the
succinate dehydrogenase from *Mytilus californianus* mantle tissue.
A. : cations - B : anions.

and Schleich, 1969 a,b : Schleich and Von Hippel, 1970; Hamabata
and Von Hippel, 1973; Von Hippel et al., 1973; Von Hippel, 1976).
In short , the net free energy stabilizing the folded form of a
protein is in most cases very small and environmental changes as
those induced for instance by addition of monovalent ions to the
medium can "push" the protein across a phase transition boundary
with the resultant formation of a different equilibrium structure.
These "distrubing" effects will of course induce modifications in
the activity of the considered proteins. These studies have been
achieved, using physical methods, on a few isolated macromolecules
taken as models (collagen - DNA - myosin). They lead to the question
of knowing if such "disrupting" effects can be observed in cells.
Two short examples from results of my laboratory will illustrate
such effects.

Between 1961 and 1974, I studied the effects of osmotic shocks on
the amino-acid pool and on several metabolic parameters in isolated
tissues of crustaceans (O_2 consumption, $C^{14}O_2$ production from labelled glucose, pyruvate and amino acids, NADH and cytochromes oxydo-
reductions level...). These results have been reviewed several times
(Gilles, 1974a, 1975, 1978, 1986) and will not be described here.
This study led us to the idea that the major monovalent ions the
concentration of which was changing during osmotic adjustment (K^+,
Cl^-, Na^+) could affect the structure and activity of enzymes implicated in the intermediary and amino-acid metabolism. Along this line
of thought, we brought between 1963 and 1974 arguments showing that
the effects of monovalent ions on the enzymes activity is directly
related to a "disturbing" effect, mostly of the anion Cl^-, on the
structure of the enzyme protein (Schoffeniels and Gilles, 1963;
Gilles, 1969; Gilles et al., 1971; Gilles, 1974b). Briefly, our
results showed that the series obtained when placing various anions
in order of increasing effectiveness (i.e. Ac^- Cl^- Br^- NO_3^-, cfr.
figure 2) is similar to the one found by Von Hippel and colleagues
and in which they disturb the structure of different macromolecules
as determined by physical methods. We have also shown that the
addition of NaCl to the incubation medium modifies the thermosensitivity curve of glutamate dehydrogenase (fig.3), a result pointing
to the fact that the anion effect must be related to a direct
"disrupting" effect on the enzyme protein configuration rather than
to some indirect effect such as association ion-substrate, or

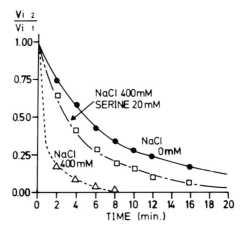

Figure 3

Effect of NaCl and serine on the thermosensitivity at 52°C of glutamate dehydrogenase from *Eriocheir sinensis* muscle tissue.

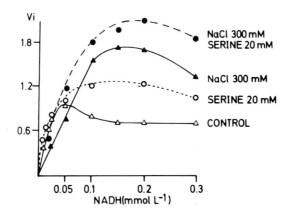

Figure 4

Effect of NaCl and serine on the activity of glutamate dehydrogenase from *Eriocheir sinensis* muscle tissue. Activity expressed in terms of O.D. units par min. per mg.protein.

electrostatic shielding. It is also worthnoticing for our present discussion that the amino acid serine can modulate the effect of NaCl on the enzyme activity (figure 4) and that it can interfere with the effect of the salt on the its thermosensitivity curve (fig.3).

The other example of disrupting effects that could be attributed to changes in intracellular amount of inorganic ions comes from a recent study we undertook on the ultrastructure of cells submitted to osmotic shocks (Delpire et al., 1985 a,b). As shown in figure 5, cultured rat PC12 cells show important ultrastructural changes when submitted to hyper or hypo-osmotic conditions. These changes essentially concern the electron density of the cytoplasmic compartment and the organization of the nucleus, chromatin appearing hypercondensated in concentrated media and decondensated in hypo-osmotic ones. These modifications can be related to changes in the ions intracellular environment rather than to changes in cell hydration. They seem also to be of general occurrence at least in mammalian cells, since we have shown them to be present in mouse Ehrlich ascited tumor cells, rat intestine and kidney tubular cells.

It is worthnoticing that changes in chromatin organization of the importance of those recorded in mammalian cells cannot be recorded in the cells of the crab *Carcinus maenas* submitted to variations in environmental osmolality in the same conditions (in vitro, same amplitude of osmotic shock, figure 5). Our crab muscle preparations, as the mammalian cells we have used, however show changes in ions concentrations during the in vitro osmotic challenges. Therefore, the fact that there is only slight changes , if any, in chromatin organization cannot be accounted for by a lack of modification in intracellular ions concentrations.

How can such a difference be interpreted ? At the moment, the possibility remains that euryhaline species such as the crab *C.maenas* might have evolved macromolecular components less sensitive to the disrupting effects of inorganic ions. Another explanation can be found in the fact that the cells of these species show large intracellular amounts of different amino-compounds (Gilles, 1975, 1986 for reviews) that could act in two different ways as "compensatory solutes". First of all, they appear to have a "stabilizing" effect

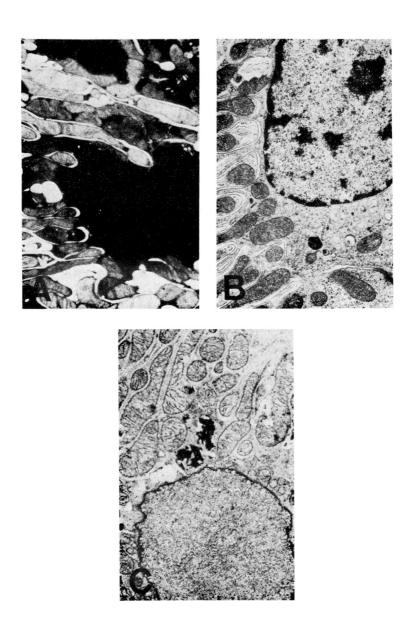

<u>Figure 5</u>

Effect of 30 minutes hypo and hyper-osmotic shocks on the ultrastructure of preparations from rat PC12 cells (A,B,C - X 10,350) and
Carcinus maenas leg muscle (D,E,F - 29,300). PC12 cells : A : hyperosmotic shock (Na$^+$ in saline : 262.8 mEq/1.) - B : control (Na$^+$ in

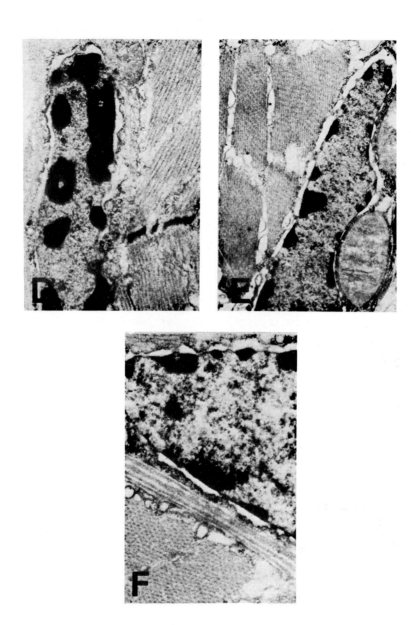

saline : 142.8 mEq/l.) - C : hypo-osmotic shock (Na$^+$ in saline :
60 mEq/l.) - *C.maenas* : D : hyperosmotic shock (Na$^+$ in saline : 960
mEq/l) - E : control (Na$^+$ in saline : 480 mEq/l.) - F : hypo-osmotic
shock (Na$^+$ in saline : 240 mEq/l.).

opposing the "disrupting" effects of the ions. In this framework,
let us recall that amino-compounds can counterbalance the effect of
monovalent ions on different macromolecular systems (see above and
also for review Yancey et al., 1982; Clark, 1985, 1986, Yancey,
1985). In intact cells however the relation amino-compounds inorganic ions is surely more complex than the one described in the
simple terms "stabilizing effects opposing to disrupting effects".
In this respect, the fact that different factors can modulate the
interplay between ions and amino-compounds as indicated in figure 4
should be taken into consideration.

How can monovalent ions disrupt the structure of macromolecules,
how can different amino-compounds oppose this effect and how can
different modulators interfere in this interplay is actually far
from being clear. Our understanding of these effects is indeed
strongly dependent on an actually lacking knowledge of 1) the equilibrium structure of water in the vicinity of proteins; 2) the
thermodynamical changes induced during the interactions ions- vicinal water-protein (see Von Hippel, 1976; Clark, 1986; Arakawa and
Timasheff, 1982, 1983, 1985 for discussion).

Amino-compounds seem also to be at work in a completely different
way. Their use as major osmotic effectors in the cells of euryhaline
invertebrates appears indeed to allow a regulation of the intracellular concentration of inorganic ions back to values close to control
during volume adjustment. In this respect, it is interesting to
consider that, though the intracellular concentration of K^+ decreases
in isolated axons and muscles of the green crab $C.maenas$ during
exposure to a hypo-osmotic saline (Kévers et al., 1979a,b; Kévers
and Gilles unpublished), there is no significant difference in the
content of this ion in the tissues of animals acclimated to sea
water or to a diluted medium (table 1). Similarly, K^+ is lost from
isolated coelomocytes of the osmoconforming worm $Glycera\ dibranchia$-
ta withstanding hypo-osmotic conditions. There is however no significant changes in the intracellular amount of this ion in the coelomocytes of animals acclimated to different salinities (Costa and
Pierce, 1982). In fact, the K^+ loss occurring in preparation of
isolated tissues from euryhaline invertebrates seems to be only
transient as indicated in the case of crustacean axons by studies

of radioactive K^+ efflux (Kévers et al., 1981) and by studies of changes in membrane potential in the case of molluscan nerve cells (Prior, 1981).

TABLE 1

K^+ level in tissues of euryhaline crustaceans after acclimation to sea water or to a diluted medium

Callinectes sapidus (a)	SW	SW/2
Axons	422 ± 16	394 ± 18
Muscle	186 ± 8	162 ± 12
Carcinus maenas (b)		
Gill	263 ± 24	214 ± 11
Hepatopancreas	288 ± 13	276 ± 17
Muscle	410 ± 27	373 ± 11
Axon	1613 ± 144	1446 ± 146

Values given in mEq/l. intracellular water for *C.sapidus* and in mEq/Kg dry weight for *C.maenas* (a) : Gérard and Gilles, 1972 (b) Kévers and Gilles, unpublished.

CONCLUSION

Euryhalinity in poecilosmotic species and its associated lack of "water intoxication" effects can thus be related to 2 different cellular mechanisms, both aiming to minimize the disrupting effects of inorganic ions on intracellular macromolecular structures. Cell of euryhaline poecilosmotic species seems indeed to have 1) the possibility to regulate the changes in the concentration of intra-cellular inorganic ions which occur during changes in blood osmola-lity. 2) The possibility to accumulate large amounts of different amino-compounds and to use them as osmotic effectors, regulating their amount with respect to the evolution of the blood osmolality.

ACKNOWLEDGMENTS

The work from my laboratory described in this review has been supported by successive grants "Crédits aux Chercheurs" from the FNRS
as well as by successive grants from the FRFC (n° 2.4516.86 for the
actual one) and a grant FRFC-IM n°130.

REFERENCES

Arakawa T, Timasheff SN (1982) Preferential interaction of proteins
    with salts in concentrated solution. Biochem. 21 : 6545-6552.
Arakawa T, Timasheff SN (1983) Preferential interaction of proteins
    with solvent components in aqueous amino acids solutions.
    Arch. Biochem. Biophys. 224 : 169-177.
Arakawa T, Timasheff SN (1985) The stabilization of proteins by
    osmolytes. Biophys. J. 47 : 411-414.
Cala PM (1983) Volume regulation by red blood cells : mechanisms
    of ion transport. Mol. Physiol. 4 : 33-52.
Cala PM (1985) Volume regulation by Amphiuma red blood cells :
    characteristics of volume-sensitive K/H and Na/H exchange.
    Mol. Physiol. 8 : 199-214.
Clark ME (1985) The osmotic role of amino acids : discovery and
    function. In : R.Gilles, M.Gilles-Baillien (eds) : Transport
    processes - Iono and Osmoregulation. Springer Verlag, Berlin,
    Heidelberg.
Clark ME (1986) Non-Donnan effects of organic osmolytes in cell
    volume changes. In : Current topics in membranes and transport.
    Academic Press, New York, London. In press.
Costa, CJ, Pierce SK (1982) Effect of divalent cations and metabolic
    inhibitors on Glycera red coelomocyte volume regulation and
    solute balance during hypoosmotic stress. J. Comp. Physiol. 151:
    133-144.
Delpire E, Duchene C, Goessens G, Gilles R (1985a) Effects of osmo-
    tic shocks on the ultrastructure of different tissues and cell
    types. Exp. Cell. Res. 160 : 106-116.
Delpire E, Gilles R, Duchene C, Goessens G (1958b) Effects of osmo-
    tic shocks on the ion content and ultrastructure of rat pheo-
    chromocytoma cells of line PC12. Mol. Physiol. 8 : 293-306.

Ellory JC, Hall AC, Stewart GW (1985) Volume-sensitive passive potassium fluxes in red cells. In : Transport processes - Iono and osmoregulation" R.Gilles and M.Gilles-Baillien (eds) Springer Verlag, Berlin, Heidelberg. pp. 401-410.

Freel RW (1978) Patterns of water and solute regulation in the muscle fibres of osmoconforming marine decapod crustaceans. J. Exp. Biol. 72 : 107-126.

Gérard, JF, Gilles R (1972) The free amino-acid pool in *Callinectes sapidus* (Rathbun) tissues and its role in the osmotic intracellular regulation. J. Exp. Mar. Biol. Ecol. 10: 125-136.

Gilles R (1969) Effect of various salts on the activity of enzymes implicated in amino acid metabolism. Arch. Internat. Physiol. Biochim. 77 : 441-464.

Gilles R (1974a) Métabolisme des acides aminés et contrôle du volume cellulaire. Arch. Internat. Physiol. Bioch. 82 : 423-589.

Gilles R (1974b) Studies on the effect of NaCl on the activity of *Eriocheir sinensis* glutamate dehydrogenase. Int. J. Biochem. 5 : 623-628.

Gilles (1975) Mechanisms of ion and osmoregulation. In : O.Kinne (ed) : Marine Ecology. Wiley Intersciences, London, New York, vol. 2, part 1, pp.259-347.

Gilles R (1978) Intracellular free amino acids and cell volume regulation during osmotic stresses. In : C.Barker Jorgensen, E.Skadhauge (eds) : Osmotic and volume regulation. A.Benzon symposium XI. Munksgaard, Copenhagen, pp.470-491.

Gilles R (1979) Intracellular organic osmotic effectors. In : R.Gilles (ed) : Mechanisms of osmoregulation in animals, chap.4 Wiley Interscience, London, New York , pp. 111-153.

Gilles R (1986) Volume regulation in cells of euryhaline invertebrates. In : Current Topics in Membrane and Transport. Academic Press, New York, in press.

Gilles R, Péqueux A (1983) Interactions of chemical and osmotic regulation with the environment. In : Vernberg FJ, Vernberg WB (eds) : The Biology of Crustacea. Vol. 8 : Environmental adaptations. Academic Press, New York, London, pp. 109-117.

Gilles R, Hogue P, Kearney EB (1971) Effect of various ions on the succinic dehydrogenase activity of *Mytilus californianus* Life Sci. 10 : Part II, 1421-1427.

Grinstein S, Goetz JD, Furuya W, Rothstein A, Gelfand EW (1985) Mechanism of regulatory volume increase in osmotically shrinken lymphocytes. Mol. Physiol. 8 : 185-198.

Hamabata A, Von Hippel PH (1973) Effects of vicinal groups on the specificity of binding of ions to amide groups. Biochem. 12 : 1264-1271.

Hoffmann EK (1977) Control of cell volume. In : Gupta BL, Moreton RB, Oschman J, Wall BJ (eds) : Transport of ions and water in animals. Academic Press, London, New York, pp. 285-332.

Hoffmann EK (1985a) Regulatory volume decrease in Ehrlich ascites tumor cells : role of inorganic ions and amino-compounds. Mol. Physiol. 8 : 167-184.

Hoffmann EK (1985b) Cell volume control and ion transport in a mammalian cell. In : R.Gilles, M.Gilles-Baillien (eds) : Transport processes - Iono and osmoregulation. Springer Verlag Berlin, Heidelberg, pp. 389-400.

Hoffmann EK (1986) Volume regulation in cultured cells. In : Current topics in membrane and transport. Academic Press, New York, London, in the press.

Kévers C, Péqueux A, Gilles R (1979a) Effects of hypo- and hyper-osmotic shocks on the volume and ions content of *Carcinus maenas* isolated axons. Comp. Biochem. Physiol. 64A : 427-431.

Kévers C, Péqueux A, Gilles R (1979b) Effects of an hypo-osmotic shock on Na^+, K^+ anc Cl^- levels in isolated axons of *Carcinus maenas*. J. Comp. Physiol. 129 : 365-371.

Kévers C, Péqueux A, Gilles R (1981) Role of K^+ in the cell volume regulation response of isolated axons of *Carcinus maenas* submitted to hypo-osmotic conditions. Mol. Physiol. 1 : 13-22.

Kregenow FM (1981) Osmoregulatory salt transporting mechanisms : control of cell volume in anisotonic media. Ann. Rev. Physiol. 43 : 493-505.

Lauf PK (1985) On the relationship between volume and thiol-stimu-lated K^+ Cl^- fluxes in red cell membranes. Mol. Physiol. 8 : 215-234.

Moran WM, Pierce SK (1984) The mechanism of crustacean salinity tolerance : cell volume regulation by K^+ and glycine effluxes. Mar. Biol. 81 : 41-46.

Potts WTW, Parry G (1964) Osmotic and ionic regulation in animals. Pergamon Press, London, New York . 423 pp.

Prior DJ (1981) Hydration related behavior and the effects of osmotic stress on motor function in the slugs. *Limax maximus* and *Limax pseudoflavus*. In : J. Salanki (ed) : Neural Biology of Invertebrates. Pergamon Press, Oxford. Adv. Physiol. Sci. 23 : 131-145.

Schleich T, Von Hippel PH (1970) Ion-induced water-proton chemical shifts and the conformational stability of macromolecules. Biochem., 9 : 1059-1066.

Schoffeniels E, Gilles R (1963) Effect of cations on the activity of L-glutamic acid dehydrogenase. Life Sci. 2 : 834-839.

Skaer HL (1974) The water balance in a serpulid polychaete, *Mercierella enigmatica* (Fauvel). I. Osmotic concentration and volume regulation. J. Exp. Biol. 60 : 351-370.

Spaargaren DH (1979) Marine and brackish water animals. In : Maloiy GMO (ed) : Comparative physiology of osmoregulation in animals. Vol. 1, Academic Press, New York, London, pp. 83-116.

Von Hippel PH (1976) Neutral salts effects on the conformational stability of biological macromolecules : model studies. In : Alfsen A, Berteaud AJ (eds) : L'eau et les systèmes biologiques. Colloques CNRS n° 246. Editions du CNRS, Paris, pp. 19-26.

Von Hippel PH, Schleich T (1969a) The effects of neutral salts on the structure and conformational stability of macromolecuels in solution. In : Fasman G, Timasheff S (eds) : Biological macromolecules. Vol. II pp. 417-574. M.Dekker , New York.

Von Hippel PH, Schleich T (1969b) Ion effects on the solution structure of biological macromolecules. Accts Chem. Res. 2 : 257-265.

Von Hippel PH, Wong KY (1963) The collagen-gelatin phase transition. Biochem. 2 ; 1387-1398.

Von Hippel PH, Wong KY (1965) On the conformational stability of globular proteins : the ribonuclease transition. J. Biol. Chem. 240 : 3909-3923.

Von Hippel PH, Peticolas V, Schack L, Karlson L (1973) Model studies on the effects of neutral salts on the conformational stability of biological macromolecules I. Ion binding to polyacrylamide and polystyrene columns. Biochemistry 12 : 1256-1264.

Wanson S, Péqueux A, Gilles R (1983) Osmoregulation in the stone
 crab *Cancer pagurus*. Mar. Biol. Let. 4 : 321-330.

Yancey PH (1985) Organic osmotic effectors in cartilaginous fishes.
 In : R.Gilles, M.Gilles-Baillien (eds) : Transport processes -
 Iono and Osmoregulation. Current Comparative approaches.
 Springer Verlag Heidelberg, New York, pp. 424-436.

Yancey PH, Clark ME, Hand SC, Bowlus RD, Somero GN (1982) Living
 with water stress : evolution of osmolyte systems. Science ,
 217 : 1214-1222.

WATER VAPOUR ABSORPTION BY ARTHROPODS: DIFFERENT SITES, DIFFERENT MECHANISMS

Mike J. O'Donnel

Department of Biology
McMaster University
Hamilton, Ontario, Canada

The capacity for water vapour absorption appears to have arisen independently in several different groups of terrestrial arthropods. Absorption occurs at localized sites in the mouth or rectum, and a number of strikingly different mechanisms are employed to reduce water activity at these sites.

More than 60 species of terrestrial arthropod can actively absorb water vapour from the atmosphere. Absorption from subsaturated humidities has been demonstrated in mites, hard and soft ticks, tenebrionid and anobiid beetles, tineid moths, desert cockroaches, firebrats, silverfish, booklice, biting lice and fleas. This unusual form of water transport not only necessitates a phase change, but also involves the movement of water against enormous thermodynamic gradients. Water vapour absorption (WVA) at the lowest humidity recorded to date, 43% RH, is equivalent to moving water against an osmotic gradient of 70 osmol l^{-1} or a hydrostatic gradient of 1400 bar (1.4×10^5 kPa). By comparison, transport by salt glands of marine reptiles and birds involves much smaller gradients of about 1 osmol kg^{-1}. The rates of transport are also striking; the biting lice (Mallophaga) for example, can increase their weight 10%/hr by means of WVA (Rudolph, 1983), although a figure of 3-4%/day is more typical for most arthropods (Machin, 1979a). Surprisingly, although WVA involves the movement of water against large gradients, the amount of energy required is quite small, usually less than 1% of the animal's daily metabolic rate (Edney, 1977).

Early studies of absorption incorrectly suggested the animal's entire body surface as a site of uptake, and a single cuticular mechanism was proposed for a wide variety of acarine and insect species (Beament 1965). An important aspect of recent work has been to demonstrate that not only are there a number of different localized uptake sites, but that several fundamentally different mechanisms are used to reduce water activity so as to permit uptake at these sites.

These discoveries of independently evolved uptake mechanisms underscore the importance of water vapour absorption to maintenance of water balance, especially in environments where the availability of free water is limited. Although integumentary,

Comparative Physiology: Life in Water and on Land. P. Dejours, L. Bolis, C.R. Taylor, E.R. Weibel (eds.) Fidia Research Series, IX-Liviana Press, Padova © 1987

respiratory, and excretory water loss can be reduced, there are finite limits to the effectiveness of such measures, particularly for small organisms whose high ratios of surface area to volume render them vulnerable to dessication.   WVA may be necessary, therefore, for long term survival in environments where the distribution of free water is temporally or spatially uneven.

This paper is concerned primarily with absorption by two species which have proven particularly amenable to physiological experiments, and which demonstrate dramatically different absorption sites and mechanisms.   The absorption mechanisms of some other arthropods are discussed with reference to these two extremes.

ABSORPTION BY MEALWORMS (TENEBRIO MOLITOR): PRODUCTION OF CONCENTRATED KCL BY EPITHELIAL CELLS OF THE RECTAL COMPLEX.

Mealworm larvae can absorb water vapour in humidities exceeding 88% RH. The site of absorption, the rectum, presents an example of the cryptonephric condition, wherein the distal ends of the Malpighian tubules are closely applied to the rectum and enclosed with it in a special chamber, the perinephric space, which is separated from the rest of the body cavity by the perinephric membrane (Grimstone et al.., 1968). This complex operates in two modes (Machin, 1979b).   The first and probably primordial mode is that of fecal dehydration. Water activity in the lumen of the tubules is reduced by concentrations of KCl as high as 2 mol l^{-1}. Water moves passively, therefore, across the rectal cuticle and epithelium into the perinephric space and from there into the tubules.   During fecal dehydration the anus is closed and the absorption mechanism keeps pace with the entry of water in fecal material passed down from the midgut. In this mode the rectal contents and the intervening compartments become fully equilibrated with the highest osmotic pressures of the complex.   In contrast, the anus is open to the atmosphere during the vapour absorption mode, and the fecal material equilibrates with ambient humidity. A steady state is established in which water activities decrease radially to their lowest values in the Malpighian tubules.

Current studies of the absorption mechanism in Tenebrio are concerned with the cellular mechanisms responsible for concentration of KCl in the tubule lumen. An isolated and perfused preparation of the rectal complex has been developed (Tupy and Machin, 1985), and preliminary measurements of electrical potential and K^+ activities in the tubule lumen and perinephric space have been obtained (O'Donnell and Machin, unpublished observations). Double-barreled microelectrodes constructed from theta glass are resistant to tip breakage as the tissue sheaths surrounding the rectal complex are penetrated. These electrodes permit successful impalement of the rectal complex compartments. One barrel is used to measure potential and the other is used either to inject dye or to measure ionic activity. Correlation of dye location with potential measurements confirm the results of Grimstone et al. (1968) indicating a tubular compartment at a potential of +50 to +70 mV relative to the bathing saline, and a surrounding perinephric compartment at a potential of -15 to -20 mV. Moreover, K^+ activities in excess of 2 mol l^{-1} have been measured in the tubule lumen. K^+ activities in the surrounding perinephric spaces are also above the equilibrium concentration, suggesting that concentration of KCl in the tubule lumen involves a two-stage pump, transporting K^+ first from the hemolymph into the perinephric space, and secondly from the perinephric space into the tubules. Future studies will examine the effect of hemolymph ion composition and rectal lumen water activity on K^+ activities in the tubule lumen.

Elevated K^+ activities in the tubule lumen confirm high osmotic pressures of tubular fluid indicated in micropuncture experiments (Grimstone et al., 1968) or analysis of frozen sections. All of these data support the view that atmospheric water vapour absorption in Tenebrio is dependent upon a reduction of water activity by production of a concentrated solution of KCl.

Several aspects of this absorption mechanism deserve comment. Firstly, the tubule cells must be capable of cell volume regulation as the flux of water through them changes, either in response to fluctuations of ambient humidity or when lumenal osmotic pressures decrease as the animal ceases absorption and prepares to moult. Secondly, sites of reduced water activity must be restricted so as to prevent a counteracting osmotic flux from the hemolymph into the tubule lumen. Most Malpighian tubules secrete iso-osmotic fluids at relatively high rates, whereas Tenebrio tubules secrete extremely hypertonic fluids at low rates. This difference is presumably due to the presence in the latter case of an impermeant perinephric membrane, which permits localized access of KCl to the tubules but restricts osmotically coupled water flux from hemolymph to lumen.

Concentrated aqueous solutions may well be involved in the oral uptake mechanisms of mites, ticks, Psocoptera and Mallophaga. However, it is only for Tenebrio that several different experimental techniques have been employed to measure the concentration or osmolality of the fluid at the source and time of its production. For other species, the evidence for production of hygroscopic solutions is circumstantial and has been inferred from the presence of high concentrations or crystalline deposits of inorganic salts on the mouthparts or associated ducts of dehydrated animals. However, dried residues that appear during dehydration may have little to do with the hygroscopic properties of the saliva during rehydration. Instead, they may be secreted there for temporary storage excretion to help maintain hemolymph ion concentrations and osmotic pressures (Needham and Teel, 1985).

Rectal sites of absorption in other insects

In firebrats and silverfish there is no evidence for the production of concentrated salt solutions by the rectal epithelium during uptake (Noble-Nesbitt, 1978). An electro-osmotic mechanism which has been proposed to explain the reduction of water activity in the epithelium of these insects suffers from serious flaws because it is based on experimental observations made under clearly non-physiological conditions (Kuppers and Thurm, 1980). Apparent electro-osmotic water transport was measured when the rectum was filled with 500 mmol 1^{-1} KCl which has a water activity (0.994) that is negligibly lower than that of hemolymph (0.995) and may actually be higher than the water activities in the hemolymph and tissues of dehydrated animals. Moreover, the presence of KCl in the lumen might promote KCl reabsorption and associated osmotic water flux in the tissues of the posterior rectum. Clearly, further evaluation of the possible significance of electro-osmotic water flow must await experiments in which the rectal lumen is filled with humid air or with a concentrated solution of non-electrolytes.

ABSORPTION BY THE DESERT COCKROACH:
ORAL UPTAKE, AND HYDROPHILIC CUTICLE

Absorption by the desert cockroach, Arenivaga investigata provides a sharp contrast to the solute-dependent rectal uptake mechanism proposed for Tenebrio. The oral absorption site of Arenivaga is utilized solely for atmospheric uptake, a transporting epithelium is not a component of the absorption system, and there is no evidence for the production of hyperosmotic solutions.

During absorption two bladder-like diverticula of the hypopharynx are protruded from the buccal cavity by inflation of the hypopharynx with hemolymph (O'Donnell, 1981a,b). Each bladder is covered with a dense mat of more than 10^4 helicoid cuticular hairs which are 160 - 180 nm in diameter over most of their length of 80 -100 um. Measurements of bladder surface temperature with micro-thermocouples indicate that the bladders are the sites of condensation during WVA (O'Donnell, 1977, 1978). During absorption a non-hygroscopic fluid is applied to the bladder surface. The fluid appears to be an ultrafiltrate of the hemolymph and is produced by a pair of structures called frontal bodies. Each spheroidal cuticle-derived frontal body is connected to the ipsilateral bladder by a groove in the epipharynx. Distortion of each frontal body by cyclical contractions of a heavy mass of frontal muscles forces fluid across a porous cuticular plate which opens into the epipharyngeal groove. Fluid moves by capillarity along the groove and onto the bladder. The frequency of muscle contraction increases with humidity, and, therefore, with the rate of condensation (O'Donnell, 1981a). This relationship suggests that sophisticated feedback mechanisms may maximize the rate of uptake by ensuring that the amount of water condensed per frontal body cycle is relatively constant (Machin et al., 1982). Applications of fluorescent or radioactive tracer solutions to the bladders indicate that the frontal body fluid and condensate moves posteriorly and laterally away from the point at the anterior edge of the labrum where the epipharyngeal groove contacts the bladder surface (O'Donnell, 1981b). The tracers appear transiently on the posterior hypopharynx and accumulate in the esophagus and crop (Figure 1).

Several types of experimental evidence indicate that the frontal body fluid is not hygroscopic. Calculations based on measurements of freezing point depression in frontal body tissues indicate osmolatities of only 645 mOsm kg^{-1}, whereas a hygroscopic solution capable of absorption at the threshold of 81% RH would require osmolalities in excess of 11,700 mOsm Kg^{-1} (O'Donnell, 1981a). Furthermore, analysis of frontal bodies and bladder tissues by neutron activation and atomic absorption spectrophotometry indicate that the

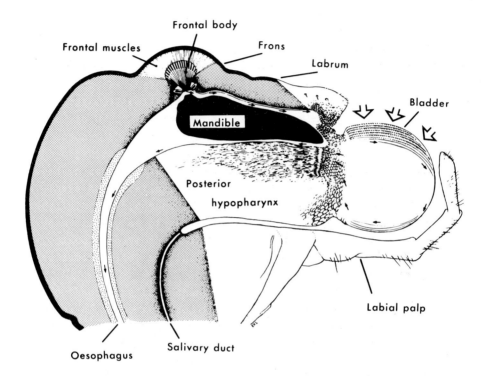

Figure 1

Drawing of the head of <u>Arenivaga</u> with portions of the head removed to show the surface features of the distal end of the epipharynx and the hypopharynx. Parts of the head in sagittal section are stippled. Solid arrows indicate fluid movements from the site of production in the frontal bodies to the oesophagus. Atmospheric water condenses on to the bladders (open arrows) and is entrained in the fluid movements towards the oesophagus.

levels of inorganic salts in the absorption system are not elevated above normal physiological levels (O'Donnell, 1982a). An observation of the bladder surfaces of absorbing animals supports this conclusion. When ambient humidity is rapidly lowered below threshold the bladder surface dries out and condensation ceases. However, no crystalline deposits are evident on the surfaces of these dried bladders, suggesting that high concentrations of inorganic salts are not present on bladders during absorption. In addition, a number of micro-methods based on fluorometric reagents have been adapted for determination of the levels of hydrophilic organic molecules such as peptides, amino acids, polyhydroxl alcohols and reducing sugars on the bladder surface. The sum of the concentrations of all of these compounds is less than 2 mmol l^{-1}. Nor do frontal bodies appear to be active sites for synthesis of hydrophilic proteins or polysaccharides, because they do not accumulate significant amounts of radiolabelled precursors from the hemolymph.

These results suggested that the frontal bodies play a subordinate or ancillary role in the absorption machanism, and that the reduction of water activity at the absorption site is dependent upon the hydrophilic nature of the cuticular hairs covering the bladder surface.

Bladder cuticle has a much greater water affinity than do unspecialized cuticles from other regions of the body of this and other species (O'Donnell, 1982b). In other words, at a given humidity, the water content of bladder cuticle (g H_2O/g dry weight) is greater than that of non-specialized cuticle. Moreover, these differences remain in washed cuticle samples, indicating that the hydrophilic properties of the bladder cuticle are not due to dissolved salts in frontal body fluid applied to the bladders in vivo. Surprisingly, the addition of salts to the bladder cuticle actually reduces water affinity. Corresponding examinations of bladder cuticle in the scanning electron microscope indicate that the hairs do, in fact, swell in conditions of low ionic strength (O'Donnell, 1982b).

Swelling of the bladder hairs in conditions of low ionic strength is an integral part of the proposed mechanism of WVA (O'Donnell, 1982b). It is well known that polyelectrolytes, such as chitin and protein, or synthetic polyacrylamide gels, swell as the ionic strength of the surrounding fluids is reduced. Much of this effect arises because increasing salt concentration reduces the mutual repulsive forces between like charges on a protein or gel structure, a greater number of cross-links, possibly of the van der Walls type, is then possible and shrinkage occurs (Katchalsky, 1954). Tanaka (1981) has summarized experiments demonstrating that in some conditions the swelling or shrinkage of a gel is discontinuous, so that an infinitesimal change in temperature, pH or ionic strength can cause a large change in volume.

Absorption produced by cyclical modulation of the water affinity of the bladder hairs may occur as follows. When the bladders are first protruded, the cockroach salivates copiously and much of this saliva evaporates. Gradually the rate of evaporation declines and the water activity of the bladder cuticle approaches equilibrium with atmospheric water activity. The addition of frontal body fluid perturbs this equilibrium. Contraction of the frontal muscles results in a short pulse of fluid of relatively high ionic strength being applied to the bladders, and moving posteriorly and laterally over the bladder surface as a zone or wave of increased ionic strength. The water affinity of the hairs is transiently lowered, and they release water and reduce their volume. Some of this fluid is swallowed.

At this point in the cycle, it is suggested that the water affinity of the hairs is reduced but highly unstable. Similar properties are apparent in synthetic polyelectrolytes over particular ranges of ionic strength or pH. Tiny thermal fluctuations and the randomness of molecular motions will result in some condensation onto the bladder surface. An infinitesimal amount of condensation will decrease the ionic strength of the fluid on the bladder hairs, and their water affinity will increase, causing further net condensation. This positive feedback upon the water affinity of the hairs continues rapidly, until the volume of the hairs is constrained by polymer elasticity or by the next addition of frontal body fluid.

Oral sites of absorption in other insects.

WVA involving the mouthparts has also been described in mites (Wharton and Furumizo, 1977), ticks (Rudolph and Knulle, 1974), the Psocoptera, or booklice (Rudolph, 1982) and the Mallophaga or biting lice (Rudolph, 1983). The absorbing systems of the latter two groups are in some ways similar to that of Arenivaga. A thin iridescent layer of fluid present on the ventral hypopharyngeal surface during absorption quickly dries on exposure to humidities below the CEH. This fluid, apparently produced by the dorsal labial glands, presumably moves across the mouthparts. It is not known if the fluid is hygroscopic.

Although the hypopharyngeal bladders of Areneivaga are utilized only for the absorption of atmospheric water, the African termite Macrotermes employs similar structures for the imbibition of water from porous surfaces (Sieber and Kokwaro, 1982). This arrangement may represent an early step in the evolution of the more dramatic uptake mechanism of the desert cockroach.

References

Beament JWL (1965) The active transport of water: evidence, models and mechanisms. Symp Soc exp Biol 19: 273-298.

Edney EB (1977) Water balance in land arthropods. Springer, Heidelberg; 288 pp.

Grimstone AV, Mullinger AM, Ramsay JA (1968) Further studies on the rectal complex of the mealworm, Tenebrio molitor, L. (Coleoptera, Tenebrionidae) Proc Roy Soc Lond B253: 343-382.

Katchalsky, A. (1954) Polyelectrolyte gels. Prog. Biophys. biophys. Chem. 4: 1-59.

Kuppers J, Thurm U (1980) Water transport by electroosmosis. In: Locke M, Smith DS (eds): Insect Biology in the Future. Academic Press, New York; pp. 125-144.

Machin, J. (1979a) Atmospheric water absorption in arthropods. Adv. Insect. Physiol. 14: 1-48.

Machin, J. (1979b) Compartmental osmotic pressures in the rectal complex of Tenebrio larvae: evidence for a single tubular pumping site. J exp Biol 82: 123-137.

Machin J, O'Donnell MJ, Coutchie PA (1982) Mechanisms of water vapor absorption in insects. J Exp Zool 222: 309-320.

Needham GR, Teel PD (1985) Water balance by ticks between bloodmeals. In: Sauer JR, Hair JA, (eds): Morphology, physiology, and behavioral biology of ticks.
Ellis Horwood Ltd., Chickester UK, in Press

Noble-Nesbitt, J.N. (1978) Absorption of water vapour by Thermobia domestica and other insects.
In: Schmidt-Nielsen K, Bolis L, Maddrell SHP (eds):Comparative Physiology - Water, Ions and Fluid mechanics. Cambridge University Press, Cambridge; pp. 53-66.

O'Donnell MJ (1977) Site of water vapor absorption in the desert cockroach, Arenivaga investigata. Proc. natn Acad. Sci. U.S.A. 74: 1757-1760.

O'Donnell MJ (1978) The site of water vapour absorption in Arenivaga investigata. In: Schmidt-Nielsen K, Bolis L, Maddnell SHP (eds):
Comparative Physiology-Water, Ions and Fluid Mechanics. Cambridge University Press, Cambridge; pp. 115-121.

O'Donnell MJ (1981a) Frontal bodies: novel structures involved in water vapour absorption in the desert burrowing cockroach, Arenivaga investigata. Tissue and Cell 13: 541-555.

O'Donnell, MJ (1981b) Fluid movements during water vapour absorption by the desert burrowing cockroach, Arenivaga investigata. J. Insect Physiol. 27: 877-887.

O'Donnell, MJ (1982a) Water vapour absorptin by the desert burrowing cockroach: evidence against a solute dependent mechanism. J. exp. Biol. 96: 251-262.

O'Donnell, MJ (1982b) Hydrophilic cuticle- The basis for water vapour absorption by the desert burrowing cockroach, Arenivaga investigata. J. exp. Biol. 99: 43-60.

Rudolph D (1982) Site, process and mechanism of active uptake of water vapour from the atmosphere in the Psocoptera. J. Insect Physiol 28: 205-212.

Rudolph D (1983) The water-vapour uptake system of the Phthiraptera. J Insect Physiol 29: 15-25.

Rudolph D, Knulle W (1974) Site and mechanism of water vapour uptake from the atmosphere in ixodid ticks. Nature 249: 84-85.

Sieber R, Kokwaro D (1982) Water intake by the termite Macrotermes michaelseni. Ent exp appl 31: 147-153.

Tanaka, T (1981) Gels. Scientific American 244 (1) 124-138.

Tupy J, Machin J (1985) Transport characteristics of the isolated rectal complex of the mealworm Tenebrio molitor. Can J Zool 63: 1897-1903.

Wharton GW, Furumizo RT (1977) Supracoxal gland secretions as a source of fresh water for Acaridei. Acarologia 19: 112-116.

OSMOREGULATION IN TERRESTRIAL AND AQUATIC INSECTS

Simon Maddrell

A.F.R.C. Unit, Department of Zoology
University of Cambridge
Downing Street, Cambridge CB2 3EJ, U.K.

It has recently been discovered that some flying insects generate water faster by metabolism than they lose it by respiration and evaporation. Depending on the level of their flight activity, then, terrestrial insects may face problems ranging from waterloading to desiccation. Insects living in aquatic environments from fresh water to highly saline waters face much the same problems. How terrestrial and aquatic insects cope with these situations is described and compared.

INTRODUCTION

Most insects live in the terrestrial environment and under most conditions (loosely where the relative humidity is less than 98%) this environment is a drying one (Edney, 1977). Insects, because they are very small relative to most other terrestrial organisms, have a higher surface area/volume ratio. This suggests that particularly for them, life on land poses acute problems of maintaining water content; they seem certain to be prone to desiccation. On the other hand, insects living in fresh water or water whose osmotic concentration is less than that of their haemolymph are likely to face exactly the opposite problem - they have to cope with a constant surplus of water as a result of osmotic influx. Understandably, then virtually all accounts of insect osmoregulation have concentrated on adaptations for water conservation in terrestrial insects and for water elimination in insects living in dilute waters.

RESPIRATORY WATER LOSS IN TERRESTRIAL INSECTS

A clear gap in our understanding of osmoregulation in terrestrial insects has been any detailed knowledge of the respiratory water loss. As I have

Comparative Physiology: Life in Water and on Land. P. Dejours, L. Bolis, C.R. Taylor, E.R. Weibel (eds.) Fidia Research Series, IX-Liviana Press, Padova © 1987

urged before (Maddrell, 1982), "What is now urgently needed are some actual determinations of rate of respiratory water loss in insects...... It is of major importance to see how the figures for terrestrial insects compare with those for vertebrates of broadly similar size, not only in respect of water loss for both groups of animals both at rest and in activity, but also in relation to the oxygen consumption. The interesting question is: How much water is lost by insects and by vertebrates when they take up similar amounts of oxygen? At the risk of unnecessary repetition, it does seem that this is the major unexplored area in the field of insect osmoregulation".

Recent work, particularly by Nicolson & Louw (1982) and by Bertsch (1984) has now explored this area and answered the interesting question. As a result the simple dichotomy between the problems of insects on land and in water is destroyed, and the way the one views osmoregulation in terrestrial insects is completely changed. Their important and surprising results show that under normal environmental conditions, flying carpenter bees and bumblebees produce water by metabolism faster than it is lost by evaporation from the respiratory system and general body surface. Given that fuel for flight in these bees is ingested as nectar (25-50% sucrose solution in water) the metabolism of the sugar from the nectar produces yet more water. The net result may be a very considerable water load for the flying insect; Bertsch's calculations (1984) suggest that for a 220 mg male bumblebee, the water load produced by the flight activity of a single day equals the whole volume of body water. The relative rates of metabolic water production and evaporation water loss in flying birds and bats are very different from those of flying insects. Both flying birds and bats lose water faster than they produce it, whereas the insects that have so far been examined produce water faster than they lose it.

These discoveries raise some new questions but also, as we shall see later, provide explanations for what had seemed surprising or unnecessary adaptations of the excretory systems of terrestrial insects.

The major question that is raised is how the insect respiratory system is adapted to provide in flight the very effective water retention that is observed. The best analysis of insect respiration and respiratory water loss is that of Kestler (1985) but his elegant work is confined to the situation of relatively inactive insects. The major adaptations he discovered are concerned with the advantages of maintaining the spiracles at or near the closed position so that inward air movements occur at high linear speeds and prevent outward diffusive water losses. Such a mechanism presumably does not help much in understanding the situation in flight. At the moment then the question of what are the water saving features of the tracheal system in flight cannot be answered.

One possibility might be not that the insect respiratory system is so effective in water retention in flight but that the systems in flying bats and birds may have evolved to be less effective in order to contribute to thermoregulation. Flying bats and birds are large relative to insects and may have problems in eliminating excess heat. If so then it would not be surprising if more water were evaporated from the respiratory system, perhaps from the surfaces not involved in gas exchange in order to provide cooling. It might be interesting to measure water loss from the smallest birds or bats flying in cool air to see if the amount of water loss per unit of oxygen taken up were smaller than in the cases so far examined.

THE OCCURRENCE OF DIURETIC HORMONES IN TERRESTRIAL INSECTS

In virtually all cases where they have been sought, insects have been shown to possess a diuretic hormone capable of stimulating a great increase in the rate of fluid secretion by the Malpighian tubules (see Table 1). This is of obvious use in blood sucking insects such as mosquitos, the tsetse fly and Rhodnius prolixus and its relatives where the need after a blood meal is rapidly to eliminate surplus fluid (Maddrell, 1980) and in insects such as newly emerged adult lepidopterans where the haemolymph volume is greatly reduced, giving an appropriate weight reduction for insects about to begin a flying phase (see Nicolson, 1976 for example). In other cases, however, the need for a diuretic hormone has not been so clear. A particularly striking example is the recent discovery of a diuretic hormone in a desert beetle (Nicolson & Hanrahan, 1986), in which one would not expect a need for rapid fluid

Table 1 Insects for which there is evidence that the rate of fluid transport by their Malpighian tubules is regulated by hormones.

Order	Species	Reference
Hymenoptera	Apis mellifera	Altmann (1956)
Diptera	Calliphora vomitoria	Knowles (1976)
	Calliphora erthyrocephala	Schwartz & Reynolds (1979)
	Glossina morsitans	Gee (1976)
	Aedes taeniorhynchus	Maddrell & Phillips (1978)
	Anopheles freeborni	Nijhout & Carrow (1978)
	Aedes aegypti	Williams & Beyenbach (1983)
Lepidoptera	Pieris brassicae	Nicolson (1976)
	Calpodes ethlius	Ryerse (1978)
	Danaus plexippus	Dores, Dallman & Herman (1979)
Coleoptera	Anisotarsus cupripennis	Nunez (1956)
	Onymacris plana	Nicolson & Hanrahan (1986)
Hemiptera	Rhodnius prolixus	Maddrell (1962)
	Triatoma infestans	Maddrell (unpublished results)
	Triatoma phyllosoma	" " "
	Dipetalogaster maxima	" " "
	Dysdercus fasciatus	Berridge (1966)
Orthoptera	Periplaneta americana	Mills (1967)
	Locusta migratoria	Cazal & Girardie (1968)
	Schistocerca gregaria	Mordue (1969)
	Carausius morosus	Pilcher (1970)

elimination. For such insects has been argued that they might need to release a diuretic hormone during activity, as in flight for example, in order to accelerate the activity of the excretory system, not so that fluid can be eliminated but so that the haemolymph can be much more rapidly filtered through the tubules to remove the metabolic wastes which accumulate more rapidly during activity (Maddrell, 1980); It was seen as a necessary concomitant to this that water, ions, and useful substances in the fast flow of primary excretory fluid would be rapidly reabsorbed in the hindgut (Maddrell, 1980); no water need be lost. However, if the results obtained from bees are typical, it is now apparent that flying insects are still likely to need to reabsorb ions and useful solutes such as sugars and amino acids but should eliminate the water.

It has often been observed that insects urinate in flight. They include such small insects as aphids (Cockbain, 1961) and others such as honeybees (Pasedach-Poeverlein, 1941), Rhodnius (Gringorten & Friend, 1979), carpenter bees (Nicolson & Louw, 1982), and bumblebees (Bertsch, 1984). The significance of this as a consequence of water generation during flight is now clear.

Since insect flight muscles are the most active tissues known, they must produce metabolic water faster than any other tissue and, as this is not all removed in respiration and by evaporation from the general body surface, the rate at which a water load accumulates may also be fast. This may explain why it is that many insects not only have diuretic hormones but when released they induce fluid secretion in the Malpighian tubules at what may seem to be unnecessarily high rates. For example, the Malpighian tubules of the blowfly Calliphora can each secrete fluid at rates of 30-40 nl min^{-1} (Knowles 1976; Schwartz & Reynolds, 1979) a rate which per unit length approaches that seen in the Malpighian tubules from bloodsucking insects which are thought to be capable of the highest known rates of fluid secretion per unit weight of tissue. Perhaps even more impressive is the recently reported finding that Malpighian tubules of the desert beetle, Onymacris plana, can when stimulated each secrete fluid at rates higher than 100 nl min^{-1} and this would otherwise seem very difficult to explain.

RECTAL ION ABSORPTION IN TERRESTRIAL INSECTS

In Phillips' pioneering studies on the physiology of the insect rectum (Phillips, 1964, 1969) it was clear that the locust and blowfly could produce urine which contained very low levels of ions, as low as 6mM Cl in the blowfly and 1mM Na, 22m M K & 5mM Cl in the locust. Of course such an ability would be useful to the insect if fed on an ion deficient diet but this unlikely in the wild. An alternative or an additional explanation is that such ion conservation would be needed during diuresis in flight.

WATER BALANCE IN TERRESTRIAL INSECTS

The above discussion concentrates on the water loading problems that terrestrial insects may face during and after flight. Their situation depends on the fact that the respiratory system loses water more slowly than it is produced by the metabolism that the respiration supports. Given the water conserving adaptations of the system that come into play in relatively quiescent insects (Kestler, 1985), it might seem that they too would have similar problems. However, insect cuticle although of low permeability to water cannot prevent a slow loss (Edney, 1977). The rate at which this occurs relative to respiratory loss will depend on the particular insect, on its level of activity and on the environmental conditions. It tsetse flies, cuticular loss is about 75% of the total in resting flies but is only 43% where the flies are active 30% of the time (Bursell, 1959) while in flying locusts cuticular losses have been estimated as being only 1/4 - 1/3 of the respiratory losses (Loveridge, 1968). In a non-active insect it must be the case that spiracular water loss is at a much lower level, in line with the much reduced metabolism. However, cuticular loss while it may be lower at what may be a lower body temperature, will still be at appreciable levels. Probably now water loss from the respiratory system and through the general body cuticle will much exceed that metabolically produced and the insect will face just the sorts of desiccating problems that nearly all accounts of insect osmoregulation have until now assumed it would face under nearly all conditions.

So if these considerations are valid, terrestrial insects may face a water load when active but net water loss when not active. Intriguing possibilities present themselves. Terrestrial insects might use periods of activity as a means of replacing water loss during a preceding inactive phase - though at the cost of using up food reserves. The success of insects in the competition with vertebrates in the terrestrial environment for niches that can be occupied by small animals (of say, less than 10g) may well have a lot to do with a superior osmoregulatory power (Maddrell, 1981; 1982), possibly now this can be attributed to an ability (so far unique to insects), to give themselves a positive water balance by activity.

Some circumstantial evidence provides support for these ideas. Flightless insects would disqualify themselves from the advantages we have discussed above. When Edney (1977) compiled a list of all the insects known to be able to absorb water vapour from unsaturated air, he noted that they were all wingless. He supposed that this was mere coincidence, but it is at least reasonable to suggest that since wingless insects could not as easily generate water by intense muscular activity as can flighted insects, there may have been more selection pressure on them to develop alternative water-gaining systems.

TERRESTRIAL AND AQUATIC INSECTS

Taking into account the likelihood that the osmoregulatory problems of terrestrial insects depends on their recent patterns of activity, we can now compare the responses they make with those that aquatic insects make to similar problems. We can, for example, consider fresh water insects and active terrestrial insects; both are likely to have to react to the presence of surplus water. On the other hand, many terrestrial insects have to cope with the drying affects of the environment as do aquatic insects living in hyperosmotic waters.

RESPONSES TO WATERLOADING

How do terrestrial and fresh-water insects deal with the problems of surplus water? It is argued above that terrestrial insects release diuretic hormone to stimulate fast secretion by their Malpighian tubules. Reabsorption of ions in the hindgut leaves a dilute urine to be eliminated. In fresh water insects the difference is that here the water load is continuous and what evidence there is suggests that the tubules secrete continuously, presumably under the control of a diuretic hormone. Since Malpighian tubule fluid is iso-osmotic with the haemolymph, here too there must be reabsorption of ions in the hindgut, and this is known to occur (Ramsay, 1950). There is an indication that the usually continuous tubule activity is under control, for in larvae of Aedes aegypti transferred to more concentrated water, the need for water elimination is reduced and the tubules fill with solid matter; plainly the fast fluid secretion had slackened (Wigglesworth, 1953).

One obvious difference between terrestrial and freshwater insects is that fresh water insects can absorb ions from their environment. Such structures as the anal papillae of mosquito larvae are used for active uptake of ions from the water (Stobbart & Shaw, 1964). Terrestrial insects presumably rely on the ion content of the food.

RESPONSES TO DRYING ENVIRONMENTS

Relatively inactive terrestrial insects may face water loss. This can be relieved by such means as feeding or drinking but we are more concerned here with the contribution that can be made by the excretory system. The Malpighian tubule secrete fluid at a relatively reduced rate and this is passed into the hindgut where the pronounced water recovery abilities of this region, especially the rectum, come into play and any fluid that is eliminated is very concentrated (Wall & Oschman, 1975). In many cases virtually all the water is recovered and only dry excreta are produced.

In comparison with this, aquatic insects in osmotically concentrated waters have developed somewhat different solutions to the problem.

At least one insect, <u>Culiseta inornata</u>, has become an osmoconformer; in water of more than about 400 mosmol l^{-1} the osmotic concentration of the haemolymph is practically the same as that of the medium up to 700 mosmol l^{-1} (Garrett & Bradley, 1984). The ionic content of the haemolymph ions varies little so that in the more concentrated waters the haemolymph are maintained at levels considerably lower than that of the medium (Garrett & Bradley, 1984). So the osmotic problems are solved in a manner previously undescribed for mosquitoes. How the necessary ionic regulation is achieved is not yet clear. Also, it will be interesting to discover what osmotically active compounds this mosquito uses to raise the osmotic concentration of its haemolymph when in concentrated waters.

Other mosquito larvae living in hyperosmotic environments maintain the osmotic concentration of their haemolymph at normal levels. How is this achieved? The urine that is produced is concentrated; one might suspect that water recovery in the hindgut was responsible just as occurs in terrestrial insects. However, the concentrated urine has another source. It is produced by a secretion of hyperosmotic fluid by an unusual posterior compartment of the rectum (Bradley & Phillips, 1977). So, far from reabsorbing fluid from the flow of fluid from the Malpighian tubules, strongly hyperosmotic fluid is added to it. Such a strategy of course depends on the plentiful supply of fluids, albeit hyperosmotic, from the environment. These animals solve their osmotic problems by drinking the environment and eliminating the salt load through a special organ. There are obvious parallals to the rectal glands of elasmobranchs and the chloride cells in fish gills and their opercula.

I would like to thank Dr Sally Corbet for drawing my attention to papers on water balance in flying insects and to Dr Paul Webb who suggested to me that flying birds and bats might need to lose extra water from the respiratory system for cooling.

REFERENCES

Altman, G. (1956) Die Regulation des Wasserhaushaltes der Honigbiene. Insectes soc. <u>3</u>, 33-40.

Berridge, M.J. (1966) The physiology of excretion in the cotton stainer, <u>Dysdercus fasciatus</u> Signoret. IV. Hormonal control of excretion. <u>J. exp. Biol</u>. <u>44</u>, 553-566.

Bertsch, A. (1984) Foraging in male bumblebees (<u>Bombus lucorum</u> L.): maximizing energy or minimizing water load? <u>Oecologia</u> <u>62</u>, 325-336.

Bradley, T.J. & Phillips, J.E. (1977) The location and mechanism of hyperosmotic fluid secretion in the rectum of the saline-water mosquito, Aedes taeniorhynchus. J. exp. Biol. 66, 111-126.

Bursell, E. (1959) Metabolic rate and water loss during flight. Ann. Rep. E. Afr. Tsetse Trypanosom. Res. Org. 1958, 32-35.

Cazal, M. & Girardie, A. (1968). Controle humoral de l'equilibre hydrique chez Locusta migratoria migratorioides. J. Insect Physiol. 14, 655-668.

Cockbain, A.J. (1961) Water relationships of Aphis fabae during tethered flight. J. exp. Biol. 38, 175-180.

Dores, R.M., Dallmann, S.H. & Herman, W.S. (1979) The regulation of post-eclosion and post-feeding diuresis in the monarch butterfly, Danaus plexippus. J. Insect Physiol. 25, 895-901.

Edney, E.B. (1977) Water Balance in Land Arthropods. Springer, Heidelberg.

Garrett, M, & Bradley, T.J. (1984) The pattern of osmotic regulation in larvae of the mosquito, Culiseta inornata. J. exp. Biol. 113, 133-141.

Gee, J.D. (1976) Active transport of sodium by the Malpighian tubules of the tsetse fly, Glossina morsitans. J. exp. Biol. 64, 357-368.

Gringorten, J.L. & Friend, W.G. (1979) Haemolymph volume changes in Rhodnius prolixus during flight. J. exp. Biol. 83, 325-333.

Kestler, P. (1985) Respiration and respiratory water loss. In Environmental Physiology and Biochemistry (ed. K.H. Hoffmann). Springer, Heidelberg.

Knowles, G. (1976) The action of the excretory apparatus of Calliphora vomitoria in handling injected sugar solution. J. exp. Biol. 64, 131-140.

Loveridge, J.P. (1968) The control of water loss in Locusta migratoria migratorioides R.&F. II. Water loss through the spriacles. J. exp. Biol. 49, 15-29.

Maddrell, S.H.P. (1962) A diuretic hormone in Rhodnius prolixus Stal. Nature, Lond. 194, 605-606.

Maddrell, S.H.P. (1980) The control of water relations in insects. In Insect Biology in the Future (eds. D.S. Smith & M. Locke). Academic Press, New York.

Maddrell, S.H.P. (1981) The functional design of the insect excretory system. J. exp. Biol. 90, 1-15.

Maddrell, S.H.P. (1982) Insects: small size and osmoregulation. In A Companion to Animal Physiology (eds. C.R. Taylor, K. Johansen & L. Bolis). Cambridge University Press, Cambridge.

Maddrell, S.H.P. and Phillips, J.E. (1978) Induction of sulphate transport and hormonal control of fluid secretion by Malpighian tubules of larvae of the mosquiteo, Aedes taeniorhynchus. J. exp. Biol. 75, 133-145.

Mills, R.R. (1967) Hormonal control of excretion in the American cockroach. I. Release of a diuretic hormone from the terminal abdominal ganglion. J. exp. Biol. 46, 35-41.

Mordue, W. (1969) Hormonal control of Malpighian tubules and rectal function in the desert locust Schistocerca gregaria. J. Insect Physiol. 15, 273-285.

Nicolson, S.W. (1976) The hormonal control of diuresis in the Cabbage white butterfly, Pieris brassicae. J. exp. Biol. 65, 565-575.

Nicolson, S.W. and Hanrahan, S.A. (1986) Diuresis in a desert beetle? Hormonal control of the Malpighian tubules of Onymacris plana (Coleoptera, Tenebrionidae). J. Comp. Physiol. B 156, 407-413.

Nicolson, S.W. and Louw, G.N. (1982) Simultaneous measurement of evaporative water loss, oxygen consumption and thoracic temperature during flight in a carpenter bee. J. exp. Zool. 222, 287-296.

Nijhout, H.F. and Carrow, G.M. (1978) Diuresis after a blood-meal in female Anopheles freeborni. J. Insect Physiol. 24, 293-298.

Nunez, J.A. (1956) Untersuchungen uber die Regelung des Wasserhaushaltes bei Anisotarsus cupripennis Germ.. Z. vergl. Physiol. 38, 341-354.

Pasedach-Poeverlein, K. (1941) Uber das "Spritzen" der Bienen und uber die Konzentrationsanderung ihres Honigblaseninhalts. Z. vergl. Physiol. 28, 197-210.

Phillips, J.E. (1964) Rectal absorption in the desert locust, Schistocerca gregaria Forskal. III. The nature of the excretory process. J. exp. Biol. 41, 69-80.

Phillips, J.E. (1969) Osmotic regulation and rectal absorption in the blowfly, Calliphora erythrocephala. Can. J. Zool. 47, 851-863.

Pilcher, D.E.M. (1970) Hormonal control of the Malpighian tubules of the stick insect, Carausius morosus. J. exp. Biol. 52, 653-665.

Ramsay, J.A. (1950) Osmotic regulation in mosquito larvae. J. exp. Biol. 27, 145-157.

Ryerse, J.S. (1978) Developmental changes in Malpighian tubule fluid transport. J. Insect Physiol. 24, 315-319.

Schwartz, L.M. and Reynolds, S.E. (1979) Fluid transport in Calliphora Malpighian tubules: a diuretic hormone from the thoracic ganglion and abdominal nerves. J. Insect Physiol. 25, 847-854.

Stobbart, R.M. and Shaw, J. (1974) Salt and water balance: excretion. In The Physiology of Insecta (ed. M. Rockstein) vol.5, 361-446. Academic Press, New York.

Wall, B.J. and Oschman, J.L. (1975) Structure and function of the rectum in insects. Fortschritte der Zoologie 23, 193-222.

Wigglesworth, V.B. (1953) The Principles of Insect Physiology, 5th edition. Methuen, London.

Williams, S.C. and Beyenbach, K.W. (1983) Differential effects of secretagogues on Na and K secretion in the Malpighian tubules of Aedes aegypti (L.). J. comp. Physiol. 149, 511-517.

NEURAL ADAPTATIONS
TO OSMOTIC AND IONIC STRESS
IN AQUATIC AND TERRESTRIAL INVERTEBRATES

John Treherne

A.F.R.C. Unit, Department of Zoology
University of Cambridge
Downing Street, Cambridge CB2 3EJ, U.K.

The cells and tissues of invertebrate animals are not
necessarily passive entities which are at the mercy of the
composition of their body fluids. This principle is
illustrated using two very different invertebrate
representatives. The first is an extreme euryhaline
osmoconformer (the serpulid worm, Mercierella enigmatica)
whose nerve cells are adapted to withstand massive changes
in the osmotic and ionic concentration of the body fluids.
The second is an osmoregulating terrestrial species (the
cockroach, Periplaneta americana) whose nervous system is,
nevertheless, subjected to considerable inconstancy due to
the cycling of the blood ions. In this case, homeostasis is
achieved not at the neuronal level, but by the provision of
a glial blood-brain barrier whose permeability properties
can, in addition, be modulated to counteract the ionic
inconstancy of the body fluids.

INTRODUCTION

One of the classic generalizations of comparative physiology is Claude Bernard's
dictum of 1865: 'La fixite du milieu interieur c'est la condition de la vie
libre'. And it makes a good deal of sense if one takes a Napoleonic view of
Nature. Animals that live in the sea get their constancy by exploiting the
remarkable stability of the marine environment: a squid, a marine worm or a
jellyfish do not need to osmoregulate; life can proceed happily at 1000
milliosmoles, with plenty of sodium and chloride and the other inorganic cations
and anions in the happy balance provided in the ocean by a provident Nature.
When animals colonized estuarine or fresh waters, or moved onto the land,
'fixite' for their cells and tissues was preserved to provide osmotic and ionic
constancy.

Comparative Physiology: Life in Water and on Land. P. Dejours, L. Bolis, C.R. Taylor, E.R. Weibel
(eds.) Fidia Research Series, IX-Liviana Press, Padova © 1987

Bernard's dictum can certainly be applied to vertebrate and many invertebrate animals. However, it is not a generalization which we can safely apply to all invertebrates. Some species employ very different physiological strategies which can involve dramatic <u>cellular</u> adjustments to large changes in the chemical composition of the body fluids. This will be demonstrated here by the results of research on two very different invertebrate animals. The first is an aquatic organism, a euryhaline estuarine worm (the most extreme osmoconformer known), and the second a terrestrial insect in which, despite relative osmotic stability, the ionic composition of the body fluids can be very far from stable.

Both of these animals illustrate the theme of this contribution. Namely, that among invertebrates, ionic and osmotic regulation can involve not only homeostasis of the body fluids (the phenomenon most commonly studied by osmoregulationists) but also very spectacular adjustments at the level of the cells and tissues, in this case, in the organ that is most sensitive to fluctuations in the chemical composition of its extracellular fluids: the central nervous system.

NEURONAL OSMOTIC AND IONIC HOMEOSTASIS IN A EURYHALINE OSMOCONFORMER

<u>Mercierella enigmatica</u> - a serpulid worm - is the most extreme osmoconformer known. It occurs in natural waters of salinities between 1 and 55 $^{o}/oo$ (Tebble, 1953) and can withstand changes in the osmotic concentration of its blood of between 84 and <u>2304</u> mOsM (Skaer, 1974; Benson & Treherne, 1978a).

Even more remarkable is the fact that the neurones are not protected from such huge fluctuations in osmotic and ionic composition by any blood-brain barrier. In <u>Mercierella</u>, the so-called giant axons are overlaid only by narrow glial processes which provide an incomplete covering of the axonal surfaces and which allows, for example, ionic lanthanum to leak into the fluid space immediately adjacent to the axons (Skaer <u>et al</u>., 1978).

An immediate physiological danger for neurones exposed to extreme osmotic challenge is the increase in membrane tension due to osmotic swelling which particularly affects the axonal spike-generating system (cf. Pichon & Treherne, 1976). Such adverse effects are however greatly reduced if the axons are very small. This is because (according to Laplace's Law) the membrane tension for a long cylinder is given by

$$T = PR$$

where P is the internal hydrostatic pressure and R is radius. Now with a small axon such as that of <u>Mercierella</u> (i.e. 15 μm radius) the tension generated by a notional excess of 100 mOsmoles is:

$$T = PR = 3.63 \text{ N m}^{-1}$$

With a large axon, such as those of a squid (say around 1 mm diameter) the equivalent figure will be about 300 times greater than that for the _Mercierella_ axon.

Nevertheless, this value of 3.63 N m^{-1} is still unrealistically large when compared with the values generally given for cellular membrane tension. For example, human erythrocytes can withstand only 0.02 N m^{-1} before haemolysis occurs (Rand, 1964).

It is therefore improbable that the _Mercierella_ axons could withstand hydrostatic pressures resulting from appreciable osmotic gradients across the membrane without structural support. These we found in the form of numerous hemidesmosome-like structures on the axon membranes which are linked to a system of intracellular filaments (Skaer et al., 1978).

The average spacing between these hemidesmosomes is about 0.25 µm. This close-spacing reduces the radius of curvature of the unrestrained portions of the membrane and, consequently, reduce the tension on the membrane as illustrated in Fig. 1. In such a system a 10% stretching would reduce the membrane tension to levels which can, for example, be tolerated by erythrocyte membranes.

There remains, however, the problem of the nature of ionic mechanisms used to maintain axonal excitability during extreme ionic dilution of the body fluids. In particular, how is it possible to run sodium-dependent action potentials (cf. Carlson & Treherne, 1977) when the external sodium concentration is drastically reduced?

That the _Mercierella_ axons are capable of functioning in the face of such ionic dilution (down to 7.5%) is demonstrated in Fig. 2. A consistent feature of this hyposmotic adaptation is a progressive hyperpolarization of the axon membrane (Benson & Treherne, 1978b). This hyperpolarization is an important factor because it partially compensates for the decline in the overshoot resulting from the dilution of external sodium ions (which carry the inward current of the action potential) and also tends to reduce sodium inactivation and thus maintain a rapid rate of rise of the action potential and a closer approach to the sodium equilibrium potential (Benson & Treherne, 1978b; Treherne, 1980, 1984b).

The hyperpolarization seems to result from two specializations: the presence of unusually high concentrations of potassium (>30 mM) in the blood of sea-water

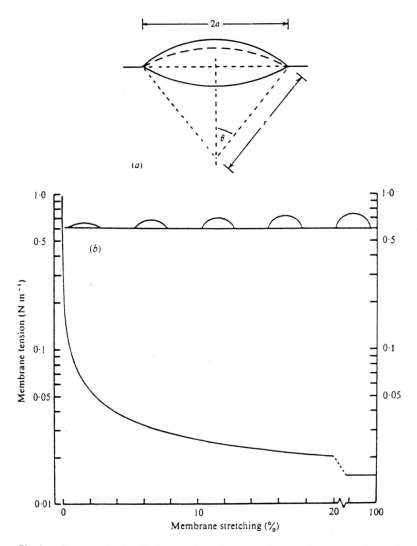

Fig. 1. a Diagram of a localised area of membrane, or 'blister', between regions of attachment of hemidesmosome-like structures on the membrane of the giant axons of the serpulid, *Mercierella enigmatica*. A national internal excess of osmotic concentration of 100 mOsmol, within the axon, corresponds to an equilibrium hydrostatic pressure (P) of 2.42×10^5 Pa, which is shown as forming localised membrane swellings between the points of attachment of the hemidesmosomes. A circular area of membrane, or radius a, originally flat, is distended to form part of a sphere, or radius r, subtending a semi-angle θ at its centre of curvature. **b** Membrane tension plotted against membrane stretching, expressed as the percentage increase in area (where $T = P_a/2 \sin \theta = 1/2 \, Pr$) and the percentage increase above the original area is $A = 100 \dfrac{1 - \cos \theta}{1 + \cos \theta}$ for an internal pressure of 2.42×10^5 Pa. Above the curve are sketched profiles of the 'blister' for various degrees of stretching. (From Skaer et al. 1978)

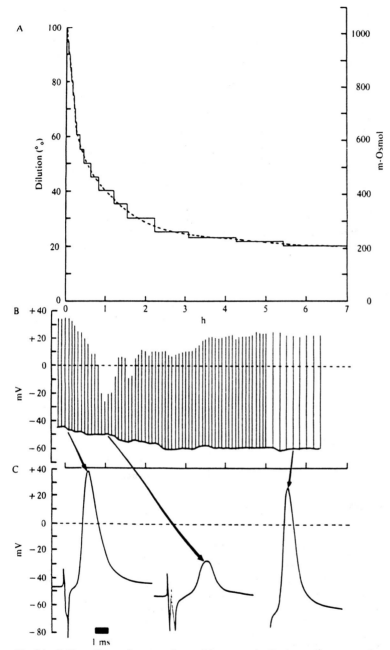

Fig. 2A–C. The effects of an experimental hyposmotic dilution regime on resting and action potentials of a *Mercierella* giant axon recording using an intracellular-located microelectrode. **A** The experimental dilution regime, shown as a series of step changes, mimics the rate of reduction in blood concentration (*broken line*) observed on transfer of a colony of *Mercierella* (in their tubes) to distilled water (data from Skaer 1974). **B** Continuous recording of the resting and action potentials during the above experimental dilution regime. **C** Action potentials recorded as successive stages of hyposmotic adaptation. (From Benson and Treherne 1978b)

adapted animals (Skaer, 1974) and, also, from a remarkably, and unusual,
sensitivity of the axon membrane potential to potassium ions in the
physiological concentration range (Benson & Treherne, 1978b).

Hyposmotic adaptation of the Mercierella axons also involves reduction of the
intracellular sodium and potassium concentrations; a dilution which does not
result from axonal swelling, for changes in axonal diameter could not be
detected with Nomarski optics (Benson & Treherne, 1978b). The decrease in
internal sodium is of some advantage in reducing the osmotic gradient and
considerable value in increasing the inward electrochemical gradient for this
cation. The latter effect was reflected in a progressive increase in overshoot
following ionic dilution of the bathing medium, a response which was abolished
in the presence of ouabain (Benson & Treherne, 1978b).

From the shifts in the Nernst slopes of sea water- and hyposmotically-adapted
axons (Benson & Treherne, 1978b) it was estimated that there is a
'non-proportional' reduction in intracellular potassium (i.e. dilution of $[K^+]_o$
to 25% resulted in approximate halving of $[K^+]_i$). This non-proportional
retention of intracellular potassium is of critical importance in hyposmotic
adaptation of the Mercierella axons because it contributes to the axonal
hyperpolarization. The hyperpolarization tends to compensate for the reduction
in overshoot of the action potential and, also, reduces sodium inactivation and,
consequently, maintains a relatively rapid rate of rise during extreme dilution
of the bathing medium.

So, in this serpulid, hyposmotic adaptation is a compromise. Particularly in
relation to intracellular potassium levels where it is necessary to retain
sufficient within the axons to produce some degree of hyperpolarization, while,
at the same time, letting enough go to reduce the osmotic imbalance.

The Mercierella and its axons are, thus, a non-Bernardian system. That is, there
is virtually no control of the 'milieur interieur', but remarkable cellular
adaptations that enable the axons to adapt and function over a very wide range
of osmotic and ionic concentrations and allow this animal to colonize 'extreme'
aquatic environments.

IONIC HOMEOSTASIS OF THE INSECT CNS
Insects conform to the Bernardian dictum in that the osmotic concentration of
the blood plasma is subject to fairly precise physiological control (vid.
Bradley, 1985). However, in one terrestrial species at least, the cockroach
(Periplaneta americana), there can be very dramatic fluctuations in the

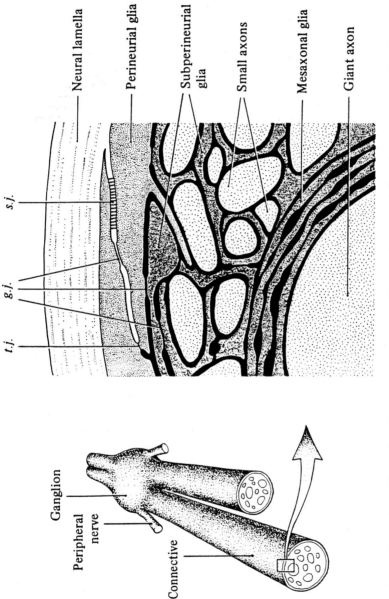

FIG. 3. Connectives of the cockroach central nervous system (left) contain an association of axons and neuroglia which is ensheathed by a layer of glial cells termed the perineurium, overlaid by the neural lamella (right). Tight junctions (t.j.) and septate junctions (s.j.) are found between perineurial cells. Gap junctions (g.j.) connect perineurial and glial cells. Giant axons are surrounded by many glial folds, the mesaxon; smaller axons have less glial investment. The sub-perineurial extracellular system, around axons and glia, is indicated in solid black. Schematic drawing, not to scale. (From Treherne, J. and Schofield, P. 1981.)

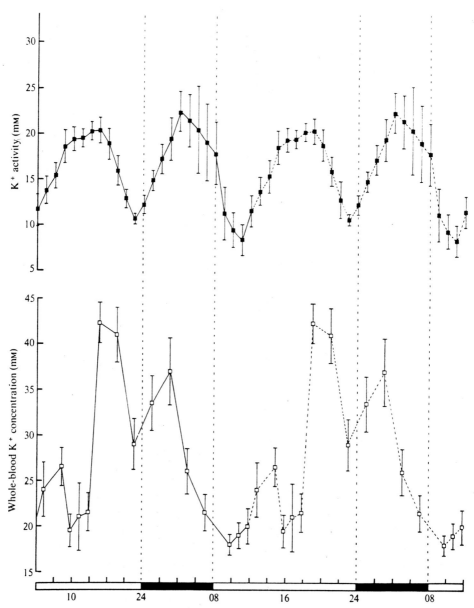

Fig. 4 (a) Composite curve of diel changes in K+ activity from 14 cockroaches, expressed in
mM. Bars are ± 1 × s.e. of the mean. (b) Diel changes in mean whole-blood K+ concentration.
Points are means of 12–24 readings from 6–14 individual males. Bars are ± 1 × s.e. of the mean.
LD 16:8. Dotted lines join points that have been plotted twice. (Lettau et al. 1977)

activities of some major blood cations (Lettau et al., 1977). This is
particularly evident for plasma potassium activity (see Fig. 4) in which there
can be more than a three-fold cycling in relation to the light-dark regime.
Equivalent changes also occur in plasma sodium activity, but in a more
idiosyncratic way, not obviously related to the light-dark regime (Lettau et
al., 1977).

Now unlike the serpulid axons, those of the cockroach CNS are protected by a
well-developed blood-brain barrier system, associated with the most superficial
layer of flattened neuroglia, the perineurium (Fig. 3). The available evidence
indicates that the restriction to the intercellular access of water-soluble ions
and molecules to the axon surfaces results from the presence of junctional
complexes at the inner ends of the narrow clefts between adjacent perineurial
cells (cf. Lane & Treherne, 1972; Schofield & Treherne, 1984; Schofield, Swales
& Treherne, 1984; Treherne, 1984a). The transperineurial resistance (calculated
from the attenuation of current pulses across the perineurium) is >900 ohmm cmγ
(Schofield & Treherne, 1984). This characterizes the cockroach perineurium as a
fairly tight epithelium, for example, roughly half that of mammalian brain
capillaries (Crone & Oleson, 1982) and very much higher than frog choroid plexus
(26 ohmm cmγ) (Zeuthen & Wright, 1981).

An important characteristic of this perineurial blood-brain barrier is that its
permeability is not a static parameter, but can apparently be modulated in
response to the fluctuating ionic conditions of the blood plasma (e.g. Fig. 4).

This modulation - which is likely to be of critical importance in maintaining
the homeostasis of the ionic environment of the neurones - appears to be
mediated by a biogenic amine, octopamine, which is known to be released into the
blood plasma of this insect (Davenport & Evans, 1984). This is shown, for
example, by the effects of octopamine on potassium-induced depolarizations of
the outwardly-directed perineurial membranes and on transperineurial resistance
(Fig. 5). That these effects are mediated by an octopamine receptor is indicated
by the effects of DL-synephrine in depressing the perineurial response to
potassium ions and the blocking by phentolamine and partial blockage by
propanolol (Schofield & Treherne, 1985, 1986).

The effects of octopamine on the permeability of the insect blood-brain
interface appears to result from a reduction in the potassium conductance of the
basolateral membrane of the perineurial glia (Schofield & Treherne, 1986). This
reduction is accompanied by a decline in the net potassium permeability of the
perineurial barrier (Schofield & Treherne, 1985), such that the rate of leakage

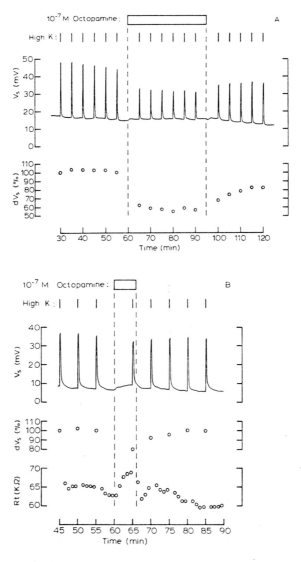

Fig. 5. Effect of 10^{-7} M DL-octopamine upon (A) interstitial potential and (B) transperineurial resistance (R_t) in cockroach central nervous connectives. A: top trace shows the continuous record of potential and K-induced changes (dV_s), while the lower graph shows dV_s as a percentage of dV_s at 5 min before octopamine. K concentration was raised from 3 to 67 mM for 15 s periods. Time scale indicates time from setting up preparation. B: experiment carried out as in A except that current was pulsed through a second electrode in the interstitial system, producing deflections in V_s (not shown for clarity), to measure R_t shown in the lower graph. (Schofield & Treherne, 1985).

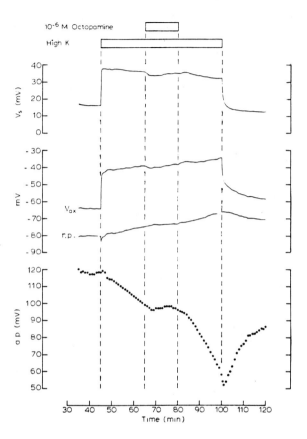

Fig. 6. Effect of 10^{-6} M DL-octopamine upon K entry into a connective, as revealed by simultaneous recording from the interstitial system and an axon. By subtracting the interstitial potential (V_s) from the potential recorded in the axon (V_{ax}), the resting potential across the axon membrane (r.p.) is obtained. Propagated action potentials (a.p.) were obtained by extracellular stimulation of adjacent connectives. Upon raising the external K concentration from 3 to 67 mM (high-K), there was a slow depolarization of r.p., and steady decline in a.p. amplitude, as K leaked into the connective. Five minutes after the introduction of 10^{-6} M octopamine into the bath, the indicated K entry was virtually abolished: in the next 10 min, r.p. depolarized by 0.2 mV, and a.p. amplitude decreased by only 1 mV. In contrast, during the 10 min before the amine, r.p. decreased by 2.9 mV and the spike fell by 10 mV. This indicates a slowing of K entry, especially when the non-linear nature of the K effects is considered. (Schofield & Treherne, 1985).

of potassium ions to the axon surfaces can be substantially reduced by the
biogenic amine (Fig. 6). This seems to be the first quantitative demonstration
of a hormonal modulation of the permeability of a glial blood-brain barrier
system. It is also conceivable that this represents a physiological response to
octopamine circulating in the blood plasma. In cockroach blood plasma,
octopamine is found at a basal level of 3×10^{-8}M: a concentration which roughly
doubles during activity at night (when there is a substantial increase in blood
potassium activity) and triples following stress (Davenport & Evans, 1984). The
highest sensitivity of the K-induced changes on the perineurium are found within
this range (Schofield & Treherne, 1986). As blood potassium (Fig. 4), rises
during the night its access to the axons would cause depolarization and
inactivation of the sodium channels. The parallel increase in the circulating
biogenic amine (Davenport & Evans, 1984) would thus counter an increased
tendency of potassium ions to cross the blood-brain barrier to the underlying
neurones.

Figs. 5 & 6 are reproduced by permission of Elsevier/North-Holland Scientific
Publishers Inc.

REFERENCES

Benson JA, Treherne JE (1978a) Axonal adaptations to osmotic and ionic stress in
    an invertebrate osmoconformer (Mercierella enigmatica Fauvel). II. Effects
    of ionic dilution on the resting and action potentials. J exp Biol 76:
    205-219.

Benson JA, Treherne JE (1978b) Axonal adaptations to osmotic and ionic stress in
    an invertebrate osmoconformer (Mercierella enigmatica Fauvel). III.
    Adaptations to hyposmotic dilution. J. exp Biol 76: 221-235.

Bernard C (1865) Introduction a l'etude de la medecine experimentale. Paris.

Bradley TJ (1985) The excretory system: structure and physiology, Vol.4 In:
    Kerkut GA, Gilbert LI (ed): Comprehensive Insect Physiology, Biochemistry
    and Pharmacology. Pergamon Press, Oxford; pp.421-465.

Carlson AD, Treherne JE (1977) Ionic basis of excitability in an extreme
    euryhaline osmoconformer, the serpulid worm Mercierella enigmatica
    (Fauvel). J. exp. Biol. 67: 205-215.

Crone C, Oleson SP (1982) Electrical resistance of brain microvascular
    epithelium. Brain Res 241: 49-55.

Davenport AP, Evans PD (1984) Stress-induced changes in the octopamine level of
    insect haemolymph. Insect Biochem 14: 135-143.

Lane NH, Treherne JE (1972) Studies on perineurial junctional complexes and the
    sites of uptake of microperoxidase and lanthanum in the cockroach central
    nervous system. Tissue Cell 4: 427-436.

Lettau J, Foster WA, Harker JE, Treherne JE (1977) Diel changes in potassium activity in the haemolymph of the cockroach Leucophaea madera. J exp Biol 71: 171-186.

Pichon Y, Treherne JE (1976) The effects of osmotic stress on the electrical properties of the axons of a marine osmoconformer (Maia squinado, Brachyura:Crustacea). J exp Biol 65: 553-563.

Rand RP (1964) Mechanical properties of the red cell membrane. II. Viscoelectric breakdown of the membrane. Biophys. J. 4: 303-316.

Schofield PK, Swales LS, Treherne JE (1984) Potentials associated with the blood-brain barrier of an insect: recordings from identified neuroglia. J exp Biol 109: 307-318.

Schofield PK, Treherne JE (1984) Localization of the blood-brain barrier of an insect: electrical model and analysis. J exp Biol 109: 319-332.

Schofield PK, Treherne JE (1985) Octopamine reduces potassium permeability of the glia that form the insect blood-brain barrier. Brain Res 360: 344-348.

Schofield PK, Treherne JE (1986) Octopamine sensitivity of the blood-brain barrier of an insect. J. exp. Biol. 123, 423-439.

Skaer H leB (1974). The water balance of a serpulid polychaete. Mercierella enigmatica (Fauvel). I. Osmotic concentration and volume regulation. J exp Biol 60: 321-330.

Skaer H leB, Treherne JE, Benson JA, Moreton RB (1978) Axonal adaptations to osmotic and ionic stress in an invertebrate osmoconformer (Mercierella enigmatica Fauvel). I. Ultrastructural and electrophysiological observations on axonal accessibility. J exp Biol 76: 191-204.

Tebble N (1953) A source of danger to harbour structures. Encrustation by a tubed marine worm. J Inst municp Engrs 80: 259-263.

Treherne JE (1980) Neuronal adaptations to osmotic and ionic stress. Comp Biochem Physiol 67B: 455-463.

Treherne JE (1984a) Blood-brain barrier. In: Kerkut GA, Gilbert LI (eds): Comprehensive Insect Physiology Biochemistry and Pharmacology, Vol.5. Pergamon Press, Oxford; pp.115-137.

Treherne JE (1984b) Neuronal adaptations to osmotic stres. In: Gilles R, Gilles-Baillien M (eds): Transport Processes, Iono- and Osmoregulation. Springer-Verlag, Berlin; pp.376-388.

Treherne JE, Schofield PK (1981) Mechanisms of ionic homeostasis in the central nervous system of an insect. J exp Biol 95: 61-73.

Zeuthen T, Wright EM (1981) Epithelial potassium transport: tracer and electrophysiological studies in choroid plexus. J Membrane Biol 60: 106-128.

SALT GLAND FUNCTION IN OSMOREGULATION IN TERRESTRIAL AND AQUATIC ENVIRONMENTS

Trevor J. Shuttleworth

Department of Biological Sciences
University of Exeter, Exeter EX4 4PS
England

The elimination of excess salt in non-mammalian vertebrates involves a variety of different salt-secreting structures which appear to use similar underlying cellular processes of secretion involving a sodium-coupled secondary active transport of chloride. Recent findings also suggest the existence of previously unsuspected common pathways in the regulation of this transport process in the different salt glands. Such common features in the transport mechanism and in its control indicate the highly conserved nature of the mechanism for the elimination of excess salt in vertebrates.

In vertebrates, an intake of salt in excess of the requirements of the body can occur under a variety of different conditions. These include feeding on a diet with a high salt content, the replenishment of water losses by drinking saline water, the diffusional uptake of salt from a hypertonic external environment, and the utilisation of mechanisms of retaining water that rely on processes of solute-linked water absorption. These seemingly rather specific circumstances in fact apply to a very wide variety of species from different vertebrate groups, inhabiting both terrestrial and aquatic environments - only amphibians seem to be excluded.

A diet high in salt is typical of the majority of vertebrate species living in or near the sea where, for example, feeding on marine invertebrates whose tissues are isotonic with the surrounding sea water will result in a significant salt load. Even those feeding exclusively on vertebrate prey may be unable to avoid the simultaneous ingestion of sea water. Similarly, several vertebrates inhabiting arid terrestrial environments feed on plants or insects which have a high content of salt. These problems of a high dietary intake of salt are frequently compounded by an associated limited access to free water which can restrict the ability of the animal to eliminate excess salt by renal means. In addition, in birds and reptiles, the overall retention of water involves a mechanism of solute-linked water reabsorbtion in the cloaca and lower gut creating a further salt load. Whilst the mobility of most birds reduces the problems of the availability of free water, this is only achieved at the expense of increased energy consumption and, more specifically, increased evaporative water losses during flight. In any event, this is

Comparative Physiology: Life in Water and on Land. P. Dejours, L. Bolis, C.R. Taylor, E.R. Weibel (eds.) Fidia Research Series, IX-Liviana Press, Padova © 1987

not an option available to the vast majority of reptiles, whether
terrestrial or marine, and also to the several oceanic bird species (e.g.
the albatross) which spend long periods of time feeding hundreds of miles
away from the nearest land.

In marine members of the primarily aquatic vertebrate groups, the
possession of body fluids with ionic concentrations significantly lower
than in the surrounding sea water leads to the additional problem of a
diffusional influx of ions across the permeable surfaces of their body,
principally the gills. In the case of marine teleosts, there is a
considerable further salt load as a result of the mechanism used in
compensating for water lost osmotically across the gills involving the
ingestion of sea water followed by the active uptake of sodium and
chloride ions by the gut. This additional salt load is avoided in
elasmobranchs because their body fluids are, in fact, slightly
hyperosmotic to the surrounding sea water as a result of the retention of
urea and trimethylamine oxide.

SALT-SECRETING TISSUES

Clearly then, the problem of an excess salt intake is widespread and
applies to many members of different vertebrate groups living in both
terrestrial and aquatic environments. It is perhaps not surprising
therefore that these different groups have evolved a range of structures
responsible for its subsequent elimination. For mammals possessing a
kidney capable of producing a urine significantly hypertonic to the body
fluids, the elimination of excess salt does not generally pose many
problems. In the remaining vertebrate groups which, with the exception of
some birds, do not have a kidney that is capable of producing a
significantly hypertonic urine, the problem of the elimination of excess
salt must be solved by extrarenal means - via the various so-called "salt
glands".

Originally, the term "salt gland" was restricted to the sodium chloride
secreting glands in the cranial region of marine birds and reptiles - the
nasal salt glands of marine birds and the lachrymal salt glands of
turtles. Over the years, this definition has been extended to include
obviously analagous structures such as the nasal salt glands of
terrestrial birds and lizards (some of which secrete potassium rather than
sodium), and even the salt-secreting rectal gland of elasmobranch fish. In
addition, a salt-secreting function has been attributed to lingual glands
in certain crocodiles (Taplin and Grigg, 1981) and to posterior sublingual
glands in sea snakes (Dunson, 1976). This discussion will also include
information, where relevant, on the branchial and opercular "chloride
cells" of marine teleosts which are the sites of the elimination of excess
salt in these animals (Foskett et al., 1983). Although these clearly do
not form a discrete gland and do not produce a fluid secretion as such,
physiologically they perform the same function and, at the cellular level,
show many similarities to the true salt glands of the other vertebrate

groups. Indeed, a principal aim of this review is to indicate that, despite the diverse sources of the excess salt load in the different animal groups, and the even more diverse nature of the tissue responsible for its elimination, there is increasing evidence of a remarkable constancy, particularly at the cellular level, in the way in which the excess salt is removed from the body fluids in order to maintain an appropriate overall hydromineral status.

In the majority of the tissues described, the principal ions secreted are sodium and chloride although, in certain cases, the major secreted cation is potassium (see later). The rate at which these ions are secreted varies from species to species and, more particularly, from group to group, and is dependent on the hydromineral status of the particular animal concerned (Table 1). In the true salt glands this secretion is seen as the

Table 1. Typical values for salt (Na^+) secretion in different groups.

	Secretion rate (mmol kg^{-1} h^{-1})	Concentration (mmol l^{-1})
Bird (nasal gland)	1 - 10	700 - 800
Turtle (lachrymal gland)	1 - 3	650 - 700
Elasmobranch (rectal gland)	0.2 - 2	500 - 550
Teleost (chloride cells)	0.5 - 3	N/A

production of a fluid with an ionic concentration that is markedly higher than that of the body fluids. In birds, the secreted fluid may reach concentrations in excess of 1100 mmol l^{-1}, equivalent to some seven or eight times the corresponding blood concentration. This illustrates the effectiveness of these organs as avenues for the elimination of excess salt with the minimum of associated water. In contrast to the other salt glands, the elasmobranch rectal gland produces a secretion that is isotonic to the body fluids. However, as this secretion contains essentially none of the urea found in the body fluids of these animals, this still represents the elimination of a fluid with a sodium chloride concentration approximately twice that of the blood. The observed rates of secretion represent extremely high levels of transport activity in these tissues. For example, the rectal gland of the dogfish Scyliorhinus is capable of secreting some 50 μmol of sodium per gram of gland every minute (Shuttleworth and Thompson, 1986) and, on the same basis, secretion rates in the avian gland may be as much as ten times this!

MECHANISM OF SECRETION

It is clear then, that all these various tissues have the ability to remove sodium and chloride ions from the blood or body fluids and secrete them against often very large concentration gradients. Although this has been known for several years, it is only relatively recently that any detailed knowledge of the actual mechanism of transport has become apparent. Much of the original work centered on the avian salt gland but the mechanism of salt secretion has been most thoroughly investigated in the elasmobranch rectal gland where direct cannulation of the secretory duct is possible and the gland can be removed and perfused in isolation. Experiments have shown that chloride secretion is dependent on the presence of sodium, and vice versa, and that secretory activity is associated with the development of a lumen-negative electrical potential such that, whilst the transepithelial movement of sodium can be explained on purely passive grounds, chloride secretion occurs against a large prevailing electrochemical gradient (Silva et al., 1977). The tissue contains high levels of Na-K-ATPase, localised on the basolateral membranes of the transporting cells, and the secretion of both sodium and chloride is inhibited by ouabain. It is also inhibited by the loop diuretics, such as furosemide or bumetanide, which are believed to act by blocking a carrier molecule involved in the coupled cotransport of sodium and chloride (together with potassium).

Overall therefore, the secretory process is an uphill movement of chloride, linked to simultaneous transport of sodium and involves a basolaterally located sodium pump and sodium-chloride coupled cotransporter. This information has been integrated into a model (Greger and Schlatter, 1984) which can be described as representing a secondary-active transport of chloride (Fig. 1). In this, the operation of the basolateral sodium pump generates a gradient for the entry of sodium from the blood into the cell. The downhill movement of sodium into the cell is coupled to the uphill movement of chloride across the same membrane and in the same direction by means of a sodium-chloride (+ potassium) cotransporter. This leads to the intracellular accumulation of chloride to a level sufficient to produce a passive exit of this ion across the apical membrane. This transepithelial movement of chloride generates a lumen-negative potential gradient which induces the passive movement of sodium into the secretory lumen via a paracellular route. For the elasmobranch rectal gland there is, by now, fairly detailed and comprehensive biochemical, pharmacological and electrophysiological evidence supporting such a model (Shuttleworth and Thompson, 1980; Hannafin et al., 1983; Greger and Schlatter, 1984) and there is increasing evidence of its applicability to the other vertebrate salt-secreting tissues.

In the marine teleost chloride cell, in the absence of any discrete secreted fluid, alternative measures of net ion secretion have had to be used largely based, in accord with the above model, on the electrogenic nature of the transepithelial transport of chloride. Thus, in isolated perfused gill preparations, measurements of the "open-circuit" transepithelial electrical potential, determined in the absence of concentration gradients, have been used to give, at least, an indirect measure of the transport rate (Shuttleworth et al., 1974). Alternatively, in certain teleost species, flat-sheet opercular or skin preparations with dense populations of chloride cells can be obtained and here it has been possible to use the much more powerful technique of short-circuit current determinations (Degnan et al., 1977). As such, transport activity has again been shown to be dependent on the presence of both serosal sodium

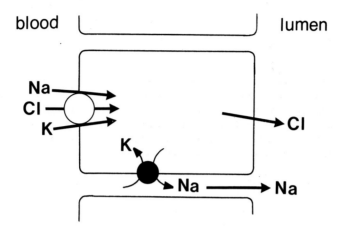

Figure 1. Model describing the secondary active transport of chloride.

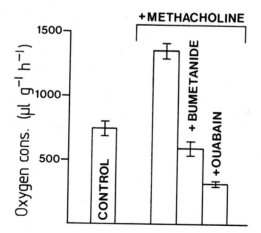

Figure 2. Oxygen consumption in <u>Malaclemys</u> lachrymal gland slices –
stimulation by methacholine and its inhibition by bumetanide and by
ouabain.

and chloride, to be inhibited by ouabain and to be sensitive to inhibition by furosemide (Shuttleworth et al., 1974; Degnan et al., 1977; Foskett et al., 1983; Davis and Shuttleworth, 1985).

The situation in the other salt glands is rather less clear. In the avian nasal gland secretory activity, as measured by tissue oxygen consumption, is inhibited by ouabain and by furosemide (Ernst and Van Rossum, 1982) or bumetanide - features that, taken together with basolateral location of the Na-K-ATPase (Ernst and Mills, 1977), are entirely consistent with the above model. The only reported measurement of the transepithelial electrical gradient during secretion indicated a lumen <u>positive</u> potential (Thesleff and Schmidt-Nielsen, 1962) which is difficult to reconcile with the model described, but there are some important methodological problems associated with these measurements and, although various alternative explanations for the overall functioning of the avian gland have been proposed (see Marshall et al., 1985), the evidence would still suggest that, at the cellular level at least, the underlying transport mechanism conforms to that described for the rectal gland and the chloride cell.

In the various reptilian glands, the picture is even more patchy. Recently, we have shown that, in the lachrymal gland of <u>Malaclemys</u>, the increase in oxygen consumption observed on stimulating secretory activity is inhibited both by ouabain and by bumetanide (Fig. 2). Apart from the existence of high levels of Na-K-ATPase in this gland, as well as in the nasal glands of the lizard <u>Dipsosaurus</u> and the sublingual glands of the sea snake <u>Pelamis</u> (Dunson and Dunson, 1975), together with the histochemical localization of this enzyme to the basolateral membrane in <u>Dipsosaurus</u> (Ellis and Goertemiller, 1974), this would appear to be the only published information available on the actual mechanism of secretion in these glands.

The development of the model referred to, owed much to earlier descriptions of sodium-coupled chloride transport in other tissues, particularly secretion in the mammalian small intestine and, as such, conforms to essentially identical descriptions of sodium-coupled chloride secretion by the mammalian trachea, cornea, and colon (Frizzell et al., 1979). Indeed, the sodium-coupled <u>absorption</u> of chloride by various tissues, including the thick ascending limb (TAL) of the loop of Henle, is similarly described by a model involving the identical components suitably "rearranged" on the appropriate membranes. Whilst much more work on the different kinds of salt gland is clearly needed, and some contradictory data notwithstanding, there would seem to be increasing evidence for a consistent mechanism underlying the transport process in all these tissues, suggesting that this particular mechanism for the coupling of the uphill transepithelial transport of chloride to the simultaneous transport of sodium has been highly conserved during vertebrate evolution.

CONTROL OF THE SECRETORY MECHANISM

If the mechanism of secretion in the various types of salt gland is based
on essentially a common plan, then, by contrast, there would appear to be
marked differences in the control of that secretory process and in the
nature of the agents responsible. Thus, there is considerable evidence
indicating that the rectal gland is principally under the control of a
peptide hormone or hormones operating via cyclic AMP, although the precise
identity of this hormone has been the subject of considerable debate.
Initially, it was claimed by Stoff et al. (1979) that vasoactive
intestinal peptide (VIP) was responsible for the regulation of secretion
but subsequent studies showed that this response only occurred in <u>Squalus</u>
and, even then, only in the isolated perfused gland preparation. More
recently a peptide moiety has been isolated from the elasmobranch
intestine that shows potent stimulatory activity in the glands from a
variety of different species (Shuttleworth and Thorndyke, 1984). The
precise identity of this peptide (provisionally named "rectin") must await
full amino acid sequencing of the molecule, but it is clear that it is not
closely related to VIP (Thorndyke and Shuttleworth, 1985). More recent
claims that an atriopeptin may be involved in the secretory response in
<u>Squalus</u> in some way (Solomon et al., 1985) have yet to be confirmed in
other species but, in any event, we have found that rectin does not show
any cross-reactivity in a radioimmunological assay for atriopeptin
(unpublished data), so again any close relationship appears unlikely.

There is also good evidence that stimulation of chloride transport by the
marine teleost chloride cell involves peptide hormones (Shuttleworth,
1985). In the chloride cell-rich opercular and skin epithelia, studies
have suggested that VIP, glucagon and possibly one of the peptides from
the caudal neurosecretory system, urotensin I, may be important
stimulators of chloride secretion (see Foskett et al., 1983). In the
perfused flounder gill, it has been possible to demonstrate a direct dose-
dependent stimulation of the transepithelial "open-circuit" potential by
the peptide hormone, glucagon (Davis and Shuttleworth, 1985). In the same
preparation, even high concentrations of VIP produced only a small effect.

If the evidence suggests a peptide-mediated stimulation of ion transport
in the salt-secreting tissues of teleost and elasmobranch fish, then the
situation is very different in the other vertebrate groups. Here, in both
the avian and, to a lesser extent, reptilian salt glands, the overwhelming
evidence is for a stimulation mediated via cholinergic neurones (see
Peaker and Linzell, 1975). A discrete "secretory nerve" can be identified
in the orbit of the bird, stimulation of which induces secretion by the
nasal salt gland. Cholinomimetics, acting via receptors of the muscarinic
type, have been shown to induce secretion when infused in vivo and, in
isolated tissues or cells, to stimulate a range of parameters related to
secretory activity, such as ouabain-sensitive and furosemide-sensitive
oxygen consumption (Ernst and Van Rossum, 1982). Similar, although less
complete, evidence has been produced for certain reptilian salt glands

where, for example, injections of cholinomimetics induce secretion in the
lachrymal glands of the turtles Caretta and Malaclemys (Schmidt-Nielsen
and Fange, 1958; Dunson, 1970), and in the nasal glands of the lizards
Ctenosaura and Sauromalus (Norris and Dawson, 1964; Templeton, 1964). We
have recently found that, in slices of the Malaclemys lachrymal salt
gland, methacholine produces marked increases in oxygen consumption that
are inhibited by either ouabain or bumetanide (Fig. 2).

The evidence would suggest therefore a marked contrast between a peptide-
mediated endocrine regulation of salt secretion in teleost and
elasmobranch fish, and a cholinergic nervous regulation of, what appears
to be, an identical secretory mechanism in the salt-secreting glands of
birds and reptiles. Whilst it is possible to speculate as to the
underlying rationale for these observed differences in control mechanism,
some recent findings suggest that they may be less clear-cut than was
previously believed. Thus, in both the nasal salt gland of the duck (Anas)
and the lachrymal salt gland of the turtle Malaclemys, we have found that
bumetanide-sensitive oxygen consumption is increased, not only by the
established methacholine-stimulated (mediated via calcium and/or cyclic
GMP), but also by a previously unsuspected pathway involving an adenylate
cyclase-cyclic AMP second messenger system (Fig. 3). On the basis of
findings in other tissues, such a system is unlikely to be involved in the
cholinergic stimulation of secretion so an additional, alternative signal
is indicated. Beta-adrenergic stimulation of the adenylate cyclase would
appear to be excluded by the finding that isoprenaline fails to elevate
oxygen consumption in either of these tissues. The interesting conclusion
is that, as in teleosts and elasmobranchs, a peptide-mediated control may
also be involved in the control of these avian and reptilian salt glands.

POTASSIUM-SECRETING SALT GLANDS

Whilst the above refers essentially to those glands responsible for the
secretion of sodium and chloride ions, several reptiles and some birds
possess salt glands which secrete large amounts of potassium (Peaker and
Linzell, 1975). Such glands are found typically in certain herbivorous
desert lizards (e.g. Sauromalus, Dipsosaurus), the marine iguana
(Amblyrhynchus) and the ostrich (Struthio). The requirement for an
extrarenal elimination of potassium results from their diet – desert
plants in the case of the ostrich and the lizards, seaweed in the case of
the iguana. Unfortunately, very little is known about the secretory
mechanism and its control in these interesting glands but, in some
preliminary work on the chuckwalla (Sauromalus), we have recently found
that when the animals are potassium-loaded, the nasal salt glands will
secrete potassium at rates up to 0.5 mmol kg^{-1} h^{-1} and will do so in
association with only negligible amounts of sodium. Most interestingly, we
found that this secretion is inhibited by the loop diuretic furosemide,
just as with the sodium-secreting glands discussed above (Fig. 4).
Furthermore, this nasal secretion of potassium can be stimulated by
methacholine and by the adenylate cyclase activator forskolin!

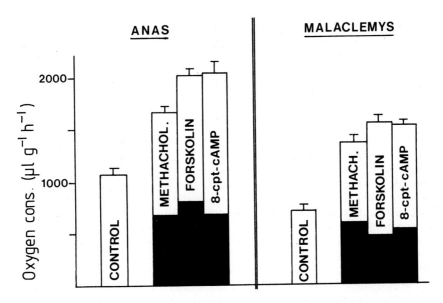

Figure 3. Oxygen consumption in tissue slices of _Malaclemys_ lachrymal gland and duck (_Anas_) nasal gland - effects of methacholine, the adenylate cyclase activator forskolin, and 8-cpt-cyclic AMP. The black segments represent the same drugs in the presence of bumetanide.

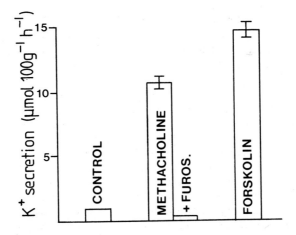

Figure 4. Potassium secretion rate in _Sauromalus_ nasal gland, measured using a perfused head preparation (Shuttleworth, unpublished data).

CONCLUSIONS

This brief review has mentioned just some of the information available on current ideas on the mechanisms of secretion and its control. Many aspects have not been covered, particularly where our knowledge is especially fragmentary - for example, the nature of the sensory signal for regulating secretory activity, the significance of changes in blood flow associated with secretion and how these changes are controlled, and the longer-term adaptive changes seen in certain salt glands in response to salt loads. Throughout this review, the theme has been that, despite the seemingly diverse phenomenon of the elimination of excess salt in vertebrates, evidence is accumulating that suggests the existence of certain common features - the repeated involvement of particular processes and cellular components. This evidence concerns the nature of the transport processes and of the specific components involved, and also the mechanisms by which these processes are regulated. These common features can be seen irrespective of the phylogenetic group to which the animal belongs, the habitat in which it lives, the origin of the excess salt load, the tissue responsible for its elimination, and even, possibly, the nature of the ion secreted.

Powerful pharmacological evidence for the validity of this "conservative theory" of vertebrate salt secretion comes from recent studies showing that specific stereoisomers of inhibitors of the sodium-coupled chloride carrier in the TAL of the mammalian loop of Henle also inhibit chloride secretion by such tissues as the elasmobranch rectal gland and teleost chloride cell (Eriksson and Wistrand, 1986), suggesting that an identical carrier molecule may be involved in these very different tissues. That apparently identical molecules are involved in, for example, the chloride-secreting rectal gland of elasmobranchs and the chloride-reabsorbing TAL of the mammalian kidney, both of which are intimately involved in the overall process of salt elimination and hydromineral balance, is convincing evidence for the conservative nature of the response of animals to a common problem. That this, and other, common threads may underly the whole, seemingly varied, phenomenon of salt secretion in vertebrates is surely a proposition that will serve to stimulate comparative and general physiologists alike.

The work reported here was supported by the U.K. Science and Engineering Research Council, the Nuffield Foundation and the University of Exeter Research Fund.

REFERENCES

Davis MS, Shuttleworth TJ (1985) Peptidergic and adrenergic regulation of electrogenic ion transport in isolated gills of the flounder (Platichthys flesus L.). J comp Physiol 155: 471-478.
Degnan KJ, Karnaky KJ, Zadunaisky JA (1977) Active chloride transport in

the in vitro opercular skin of a teleost (<u>Fundulus</u> <u>heteroclitus</u>), a gill-like epithelium rich in chloride cells. J Physiol 271: 155-191.

Dunson WA (1970) Some aspects of electrolyte and water balance in three estuarine reptiles, the diamondback terrapin, American and "saltwater" crocodiles. Comp Biochem Physiol 32: 161-174.

Dunson WA (1976) Salt glands in reptiles. In: Gans C, Dawson WR (eds): Biology of the Reptilia Vol 5. Academic Press; pp. 413-445.

Dunson MK, Dunson WA (1975) The relation between plasma Na concentration and salt gland Na-K ATPase content in the diamondback terrapin and the yellow-bellied sea snake. J comp Physiol 101: 89-97.

Ellis RA, Goertemiller CC (1974) Cytological effects of salt-stress and localization of transport adenosine triphosphatase in the lateral nasal glands of the desert iguana <u>Dipsosaurus</u> <u>dorsalis</u>. Anal Rec 180: 285-297.

Eriksson O, Wistrand PJ (1986) Chloride transport inhibition by various types of 'loop' diuretics in fish opercular epithelium. Acta Physiol Scand 126: 93-101.

Ernst SA, Mills JW (1977) Basolateral plasma membrane localization of ouabain-sensitive sodium transport sites in the secretory epithelium of the avian salt gland. J cell Biol 75: 74-94.

Ernst SA, Van Rossum GDV (1982) Ions and energy metabolism in duck salt-gland: possible role of furosemide-sensitive co-transport of sodium and chloride. J Physiol 325: 333-352.

Foskett JK, Bern HA, Machen TE, Conner M (1983) Chloride cells and the hormonal control of teleost fish osmoregulation. J exp Biol 106: 255-281.

Frizzell RA, Field M, Schultz SG (1979) Sodium-coupled chloride transport by epithelial tissues. Am J Physiol 236: F1-F8.

Greger R, Schlatter E (1984) Mechanism of NaCl secretion in the rectal gland of spiny dogfish (<u>Squalus</u> <u>acanthias</u>) I. Experiments in isolated in vitro perfused rectal gland tubules. Pflugers Arch 402: 63-75.

Hannafin J, Kinne-Saffran E, Friedman D, Kinne R (1983) Presence of a sodium-potassium chloride cotransport system in the rectal gland of <u>Squalus acanthias</u>. J Membr Biol 75: 73-83.

Marshall AT, Hyatt AD, Phillips JG, Condron RJ (1985) Isosmotic secretion in the avian nasal salt gland : X-ray microanalysis of luminal and intracellular ion distributions. J comp Physiol 156: 213-227.

Norris KS, Dawson WR (1964) Observations on the water economy and electrolyte excretion of chuckwallas (Lacertilia, <u>Sauromalus</u>). Copeia 4: 638-646.

Peaker M, Linzell JL (1975) Salt glands in birds and reptiles, Cambridge Univ Press, Cambridge.

Schmidt Nielsen K, Fänge R (1958) Salt glands in marine reptiles. Nature 182: 783-785.

Shuttleworth TJ (1985) Peptidergic control of transport in the teleost gill. In: Gilles R, Gilles-Baillien M (eds): Transport processes, iono- and osmoregulation. Springer Verlag, Berlin; pp. 194-203.

Shuttleworth TJ, Potts WTW, Harris JN (1974) Bioelectric potentials in the gills of the flounder <u>Platichthys</u> <u>flesus</u>. J comp Physiol 94: 321-329.

Shuttleworth TJ, Thompson JL (1980) The mechanism of cyclic AMP stimulation of secretion in the dogfish rectal gland. J comp Physiol 140:

209-216.

Shuttleworth TJ, Thorndyke MC (1984) An endogenous peptide stimulates secretory activity in the elasmobranch rectal gland. Science 225: 319-321.

Shuttleworth TJ, Thompson JL (1986) Perfusion-secretion relationships in the isolated elasmobranch rectal gland. J exp Biol, in press.

Silva P, Stoff J, Field M, Fine L, Forrest JN, Epstein FH (1977) Mechanism of active chloride secretion by shark rectal gland : role of Na-K-ATPase in chloride transport. Am J Physiol 233: F298-F306.

Solomon R, Taylor M, Dorsey D, Silva P, Epstein FH (1985) Atriopeptin stimulation of rectal gland function in Sqaulus acanthias. Am J Physiol 249: R348-R354.

Stoff JS, Rosa R, Hallac R, Silva P, Epstein FH (1979) Hormonal regulation of active transport in the dogfish rectal gland. Am J Physiol 237: F138-F144.

Taplin LE, Grigg GC (1981) Salt glands in the tongue of the estuarine crocodile Crocodylus porosus. Science 212: 1045-1047.

Templeton JR (1964) Nasal salt excretion in terrestrial lizards. Comp Biochem Physiol 11: 223-229.

Thesleff S, Schmidt-Nielsen K (1962) An electrophysiological study of the salt gland of the herring gull. Am J Physiol 202: 597-600.

Thorndyke MC, Shuttleworth TJ (1985) Biochemical and physiological studies on peptides from the elasmobranch gut. Peptides 6, Suppl 3: 369-372.

LIST OF PARTICIPANTS

Ahearn, Gregory - Department of Zoology, 2538 The Mall, University of Hawaii at Manoa, Honolulu, HI 96822, USA

Ar, Amos - Department of Zoology, The George S. Wise Faculty of Life Sciences, Tel-Aviv University, Ramat-Aviv, P.O.B. 39040, Tel-Aviv 69978, Israel

Atema, Jelle - Boston University Marine Program, Marine Biological Laboratory, Woods Hole, MA 02543, USA

Bartholomew, George A. - Department of Biology, University of California, Los Angeles, CA 90024, USA

Bennett, Albert F. - Department of Cell and Developmental Biology, University of California, Irvine, CA 92717, USA

Benson, Andrew A. - Marine Biology Research Division A-002, Scripps Institution of Oceanography, La Jolla, CA 92093, USA

Block, Barbara - Department of Zoology, Duke University, Durham, NC 27706, USA

Bolis, Liana C. - General Biology, University of Milan, Via Balzaretti, 20133 Milan, Italy

Canciglia Paolo - Università di Messina, Istituto di Fisiologia Generale, Via di Verdi 85, 98100 Messina, Italy

Childress, James J. - Department of Biology, University of California, Santa Barbara, CA 93106, USA

Clegg, James S. - Bodega Marine Laboratory, University of California, Bodega Bay, CA 94923, USA

Daniel, Thomas L. - Department of Zoology, University of Washington, Seattle, WA 98195, USA

Dejours, Pierre - Laboratoire d'Etude des Régulations Physiologiques, Centre National de la Recherche Scientifique, 23 rue Becquerel, 67087 Strasbourg, France

Diamond, Jared M. - Department of Physiology, Center for the Health Sciences, University of California, Los Angeles, CA 90024, USA

Evans, David H. - Department of Zoology, University of Florida, Gainesville, FL 32611, USA

Fidanza, Alberto - Università di Roma, Istituto di Fisiologia Generale, Città Universitaria, 00100 Roma, Italy

Gilles, Raymond - Laboratoire de Physiologie Animale, Université de Liège, Quai Van Beneden 22, 4020 Liège, Belgium

Hamner, William M. - Department of Biology, University of California, Los Angeles, CA 90024, USA

Hess, Ernst - Institut de Zoologie, Chemin de Chantemerle 22, 2000 Neuchâtel 7, Switzerland

Jackson, Donald C. - Division of Biomedical Sciences, Brown University, Box G, Providence, RI 02912, USA

Karasov, William - Department of Wild Life Ecology, University of Wisconsin, 1630 Linden Drive, Madison, WI 53706, USA

Keynes, Richard D. - Department of Physiology, Cambridge University, Downing Street, Cambridge CB2 3EJ, UK

Kooyman, Gerald L. - Physiological Research Laboratory, Scripps Institution of Oceanography, La Jolla, CA 92903, USA

Land, Michael F. - Neurophysiology and Ethology Group, University of Sussex, Brighton, Sussex BN1 9QG, UK

Linzen, Bernt - Zoologisches Institut der Universität, Luisenstrasse 14, 8000 München 2, FRG

Louw, Gideon N. - Department of Zoology, University of Cape Town, Private Bag, 7700 Rondebosch, South Africa

Maddrell, Simon H.P. - Department of Zoology, University of Cambridge, Downing Street, Cambridge CB2 3EJ, UK

Maloiy, Ole G.M. - College of Agriculture and Veterinary Sciences, P.O. Box 29053, Kabete, Kenya

Nachtigall, Werner - Zoologisches Institut der Universität des Saarlandes, 6600 Saarbrucken, FRG

Nagy, Kenneth A. - Laboratory of Biochemical and Environmental Sciences, University of California, Los Angeles, CA 90024, USA

O'Donnell, Mike J. - Department of Biology, Faculty of Science, 4700 Keele Street, Downsview, Ontario M3J 1P3, Canada

Pennycuik, Colin J. - Department of Biology, University of Miami, P.O. Box 249118, Coral Gables, FL 33124, USA

Powell, Mark A. - Scripps Institute of Oceanography, University of California, La Jolla, CA 92093, USA

Prinetti, Alessandro - General Biology, University of Milan, Via Balzaretti, 20133 Milano, Italy

Rahmann, Hinrich - Institut für Zoologie (220), Universität, Stuttgart-Hohenheim, 7000 Stuttgart 70, FRG

Rahn, Hermann - Department of Physiology, State University of New York, 120 Sherman Hall, Buffalo, NY 14214, USA

Rieppel, Olivier - Paläontologisches Institut und Museum der Universität, Künstlergasse 16, 8006 Zürich, Switzerland

Scheid, Peter - Institut für Physiologie, Lehrstuhl 1, Ruhr-Universität Bochum, Universitätsstrasse 150, Postfach 10 21 48, 4630 Bochum 1, FGR

Schmidt-Nielsen, Knut - Department of Zoology, Duke University, Durham, NC 27706, USA

Seymour, Roger S. - Department of Zoology, University of Adelaide, G.P.O. Box 498, Adelaide, South Australia

Shkolnik, Amiram - Department of Zoology, The George S. Wise Faculty of Life Sciences, Tel-Aviv University, Ramat-Aviv, P.O.B. 39040, Tel-Aviv 69978, Israel

Shoemaker, Vaughan H. - Department of Biology, University of California, Riverside, CA 92521, USA

Shuttleworth, Trevor J. - Department of Biological Sciences, University of Exeter, Hattlerly Laboratories, Prince of Wales Road, Exeter, EX4 4PS, UK

Somero George N. - Scripps Institution of Oceanography, University of California, La Jolla, CA 92093, USA

Taylor, C. Richard - Concord Field Station, Museum of Comparative Zoology, Harvard University, Old Causeway Road, Bedford, MA 01730, USA

Toulmond, André - Biologie et Physiologie des Organismes Marins, Université Pierre et Marie Curie, 4 place Jussieu, 75230 Paris 05, France

Treherne, John E. - Department of Zoology, University of Cambridge, Downing Street, Cambridge CB2 3EJ, UK

Truchot, Jean-Paul - Station Marine, Laboratoire de Neurobiologie et de Physiologie Comparées, 2 rue du Professeur Jolyet, 33120 Arcachon, France

Webb, Paul - School of Natural Resources, University of Michigan, Ann Arbor, MI 48108, USA

Wehner, Rüdiger - Department of Zoology, University of Zürich, 8057 Zürich, Switzerland

Weibel, Ewald R. - Anatomisches Institut, Universität Bern, Bühlstrasse 26, 3000 Bern, Switzerland

White, Fred N. - Physiological Research Laboratory, Scripps Institution of Oceanography A-004, University of California, La Jolla, CA 92093 USA

Wood, Stephen C. - Department of Physiology, University of New Mexico, Albuquerque, NM 87131, USA

Wright, Stephen H. - Department of Physiology, College of Medicine, University of Arizona, Tucson, AZ 85724, USA

AUTHOR INDEX

SUBJECT INDEX

Finito di stampare
nel mese di Maggio del 1987
dalla tipolitografia «La Grafica & Stampa Editrice s.r.l.» di Vicenza
per conto della Liviana Editrice s.p.a. di Padova